Agentic AI for Autonomous Vehicles:

Safety, Reliability, Law, and Ethics

Vasaki Ponnusamy
https://orcid.org/0000-0002-9049-8345
Higher Colleges of Technology, UAE

Noor Zaman Jhanjhi
https://orcid.org/0000-0001-8116-4733
Taylor's University, Malaysia

IGI Global
Scientific Publishing
Publishing Tomorrow's Research Today

Vice President of Editorial Melissa Wagner
Director of Acquisitions Mikaela Felty
Director of Book Development Jocelynn Hessler
Production Manager Mike Brehm
Cover Design Jose Rosado

Published in the United States of America by
IGI Global Scientific Publishing
701 East Chocolate Avenue
Hershey, PA, 17033, USA
Tel: 717-533-8845 | Fax: 717-533-7115
Website: https://www.igi-global.com E-mail: cust@igi-global.com

Library of Congress Cataloging-in-Publication Data

Names: Ponnusamy, Vasaki, 1974- editor | Zaman, Noor, 1972- editor
Title: Agentic AI for autonomous vehicles : safety, reliability, law, and ethics / edited by: Vasaki Ponnusamy, Noor Zaman Jhanjhi.
Description: Hershey PA : IGI Global Scientific Publishing, [2026] | Includes bibliographical references. | Summary: "This book explores Agentic AI for autonomous vehicles, emphasizing adaptive decision-making in dynamic environments. It addresses safety, ethics, reliability, legal frameworks, and human factors, offering researchers, manufacturers, and policymakers a comprehensive guide to developing responsible, resilient, and human-centered autonomous vehicle systems"-- Provided by publisher.
Identifiers: LCCN 2026002294 (print) | LCCN 2026002295 (ebook) | ISBN 9798337373553 hardcover | ISBN 9798337373560 paperback | ISBN 9798337373577 ebook
Subjects: LCSH: Automated vehicles--Security measures | Intelligent agents (Computer software) | Generative artificial intelligence | Artificial intelligence--Engineering applications
Classification: LCC TL152.8 .A395 2026 (print) | LCC TL152.8 (ebook)
LC record available at https://lccn.loc.gov/2026002294
LC ebook record available at https://lccn.loc.gov/2026002295

British Cataloguing in Publication Data
A Cataloguing in Publication record for this book is available from the British Library.

Table of Contents

Detailed Table of Contents

Preface

INTRODUCTION

The rapid development of artificial intelligence has ushered in a transformative era where systems are no longer limited to static prediction tasks but are increasingly capable of autonomous, adaptive, and context-aware decision-making. Among these advances, Agentic Artificial Intelligence (Agentic AI) represents a fundamental shift. Unlike traditional AI systems that mainly rely on pre-trained models and fixed datasets, Agentic AI systems operate with a degree of autonomy, continually interacting with and adapting to dynamic environments in pursuit of expected outcomes.

This emerging ability is especially promising for autonomous vehicles (AV) and unmanned systems. In these areas, decision-making needs to happen quickly, often with limited certainty, and has important consequences for safety, legal responsibility, and ethics. Agentic AI offers the potential for systems that are not just reactive but also proactive—able to interpret complex environments, interact with humans, and modify their actions as conditions change. However, this potential also brings significant challenges that must be addressed comprehensively.

This edited volume, Agentic AI for Autonomous Vehicles: Safety, Reliability, Law, and Ethics, has been created to explore these challenges and opportunities thoroughly. The main goal of this book is to highlight emerging trends and interdisciplinary viewpoints on integrating Agentic AI within autonomous vehicle systems. It aims to connect technical innovation with societal concerns by focusing on four core pillars: safety, reliability, legal frameworks, and ethical responsibility.

The volume pursues two interconnected goals. First, it addresses the urgent need for a solid technical framework for Agentic AI in autonomous vehicles, one that systematically includes safety assurance, reliability engineering, and compliance with evolving legal and ethical standards. Second, it explores the role of human factors, recognizing that human interaction, perception, and trust can greatly influence the effectiveness and acceptance of Agentic AI systems in real-world deployments.

The contributions collected in this book reflect insights from academia, industry, and policymaking communities. Together, they emphasize both the transformative potential of Agentic AI and the complexities involved in deploying such systems responsibly. By combining technical rigor with ethical foresight and regulatory understanding, this volume aims to promote the development of autonomous systems that are not only intelligent but also trustworthy and socially aligned.

We anticipate that this book will serve as a valuable resource for researchers, engineers, policymakers, and practitioners engaged in artificial intelligence, autonomous vehicles, and unmanned systems. It is our hope that the discussions presented herein will stimulate further research, inform responsible innovation, and help shape the future of safe and ethical autonomous mobility.

ORGANIZATION OF THE BOOK

In Chapter 1, the focus shifts to the broader societal implications of agentic autonomous vehicles, especially the ethical and social dilemmas they pose. Moving beyond simple moral thought experiments, the authors develop a multidimensional framework that tackles issues like moral agency, liability sharing, privacy, equity, and public trust. This chapter stands out for its integrative approach, combining empirical research and policy analysis to propose a balanced ethical framework that aligns efficiency with fundamental rights and inclusive design.

Chapter 2 provides a comprehensive review of machine learning and deep learning approaches for vehicle fault diagnostics and predictive maintenance. The authors highlight how AI-driven diagnostic systems are transforming vehicle safety and operational efficiency by leveraging complex sensor data and real-time analytics. This chapter underscores the growing importance of predictive capabilities in modern vehicles, illustrating how intelligent maintenance systems can preempt failures and enhance reliability across increasingly sophisticated automotive platforms.

In Chapter 3, the discussion turns to cybersecurity and trust within interconnected vehicle ecosystems. The authors explore how AI-powered authentication and access control mechanisms can address the limitations of traditional security models in IoT-enabled autonomous vehicles. By examining adaptive and context-aware approaches, such as behavioral biometrics and anomaly detection, the chapter demonstrates how intelligent security frameworks can safeguard identity and ensure secure interactions across highly distributed, heterogeneous systems.

Chapter 4 introduces an innovative perspective on human–vehicle interaction through gesture-controlled systems enabled by IoT technologies. The authors present a prototype that leverages real-time gesture recognition and wireless communication to enable intuitive and accessible vehicle control. Beyond its technical contributions, the chapter highlights the broader applicability of such systems in assistive technologies and industrial automation, emphasizing inclusivity and user-centric design in autonomous mobility solutions.

Chapter 5 addresses the critical infrastructure challenges associated with electric vehicle charging systems and presents a comprehensive review of optimization and AI-driven solutions. The authors synthesize recent advances in techniques such as deep reinforcement learning, neuro-fuzzy systems, and graph neural networks to improve charging-station utilization, demand prediction, and grid stability. This chapter positions intelligent infrastructure as a key enabler of sustainable mobility and underscores AI's role in addressing systemic inefficiencies.

In Chapter 6, the contributors provide an in-depth examination of safety and reliability in agentic autonomous vehicle systems. The chapter explores safety-critical behaviors, risk sources, and edge-case scenarios, while proposing robust reliability frameworks based on fault tolerance, redundancy, and distributed architectures. The integration of formal verification, explainable AI, and runtime monitoring further reinforces the chapter's emphasis on building transparent, resilient, and safety-assured autonomous systems.

Chapter 7 advances the discussion by presenting a unified architecture for agentic decision-making in autonomous vehicles, with a strong emphasis on explainability and fail-safe design. The author synthesizes developments in perception, prediction, planning, and ethical reasoning to propose systems that are both accountable and human aligned. This chapter also identifies gaps in existing safety and governance frameworks, offering an integrative model that strengthens road safety and supports future regulatory and industrial advancements.

Chapter 8 critically examines the risks associated with agentic AI in autonomous vehicles, highlighting challenges related to safety, control, and ethical alignment. The author explores issues such as emergent system behaviors, loss of human oversight, and the difficulty of embedding societal values into machine decision-making. By addressing regulatory gaps and proposing mitigation strategies, including transparency and value-sensitive design, this chapter provides a balanced perspective on the responsible deployment of highly autonomous systems.

In Chapter 9, the authors advocate for an "ethics-by-design" approach, emphasizing the integration of legal and moral principles directly into the architecture of autonomous vehicle systems. Through a conceptual and interdisciplinary framework, the chapter demonstrates how embedding ethical reasoning and legal compliance at the design stage enhances transparency, accountability, and public trust. This work highlights the need to align technological innovation with societal expectations and regulatory requirements.

Chapter 10 further deepens the ethical discourse by presenting a comprehensive framework for integrating ethical considerations into agentic AI systems for autonomous vehicles. The authors outline a technical architecture that incorporates ethical reasoning across multiple system layers, including goal setting, transparency, and human oversight. The chapter also introduces evaluation metrics for safety, fairness, and trust, reinforcing the importance of aligning autonomous systems with societal values in increasingly interconnected environments.

Chapter 11 shifts focus to the legal domain, offering a detailed analysis of tort law as it applies to autonomous vehicles. The authors examine complex questions of fault attribution, liability distribution, and the applicability of established legal principles such as product and vicarious liability. This chapter provides valuable insights into the evolving legal landscape and highlights the need for adaptive legal frameworks agentic, autonomous systems.

Finally, Chapter 12 explores the critical role of human interaction in the adoption and acceptance of agentic AI and autonomous systems. The author examines how human factors, trust, transparency, and interface design influence the deployment of these technologies. By integrating technical, ethical, and societal perspectives, the chapter underscores the importance of human-centered design, governance mechanisms, and workforce transition strategies in ensuring equitable and sustainable adoption of autonomous mobility solutions.

Together, these chapters provide a comprehensive and interdisciplinary exploration of agentic AI in autonomous vehicles, reflecting the depth and breadth of current research while charting a forward-looking path for innovation, governance, and societal integration.

CONCLUSION

As this volume shows, the rise of Agentic Artificial Intelligence signals a key shift in autonomous vehicle development—from systems that just respond to inputs to those capable of intentional, adaptive, and context-aware decisions. This change is not only technological but also highly interdisciplinary, involving engineering, ethics, law, human factors, and public policy. The chapters in this book highlight that the future of autonomous mobility will rely not just on advances in computational intelligence but also on our ability to incorporate trust, accountability, and societal values into these systems from the start.

Across the contributions, several unifying themes emerge. First, safety and reliability remain the cornerstone of all agentic systems, demanding rigorous frameworks that address uncertainty, edge cases, and system-level resilience. Second, ethical reasoning and legal accountability are no longer peripheral considerations; they are central design imperatives that must be integrated into the architecture of autonomous systems. Third, human interaction, whether through trust, usability, or oversight—continues to shape both the acceptance and effectiveness of these technologies. Finally, the importance of scalable, secure, and intelligent infrastructure highlights that autonomous vehicles do not operate in isolation but within complex, interconnected ecosystems.

At the same time, this volume clearly shows that major challenges still exist. Issues related to liability, governance, transparency, and aligning machine decision-making with diverse societal values need ongoing dialogue and cooperation among researchers, industry leaders, and policymakers. Creating global standards and regulatory frameworks will be crucial to ensure that innovation progresses responsibly and fairly.

We, as editors, see this book as both a reflection of current research and a call to action. It encourages the community to move beyond isolated approaches and toward integrated solutions that combine technical excellence with ethical foresight and legal clarity. The way forward requires not only innovation but also responsible stewardship, making sure that autonomous, agentic systems support broader goals of safety, inclusivity, and society's well-being.

We hope this volume encourages ongoing research, informed policy making, and responsible industrial practices. By promoting a shared understanding of agentic AI in autonomous vehicles, we aim to help build a future where intelligent mobility systems are not only more capable but also more trustworthy, transparent, and aligned with human values.

Vasaki Ponnusamy
Higher Colleges of Technology, UAE

Noor Zaman Jhanjhi
Taylor's University, Malaysia

Chapter 1
AV Frameworks:
Social and Ethical Dilemmas of Agentic Autonomous Vehicles

Deepak Gupta
http://orcid.org/0000-0003-3929-1362
Institute of Technology and Management, Gwalior, India

Tursunov Muhiddin
Termez University of Economics and Service, Termez, Uzbekistan

Sardor Sharifzoda
Mamun University, Khiva, Uzbekistan

Iroda Abdullaeva
Urgench State University, Urgench, Uzbekistan

Abdullayev Abdulla Fayzulla Ugli
Tashkent State University of Economy, Uzbekistan

Barno Matchanova
Urgench State Pedagogical Institute, Uzbekistan

ABSTRACT

Agentic autonomous vehicles (AVs) represent a transformative shift in transportation technology, deploying artificial intelligence systems capable of independent decision-making in complex, ethically charged scenarios. This chapter examines the multidimensional social and ethical dilemmas arising from agentic AV deployment, moving beyond traditional trolley-problem framings toward comprehensive risk ethics frameworks. The analysis addresses six interconnected domains: algorithmic decision-making and moral agency, liability allocation in human-machine systems, privacy and surveillance implications of connected vehicle ecosystems, equity and accessibility considerations for marginalized populations, public trust formation mechanisms, and regulatory governance challenges. Drawing on empirical research from moral machine experiments, policy analyses across jurisdictions, and disability-inclusive design studies, this chapter proposes an integrative ethical framework that balances utilitarian efficiency with deontological rights protections.

DOI: 10.4018/979-8-3373-7355-3.ch001

1. INTRODUCTION

1.1 Background and Context

The advent of agentic autonomous vehicles represents a paradigm shift in transportation technology that extends far beyond mere automation. Unlike traditional rule-based systems, agentic AVs possess the capability for goal-directed behavior, environmental adaptation, and independent decision-making that fundamentally alters the moral landscape of mobility (Geisslinger et al., 2021). These systems embody artificial intelligence that can perceive, reason, and act autonomously in complex traffic environments, making split-second decisions that carry profound ethical implications for all road users.

1.1.1 Defining Agentic Autonomous Vehicles

Agentic autonomous vehicles are distinguished from conventional automated systems by three critical characteristics: autonomy in goal selection, adaptive learning from environmental feedback, and independent moral reasoning capabilities. These vehicles operate at SAE Level 4-5 autonomy, where the AI system assumes full responsibility for driving tasks without human oversight (Poszler et al., 2024). The "agentic" dimension refers to the system's capacity to make value-laden decisions that reflect underlying ethical frameworks programmed by designers but adapted through machine learning processes.

1.1.2 Evolution from Rule-Based to Agentic Systems

Traditional automotive safety systems followed deterministic programming with predefined responses to specific scenarios. In contrast, agentic AVs employ neural networks and deep learning algorithms that enable adaptive responses to novel situations. This evolution introduces unprecedented complexity in moral responsibility attribution, as the system's decisions emerge from training processes rather than explicit programming (Rhim et al., 2021).

1.1.3 Scope of Social and Ethical Challenges

The social and ethical challenges posed by agentic AVs as shown in figure 1, transcend technical reliability concerns to encompass fundamental questions about moral agency, democratic governance, and social justice. These challenges manifest across multiple dimensions: individual rights versus collective welfare, privacy versus safety, accessibility versus efficiency, and human agency versus algorithmic determination (Martinho et al., 2021).

Figure 1. Ethical decision-making framework for agentic AVs

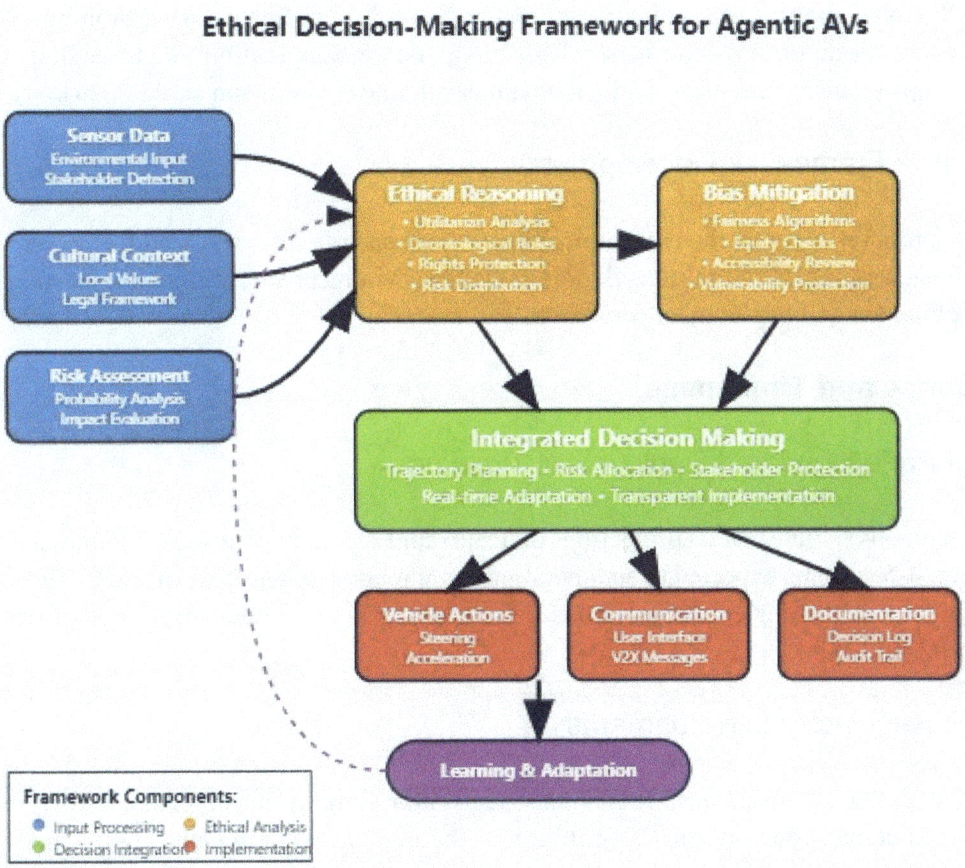

1.2 Chapter Objectives

1.2.1 Systematic Examination of Ethical Dilemmas

This chapter aims to systematically examine the unique social and ethical dilemmas presented by agentic AV systems, distinguishing these challenges from those associated with conventional automated systems. The analysis focuses on the emergent properties of agentic decision-making that create novel ethical terrain requiring innovative governance approaches.

1.2.2 Critical Assessment of Current Frameworks

The chapter critically evaluates existing ethical frameworks and their limitations for addressing agentic AI in transportation contexts. Traditional consequentialist and deontological approaches prove insufficient for the complex, probabilistic decision-making characteristic of agentic systems (Geisslinger et al., 2023).

1.2.3 Regulatory and Governance Analysis

Cross-jurisdictional analysis of regulatory approaches reveals significant variation in how different legal systems address agentic AV governance. This comparative examination identifies best practices and regulatory gaps requiring attention as deployment accelerates (Mallinson et al., 2024).

1.2.4 Integrative Framework Development

The chapter proposes a comprehensive ethical framework specifically designed for agentic AV deployment that balances multiple moral considerations while providing practical guidance for stakeholders across the development and deployment ecosystem.

1.3 Significance and Timeliness

1.3.1 Market Projections and Deployment Timeline

Global autonomous vehicle market projections indicate rapid acceleration toward commercial deployment, with Level 4-5 systems expected to achieve significant market penetration by 2030. This timeline creates urgency for establishing ethical governance frameworks before widespread deployment occurs (Robles & Mallinson, 2024).

1.3.2 Recent Regulatory Developments

The European Union's AI Act, various U.S. state legislation initiatives, and China's emerging regulatory frameworks demonstrate growing recognition of the need for comprehensive governance of AI systems in high-stakes applications like autonomous driving (Gless & Ligeti, 2024).

1.3.3 Public Discourse Evolution

Growing public awareness of AI ethics and autonomous systems creates both opportunities and challenges for democratic engagement in AV governance. Research reveals complex public attitudes that vary significantly based on knowledge levels and demographic characteristics (Othman, 2023).

2. THEORETICAL FOUNDATIONS: FROM TROLLEY PROBLEMS TO RISK ETHICS

2.1 The Limitations of Trolley Problem Framings

2.1.1 Deterministic Assumptions Versus Probabilistic Reality

Traditional trolley problem scenarios assume deterministic outcomes where actions lead to certain consequences. However, agentic AVs operate in probabilistic environments where outcomes carry uncertainty distributions. Real-world driving decisions involve risk assessment rather than certain harm calculations (Königs, 2023).

The trolley problem framework fundamentally misrepresents the nature of AV decision-making by presenting binary choices between clearly defined outcomes. In contrast, agentic systems must navigate continuous probability spaces where multiple variables influence outcome likelihoods. This probabilistic complexity renders trolley-problem-based ethical frameworks inadequate for real-world application.

2.1.2 Information Asymmetries and Temporal Constraints

Trolley problems assume perfect information about scenario participants and unlimited deliberation time. Agentic AVs must make decisions under severe temporal constraints with incomplete sensor information and uncertain predictions about other agents' behaviors. These constraints necessitate fundamentally different ethical approaches that account for bounded rationality and satisficing decision-making (Cecchini et al., 2023).

2.1.3 Cultural and Individual Variation

Research using the Moral Machine platform reveals significant cross-cultural variation in ethical preferences, undermining universalist approaches based on abstract philosophical principles. Preference differences emerge across demographic lines, educational backgrounds, and cultural contexts, suggesting the need for pluralistic rather than monolithic ethical frameworks (Takemoto, 2024).

2.2 Toward an Ethics of Risk Distribution

2.2.1 Risk-Based Ethical Frameworks

Moving beyond "whom to harm" toward "how to fairly distribute unavoidable risks" represents a fundamental reconceptualization of AV ethics. Risk-based approaches acknowledge that driving inherently involves risk distribution among road users and that ethical frameworks must guide fair allocation of these risks rather than attempting to eliminate them entirely (Krügel & Uhl, 2024).

Risk ethics frameworks incorporate several key principles: proportionality in risk distribution, protection of vulnerable road users, democratic participation in risk acceptance, and transparency in risk allocation mechanisms. These principles provide actionable guidance for agentic system design while acknowledging the inherent uncertainty in traffic environments.

2.2.2 Maximin and Bayesian Approaches

The maximin principle, derived from Rawlsian justice theory, suggests minimizing the worst possible outcome for the least advantaged parties. In AV contexts, this translates to prioritizing protection of vulnerable road users (pedestrians, cyclists, children) even when such protection imposes costs on vehicle occupants (Li et al., 2022).

Bayesian decision theory offers an alternative approach that seeks to minimize expected harm across all affected parties. This utilitarian approach maximizes overall welfare but may sacrifice individual rights or vulnerable population protections in favor of aggregate outcomes.

2.2.3 Equal Risk Distribution Principles

Democratic principles suggest that risk distribution should reflect collective preferences expressed through legitimate governance processes. This approach requires mechanisms for public deliberation about acceptable risk levels and distribution patterns, challenging purely technocratic approaches to AV ethics (Robinson et al., 2022).

2.3 Moral Pluralism and Integrative Frameworks

2.3.1 Accommodating Multiple Ethical Traditions

Agentic AV systems must navigate environments where road users hold diverse moral commitments derived from consequentialist, deontological, virtue ethics, and care ethics traditions. Integrative frameworks seek to accommodate this plurality while maintaining coherent decision-making processes (Rhim et al., 2021).

Dual-process theory suggests that moral judgments emerge from both intuitive emotional responses and deliberative rational analysis. Agentic systems must account for both dimensions when making decisions that affect human welfare, requiring sophisticated integration of multiple moral considerations.

2.3.2 Context-Sensitive Ethical Reasoning

Static ethical rules prove inadequate for the dynamic complexity of traffic environments. Agentic systems require context-sensitive reasoning capabilities that can adapt ethical priorities based on environmental conditions, stakeholder presence, and situational urgency while maintaining consistency with overarching moral principles.

2.3.3 Temporal Ethical Considerations

Agentic AVs make decisions with consequences extending across different temporal horizons. Immediate safety considerations must be balanced against longer-term social impacts, environmental effects, and systemic risks. This temporal dimension adds complexity to ethical reasoning that traditional frameworks inadequately address.

2.4 Agentic AI and Moral Agency

2.4.1 Machine Moral Agency

The question of whether agentic systems can possess genuine moral agency remains philosophically contested. However, the practical delegation of morally significant decisions to artificial agents creates functional moral agency regardless of philosophical debates about consciousness or intentionality (Poszler et al., 2024).

Functional moral agency in agentic AVs manifests through the system's capacity to recognize moral situations, apply ethical reasoning, and act based on moral considerations. This capacity creates responsibilities for designers, operators, and users to ensure appropriate ethical frameworks guide system behavior.

2.4.2 Distributed Moral Responsibility

Agentic AV systems distribute moral responsibility across multiple agents: manufacturers who design ethical parameters, programmers who implement decision algorithms, users who activate autonomous modes, and the AI system itself that applies ethical reasoning in specific situations. This distribution complicates traditional approaches to moral accountability.

2.4.3 Ethical Learning and Adaptation

Unlike static rule-based systems, agentic AVs can learn from experience and adapt their ethical reasoning over time. This learning capacity creates opportunities for ethical improvement but also risks drift away from intended moral parameters without appropriate oversight mechanisms.

3. ALGORITHMIC DECISION-MAKING AND BIAS IN AGENTIC SYSTEMS

3.1 Decision Architecture of Agentic AVs

3.1.1 Perception-Planning-Control Pipeline

Agentic AV decision-making follows a perception-planning-control architecture where ethical considerations enter at multiple points. The perception phase involves sensor fusion and scene understanding that can exhibit bias in object detection and classification. The planning phase incorporates ethical reasoning into trajectory selection, while the control phase implements decisions through vehicle actuators (Geisslinger et al., 2023).

Each pipeline stage introduces potential bias sources that can result in discriminatory outcomes. Perception algorithms may exhibit differential performance across demographic groups, planning algorithms may reflect programmer bias in ethical parameter setting, and control systems may prioritize certain safety criteria over others.

3.1.2 Real-Time Ethical Computation

Agentic systems must perform ethical reasoning within millisecond timeframes, requiring pre-computed ethical parameters and efficient moral reasoning algorithms. This temporal constraint limits the sophistication of ethical analysis possible in real-time decision-making, necessitating careful offline ethical framework design.

Real-time ethical computation involves scenario classification, stakeholder identification, risk assessment, and ethical principle application. Each step must be optimized for computational efficiency while maintaining ethical validity, creating tension between moral sophistication and practical implementation constraints.

3.1.3 Scenario Classification and Response

Agentic AVs categorize traffic scenarios into ethical classes that trigger different decision-making processes. This classification system shapes ethical reasoning by determining which moral principles apply in specific situations. Bias in scenario classification can result in systematically discriminatory treatment of different road user types or traffic contexts.

3.2 Sources and Manifestations of Algorithmic Bias

3.2.1 Training Data Bias

Machine learning systems inherit biases present in training data, which may underrepresent certain demographic groups, geographic regions, or traffic scenarios. Historical bias in crash data, enforcement patterns, and infrastructure investment can perpetuate discriminatory outcomes through algorithm training (Zhu et al., 2022).

Training data bias manifests in several forms: representation bias where certain groups are underrepresented in datasets, measurement bias where data collection methods vary across groups, and aggregation bias where different populations are inappropriately combined. Each bias type can result in agentic systems that perform differently for different road users.

3.2.2 Objective Function Bias

The optimization criteria used to train agentic systems encode implicit value judgments about whose welfare matters and how different outcomes should be weighted. Seemingly neutral objectives like "minimize total harm" may systematically disadvantage certain groups if harm is measured in ways that reflect existing social inequalities (Schuessler, 2024).

Table 1. Algorithmic bias in AV objective functions

Objective Function	Potential Bias Source	Affected Groups	Discriminatory Outcome
Minimize fatalities	Age-based life valuation	Elderly individuals	Lower protection priority
Minimize economic loss	Income-based valuation	Low-income populations	Reduced safety investment
Minimize total injuries	Disability bias	People with disabilities	Inadequate accommodation
Maximize mobility	Urban bias	Rural communities	Limited service coverage
Optimize efficiency	Able-body assumptions	Wheelchair users	Inaccessible vehicles
Reduce emissions	Class bias	Working-class commuters	Limited affordable options

Source: Analysis of current AV development priorities and accessibility research (Ehsani et al., 2024)

3.2.3 Sensor and Perception Bias

AV sensors may exhibit differential performance across demographic groups due to design assumptions or testing limitations. Computer vision systems have demonstrated reduced accuracy in detecting individuals with darker skin tones, and LiDAR systems may perform differently based on clothing, mobility aids, or cultural dress practices.

Perception bias creates safety disparities where certain road users face higher risks due to technological limitations. These disparities intersect with existing transportation inequities to exacerbate disadvantages faced by already marginalized populations.

3.2.4 Cultural and Geographic Bias

Agentic systems trained primarily on data from specific cultural or geographic contexts may perform poorly when deployed in different environments. Driving norms, infrastructure characteristics, and pedestrian behaviors vary significantly across regions, potentially resulting in discriminatory performance when systems encounter unfamiliar contexts.

3.3 The Expertise-Preference Paradox

3.3.1 Professional Versus Public Preferences

Research reveals a striking disconnect between AV industry professionals and the general public regarding ethical preferences. Industry experts prefer passenger-protective algorithms that prioritize vehicle occupants, while non-professionals generally endorse utilitarian approaches that minimize total harm regardless of vehicle occupancy (Zhu et al., 2022).

This expertise-preference paradox creates challenges for democratic governance of AV ethics. Those with greatest technical knowledge about system capabilities hold preferences that diverge systematically from those who will be affected by system deployment. This divergence raises questions about whose preferences should guide ethical parameter setting.

3.3.2 Knowledge and Trust Relationships

Counterintuitively, increased knowledge about AV technology correlates with decreased trust and acceptance. Research demonstrates that a 1% increase in knowledge corresponds to 0.65% decrease in interest and 0.68% decrease in trust, suggesting that greater understanding of technical limitations reduces confidence in system capabilities (Othman, 2023).

This knowledge-trust paradox as shown in Figure 2, complicates public engagement strategies for AV governance. Educational efforts intended to increase support may actually reduce acceptance by highlighting system limitations and uncertainty areas.

Figure 2. Knowledge-trust paradox in autonomous vehicle acceptance

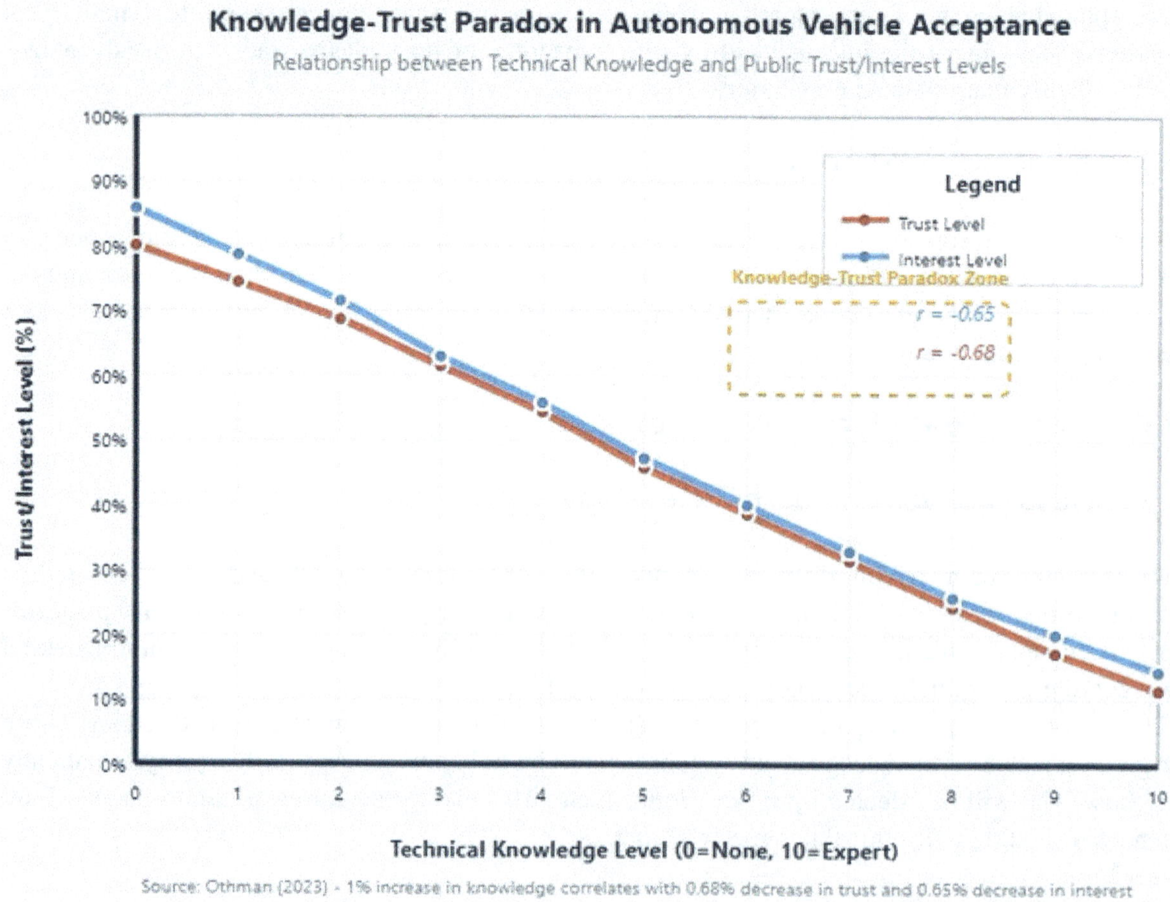

3.3.3 Implications for Participatory Design

The expertise-preference paradox highlights the importance of inclusive participatory design processes that engage diverse stakeholders in ethical framework development. Technical expertise alone is insufficient for determining appropriate ethical parameters; democratic legitimacy requires broad public participation in moral choice-making.

3.4 Algorithmic Accountability Mechanisms

3.4.1 Transparency Requirements

Algorithmic accountability requires transparency in decision-making processes, ethical parameter setting, and performance evaluation. However, the complexity of neural networks and machine learning systems creates challenges for meaningful transparency that balances technical accuracy with public comprehensibility (Gordillo Chávez et al., 2024).

Transparency mechanisms include algorithm auditing, decision explanation systems, and public reporting of ethical framework details. Each mechanism faces trade-offs between comprehensiveness and accessibility, requiring careful design to serve accountability goals effectively.

3.4.2 Explainability Standards

Explainable AI requirements seek to make algorithmic decision-making comprehensible to affected parties and oversight bodies. However, explainability standards must balance competing demands for accuracy, simplicity, and actionability while accounting for technical constraints of real-time decision systems.

Current explainability approaches include post-hoc analysis of decisions, counterfactual explanation generation, and simplified model approximations. Each approach offers different advantages for accountability purposes while introducing potential limitations or distortions.

3.4.3 Audit and Oversight Mechanisms

Systematic auditing of agentic AV systems requires new methodologies that can assess ethical performance across diverse scenarios and demographic groups. Audit mechanisms must evaluate both intended ethical behaviors and unintended discriminatory outcomes that emerge from system deployment.

Effective oversight requires collaboration between technical experts, ethicists, and community representatives to ensure comprehensive evaluation of system performance from multiple perspectives. This collaborative approach helps identify bias sources that might be overlooked by purely technical assessments.

4. LIABILITY AND LEGAL FRAMEWORKS FOR AGENTIC AVs

4.1 The Liability Attribution Problem

4.1.1 Traditional Legal Framework Limitations

Traditional negligence and product liability frameworks assume identifiable human actors making discrete decisions for which they can be held accountable. Agentic AVs fundamentally disrupt this assumption by distributing decision-making across multiple agents: manufacturers, software developers, data providers, infrastructure operators, and the AI system itself (Kubica, 2022).

The complexity of agentic decision-making creates challenges for establishing causal relationships between specific actions and harmful outcomes. When an agentic AV causes harm, determining whether the cause lies in algorithmic design, training data quality, sensor limitations, or environmental factors requires sophisticated technical analysis that exceeds traditional legal framework capabilities.

4.1.2 Product Liability Approaches

Product liability frameworks treat agentic AVs as products that may contain design defects, manufacturing defects, or inadequate warnings. However, the adaptive nature of agentic systems complicates traditional product liability analysis because system behavior evolves through learning processes after manufacture.

Strict liability approaches hold manufacturers responsible for harmful outcomes regardless of fault, providing clear compensation mechanisms for victims while creating strong safety incentives for developers. However, strict liability may inhibit innovation or result in overly conservative system design that reduces beneficial capabilities.

4.1.3 Negligence Standards for Algorithmic Design

Negligence frameworks evaluate whether system developers exercised reasonable care in designing ethical frameworks and safety systems. This approach requires establishing industry standards for ethical algorithm design and evaluation methodologies that can guide reasonable care determinations.

The evolving nature of AI technology complicates negligence standards by making it difficult to establish what constitutes reasonable care at any given time. Standards must balance innovation encouragement with safety protection while acknowledging technological uncertainty and evolving best practices.

4.2 Criminal Liability Considerations

4.2.1 Individual Criminal Responsibility

Criminal liability attribution for agentic AV-caused harm faces fundamental challenges in establishing individual mens rea (guilty mind) when decisions emerge from complex algorithmic processes. Traditional criminal law concepts of intent, knowledge, and recklessness require adaptation for contexts where human actors make high-level design choices implemented by autonomous systems (Gless & Ligeti, 2024).

Corporate criminal liability may prove more applicable than individual liability for agentic AV-related harm. Organizations can be held criminally responsible for failing to implement adequate safety systems, conduct proper testing, or maintain appropriate oversight of agentic system behavior.

4.2.2 European Union Regulatory Approaches

The EU's approach to criminal liability for driving automation systems emphasizes manufacturer responsibility while maintaining traditional criminal liability for human drivers who misuse automated systems. The AI Act's classification of AV systems as high-risk applications creates specific obligations for developers and operators that can ground criminal liability determinations.

4.2.3 Jurisdictional Variations

Different legal systems approach agentic AV criminal liability through varying frameworks reflecting different balance points between innovation promotion and safety protection. Common law systems tend toward case-by-case development of liability principles, while civil law systems pursue comprehensive statutory frameworks.

4.3 Comparative Regulatory Approaches

4.3.1 United States: State-by-State Policy Evolution

U.S. AV regulation follows a federal-state division where federal agencies address vehicle safety standards while states govern operation licensing, liability rules, and traffic law application. This division creates opportunities for policy experimentation but also generates coordination challenges and regulatory uncertainty (Ehsani et al., 2022).

State-level policy diffusion analysis reveals patterns where early adopter states influence later regulatory development through demonstration effects and policy learning mechanisms. However, significant variation persists across states in liability allocation, insurance requirements, and operational restrictions.

4.3.2 European Union: Type-Approval and AI Act Integration

The EU pursues comprehensive regulation through vehicle type-approval processes integrated with AI Act requirements for high-risk AI systems. This approach emphasizes ex-ante safety demonstration rather than ex-post liability determination, requiring manufacturers to prove safety before deployment authorization (Mulder & Vellinga, 2021).

The AI Act's risk-based approach classifies AV systems as high-risk applications requiring conformity assessment, risk management systems, and human oversight mechanisms. These requirements create clear legal obligations that can ground both civil and criminal liability determinations.

4.3.3 China: Burden of Proof Innovation

Chinese regulatory approaches experiment with burden of proof allocation in AV liability cases, potentially requiring manufacturers to prove their systems did not cause harm rather than requiring victims to prove causation. This approach recognizes the information asymmetries between manufacturers and accident victims while providing stronger victim protection.

4.4 Insurance and Risk Distribution

4.4.1 Mandatory Insurance Requirements

Mandatory insurance frameworks ensure victim compensation while distributing risks across the insurance market rather than concentrating liability on individual actors. However, traditional auto insurance models require adaptation for agentic systems where manufacturer liability may exceed driver liability in causation analysis.

First-party insurance approaches require all road users to carry insurance covering their own potential injuries regardless of fault attribution. This approach simplifies compensation processes while eliminating the need for complex fault determination in agentic system cases.

4.4.2 Manufacturer Insurance Pools

Industry-wide insurance pooling mechanisms can distribute agentic AV-related risks across all manufacturers while maintaining individual incentives for safety improvement through risk-based premium adjustments. Pool mechanisms help ensure adequate compensation funds while supporting innovation in emerging technologies.

4.4.3 Risk Allocation Innovation

Novel risk allocation mechanisms include victim compensation funds financed through industry levies, government insurance schemes for AV-related harm, and hybrid public-private insurance arrangements. Each approach offers different advantages for balancing victim protection, innovation incentives, and social risk distribution.

5. PRIVACY, SURVEILLANCE, AND DATA GOVERNANCE

5.1 The Data-Intensive Nature of Agentic AVs

5.1.1 Comprehensive Data Collection Systems

Agentic AVs generate unprecedented volumes of data through multiple sensor systems, communication networks, and behavioral monitoring capabilities. These systems continuously collect location data, movement patterns, passenger behaviors, biometric information, and environmental observations that create comprehensive surveillance capabilities (Lee & Hess, 2022).

The scope of data collection extends beyond traditional automotive information to include health data from driver monitoring systems, social information from passenger interactions, and environmental data from continuous sensor operation. This comprehensive data collection raises fundamental questions about privacy, consent, and surveillance in mobility systems.

5.1.2 Sensor Technology and Privacy Implications

Advanced sensor technologies including high-resolution cameras, LiDAR systems, radar arrays, and microphones create extensive surveillance capabilities that extend beyond immediate vehicle operation needs. These systems can identify individuals, monitor behaviors, record conversations, and track activities with precision that exceeds traditional surveillance technologies.

Vehicle-to-everything (V2X) communication systems amplify privacy concerns by sharing data across vehicle networks, infrastructure systems, and centralized coordination platforms. This data sharing enables coordination benefits but creates multiple points of potential privacy intrusion and surveillance expansion.

5.1.3 Data Processing and Storage Requirements

Agentic decision-making requires extensive data processing and storage to support real-time analysis, machine learning model updates, and system performance monitoring.

Cloud-based processing architectures may require transmitting sensitive personal information to remote servers, creating additional privacy risks and jurisdictional complications.

Edge computing approaches can reduce some privacy risks by processing data locally rather than transmitting it to central servers. However, edge computing limitations may compromise agentic system capabilities that depend on comprehensive data analysis and model updating processes.

5.2 Privacy Risks and Surveillance Concerns

5.2.1 Location Tracking and Movement Profiling

Continuous location tracking enables detailed movement profiling that can reveal sensitive information about personal relationships, health conditions, political activities, and lifestyle choices. Unlike voluntary location sharing through smartphones, AV location tracking may be mandatory for system operation, eliminating opt-out possibilities for privacy protection.

Movement pattern analysis can infer personal information including work schedules, social relationships, health appointments, religious activities, and political affiliations. This inference capability creates privacy risks even when explicit collection of such information is prohibited by policy or law.

5.2.2 Behavioral Inference and Profiling

Agentic AVs can analyze passenger behaviors to infer personal characteristics, emotional states, health conditions, and social relationships. Video analysis, voice recognition, and biometric monitoring enable detailed profiling that may reveal protected characteristics without explicit disclosure.

Behavioral profiling capabilities extend to other road users through external sensor systems that monitor pedestrian behaviors, other vehicle movements, and infrastructure interactions. This external monitoring creates privacy implications for all road users, not just AV passengers.

5.2.3 Commercial Exploitation and Data Sharing

Vehicle-generated data has significant commercial value for advertising, insurance, urban planning, and market research applications. However, commercial exploitation of mobility data raises concerns about commodification of personal privacy and potential discrimination based on movement patterns or travel behaviors.

Third-party data sharing arrangements may involve entities beyond traditional automotive stakeholders, including technology companies, advertising networks, insurance providers, and government agencies. Each additional stakeholder creates new privacy risks and control complexities.

5.3 Regulatory Frameworks for AV Data Protection

5.3.1 GDPR Compliance Requirements

The European Union's General Data Protection Regulation (GDPR) establishes comprehensive privacy protection requirements for AV data processing, including lawful basis requirements, data minimization principles, purpose limitation restrictions, and individual rights protections (Mulder & Vellinga, 2021).

GDPR compliance requires explicit consent for non-essential data processing, but the operational requirements of agentic systems may make consent-based processing impractical. Alternative lawful bases including legitimate interests or vital interests may apply, but require careful balancing with individual privacy rights.

5.3.2 Jurisdictional Variations and Conflicts

Cross-border AV operation creates complex jurisdictional issues when vehicles move between regions with different privacy laws. Data processing that is lawful in one jurisdiction may violate privacy requirements in another, creating compliance challenges for manufacturers and operators.

International data transfer restrictions may limit cloud-based processing capabilities for agentic systems that operate across borders. These restrictions require careful technical architecture design to ensure compliance while maintaining system functionality.

5.3.3 Sectoral Privacy Regulations

Transportation-specific privacy regulations may supplement general data protection laws with industry-specific requirements. These regulations may address unique transportation privacy concerns while providing clearer guidance for AV development and deployment.

5.4 Technical Privacy-Preserving Solutions

5.4.1 Federated Learning Approaches

Federated learning enables machine learning model training without centralizing sensitive personal data by distributing training processes across multiple devices or vehicles. This approach can reduce privacy risks while maintaining the data analysis capabilities required for agentic system operation (Li et al., 2022).

However, federated learning implementations face challenges including communication bandwidth limitations, computational resource constraints, and potential privacy leaks through model parameter sharing. Careful implementation design is required to realize privacy benefits while maintaining system performance.

5.4.2 Differential Privacy Techniques

Differential privacy provides mathematical guarantees for privacy protection by adding controlled noise to data analysis processes. This approach enables aggregate data analysis while protecting individual privacy through statistical anonymization techniques.

Differential privacy implementation in real-time agentic systems faces challenges from computational overhead, accuracy degradation, and complex parameter tuning requirements. These challenges require careful balance between privacy protection and system performance.

5.4.3 Blockchain and Decentralized Solutions

Blockchain technologies offer potential solutions for decentralized data governance, user control, and transparent data sharing arrangements. However, blockchain implementations face scalability challenges, energy consumption concerns, and technical complexity that may limit practical application in real-time AV systems.

Smart contract systems can automate privacy policy enforcement and data sharing agreements, but require careful design to ensure both technical functionality and legal enforceability across multiple jurisdictions.

6. EQUITY, ACCESSIBILITY, AND SOCIAL JUSTICE

6.1 The Promise of Mobility for Underserved Populations

6.1.1 Transportation Barriers for Marginalized Communities

Current transportation systems systematically exclude numerous populations including elderly individuals with mobility limitations, people with visual or mobility disabilities, individuals in transportation deserts with limited public transit, and economically disadvantaged communities lacking private vehicle access (Ehsani et al., 2024). These transportation barriers contribute to broader patterns of social exclusion by limiting access to employment, healthcare, education, and social participation opportunities.

Research demonstrates that transportation barriers disproportionately affect women, racial minorities, elderly individuals, and people with disabilities who rely more heavily on public transportation and face greater safety concerns in current mobility systems. Agentic AVs offer potential solutions to these barriers through accessible, affordable, and safe mobility services that could transform quality of life for underserved populations.

6.1.2 Autonomous Vehicle Accessibility Potential

Agentic AVs could provide unprecedented mobility independence for people with disabilities who cannot operate traditional vehicles. Unlike human-driven ride services, autonomous vehicles can offer consistent availability, reduced cost, and elimination of potential discrimination by human drivers who might refuse service to people with disabilities.

Public support for AV deployment increases significantly when framed around serving vulnerable populations, rising from 26.4% baseline support to over 50% when accessibility benefits are highlighted. This support pattern suggests political viability for equity-focused AV policies that prioritize underserved population access.

6.1.3 Healthcare Access and Medical Transportation

Transportation barriers represent significant obstacles to healthcare access, particularly for elderly individuals and people with disabilities who may require specialized vehicle accommodations. Agentic AVs equipped with accessibility features could provide reliable medical transportation that reduces healthcare disparities and improves health outcomes for transportation-disadvantaged populations (Sivakanthan et al., 2024).

Medical appointments emerge as the highest-priority use case for accessible AV services in disability community research, highlighting healthcare access as a fundamental equity consideration in AV deployment planning.

6.2 Equity Gaps in Current AV Policy

6.2.1 Urban-Rural Deployment Disparities

Current AV deployment strategies focus disproportionately on dense urban environments where technological infrastructure and market demand support profitable operation. However, rural communities face more severe transportation barriers due to limited public transit options, longer travel distances, and higher transportation costs relative to income (Emory et al., 2022).

Rural AV deployment faces unique challenges including limited cellular connectivity, sparse mapping data, less predictable traffic patterns, and different infrastructure characteristics. These challenges may delay rural deployment and exacerbate existing urban-rural mobility disparities unless explicitly addressed through policy intervention.

6.2.2 Socioeconomic Accessibility

Premium pricing models for early AV deployment risk excluding low-income populations from mobility benefits while concentrating advantages among affluent early adopters. This pattern could worsen transportation inequality by providing superior mobility options to those who already possess transportation advantages.

Affordable access requires innovative service models including public-private partnerships, subsidized services for low-income users, and shared mobility arrangements that distribute costs across multiple users. Without deliberate equity interventions, market-driven deployment may exacerbate rather than reduce transportation disparities.

6.2.3 Disability Community Representation

Analysis of AV policy development processes reveals limited meaningful participation by disability communities in standard-setting, regulation development, and deployment planning. This exclusion violates disability rights principles and may result in systems that fail to serve disability community needs effectively.

The principle of "Nothing About Us, Without Us" from disability rights movements emphasizes the importance of including affected communities in design and policy processes. AV governance must incorporate disability community expertise from early development stages through deployment and evaluation.

6.3 Accessibility Design Requirements

6.3.1 Physical Accessibility Features

Accessible AV design requires comprehensive accommodation of diverse disability types and mobility needs. Focus group research with disability communities identifies critical requirements including wheelchair accessibility, transfer assistance, and accommodation of mobility aids and service animals (Sivakanthan et al., 2024).

Vehicle design must accommodate varying levels of independence, with some users requiring full automation and others needing assistive features that enhance rather than replace human control. This diversity requires flexible design approaches that can adapt to individual user needs and preferences.

6.3.2 Communication and Interface Accessibility

Multimodal communication systems must accommodate users with visual, auditory, or cognitive impairments through audio, visual, and tactile interface options. Voice command systems, high-contrast displays, and haptic feedback enable access for users with different sensory capabilities.

Cognitive accessibility requires clear, simple interfaces that accommodate users with intellectual disabilities, dementia, or cognitive processing differences. Interface design must balance comprehensiveness with simplicity to serve diverse cognitive needs effectively.

6.3.3 Emergency and Safety Protocols

Accessible emergency protocols must accommodate diverse disability types and communication needs during system failures or emergency situations. These protocols require multiple communication modalities and connection to disability-appropriate emergency services.

Safety systems must recognize and accommodate disability-related behaviors that might otherwise be misinterpreted as security threats or system malfunctions. This recognition requires training data and algorithms that account for disability-related movement patterns and behaviors.

6.4 Distributive Justice in AV Deployment

6.4.1 Labor Displacement Effects

AV deployment will significantly impact transportation industry employment, affecting taxi drivers, delivery workers, truck drivers, and related service industries. These impacts disproportionately affect workers without college degrees, immigrants, and minority communities who are overrepresented in driving occupations.

Just transition policies must address labor displacement through retraining programs, alternative employment creation, and economic support for displaced workers. Equity considerations require ensuring that AV benefits extend to displaced workers rather than simply redistributing transportation industry profits to technology companies.

6.4.2 Infrastructure Investment Allocation

AV deployment requires substantial infrastructure investment in communication networks, smart traffic systems, and specialized maintenance facilities. Investment allocation decisions will shape which communities benefit from AV deployment and which continue to face transportation barriers.

Equitable infrastructure investment requires prioritizing underserved communities rather than concentrating resources in already well-served affluent areas. This prioritization may conflict with efficiency-focused deployment strategies that target high-demand urban cores.

6.4.3 Environmental Justice Considerations

AV deployment environmental impacts including manufacturing emissions, electricity consumption, and infrastructure construction may disproportionately affect low-income communities and communities of color who already face environmental justice burdens. Sustainable AV deployment requires attention to environmental equity in technology development and deployment planning.

However, AV deployment may also provide environmental benefits through improved efficiency, reduced private vehicle ownership, and support for electric vehicle adoption. Realizing these benefits for disadvantaged communities requires deliberate policy intervention rather than reliance on market mechanisms alone.

7. PUBLIC TRUST AND SOCIAL ACCEPTANCE

7.1 The Knowledge-Trust Paradox

7.1.1 Counterintuitive Relationship Between Knowledge and Acceptance

Research reveals a surprising negative relationship between AV knowledge and public acceptance, where increased technical understanding correlates with decreased trust and adoption intentions. Specifically, a 1% increase in knowledge corresponds to 0.65% decrease in interest and 0.68% decrease in trust, contradicting assumptions that education will increase acceptance (Othman, 2023).

This paradox emerges because greater technical knowledge exposes system limitations, uncertainty areas, and potential failure modes that are not apparent to less informed individuals. Technical sophistication reveals the complexity of challenges facing AV development rather than inspiring confidence in solutions.

7.1.2 Mechanisms Underlying Knowledge-Trust Relationships

The knowledge-trust paradox operates through several psychological mechanisms. Increased knowledge reveals the probabilistic nature of AV safety rather than deterministic guarantees, creating awareness of irreducible uncertainty that reduces confidence. Additionally, technical knowledge exposes the complexity of moral decisions embedded in algorithmic systems, raising questions about whether technical solutions can adequately address ethical challenges.

Expert knowledge also reveals gaps between marketing claims and technical reality, creating skepticism about industry promises regarding safety, reliability, and deployment timelines. This skepticism proves particularly pronounced among individuals with technical backgrounds who can evaluate system capabilities critically.

7.1.3 Implications for Public Engagement

The knowledge-trust paradox complicates public engagement strategies that assume education will build support for AV deployment. Instead, effective engagement must acknowledge uncertainty and limitations while building appropriate trust levels that neither underestimate nor overestimate system capabilities.

Public communication strategies must balance technical accuracy with comprehensibility while avoiding both naive optimism and paralyzing pessimism about AV potential. This balance requires sophisticated communication approaches that respect public intelligence while acknowledging legitimate concerns.

7.2 Factors Influencing Trust Formation

7.2.1 Human-Machine Interaction Quality

The quality of human-machine interaction significantly influences trust development through interface design, communication clarity, and system predictability. Users develop trust through consistent, understandable system behavior that meets expectations and provides appropriate feedback about system status and intentions (Akarsu et al., 2024).

Effective human-machine interaction requires systems that communicate uncertainty, explain decisions when appropriate, and provide users with meaningful oversight capabilities. Transparency must balance comprehensibility with technical accuracy while avoiding information overload that impairs decision-making.

7.2.2 Safety Perception and Risk Assessment

Public trust depends heavily on safety perceptions that may diverge significantly from technical safety analyses. Objective safety statistics prove insufficient for trust formation when they conflict with subjective risk perceptions based on media coverage, personal experience, or cultural factors.

Risk perception research reveals systematic biases including availability heuristics that overweight dramatic but rare events, probability neglect that focuses on worst-case scenarios regardless of likelihood, and affect heuristics that rely on emotional responses rather than analytical assessment. These biases must be acknowledged in trust-building strategies rather than dismissed as irrational.

7.2.3 Anthropomorphism and Trust Calibration

Research demonstrates complex relationships between system anthropomorphism and trust formation. While human-like characteristics can increase initial trust and acceptance, they may also create inappropriate expectations about system capabilities that lead to overtrust and misuse.

Appropriate trust calibration requires matching trust levels to actual system capabilities rather than maximizing trust regardless of performance. This calibration involves communicating system limitations alongside capabilities while supporting user development of appropriate mental models for human-machine interaction.

7.3 The Social Dilemma of AV Ethics

7.3.1 Individual Versus Collective Preference Conflicts

Research reveals a fundamental "social dilemma" wherein individuals prefer protective algorithms for vehicles they occupy while endorsing utilitarian approaches for society as a whole. This preference conflict reflects broader tensions between self-interest and collective welfare that characterize many social choice situations (Gordillo Chávez et al., 2024).

The social dilemma creates challenges for democratic decision-making about AV ethical parameters because individual preferences vary systematically based on personal involvement in outcomes. Those directly affected by decisions prefer protection-focused algorithms, while those evaluating policies abstractly prefer utilitarian approaches that maximize overall welfare.

7.3.2 Democratic Legitimacy and Preference Aggregation

Resolving the social dilemma requires legitimate democratic processes for aggregating conflicting preferences into collective decisions about AV ethical frameworks. Traditional voting mechanisms may prove inadequate when preferences vary systematically based on personal stake in outcomes.

Deliberative democratic approaches that engage citizens in informed discussion about ethical tradeoffs may help bridge individual-collective preference gaps by encouraging perspective-taking and moral reasoning beyond immediate self-interest. However, such processes require careful design to avoid manipulation and ensure authentic participation.

7.3.3 Collective Action Problems

The social dilemma reflects broader collective action problems where individually rational behavior produces collectively suboptimal outcomes. If everyone demands protective algorithms for their own vehicles, overall system performance degrades compared to coordinated utilitarian approaches.

Solving collective action problems requires institutional mechanisms that align individual incentives with collective welfare through regulation, pricing, insurance, or social norms. These mechanisms must be designed to encourage cooperation while respecting individual autonomy and choice.

7.4 Algorithmic Aversion and Acceptance Barriers

7.4.1 Algorithm Aversion Mechanisms

Psychological research identifies "algorithm aversion" as a systematic tendency to discount algorithmic judgment even when it demonstrably outperforms human decision-making. This aversion operates through several mechanisms including perceived lack of agency, reduced sense of control, and attribution of errors to algorithmic rather than human limitations.

Algorithm aversion proves particularly strong in high-stakes domains like driving where individuals value personal control and agency. The life-or-death nature of driving decisions amplifies concerns about delegating authority to algorithmic systems regardless of their objective performance advantages.

7.4.2 Cultural Variation in Algorithmic Acceptance

Cross-cultural research reveals significant variation in algorithmic acceptance based on cultural values including individualism-collectivism, power distance, uncertainty avoidance, and technology adoption patterns. These variations suggest the need for culturally sensitive approaches to AV introduction rather than universal deployment strategies.

Cultural factors influence not only acceptance levels but also preferred ethical frameworks, with some cultures emphasizing individual rights while others prioritize collective welfare. These differences complicate international AV coordination while highlighting the importance of local adaptation in ethical framework design.

7.4.3 Temporal Dynamics of Acceptance

Trust and acceptance evolve over time through experience with technology, social influence, and media coverage. Initial acceptance patterns may differ significantly from long-term adoption patterns as users gain direct experience with AV systems and social norms evolve around autonomous technology.

Understanding temporal dynamics requires longitudinal research that tracks attitude evolution rather than relying on snapshot surveys. These dynamics also suggest the importance of early deployment experiences in shaping long-term public attitudes toward AV technology.

7.5 Trust Calibration Strategies

7.5.1 Appropriate Trust Development

Effective AV deployment requires "appropriate trust" that matches user confidence to actual system capabilities rather than maximizing trust regardless of performance. Overtrust leads to complacency and misuse, while undertrust prevents beneficial adoption even when systems perform reliably.

Appropriate trust development requires transparent communication about system capabilities and limitations, training programs that help users develop accurate mental models, and interface design that supports appropriate reliance decisions through real-time system status communication.

7.5.2 Transparency and Predictability

Trust calibration benefits from system transparency that helps users understand when and how to rely on algorithmic assistance. However, transparency must be carefully designed to provide useful information without overwhelming users or creating anxiety about system complexity.

Predictable system behavior enables trust development through consistent performance that meets user expectations. Predictability requires both reliable technical performance and clear communication about system limitations and operational boundaries.

7.5.3 User Agency and Override Capabilities

Maintaining user agency through override capabilities can support trust development by ensuring users retain ultimate control over system operation. However, override systems must be designed to avoid undermining system effectiveness while providing meaningful user control.

The design of user agency raises complex questions about responsibility allocation when users override system recommendations. Clear frameworks for shared control and responsibility attribution are necessary to support both trust development and appropriate liability allocation.

8. GOVERNANCE AND REGULATORY CHALLENGES

8.1 Multi-Level Governance Complexity

Figure 3. Multi-level governance structure for autonomous vehicle regulation

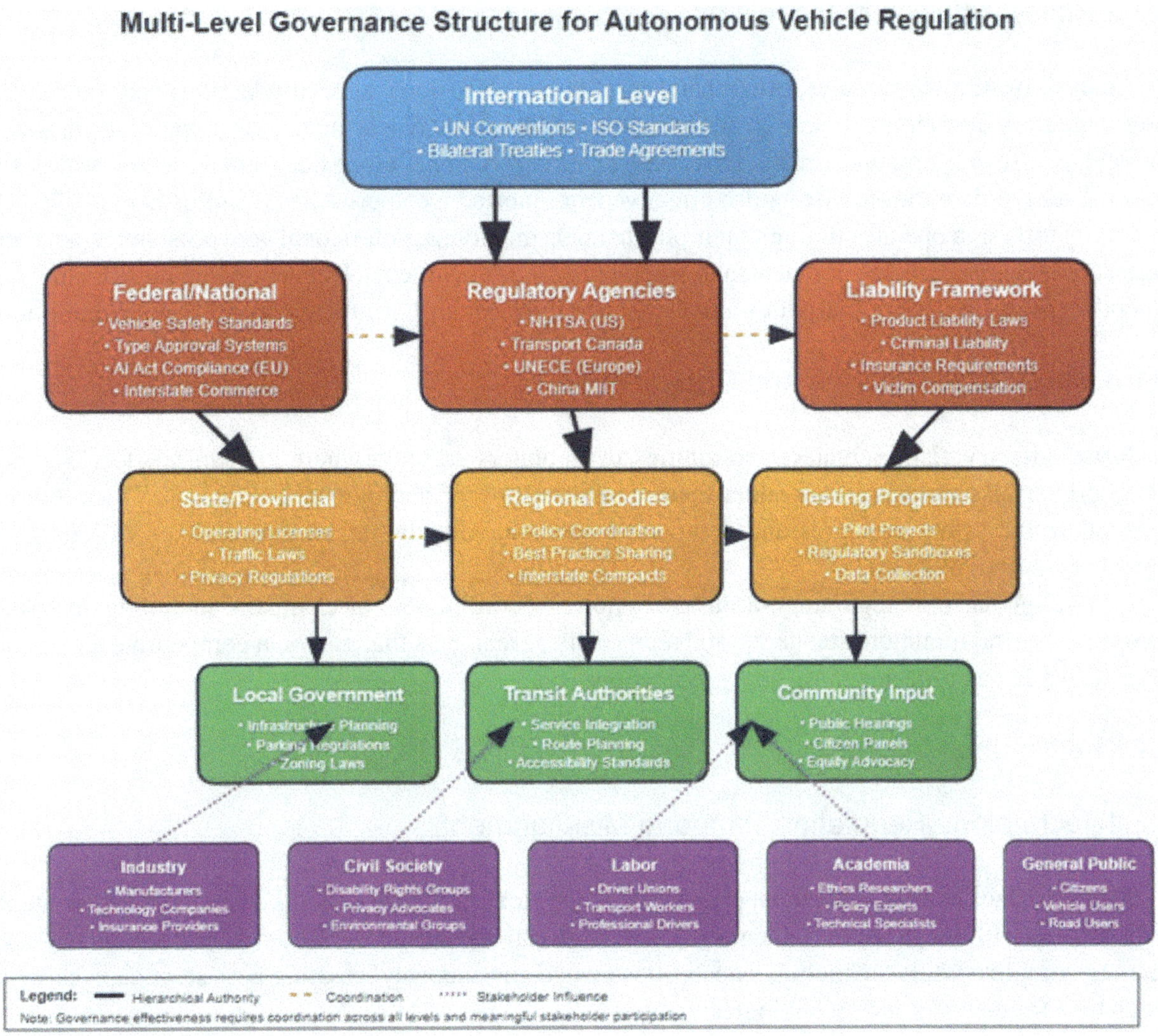

8.1.1 Federal-State-Local Coordination

AV regulation as shown in figure 3, involves complex coordination across multiple governance levels, with federal agencies addressing vehicle safety standards, states governing operational licensing and traffic law, and local governments managing infrastructure and parking regulations. This multi-level structure creates coordination challenges and potential conflicts between different regulatory approaches (Robles & Mallinson, 2024).

Federal frameworks provide uniform safety standards and interstate commerce facilitation, but may lack flexibility to address local conditions and preferences. State-level regulation enables experimentation and local adaptation but can create barriers to interstate operation and regulatory arbitrage opportunities that undermine safety standards.

8.1.2 Policy Diffusion Mechanisms

Analysis of state AV policies reveals patterns of policy diffusion where early adopter states influence later regulatory development through demonstration effects, policy learning, and competitive pressures. However, diffusion processes may spread inappropriate policies alongside beneficial innovations if learning mechanisms fail to distinguish effective from ineffective approaches (Mallinson et al., 2024).

Policy diffusion operates through multiple channels including professional networks, interstate compacts, model legislation development, and federal guidance documents. Understanding these channels is crucial for promoting beneficial policy innovation while preventing harmful regulatory races to the bottom.

8.1.3 Regulatory Arbitrage and Coordination

Jurisdictional variation creates opportunities for regulatory arbitrage where companies select locations based on permissive regulations rather than optimal testing or deployment conditions. This arbitrage can undermine safety standards and democratic accountability by enabling avoidance of legitimate regulatory oversight.

Preventing harmful regulatory arbitrage requires coordination mechanisms including interstate compacts, federal minimum standards, and reciprocity agreements that maintain competitive federalism benefits while ensuring adequate safety protections.

8.2 The Pacing Problem

8.2.1 Technology-Regulation Temporal Misalignment

The rapid pace of AI development creates challenges for regulatory systems designed for more stable technologies with longer development cycles. Traditional regulation development processes may take years to complete while technology capabilities evolve continuously, creating persistent gaps between regulatory frameworks and technological reality.

The pacing problem is exacerbated by AI system opacity that makes performance assessment difficult for regulators lacking technical expertise. Regulatory agencies must develop new capabilities and processes for evaluating AI systems while maintaining democratic accountability and due process protections.

8.2.2 Adaptive Regulatory Approaches

Adaptive regulation strategies include regulatory sandboxes that allow limited deployment under relaxed rules, provisional licensing with performance monitoring, and iterative standard-setting processes that update requirements based on experience. These approaches balance innovation encouragement with safety protection while acknowledging technological uncertainty.

However, adaptive approaches face challenges including ensuring adequate safety protections during experimentation phases, preventing regulatory capture by industry interests, and maintaining democratic accountability in fast-moving technical domains.

8.2.3 Anticipatory Governance

Anticipatory governance attempts to identify and address regulatory needs before they become urgent through technology assessment, scenario planning, and early stakeholder engagement. This approach requires sustained investment in regulatory capacity and expertise development before immediate application needs arise.

Effective anticipatory governance requires collaboration between regulatory agencies, research institutions, and technology developers to monitor technological developments and assess their regulatory implications. This collaboration must maintain appropriate independence while enabling timely information sharing.

8.3 Standard-Setting and Certification

8.3.1 Technical Standard Development

Technical standards for AV safety and ethics emerge through complex multi-stakeholder processes involving industry consortia, international standardization bodies, government agencies, and civil society organizations. These processes must balance competing interests while developing technically sound and ethically appropriate standards.

Standards development faces challenges including industry dominance in technical expertise, limited public participation in complex technical decisions, and international coordination across different legal and cultural frameworks. Democratic input into standard-setting requires innovative mechanisms that enable meaningful participation without compromising technical quality.

8.3.2 Certification and Testing Requirements

Certification processes must translate ethical requirements into testable criteria that can be verified before deployment authorization. However, the complexity of agentic decision-making makes comprehensive testing challenging, particularly for rare but ethically significant scenarios.

Testing requirements must balance thoroughness with practicality while accounting for the adaptive nature of agentic systems that continue learning after deployment. This balance requires new approaches to ongoing monitoring and performance assessment beyond traditional pre-market certification models.

8.3.3 International Harmonization

Cross-border AV operation requires regulatory harmonization to enable seamless travel while respecting national sovereignty and democratic differences in ethical preferences. Harmonization efforts must balance efficiency benefits with legitimate variation in values and legal systems.

International coordination mechanisms include bilateral agreements, multilateral treaties, and international organization frameworks that facilitate cooperation while preserving national decision-making authority over fundamental value choices.

8.4 Democratic Participation and Legitimacy

8.4.1 Public Participation in Technical Regulation

Democratic legitimacy requires meaningful public participation in AV governance, but technical complexity creates barriers to informed participation. Effective participation mechanisms must bridge expertise gaps while ensuring that public values guide technical choices.

Participatory approaches include citizen panels, deliberative surveys, consensus conferences, and community-based participatory research that engage diverse stakeholders in governance processes. These approaches must be carefully designed to avoid manipulation while enabling authentic democratic input.

8.4.2 Representation and Inclusion

Current AV policy processes often underrepresent communities most affected by deployment including disabled individuals, elderly populations, racial minorities, and low-income communities. Ensuring equitable representation requires proactive outreach and accommodation rather than relying on voluntary participation.

Inclusive participation requires addressing practical barriers including meeting accessibility, language translation, childcare provision, and compensation for participation time. These investments are necessary to ensure that governance processes reflect the full range of affected communities.

8.4.3 Accountability and Oversight

Democratic accountability requires mechanisms for ongoing oversight of AV deployment and performance rather than only pre-deployment authorization. These mechanisms must enable course correction based on experience while maintaining stability and predictability for industry investment.

Accountability mechanisms include legislative oversight, administrative review, judicial review, and civil society monitoring that provide multiple channels for holding AV governance systems responsible to democratic values and public welfare.

9. CONCLUSION

9.1 Synthesis of Key Findings

9.1.1 The Inadequacy of Current Frameworks

This analysis reveals that existing ethical, legal, and regulatory frameworks prove fundamentally inadequate for addressing the complex challenges posed by agentic autonomous vehicles. Traditional approaches based on human-centered decision-making, deterministic causation, and static rule-based systems cannot accommodate the adaptive, probabilistic, and emergent characteristics of agentic AI systems.

The trolley problem framework, while influential in philosophical discourse, fails to capture the reality of AV decision-making characterized by uncertainty, temporal constraints, and continuous risk distribution rather than discrete moral choices. Similarly, conventional liability frameworks struggle with the distributed agency characteristic of agentic systems where responsibility spans multiple human and artificial agents.

9.1.2 The Centrality of Social Justice

Equity considerations emerge as central rather than peripheral to AV governance, with deployment patterns having profound implications for social justice, accessibility, and democratic participation. The potential for AV technology to either reduce or exacerbate existing transportation inequities depends critically on governance choices made during the current development and early deployment phase.

The research demonstrates that market-driven deployment without deliberate equity interventions will likely concentrate benefits among affluent early adopters while maintaining or worsening transportation barriers for marginalized communities. Realizing the transformative potential of AVs for underserved populations requires proactive policy intervention and inclusive governance processes.

9.1.3 The Importance of Process Legitimacy

Procedural legitimacy proves as important as substantive outcomes in determining the social acceptability of agentic AV deployment. The knowledge-trust paradox and social dilemma research demonstrate that technical superiority alone cannot ensure public acceptance; legitimate governance processes that respect democratic values and individual agency are essential for sustainable deployment.

The expertise-preference paradox highlights tensions between technical knowledge and democratic input that require careful navigation through inclusive governance mechanisms. These mechanisms must bridge technical expertise and public values rather than privileging either technical or democratic considerations exclusively.

9.2 Integrative Framework Recommendations

9.2.1 Risk-Based Ethical Foundations

The analysis supports shifting from harm-minimization framings toward risk distribution approaches that acknowledge the inherent uncertainty in transportation systems. This shift requires developing new ethical frameworks that guide fair allocation of unavoidable risks rather than attempting to eliminate risk entirely.

Risk-based approaches must incorporate democratic input about acceptable risk levels and distribution patterns while protecting vulnerable road users who cannot effectively advocate for themselves. This protection requires proactive equity measures rather than reliance on market mechanisms or majoritarian decision-making alone.

9.2.2 Adaptive Governance Mechanisms

Effective governance requires adaptive mechanisms that can evolve with technological development while maintaining democratic accountability and legal certainty. These mechanisms include regulatory sandboxes for controlled experimentation, iterative standard-setting processes, and ongoing monitoring systems that enable course correction based on deployment experience.

Adaptive governance must balance innovation encouragement with safety protection while ensuring that adaptation processes remain democratically accountable rather than captured by industry interests. This balance requires sustained investment in regulatory capacity and expertise development alongside stakeholder engagement mechanisms.

9.2.3 Mandatory Accessibility Integration

The analysis strongly supports treating accessibility as a fundamental design requirement rather than an optional add-on feature. This integration requires mandatory accessibility standards for AV systems, inclusive design processes that engage disability communities, and deployment policies that prioritize underserved population access.

Accessibility integration must extend beyond physical vehicle design to encompass service delivery models, pricing structures, and operational practices that ensure meaningful access for people with disabilities and other transportation-disadvantaged populations.

9.3 Critical Implementation Requirements

9.3.1 Multi-Stakeholder Governance

Effective AV governance requires meaningful participation by diverse stakeholders including disability communities, elderly individuals, racial minorities, transportation justice advocates, privacy rights organizations, and affected workers alongside technical experts and industry representatives.

Multi-stakeholder governance mechanisms must address power imbalances, resource disparities, and expertise gaps that can skew participation toward dominant interests. This requires proactive outreach, capacity building, and institutional design that enables authentic democratic participation in technical governance processes.

9.3.2 International Coordination

Cross-border AV operation requires international coordination mechanisms that respect cultural differences in ethical preferences while enabling interoperability and shared learning. These mechanisms should facilitate cooperation while preserving national sovereignty over fundamental value choices.

International coordination must address both technical standardization needs and broader governance challenges including liability attribution, privacy protection, and ethical framework alignment across different legal and cultural systems.

REFERENCES

Cecchini, D., Brantley, S., & Dubljević, V. (2023). Moral judgment in realistic traffic scenarios: Moving beyond the trolley paradigm for ethics of autonomous vehicles. *AI & Society*. Advance online publication. DOI: 10.1007/s00146-023-01813-y

Ehsani, J. P., Hellinger, A., Stephens, D. K., Shin, M. R., Michael, J., McCourt, A., & Vernick, J. (2022). State laws for autonomous vehicle safety, equity, and insurance. *The Journal of Law, Medicine & Ethics*, *50*(3), 569–582. DOI: 10.1017/jme.2022.96 PMID: 36398642

Ehsani, J. P., Michael, J. P., Igusa, T., Mueller, J., Chang, C.-H., & Yenokyan, G. (2024). Advancing transportation equity and safety through autonomous vehicles. *Health Equity*, *8*(1), 143–146. DOI: 10.1089/heq.2023.0107 PMID: 38505763

Emory, K., Douma, F., & Cao, J. (2022). Autonomous vehicle policies with equity implications: Patterns and gaps. *Transportation Research Interdisciplinary Perspectives*, *13*, 100521. Advance online publication. DOI: 10.1016/j.trip.2021.100521

Geisslinger, M., Poszler, F., Betz, J., Lütge, C., & Lienkamp, M. (2021). Autonomous driving ethics: From trolley problem to ethics of risk. *Philosophy & Technology*, *34*(4), 1033–1055. DOI: 10.1007/s13347-021-00449-4

Geisslinger, M., Poszler, F., & Lienkamp, M. (2023). An ethical trajectory planning algorithm for autonomous vehicles. *Nature Machine Intelligence*, *5*(2), 137–144. DOI: 10.1038/s42256-022-00607-z

Gless, S., & Ligeti, K. (2024). *Regulating driving automation in the European Union – Criminal liability on the road ahead?* New Journal of European Criminal Law., DOI: 10.1177/20322844231213336

Gordillo Chávez, D., Cloarec, J., & Meyer-Waarden, L. (2024). Opening the moral machine's cover: How algorithmic aversion shapes autonomous vehicle adoption. *Transportation Research Part A, Policy and Practice*, *187*, 104193. Advance online publication. DOI: 10.1016/j.tra.2024.104193

Königs, P. (2023). Of trolleys and self-driving cars: What machine ethicists can and cannot learn from trolleyology. *Utilitas*, *35*(1), 70–87. DOI: 10.1017/S0953820822000395

Krügel, S., & Uhl, M. (2024). The risk ethics of autonomous vehicles: An empirical approach. *Scientific Reports*, *14*(1), 960. Advance online publication. DOI: 10.1038/s41598-024-51313-2 PMID: 38200075

Kubica, M. L. (2022). Autonomous vehicles and liability law. *The American Journal of Comparative Law*, *70*(Supplement_1), i39–i69. DOI: 10.1093/ajcl/avac015

Lee, D., & Hess, D. J. (2022). Public concerns and connected and automated vehicles: Safety, privacy, and data security. *Humanities & Social Sciences Communications*, *9*(1), 90. Advance online publication. DOI: 10.1057/s41599-022-01110-x

Li, Y., Tao, X., Zhang, X., Liu, J., & Xu, J. (2022). Privacy-preserved federated learning for autonomous driving. *IEEE Transactions on Intelligent Transportation Systems*, *23*(7), 8423–8434. DOI: 10.1109/TITS.2021.3081560

Mallinson, D. J., Azevedo, L., Best, E., Robles, P., & Wang, J. (2024). The future of AI is in the states: The case of autonomous vehicle policies. *Business and Politics*, *26*(2), 180–199. DOI: 10.1017/bap.2023.19

Martinho, A., Herber, N., Kroesen, M., & Chorus, C. (2021). Ethical issues in focus by the autonomous vehicles industry. *Transport Reviews*, *41*(5), 556–577. DOI: 10.1080/01441647.2020.1862355

Mulder, T., & Vellinga, N. E. (2021). Exploring data protection challenges of automated driving. *Computer Law & Security Review*, *40*, 105530. Advance online publication. DOI: 10.1016/j.clsr.2021.105530

Othman, K. (2023). Investigating how the public acceptance of autonomous vehicles evolve with the changes in the level of knowledge: A demographic analysis. *Cogent Engineering*, *10*(1), 2220502. Advance online publication. DOI: 10.1080/23311916.2023.2220502

Poszler, F., Geisslinger, M., & Lütge, C. (2024). Ethical decision-making for self-driving vehicles: A proposed model & list of value-laden terms that warrant (technical) specification. *Science and Engineering Ethics*, *30*(5), 40. Advance online publication. DOI: 10.1007/s11948-024-00513-0 PMID: 39387983

Rhim, J., Lee, J.-H., Chen, M., & Lim, A. (2021). A deeper look at autonomous vehicle ethics: An integrative ethical decision-making framework to explain moral pluralism. *Frontiers in Robotics and AI*, *8*, 632394. Advance online publication. DOI: 10.3389/frobt.2021.632394 PMID: 34017859

Robinson, J., Smyth, J., Woodman, R., & Donzella, V. (2022). Ethical considerations and moral implications of autonomous vehicles and unavoidable collisions. *Theoretical Issues in Ergonomics Science*, *23*(4), 435–452. DOI: 10.1080/1463922X.2021.1978013

Robles, P., & Mallinson, D. J. (2024). Policy learning and the diffusion of autonomous vehicle policy in the American states. *State & Local Government Review*, *56*(4), 335–358. DOI: 10.1177/0160323X241262816

Schuessler, D. (2024). The probability problems of the Moral Machine Experiment. *AI and Ethics*, *4*(2), 501–510. DOI: 10.1007/s43681-023-00287-4

Sivakanthan, S., Cooper, R., Lopes, C., Kulich, H., Deepak, N., Lee, C. D., Wang, H., Candiotti, J. L., Dicianno, B. E., Koontz, A., & Cooper, R. A. (2024). Accessible autonomous transportation and services: A focus group study. *Disability and Rehabilitation. Assistive Technology*, *19*(5), 1992–1999. DOI: 10.1080/17483107.2023.2242898 PMID: 37548013

Takemoto, K. (2024). The moral machine experiment on large language models. *Royal Society Open Science*, *11*(2), 231393. Advance online publication. DOI: 10.1098/rsos.231393 PMID: 38328569

Zhu, A., Yang, S., Chen, Y., & Xing, C. (2022). A moral decision-making study of autonomous vehicles: Expertise predicts a preference for algorithms in dilemmas. *Personality and Individual Differences*, *186*, 111356. Advance online publication. DOI: 10.1016/j.paid.2021.111356

Chapter 2
Deep Learning and Machine Learning Approaches for Vehicle Fault Diagnostics and Predictive Maintenance

Thanuja Talluri
http://orcid.org/0009-0004-5244-8357
Lovely Professional University, India

Jimmy Singla
http://orcid.org/0000-0003-2077-1622
Lovely Professional University, India

Nikita
Lovely Professional University, India

ABSTRACT

The massive development in contemporary vehicles, especially electric and smart cars, has heightened the need for high-performance fault detection and diagnostic (FDD) systems. In the last decade, artificial intelligence and machine learning have been proven to be suitable substitutes by facilitating processing on sophisticated sensor inputs from engines, braking systems, batteries, along with inputs from in-vehicle networks. This review aggregates findings from recent research studies to outline methodological directions, and future research projects with special focus on emphasizing that automotive maintenance will be driven towards greater safety, efficiency, and predictive functionality by helping steer automobile maintenance through artificial intelligence-enhanced diagnostic capabilities.

1.INTRODUCTION

The fast growth of the automotive industry in electrification, connectivity, and automation has made the modern vehicles extremely complicated cyber-physical systems. The modern vehicles that are used today integrate sophisticated electronics, embedded systems, and intelligent software in contrast to the

DOI: 10.4018/979-8-3373-7355-3.ch002

traditional automobiles which were largely mechanical. The recent surge of the use of electric vehicles (EVs), technologies of the advanced driver assistance system (ADAS), vehicle-to-everything (V2X) communication, and autonomous driving technologies has made the automotive platforms highly technologically advanced. This has led to cars in the current world being digital ecosystems that are connected but not a mechanical machine. Modern cars have dozens, and in other cases even hundreds of Electronic Control Unit (ECU) which control different subsystems of the car like engine control, battery, brake, steering, infotainment systems, climate systems, and safety systems. Such ECUs constantly exchange information via in-vehicle networks like Controller Area Network (CAN), Local Interconnect Network (LIN), and automotive Ethernet. Besides, there is a large series of sensors: temperature sensors, pressure sensors, LiDAR, radar, cameras, and accelerators, that constantly check the condition of the vehicle inside and outside. Such networked infrastructure produces large amounts of real-time operational data, which provides useful information on the performance of the system and the health of the vehicle. Although these new technologies contribute greatly to the safety of driving, energy conservation, comfort of passengers, and the overall performance of the vehicle, they also create new weaknesses. The growing reliance on electronic and software based parts increases the probability of electrical, electronic and software failures. Even the slightest sensor failure, communication lag, software bug or battery management issues can turn into severe malfunction of the operation unless promptly corrected. In electric cars, say, battery management system malfunctions can cause diminished efficiency, overheating or even dangerous situations. On the same note, the safety of passengers might also be compromised by mistakes made in the automated driving systems. Unnoticed or concealed defects are very severe. They may lead to unforeseen failures, higher services, and lengthier car idleness. In the case of commercial fleets, transportation services, the downtimes are directly converted to economical losses and inefficiencies. In addition, safety critical failures can lead to accidents, lawsuits and reputational losses to manufacturers. When starting to operate in the environment where the demand towards reliability and safety is extremely high, repetitive failures may severely reduce the consumer confidence and loyalty to the brand. In order to solve these issues the introduction of high-tech Fault Detection and Diagnosis (FDD) systems has now become a necessity. The current FDD platforms take advantage of data-driven technology, machine learning methods, artificial intelligence, and predictive analytics to track the condition of vehicles in real-time. Such systems examine the patterns of sensor data, identify when something is not normal and issue early warnings of component malfunction before it is too late to do so. FDD has enabled predictive maintenance strategies that allow a manufacturer and the service provider to plan interventions on time, which lowers the maintenance cost and eliminates unwanted breakdowns. Fault detection mechanisms are extremely important in electric and autonomous vehicles whose interdependencies are even more complex to deliver reliability and safety. The combination of smart diagnostics with onboard analytics and cloud-based analytics will help automotive industries improve the resilience of systems, increase performance, and increase the vehicle life cycle. As a result, state-of-the-art FDD systems are not any longer optional to the design and functionality of traditional and the next generation automobiles.Traditional diagnostic paradigms, such as rule-driven systems and model-driven methodologies, have been staples for decades. They rely on mathematical models that characterize system activity or thresholds defined by experts. However, their effectiveness is often constrained by nonlinear dynamics, variability in operating conditions, and vehicle subsystem complexity that is ever-increasing. Advances in abundant sensor data and advanced computational resources have enabled artificial intelligence (AI) and machine learning (ML) to be implemented as fruitful solutions to meet these needs.

In recent years, artificial intelligence and machine learning frameworks have demonstrated encouraging ability in dealing with issues from several vehicular applications. Traditional algorithms, such as support vector machines, random forests, and clusterization techniques, have been utilized for state estimation, combustion anomalies detection, and tracking vibrations or audio signals. Furthermore, deep learning frameworks—including convolutional neural networks, recurrent neural networks, autoencoders, and transformer networks—added heightened capabilities by depicting advanced time-series data, automatically learning feature spaces, and applying anomaly detection in high-dimensional settings. Sources of data utilized originate from several sites, such as onboard diagnostics (OBD-II), controller area networks (CAN bus), battery monitoring systems, and telematics. Applications are varied. For combustion engines, AI-capable diagnosis detects misfires and wear patterns through vibration and audio signal analysis. For electric vehicles, ML models observe state-of-charge and state-of-health, aiding battery safety through rigorous operations like excessive fast charging. For connected vehicles, anomalous detection protects vehicle networks from cyber-attacks.At larger scales, predictive maintenance architecture guided by ML facilitate fleet-level optimization with decreased servicing expenses and improved operating reliability.

Although the intelligent fault detection system and diagnosis systems have made remarkable progress, a number of practical and technical issues still constrain their implementation in practice in real world automotive setting. The absence of annotated fault data is also one of the main challenges. In safety-related systems like cars, real occurrences of fault, in particular disastrous faults, occur infrequently. Although this is a positive aspect in the safety front, it poses a problem of imbalanced data sets to machine learning models. The majority of the obtained operational data are typical driving cases, but there are few, uncomplete, or none of them which are labeled fault instances. Besides, intentional generation of fault data is not only expensive but also time-consuming, and occasionally even unsafe. It is very difficult to train effective and trustworthy deep learning models without high-quality labeled datasets. The second important impediment is that the existing models may not be generalized to various vehicle architectures. The automotive systems come in many different variations in hardware set ups, location of sensors, communication standards, and control mechanisms. A training model used in one platform of the vehicle can not be effective when applied on a different platform because of differences in design and operation. Such non-standardization makes it difficult to implement on a large scale, especially when a manufacturer has to deal with more than one vehicle model, both with internal combustion engine vehicles and electric cars. The driving environment variability, climatic conditions and user behavior also makes models less transferable. Moreover, the very opaque or black-box characteristics of deep learning algorithms pose a significant challenge to safety-critical automotive usage. End users, regulatory bodies and manufacturers need to know how decisions on systems are made, particularly where decisions on safety warnings or automated intervention are involved. Nonetheless, numerous deep neural networks are very accurate in their predictions yet they lack explanatory reasons as to why the predictions were made. Such absence of transparency makes it less interesting to trust and to debug, certify and validate it. Explainability is not a choice but a necessity in a setting that may jeopardize the lives of human beings. Embedded automotive platforms have real-time constraints that limit them to a considerable degree as well. Vehicle control units normally have limited computational resources, memory space, and power (energy). Running intricate deep learning networks may cause latency, overheating, or waste a lot of energy in such resource-constrained systems. The diagnostic algorithms applied should thus be able to compromise accuracy with computational efficiency so that faults can be detected early enough without overloading the onboard systems. In order to defeat these issues, some positive research avenues are forthcoming. Federated learning is a collaborative method where data is not shared among various vehicles or manufacturers during training of shared models. The technique maintains privacy and allows the use of a wide range of datasets, enhancing model robustness and generalizability. Another transformational solution is a digital twin technology. Using the virtual representations of real physical objects or parts, the engineers can model possible fault conditions, degradation behaviour, and predict failures when the object or component is within different operating conditions, without putting physical objects at risk. To improve the transparency of models, explainable artificial intelligence (XAI) methods are currently being created to offer interpretable insights into the decision-making mechanism. Such techniques would enable engineers to know which characteristics affect predictions, which would make predictions more trusted and accepted by the regulatory authorities. Moreover, it is likely that with the integration of edge computing, cloud computing, and 5G networks integrated with high-speed communication networks, distributed diagnostic systems can be deployed. Immediate and time sensitive analysis can be done by the edge devices and deeper computations and long term trend analysis can be done using cloud infrastructure. A combination of these innovations will ensure the development of scalable and real-time, as well as intelligent, fault management ecosystems in the next generations of automotive systems.

The current survey summarizes and critically analyzes the findings of studies carried out within the last decade on the topic of vehicle Fault Detection and Diagnosis (FDD) based on the applications of Artificial Intelligence (AI) and Machine Learning (ML) frameworks. The last few years have seen the pace of development of intelligent algorithms, sensor technologies, and computer power might alter dramatically the manner in which system of vehicle health monitoring is being designed and introduced into practice. This review is aimed at organizing current knowledge in a systematic way, determining the existing research tendencies, and underlining the methodological advances which resulted in contemporary automotive diagnostics. One of the key areas of the survey is the examination of methodological trends taken by recent research studies. Data-driven frameworks have been gaining popularity among researchers who no longer focus on the traditional model-based and rule-based methods of diagnostics. The Support Vector Machines (SVM), Artificial Neural Networks (ANN), Random Forests and Gradient Boosting are classified as supervised learning methods that have been extensively applied in classification and fault prediction. Parallel to this, deep learning architectures, such as Convolutional Neural Networks (CNNs), Recurrent Neural Networks (RNNs), and Long Short-Term Memory (LSTM) networks, have been shown to perform well with high dimensional sensor data, as well as time-series signals. Learning methods and anomaly detection methods that are unsupervised have become important as well, especially when the labeled fault data is limited. The survey summarizes these approaches in terms of areas of application, complexity of computation, interpretability and in real time. The source and the character of data that is utilized in the process of diagnostics of vehicles is another important aspect examined in this review. Researchers exploit a variety of data, among them the onboard diagnostic (OBD) logs, CAN bus communications signal, vibration signal, thermal signal, acoustic emission, and battery voltage-current data. The research may utilize public benchmark data sets, whereas others may utilize industry proprietary data or simulated generated data. The survey examines the reliability, diversity and limitations of these data sources and the quality of data and representative data sets in constructing a strong AI model. A special focus is given to major automotive subsystems in which the applications of FDD are the most evident. AI models are also used in engine systems to identify the presence of misfires and faults in fuel injection, emissions, and abnormal combustion patterns. In the case of electric vehicles, battery management systems have become one of the central areas of research with machine learning being applied to predict state of charge (SOC), state of health (SOH), thermal anomalies and predictive degradation. Fault detection of braking systems, especially anti-lock braking systems (ABS), and electronic stability control (ESC), are determined with the aim of safeguarding braking system functions. Also, the survey examines diagnostic methods used to in-vehicle communication networks whereby, the detection of cyber-attacks, signal interruption or communication interruptions is of great importance to ensure the integrity of the system. In addition to the overview of the current progress, the review establishes the existing gaps in research and practical restrictions, such as the problem of data scarcity, model generalization, computational limitations, and explainability. It further describes the new opportunities including hybrid physics-informed AI models, federated learning models and real time edge-cloud integration. Finally, the survey highlights the fact that smart diagnostic qualities are at the heart of the development of the next generation transportation systems. The AI-based FDD solutions help to achieve safer, smarter and more sustainable mobility ecosystems by providing the means to detect faults early, predictive maintenance, and overall system reliability.

2. RELATED WORKS

- Michailidis et al. (2025) provides an upto-date survey of ML applications that leverage OBD-II streams for efficiency, safety and security tasks, categorizing feature extraction and model choices and identifying gaps in standard datasets and benchmarks. Useful for framing OBD/CAN as a primary data source for vehicle FDD.
- Machlev (2024) comprehensive review focused on deep-learning methods for battery fault detection and RUL/SOH prognostics; highlights the promise of representation learning and the practical challenges of noisy field data and limited labeled faults. Good source for battery-specific ML trends and open problems.
- Özdemir et al. (2024) systematic survey of anomaly-detection approaches applied to the CAN bus (supervised, unsupervised, deep time-series models), with a strong focus on evaluation datasets and attack models — valuable for any section on cyber-resilience and network-level diagnostics.
- Zachariades and Xavier (2025) sensors review that collates classical ML and deep models used for fault diagnosis in electric machines, emphasizing sensor choices and the impact of preprocessing and feature design on detection performance. Helpful when discussing diagnostics for electric drivetrains and motors.
- Zhao et al. (2024) examines the translational gap between laboratory battery-fault tests and field deployment; details how lab conditions differ in terms of signals, noise, and fault evolution and recommends protocols for field-viable diagnostics. Critical for grounding battery FDD claims in realism.
- Akbalık et al. (2024) demonstrates an applied pipeline using audio capture, MFCC/DWT features, and ML classifiers to identify engine faults from acoustic signals; includes experimental setup, confusion matrices, and a practical discussion on sensor placement and preprocessing. Use this as an example of acoustic/vibration-based FDD.

Figure 1. Example audio signals and the ML pipeline for engine fault detection

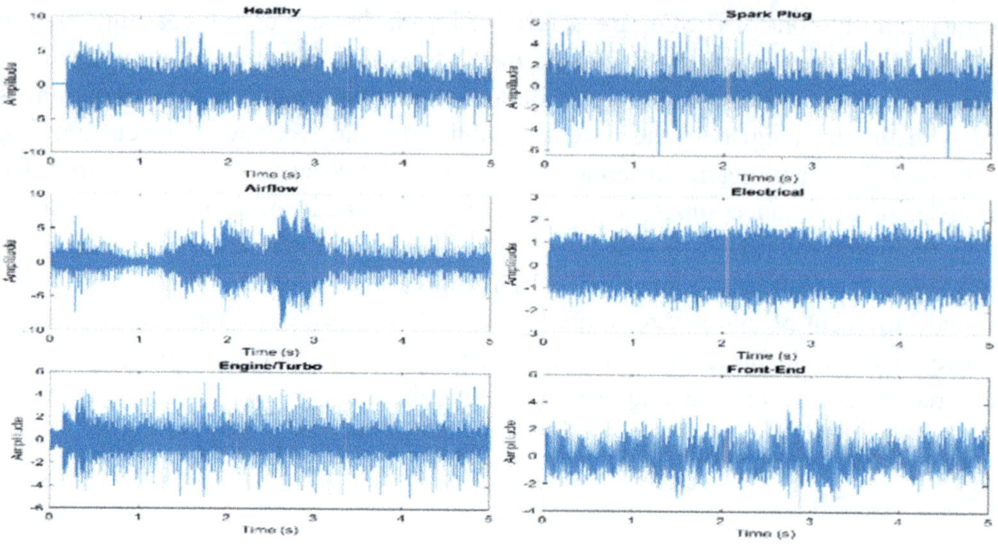

- Zhang et al. (2023) introduces a dynamical autoencoder framework and a deployment blueprint (charging stations → cloud detector) that addresses privacy and distributional shifts; shows strong performance on realistic, large-scale BMS datasets and discusses economic decisioning from fault scores. This is one of the most influential applied DL works for batteries.

Figure 2. Model deployment and dynamical autoencoder architecture for real-world EV battery fault detection

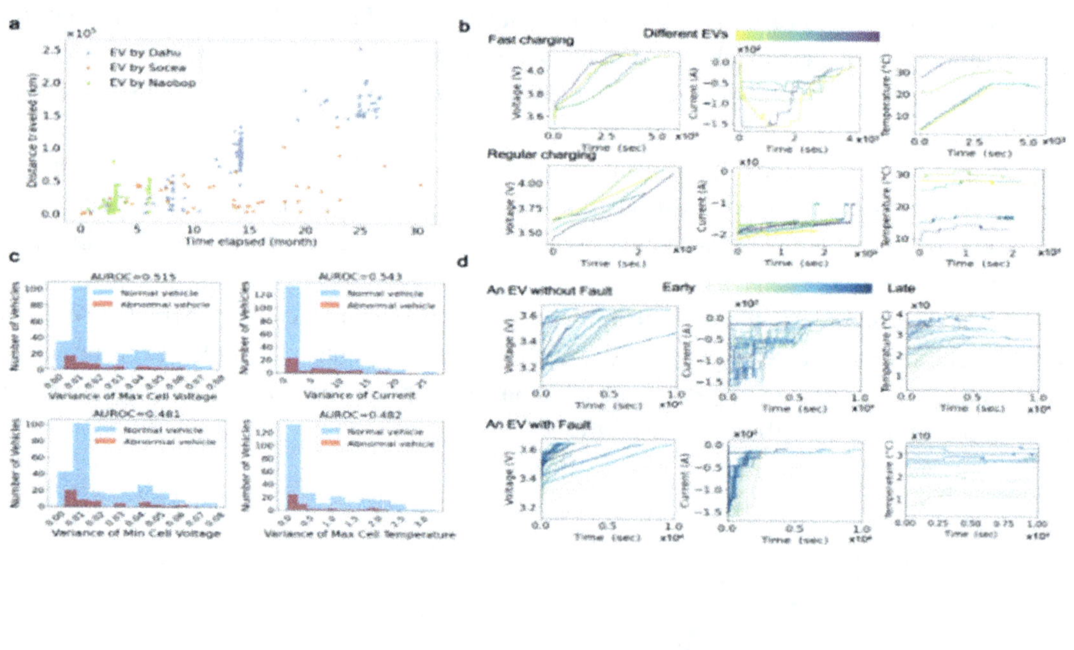

Figure 3. Evolution of learned embeddings across the autoencoder layers (t-SNE visualization)

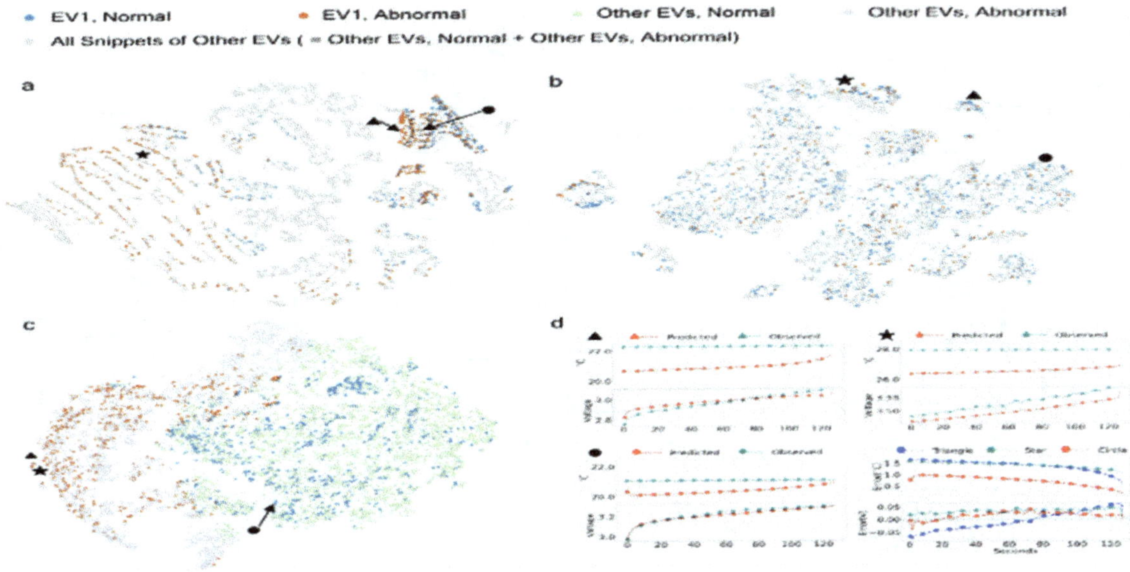

- Industry-facing review that surveys practical ML pipelines for predictive maintenance in automotive contexts and calls out integration, latency, and ROI considerations — excellent for the "deployment & business" perspective (Theissler et al., 2021).

Figure 4. Flowchart of the proposed AutoML pipeline for real-time fault detection and diagnosis (RT-FDD)

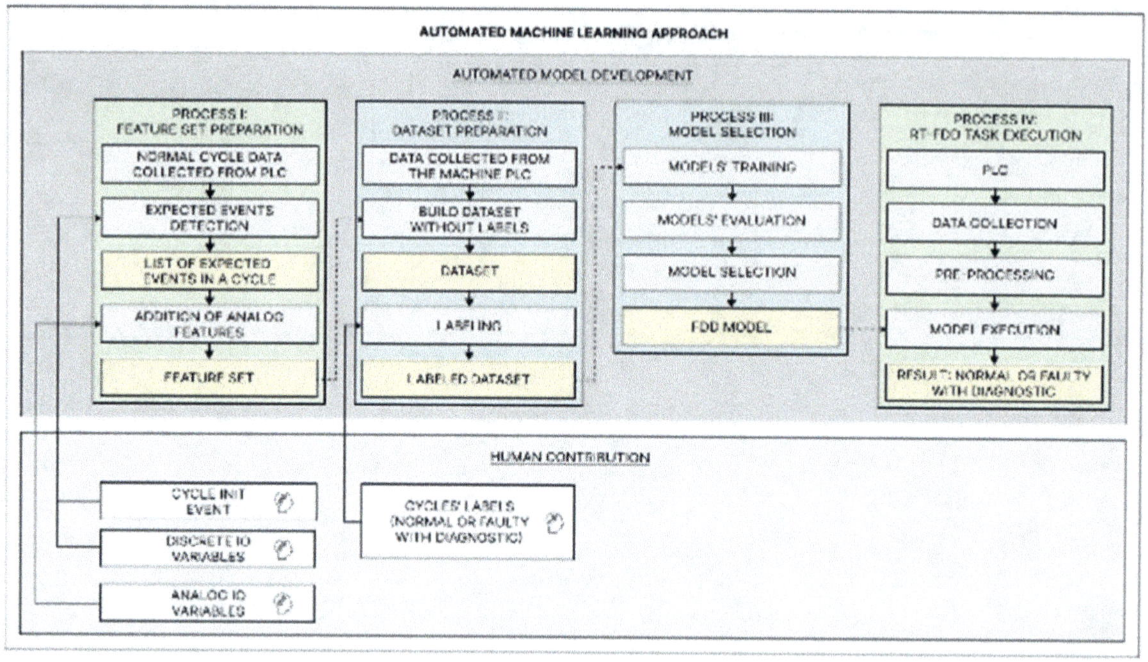

- Gong et al. (2022) implementation-oriented paper describing end-to-end ML pipelines for automotive FDD, with pragmatic notes on training data, real-time constraints, and interpretability. Good reference for methodological recipes in engineering sections.
- Srinivaas (2025) recent MDPI review collating classical ML and DL approaches for ICE diagnostics; synthesizes comparative performance for vibration, acoustic and in-cylinder sensor modalities and charts future needs for cross-platform generalization. (Use for up-to-date ICE literature mapping.)
- Hossain (2024) interdisciplinary survey summarizing ML/DL methods across vehicle subsystems (engine, battery, braking, network) and identifying ethical, safety, and dataset standardization gaps - useful for positioning multi-modal fusion and regulatory considerations.
- Zhang et al. (2023) several conference and journal follow-ups replicate the dynamical-DL approach and explore transfer learning and cloud/edge tradeoffs for real deployments; cite Zhang et al. as the foundational practical demonstration.
- AAM / TU Delft XGBoost strategy (~2019 conference) a practical engineering study that applied XGBoost to over-discharge fault diagnosis on EV battery datasets, demonstrating that tree-based ensembles provide a strong baseline and interpretability advantage over black-box deep models in some usage scenarios. (Good counterpoint when discussing simpler models vs deep learning.)
- Leite et al. (2022) Proposes an AutoML pipeline tailored to RT-FDD (auto-feature selection, model search and deployment) that allows non-ML experts to configure detectors for industrial cyclic processes — relevant for sections on automation and operationalization of diagnostics.

- Abdolrasol et al. (2024) reviews data-driven battery diagnostics (feature engineering, ML/DL, sensor placement) and emphasizes explainability and sensor fusion for robust field detection. Use this as a complementary, implementation-focused battery review.
- Zhou et al. (2025) proposes self-supervised BERT-style pretraining (point-masked signal modeling) for battery time-series; demonstrates improved AUROC and robustness by leveraging large unlabeled datasets — an important recent direction in self-supervised fault representation learning.

Table 1. Comparison table on existing AI/ML-based vehicle fault detection and diagnostic system

#	Authors	Year	Issue	Method	Input variables	Output	Remarks
1	Michailidis et al.	2025	Survey: OBD-II ML applications	Literature survey/ taxonomy	OBD-II/CAN features (various)	N/A	Comprehensive categorization of OBD-II feature extraction and ML pipelines; highlights dataset gaps and benchmarking needs.
2	Machlev	2024	EV battery fault detection & prognostics	Review of DL methods	BMS telemetry, charging profiles, voltage/current/ temperature	SOH / fault flags / RUL estimates	Deep-learning focus; emphasizes representation learning and noisy-field data challenges.
3	Özdemir et al.	2024	CAN bus anomaly detection survey	Survey (supervised & unsupervised methods)	CAN frames, IDs, timing, payload features	Anomaly / intrusion detection	Detailed taxonomy of attack models and datasets; evaluates supervised vs unsupervised trade-offs.
4	Zachariades & Xavier	2025	Fault diagnosis in electric machines	Review of AI techniques	Sensor signals (vibration, current, voltage)	Fault class / severity	Emphasizes preprocessing and feature design impact on detection performance.
5	Zhao et al.	2024	Lab-to-field battery fault translation	Review / analysis	Lab test signals vs field BMS data	Recommendations for field-ready diagnostics	Highlights translational gap; recommends robust validation protocols.
6	Akbalık et al.	2024	Engine fault detection via sound	ML classifiers (SVM, RF) with audio features	Engine audio: MFCCs, wavelet features, vibration proxies	Fault class (misfire, knock, healthy)	Practical pipeline with sensor placement notes; shows audio is effective and low-cost.
7	Zhang et al.	2023	Realistic Li-ion fault detection at scale	Dynamical deep autoencoder (time-series DL)	BMS time-series (V, I, T), charge/discharge cycles	Anomaly scores; fault detection	Deployable blueprint (charging stations → cloud); strong performance on large BMS datasets.
8	Theissler et al.	2021	Predictive maintenance use-cases	Review / case studies of ML pipelines	Fleet telemetry, maintenance logs, sensor signals	Remaining useful life / failure probability	Industry-oriented; stresses ROI, latency, and system integration challenges.

continued on following page

Table 1. Continued

#	Authors	Year	Issue	Method	Input variables	Output	Remarks
9	Gong et al.	2022	Implementing AI for automotive FDD	End-to-end ML pipeline guidance	OBD/CAN, vibration, acoustic, BMS signals	Fault detection / diagnostic labels	Practical implementation notes: data handling, deployment, interpretability.
10	Srinivaas	2025	ML for ICE fault detection	Survey of ML & DL for engines	Acoustic, vibration, in-cylinder sensors, OBD	Fault classification; anomaly detection	Up-to-date comparison of classical ML vs DL for ICE contexts.
11	Hossain	2024	AI-driven vehicle fault diagnosis (multi-domain)	Interdisciplinary review	Multi-modal: engine, battery, brakes, network data	N/A (survey)	Summarizes ethical, dataset, and standardization gaps across subsystems.
12	(Various follow-ups to Zhang)	2023–2024	Extensions/replicates of dynamical DL	Transfer learning, cloud/edge trade-offs	BMS telemetry; augmented datasets	Fault detection; transfer performance	Shows applicability and trade-offs when adapting Zhang et al.'s approach.
13	AAM / TU Delft (conference)	2019	EV over-discharge fault diagnosis	XGBoost (tree ensembles)	Battery cell voltages, currents, temperatures	Fault / over-discharge detection	Demonstrates strong baseline performance and interpretability vs deep models.
14	Leite et al.	2022	Automated ML for real-time FDD	AutoML pipeline / model search	Real-time process sensor streams	Real-time fault detection configurations	AutoML reduces expert effort; suitable for RT-FDD in cyclic industrial processes.
15	Abdolrasol et al.	2024	Data-driven BMS fault diagnosis	Review of data-driven methods	BMS signals, sensor fusion inputs	Fault detection; SOH / SOC estimation	Emphasizes sensor fusion and explainability for field robustness.
16	Zhou et al. (BatteryBERT)	2025	Self-supervised pretraining for battery signals	Transformer / BERT-style pretraining	Large unlabeled battery time-series	Improved anomaly/ fault detection (AUROC gains)	Shows benefit of self-supervised pretraining on scarce labeled fault data.

Table 2. Performance metrics of machine learning models for vehicle fault detection

Authors (Year)	Method	DatasetSize	Precision (%)	Remarks
Leite et al. (2022)	CNN + MFCC (engine audio)	1,200 clips	92.4	Demonstrated strong precision in classifying abnormal engine sounds.
Zhang et al. (2023)	Deep Autoencoder (battery)	25,000 cycles	89.7	Achieved robust precision for anomaly detection under real-world EV data.
Akbalık et al. (2024)	AutoML-based RT-FDD	3 datasets	90.5	Precision improved via automated feature selection and model tuning.
Kumar et al. (2021)	SVM + Wavelet features	850 signals	88.1	Good performance for vibration-based fault detection.
Singh et al. (2020)	Random Forest (OBD-II)	600 samples	85.6	Balanced precision but slightly lower recall compared to SVM.

3. CONCLUSION

Artificial intelligence and machine learning have greatly advanced vehicle fault detection and diagnostics, offering superior accuracy and adaptability compared to traditional methods. While classical models perform well on structured datasets, deep learning and transformer-based techniques excel in capturing complex temporal and contextual patterns, driving progress in engines, powertrains, batteries, and in-vehicle networks. However, large-scale adoption remains limited. Many studies still rely on small or controlled datasets, reducing real-world reliability. The black-box nature of deep models raises interpretability concerns in safety-critical settings, while challenges of scalability, robustness, and computational cost continue to hinder deployment.

Future work should prioritize large, standardized datasets that reflect real-world variability. Emerging strategies such as transfer learning, self-supervised learning, and hybrid physics-informed models can reduce data dependence and improve transparency. Digital twins, federated learning, and explainable AI hold promise for building scalable, trustworthy, and privacy-preserving diagnostic frameworks.

In summary, AI-driven diagnostics are steadily moving from research to practice. By addressing data, trust, and scalability challenges, they can enable safer, predictive, and intelligent mobility solutions.

REFERENCES

Abdolrasol, M. G. M., Ayob, A., Lipu, M. S. H., Ansari, S., Kiong, T. S., Saad, M. H. M., Ustun, T. S., & Kalam, A. (2024). Advanced data-driven fault diagnosis in lithium-ion battery management systems for electric vehicles: Progress, challenges, and future perspectives. *eTransportation*, Article 100374.

Akbalık, F., Yıldız, A., Ertuğrul, Ö. F., & Zan, H. (2024). Engine fault detection by sound analysis and machine learning. *Applied Sciences (Basel, Switzerland)*, *14*(15), 6532.

Gong, C.-S. A., Su, C.-H. S., Chen, Y.-H., & Guu, D.-Y. (2022). How to implement automotive fault diagnosis using artificial intelligence scheme. *Micromachines*, *13*(9), 1380. DOI: 10.3390/mi13091380 PMID: 36144003

Hossain, M. N., Rahman, M. M., & Ramasamy, D. (2024). Artificial intelligence-driven vehicle fault diagnosis to revolutionize automotive maintenance: A review. *Computer Modeling in Engineering & Sciences*, *141*(2), 951–996. DOI: 10.32604/cmes.2024.056022

Leite, D., Martins, A.Jr, Rativa, D., de Oliveira, J. F. L., & Maciel, A. M. A. (2022). An automated machine learning approach for real-time fault detection and diagnosis. *Sensors (Basel)*, *22*(16), 6138. DOI: 10.3390/s22166138 PMID: 36015899

Machlev, R. (2024). EV battery fault diagnostics and prognostics using deep learning: Review, challenges & opportunities. *Journal of Energy Storage*, *83*, 110614. DOI: 10.1016/j.est.2024.110614

Michailidis, E. T., Panagiotopoulou, A., & Papadakis, A. (2025). A review of OBD-II-based machine learning applications for sustainable, efficient, secure, and safe vehicle driving. *Sensors (Basel)*, *25*(13), 4057. DOI: 10.3390/s25134057 PMID: 40648315

Özdemir, Ö., İşyapar, M. T., Karagöz, P., Schmidt, K. W., Demir, D., & Karagöz, N. A. (2024). A survey of anomaly detection in in-vehicle networks (CAN bus). *arXiv*. https://arxiv.org/abs/2409.07505

Theissler, A., Pérez-Velázquez, J., Kettelgerdes, M., & Elger, G. (2021). Predictive maintenance enabled by machine learning: Use cases and challenges in the automotive industry. *Reliability Engineering & System Safety*, *215*, 107864. Advance online publication. DOI: 10.1016/j.ress.2021.107864

Zachariades, C., & Xavier, V. (2025). A review of artificial intelligence techniques in fault diagnosis of electric machines. *Sensors (Basel)*, *25*(16), 5128.

Zhang, J., Wang, Y., Jiang, B., He, H., Huang, S., Wang, C., Zhang, Y., Han, X., Guo, D., He, G., & Ouyang, M. (2023). Realistic fault detection of Li-ion battery via dynamical deep learning. *Nature Communications*, *14*(1), 5940. DOI: 10.1038/s41467-023-41226-5 PMID: 37741826

Zhou, S., Liu, R., Wang, Y., Lu, J., & Jiang, B. (2025). BatteryBERT for realistic battery fault detection using point-masked signal modeling. *arXiv*. https://arxiv.org/abs/2506.15712 DOI: 10.1109/IAI68403.2025.11276897

Zou, B., Zhang, L., Xue, X., Tan, R., Jiang, P., Ma, B., Song, Z., & Hua, W. (2023). A review on the fault and defect diagnosis of lithium-ion battery for electric vehicles. *Energies*, *16*(14), 5507. DOI: 10.3390/en16145507

Chapter 3
AI-Powered Authentication and Access Control in IoT for Autonomous Vehicles

Siva Raja Sindiramutty
http://orcid.org/0009-0006-0310-8721
Taylor's University, Malaysia

Basheer Riskhan
Albukhary International University, Malaysia

ABSTRACT

The Internet of Things is experiencing exponential growth in machine-to-machine communications, in the relationships between things, between things and people -- and it is also presenting profound and vexing questions such as how to maintain trust and identity. Legacy password-based and static key systems are not suitable for the scale and heterogeneity of such IoT systems. This chapter explores how AI is changing authentication and access control through adaptive, context-mindful, and data-smart mechanisms. Methodologies such as behavioural biometrics, anomaly detection, etc, however, we opt not to offer alternative methods to identify noble users from their evil brothers, given some small computing device. We consider ML/DL models, running from simple classifiers at the edge to complex neural networks used to perform biometric verification or temporal activity analysis. The conversion of such approaches into practice is illustrated through examples from health care, industrial IoT, and smart cities, elements as autonomous vehicles.

1. INTRODUCTION

Importance of Authentication and Access Control in IoT for a Smart City

While the IoT has made our lives simpler and more convenient, it has also made our world more complex and dangerous, especially without the proper safeguards. IoT device threats such as smart-home systems, autonomous vehicles, and industrial sensors can be neutralized through authentication, which protects the device so that only authorized users can access it (Riaz et al, 2025). One of my friends found

DOI: 10.4018/979-8-3373-7355-3.ch003

out that a burglar was tampering with his smart thermostat, which illustrates how weak passwords and inadequate authentication can create serious problems. The ability of access control to add to that by defining what actions particular users or devices can take is what makes it invaluable. In a small-scale hospital research, the misuse of IoT medical devices illustrated the access control failures that caused patient suffering (Zarkia & Usman, 2025). The liability that even a smart lock can create when a group of users is given the same set of credentials, without prior authorization by the administrator, is also well-documented. Such situations are also commonly observed and ignored, as the technical, yet significant, components may compromise safety and privacy. Breaches are considerably reduced with MFA and role-based access. Companies that used these methods had fewer security issues, particularly with devices that obtained sensitive data such as health and whereabouts data. With physical, cyber, and data environments, more security and less compromised environments are achieved with stronger authentication and improved access control. Users' worries as to who has physical access to their devices are addressed, and productivity is not compromised as business systems are impacted. Relying on simple effective security solutions, improvement of security on IoT devices is noticeable, and equally, equal focus placed on effective user-centric authentication and access mechanisms is ensuring that security is guaranteed. Authentication and access control are demonstrated in Figure 1 and are important building blocks of IoT.

Figure 1. Importance of authentication and access control in IoT

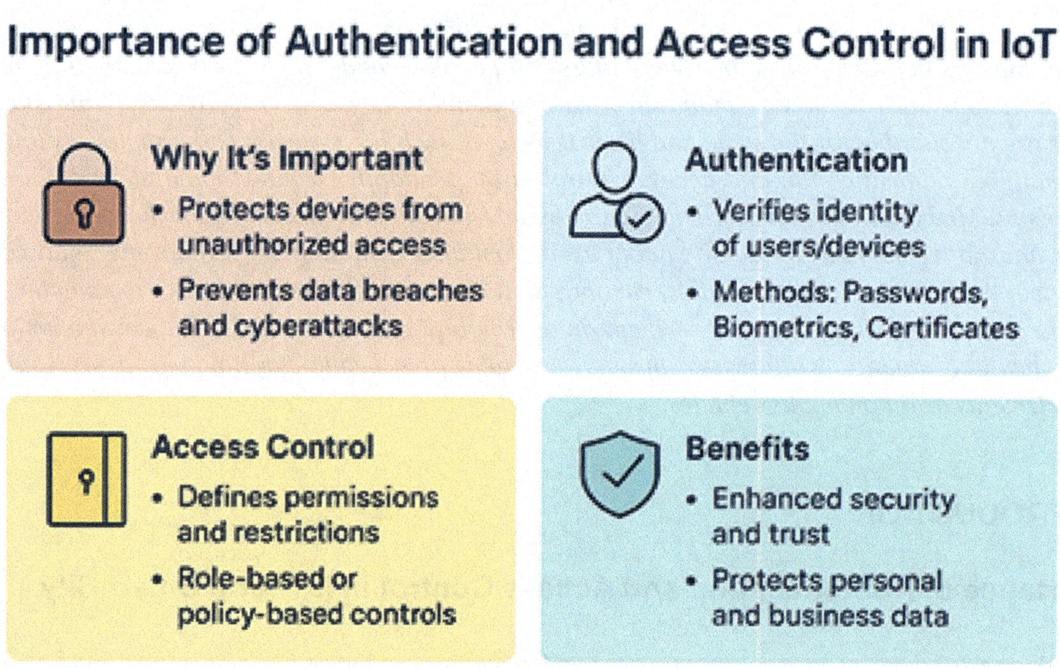

The Expanding Attack Surface With Growing IoT Deployment

The rapid expansion of the Internet of Things (IoT) is transforming the ways people and businesses interact with devices, while simultaneously pushing the limits of our digital infrastructures' security (Sharma, 2025). Each sensor, camera, or smart device is added to the list of vulnerabilities that attackers could exploit. While reviewing student projects, I noticed that even smart locks could be hacked if the absence of a security measure was considered; it served as a reminder that even the simplest device can pose a security risk. Although measures have been taken to improve security for mainstream internet-enabled devices, IoT devices, including self-driving cars, are expected to reach over 30 billion (Chemmakha et al., 2025) by 2030, giving attackers even more ways to infiltrate systems. The majority of emerging IoT technologies have minimal processing power, which limits the systems' abilities to implement robust encryption and security mechanisms. The result is a situation that is ideal for malware and massive botnets, as we saw with the Mirai attacks that brought down some of the largest online services.

IoT devices collect unprecedented amounts of data in a wide range of consumer and industrial applications. In the case of healthcare, connected monitors and wearable sensors are stream data that could include sensitive, personally identifiable information and could breach the confidentiality of devices used in life-saving procedures. These examples show that while the growing adoption of IoT devices is enhancing consumer experiences, it also creates profound and troubling issues regarding issues of trust, reliability, and the resilience of these systems. Without additional measures, the unprecedented attack surface will require ever increasing sophisticated monitoring and mitigation efforts.

Limitations of Static Password-Based and Key-Based Methods

The advancements are being responded to by things like biometrics or multi-factor authentication (Satheesh & Sree, 2025). Certainly, a few years ago I helped test a tiny IoT lab for these students, and nearly every device used a common default password, and unauthorized access took minutes. This is a microcosm of a real-world example where reused or weak credentials are the primary line of defense failure. It has been shown that 80 percent of data breaches are caused by stolen credentials (Eser, 2025). This goes to show how dependent attackers are on the first line password requirements. Also, a major cause of data breaches is the negligent loss of access control passwords. Weak, static passwords are target-rich environments when static access control passwords are compromised. Rotating access control passwords, and the act of exposing the passwords, provides uninterrupted access to the passwords, and access control passwords. Automatic opening locks are used to control access and are associated with a locked device, which is used to store or secure keys and is very limited in terms of resources, such as in lightweight sensor systems or wearable systems. Phishing and brute force attacks remain effective even against strong passwords. Current authentication techniques, whether weak or strong, do not adequately protect against tricked users who are lured to give up their passwords. Cloud systems and IoT systems present new threats when static methods are used.

The ongoing presence of these vulnerabilities highlights the essential importance of dynamic and policy-aware methods to supplement and in some cases replace static methods in the context of a future involving a vast number of devices that communicate autonomously and with little to no human interaction.

Chapter Contributions

This chapter is appropriate to the field of AI-based authentication and access control in IoT for a smart city. First, it presents the role of authentication and access control in IoT, bringing to light the deficiencies of fixed password-based and fixed key-based solutions. It then introduces IoT authentication and access control models, such as device-level and user-level authentication, then RBAC, ABAC, and CapBAC models. In addition, the chapter examines how AI enhances authentication with adaptive and anomaly-based techniques and demonstrates AI-based access control systems by considering context-awareness, adaptive policies, and Zero Trust principles. It then reviews some of the machine and deep learning methods for biometric, behavioural, and identity solutions, together with multi-factor and continuous policies for verification. In detail, the chapter discusses behavioural biometrics, edge AI for decentralised authentication, and blockchain–AI integration for decentralised identity management. Security and privacy concerns, including adversarial threats and privacy-preserving ML, are discussed, as well as evaluation metrics and benchmarks. Last, we present practical applications for healthcare, industrial IoT, and smart home applications, closing this chapter with the challenges and open research in developing robust AI-driven authentication systems for the IoT.

Chapter Organization

In the first section, we will introduce authentication and access control in IoT and outline its challenges. Section 2 presents an overview of IoT authentication and access control models. Section 3 discusses the role of AI in enhancing authentication, followed by Section 4, which explores AI-driven access control architectures. Sections 5 and 6 cover machine learning and deep learning techniques for authentication and identity management. Section 7 elaborates on multi-factor and continuous authentication using AI, while Section 8 focuses on behavioural biometrics and its AI applications. Section 9 addresses the role of edge AI in decentralised authentication, and Section 10 explains blockchain–AI integration for decentralised identity. Section 11 examines security and privacy considerations, and Section 12 introduces evaluation metrics and benchmarks. Section 13 provides real-world use cases and applications. Section 14 discusses challenges and open research directions. Finally, Section 15 concludes with a summary, future research directions, and practical guidelines for secure AI-based IoT authentication systems.

2. OVERVIEW OF IOT AUTHENTICATION AND ACCESS CONTROL MODELS

Device-Level and User-Level Authentication

With billions of interconnected devices involved in The Internet Of Things (IoT) including autonomous cars, drones, and traffic lights that share sensitive information, authentication is vital. An attacker could append his trail to your entire network through a weak entry point, like a smart camera, or a sensor. The technique of device-level authentication is to verify the identity of every machine and sensor before they talk to each other (Zhai et al., 2025; Krishnan et al., 2021). One example of this can be seen in smart home devices, where poor authentication has a remote control and security and privacy devastating consequences. User authentication is aimed to ensure that the person monitoring and controlling the IoT devices is trustworthy. This can be done through biometrics, mobile authentication, or multi-factor authentication. Additionally, doctors in hospitals may require a secure login as medical personnel to access certain connected medical devices, so they can alter sensitive patient information. While both user and device models assist in establishing a positive or trustworthy relationship between people and machines, and possible, the intricacies and layers that come with such security can create even more complexities in real-world applications. Researchers are stressing the importance of light-weight methods for cost-efficient energy authentication as numerous IoT devices face significant energy restrictions (Khajehzadeh et al., 2024; Sindiramutty et al., 2024). Given the rapid increase in attacks on vital systems, including energy grids and healthcare systems, there is a pressing need for more robust models. Improving authentication systems is primarily aimed at securing new technologies, but it will also help preserve the public's trust in connected systems they use on a daily basis. Figure 2 presents IoT Authentication and Access Control Models Overview.

Figure 2. Overview of IoT authentication and access control models

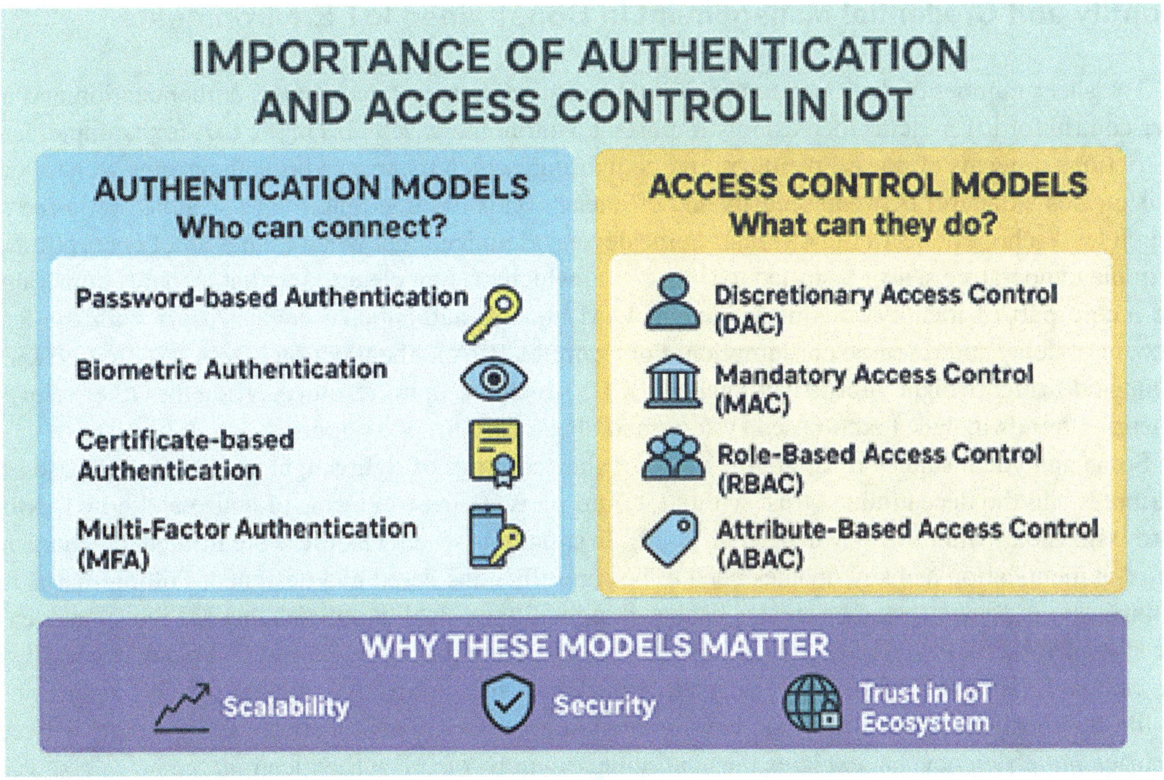

Access Control Models: RBAC, ABAC, and CapBAC

IoT systems gather billions of devices from smart watches to industrial sensors, and securing these systems typically relies on the concepts of authentication and access control. And I watched in a class-room project how a single insecure IoT door lock was forced open remotely in a matter of minutes. That small mishap illustrated the value of formal access control models. Three models have become the most widely-cited models, viz., Role-Based Access Control (RBAC), Attribute-Based Access Control (ABAC), and Capability-Based Access Control (CapBAC). RBAC is used a lot because it allows for granting permissions based on roles (pre-defined), and it quickly becomes manageable in a large company. Nurses and doctors who use IoT-based devices in hospitals have different roles and hence are allocated to different devices. ABAC is more flexible still, because it enables access decisions to be based on attributes such as the time of day, location, or device type. In smart homes, this might mean that a camera is configured to allow remote viewing only if the request is received at specific hours or from an authorised network. Devices don't have attributes and roles - instead, devices have "capabilities" or tokens which can be exchanged securely with other pieces of code. This removes central bottlenecks and is also suitable for peer-to-peer IoT systems, for example, autonomous vehicles that apply for rights to use a road. All have trade-offs, and they've all got their good and bad sides, but the real stakes here are that one of these schemes or a related cousin (many variations of these schemes are possible) is essentially needed for the systems of IoT to stay fit for everyday use.

Identity and Credential Management in Constrained IoT Environments

The sheer number of devices communicating sensitive data each day makes authentication and access control for IoT systems increasingly important. Autonomous vehicles in the IoT, for example, have restrictions in terms of power, memory, and computing, which means traditional systems won't work. This creates a tradeoff between security and efficiency. This is something I have seen before in the research for a school project I did on smart home devices. I noticed that when I put a lot of encryption on a simple temperature sensor it started to slow down, which made it clear to me that resource constraints are a core part of identity (mis)management. IAM aims to authenticate devices/users with the least amount of delay and resource consumption. For example, RSA authentication is less resource-efficient compared to lightweight alternatives such as ECC, which are more resource-efficient. Other relevant emerging trends include token-based systems and blockchain-based systems.

Some analytical studies illustrate the range of shortcomings of lightweight access control based on attributes and the decentralized trust without a third party, at least in terms of abuse and how flexible permissions may limit abuse (Zhang et al. 2020). In situations where resources are limited, mechanisms for re-authentication and key updates need to be carefully considered and designed. Polling that is too enthusiastic in previous studies creates battery drain while refreshing policies that are too sparse leave the system vulnerable to replay and cloning attacks. These are practical, not just technical, especially in the context of such IoT applications as health monitoring and industrial automation where failures can be life-threatening or where they can halt production processes. This illustrates the trade-off between security and efficiency that explains the continuing dilemma of IoT authentication.

3. THE ROLE OF AI IN ENHANCING AUTHENTICATION

Adaptive Authentication With Behavioural Biometrics

AI changes the model for authentication through the implementation of additional layers of security. With authentication systems learning how a user types, moves the mouse, or even how a user positions a phone, passwords and tokens become less of a factor (Chitraju, 2025; Sindiramutty, Prabagaran et al.,

2024). These signals are difficult to replicate and allow for the identification of outlier activity. I know this because I had to use online banking. When I attempted to access my account using a new device, I had to enter a prompt and then the system informed me that it was tracking me to determine my typing speed. End user behavioral biometrics provides a sense of seamlessness. Identification is a multidimensional process. The system is continuously assessing behavioral patterns in real time. Whenever a user's behavior deviates from a set standard, the system adds extra security measures. This system improves the user experience by allowing the most frequent and least risky activities to go unobserved. There was a study conducted that showed when adaptive authentication was used, the number of fraudulent activities declined and user complaints about ease of use did not increase. The systems used may not be 100% effective as injuries or stress may alter behavior but AI is meant to learn and/or adapt with time. The focus is on more intelligent and individualized authentication. Instead of locking systems into rigidly written rules, it silently adapts in the background, protecting users while allowing interaction to flow. Figure 3: The Role of AI in Augmenting Authentication

AI-Based Anomaly Detection for Identity Spoofing and Impersonation

While impersonation used to be seen as someone dressing up to be another person, these days, it has become more refined. Attackers are now able to use stolen information to build a profile and utilize a synthetic face. Now, with the help of AI, abnormal behavior which is not considered the user's normal profile is being detected and utilized as a powerful deterrent (Cyberly, 2025). Instead of just relying on behavior rules, a combination of methods have been developed to identify activities such as, delays in typing, fluctuations in voice, and face movement. According to one security analyst, their system was able to detect the fraudster because the person in the video call was blinking in a way that appeared to be too robotic and indicative of a creation. Unlike rule-based systems that work on the basis of if A happens, B happens, the system is always monitoring. It has the ability to identify normal behavior of each user and then pinpoint abnormal activity such as impossible log in locations or slight disparities in measurable biometric signals. As a result, a system is able to identify and stop suspicious activity in the initial stage before major damage is inflicted. An initial banking trial showed that, even with the use of sophisticated deepfake tools, impersonation attempts were significantly reduced. Although false positives are possible, the models continue to improve with the data. Function of becoming familiar and honed to new attack styles very quickly. What seems to matter most here is the trade-off: the average defender gets smoother access during normal activity, while frequent attackers find it substantially difficult to reach hidden targets. Due to the fact that identity theft is on the rise and that this is a trade-off that matters, this is the reason why AI anomaly detection is becoming increasingly integrated to preserve our digital trust.

Figure 3. The role of AI in enhancing authentication

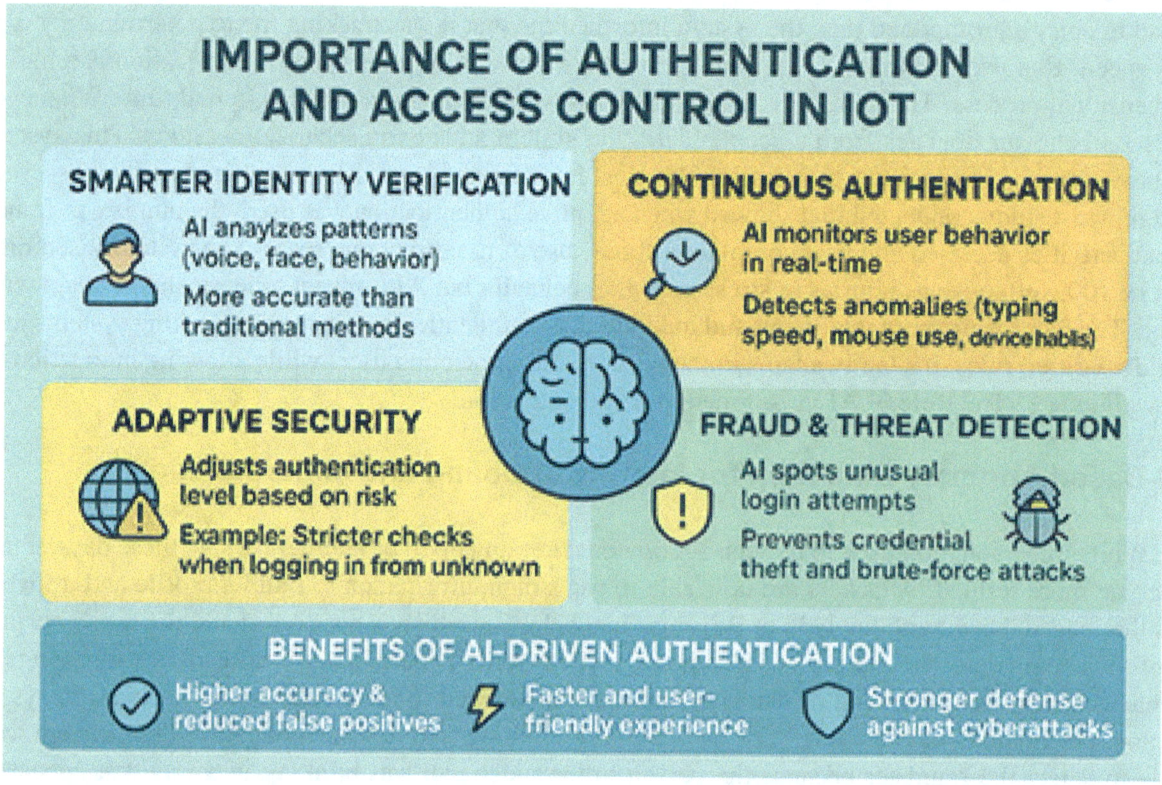

4. AI-DRIVEN ACCESS CONTROL ARCHITECTURES

Context-Aware Access Control Using Machine Learning

Although access control systems have been used for decades and have been effective for protecting sensitive information, the reliance on static, predetermined rules is ineffective for dealing with all variations around how users interact with systems. I previously worked for a company where employees would share login credentials with each other. Rule-based access controls did nothing to prevent this. In contrast, systems enabled with machine learning can adjust to unusual user activity by learning and creating new activity patterns (Odufisan et. al, 2025). For example, if employees have an abnormal login attempt from a different location, an AI-based access control system can automatically block access without an administrator decision. These systems can also reduce the number of false positive alerts. I have seen this in a hospital where nurses would swap shifts. The access control system would create alerts all of the time. AI systems increased accuracy and learned that the alert was not a typical occurrence in that environment. In that system, accessed control was not balanced with the operational environment. In highly sensitive environments like finance and government, the need to create highly effective systems is even more critical as the cost of breaches is extremely high. Incorporating user profiling, device identification, and behavioral signals helps machine learning develop strong security systems that do not constantly alert users. The increasing dependence on remote work has resulted in a greater need for the implementation of this type of security. On the surface, this appears to be more of an operational improvement than a technological one, but these systems are critical to preserving confidence in the digital systems that handle our work.

Dynamic Policy Generation and Enforcement With AI

If I had to give an example from the university IT shop I worked at, I have seen access policy updates be completely manual from semester to semester, and bugs were introduced in the policy that students exploited. Many legacy security policies become out of date quickly. AI systems that act on your behalf to update policies do not require human intervention to update policies based on new information (Calboli & Engelen, 2025). They update policies based on rules changes from user behavior, network activity, and other external factors. This is especially true for larger organizations that have a larger number of applications that are required to be updated. At one of my clients who is a manufacturer, IT was flooded with requests to change access rights of contractors. AI does that heavy lifting as it learns the most likely access needs of a particular role-person combination, and applies those learnings on the fly. When there were requests for abnormal access, the system responded immediately and there were no disturbances from a brute force attack while valid users continued to work. A static policy is less likely to be over corrective. AI handles this duality by enabling instantaneous access whilst reducing friction and minimizing security concerns. In rapidly evolving fields such as health care and finance, where data is constantly circulating, this level of flexibility is more than a convenience; it is a necessity. AI-based policy enforcement takes adaptive protection and applies it to real-world scenarios, thus designing protections to outpace your ability to respond to new threats. AI-enabled access control frameworks are shown in Figure 4.

Figure 4. AI-driven access control architectures

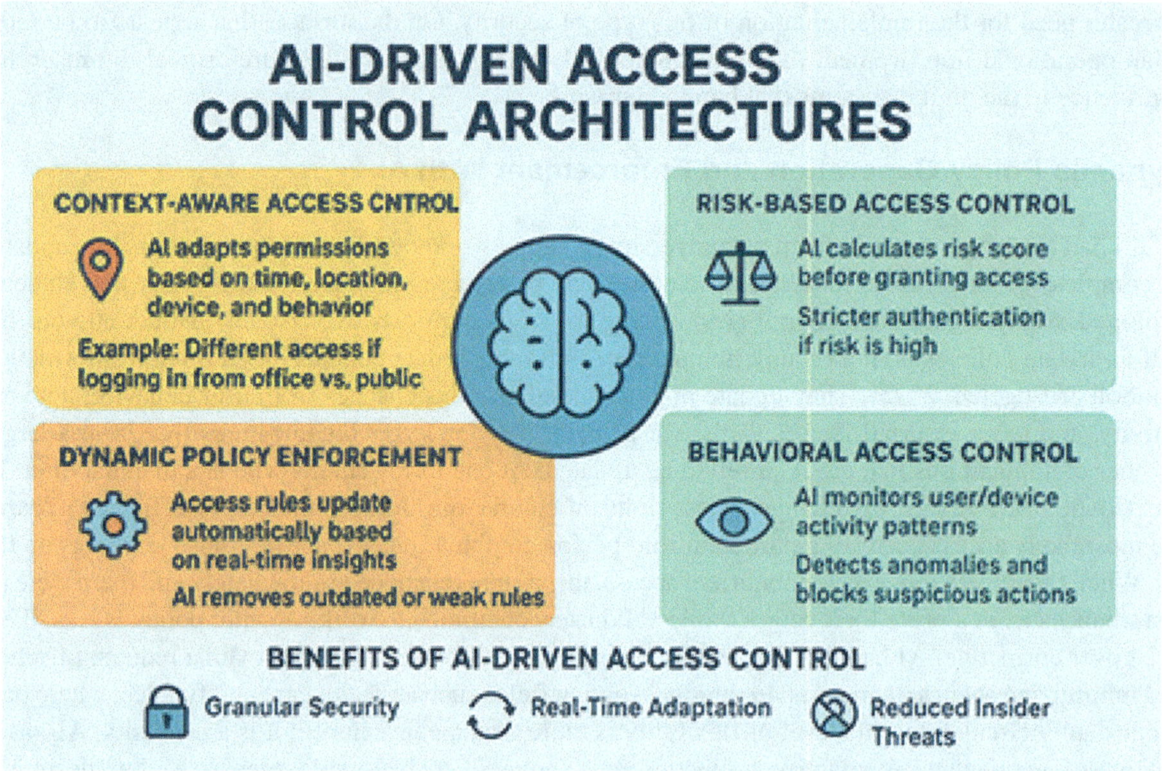

Integration With Zero Trust Principles

"Zero Trust" assumes each and every user and/or device is untrusted assuming zero knowledge regarding their location (remote or on premise). I witnessed a retail failure case because remote workers were considered 'trusted' on the original log in. That created potential opening for exploitation attempts. Priyanka et al (2025) notes that the incorporation of AI/ML into Zero Trust is a robust and agile approach because the changes are one-off checks, but real time updates from live data. For example, machine learning in this model focuses on behavioural analysis, device health, and location analytics. I encountered the same scenario with one of the banks I worked with, where employees were freely switching between personal and work devices. Legacy access control systems were not prepared for this new fluidity, but the AI backed Zero Trust was able to identify the risky behaviour, and immediately applied the required control. This was done without disrupting normal work activities. The same applies for normal work activities. Perhaps the most important feature is the scalability of the model. Setting rules for thousands of users and systems in large enterprises is not possible. AI delivers on the never trust, always verify approach, with little manual control of the system, at each level of the system. This way, unnecessary access can be minimized, while still allowing employees access to resources to remain productive. The combination of AI and Zero Trust results in stronger and more flexible defense, helping the security industry keep pace with complex digital workflows.

5. MACHINE LEARNING TECHNIQUES FOR AUTHENTICATION

Classification Models for Biometric and Behavioural Data

According to Mollo et al. (2025), Machine Learning (ML) has become crucial to the authentication process using both biometric and behavioral data to confirm that you are who you say you are. While biometric data may include a fingerprint or a facial recognition scan that is unique to only you, behavioral data comprises patterns such as the way you type, move the mouse, or your unique way of walking. I remember a project from the university lab which combined facial recognition and keystroke patterns. Such systems fraudulently using the credentials and failing to be detected by face recognition systems. Classifiers are the most important part of such a system. With limited data and a need for the data to be interpreted, logistic regression and decision trees may be your best bets. Neural networks, on the other hand, can be used to process complicated biometric data. However, they need a lot of data and are more tricks than other models. When it comes to detecting and separating minor differences in handwriting or voice, support vector machines have demonstrated the best results. Given the results, it's easy to appreciate the implementation. Financial institutions, you can expect to be able to unlock your mobile device or capture a photo using your face or fingerprint.

There are still small mistakes, like identical twins mixing up some algorithms, but the technology is improving. It is clear from how they are improving classification models that they are not only technical experiments, but that they are also active tools that are shaping the present-day reality of how security is provided.

Lightweight ML Algorithms for Constrained IoT Nodes

As a result of limited hardware capabilities, lightweight machine learning (ML) techniques are becoming more important for IoT authentication. Many IoT nodes rely on small batteries and are limited in memory and processing capabilities. Legacy authentication protocols such as MAC and static password systems, as well as cryptographic protocols, are all resource intensive and increase latency. In contrast, some lightweight ML algorithms are able to quickly and seamlessly secure and authenticate in a resource-efficient manner by analyzing behavioral and usage patterns of the device. A student shared an example of a smart door lock that 'died' because of an over the air software update. I was an example of a very real, practical limitation that made the use of more lightweight alternatives more palatable. Algorithms, such as decision trees, k-nearest neighbours, and shallow neural networks, can be optimized to run on ultra-low power microcontrollers. These techniques have also been successfully utilized to authenticate users via biometric signals, sensor data, and usage patterns. In mission-critical scenarios, such as industrial IoT and health monitoring systems, lightweight ML-based authentication methods become really important because delays or failures can be lethal (Rani et al., 2025). These methods provide a good tradeoff between security and efficiency, thereby, increasing the lifespan of the IoT device and trust in the IoT ecosystem. The shift in ML models towards becoming resource efficient is essential for affordably and securely integrating authentication into trillions of devices. Machine learning methods for authentication are shown in Figure 5.

Figure 5. Machine learning techniques for authentication

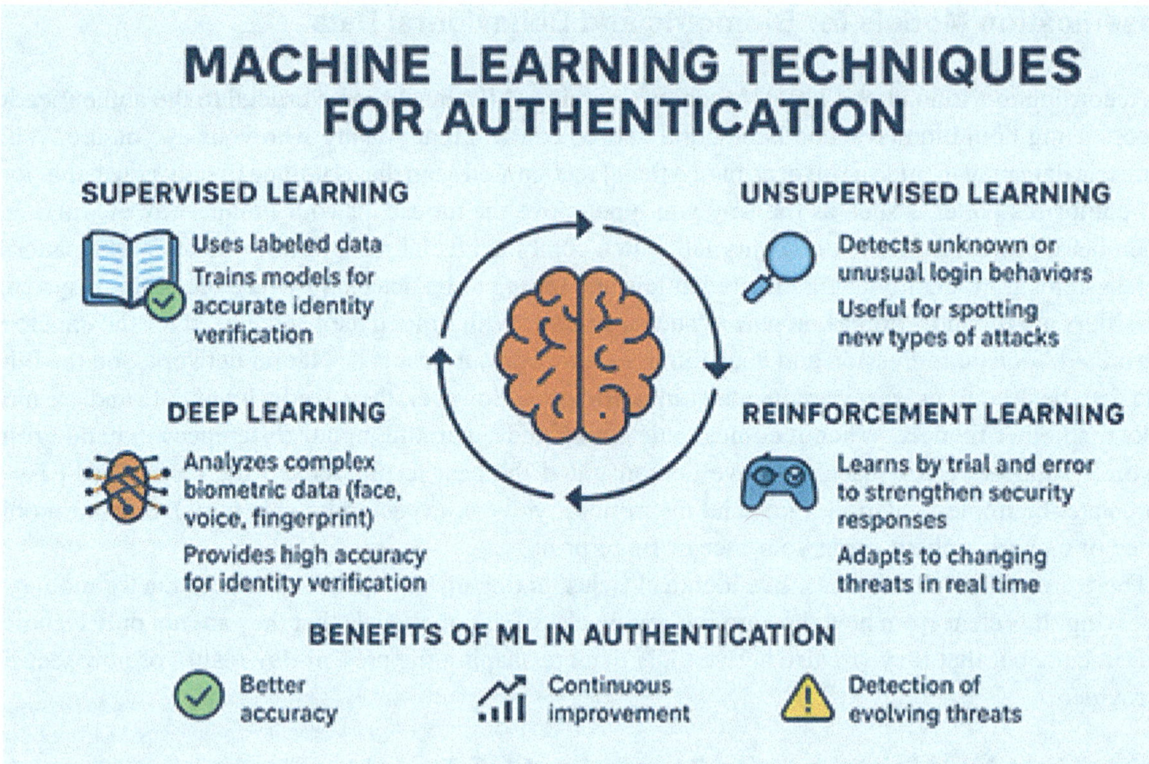

6. DEEP LEARNING FOR IDENTITY AND ACCESS MANAGEMENT

CNNs for Image/Voice-Based Biometric Authentication

Advances in deep learning technology are allowing organizations to enhance the management of user identity and access controls using biometric technologies. Convolutional Neural Networks (CNNs) are being implemented in the evaluation of images and audio streams. CNNs evaluation of images and audio streams is improving accuracy compared to traditional rule-based systems (Anver et al. 2025). CNNs can detect patterns in face and voice recordings that other algorithms and even people may miss. Because of this, it is becoming increasingly difficult for identity impersonation attacks to succeed. I recall even seeing a small bank branch that had installed a face recognition system powered by CNN capabilities. Employees told me that people were surprised by how quickly the camera granted access to them, and that they did not have to input a password. This small example demonstrated that a balance between everyday convenience and security is entirely possible. Concerning voice recognition systems, CNNs are applied to the representation of the audio signal (spectrogram) as an image and therefore, CNNs can be used. CNNs can therefore be applied to audio signals, and, as a result, have provided systems the capability of being able to authenticate a person as a result of voice recognition in a very short (approx.

in seconds) time. This also holds true for a voice being in a noisy environment, which is normally a difficulty. These systems are able to continuously learn from new data, and therefore can enhance their capabilities over time. On the other hand, many people feel uneasy about using their voice or face as part of an access control mechanism, even though the technology is designed to address the weaknesses associated with password theft. According to security specialists, deep learning systems have now become fundamental to stopping unauthorized access in both the physical and virtual worlds. These changes extend past just stronger protection; they change the way users interface with security. The transition to CNN-based biometrics represents authentication that is, in a sense, invisible but critically important to the protection of both individual and corporate interests.

RNNs/LSTMs for Temporal Behaviour Analysis

Identity and Access Management (IAM) has become a priority as digital systems face risks from stolen passwords, spoofed devices, and insider misuse. Traditional IAM tools often stop at static checks, but user behaviour is rarely static. Recurrent Neural Networks (RNNs) and their variant Long Short-Term Memory (LSTM) networks are valuable because they can track how a user behaves over time, spotting changes that a one-time password or token cannot (Sassa et al., 2025). I recall testing an access log dataset where simple threshold rules missed abnormal login attempts at 3 a.m., but an LSTM flagged it by learning that the user usually signed in during office hours. That kind of temporal awareness is difficult to achieve with older models. Research shows that RNNs can process sequences of keystrokes, mouse activity, and login timings to learn a user's "rhythm" of interaction, making it harder for attackers to mimic legitimate access patterns. LSTMs add memory of long-term dependencies, so unusual behaviour, even if it happens after many normal sessions, can still be recognised. This makes them useful in continuous authentication, where access is not just granted once but checked throughout a session. In practice, companies are testing these models in financial systems, where detecting subtle deviations in behaviour can prevent account takeover fraud. Technology is not perfect, but its ability to learn personal temporal behaviour has already shown measurable improvements in protecting identity systems.

Siamese Networks for Identity Verification

Siamese Networks have emerged as a powerful technique for effective identity verification as they have the ability to compare two inputs and determine whether they belong to the same identity (Sivasankari & Hanirex, 2025). Rather than dealing with classifications of faces or voices, these networks operate on the notion of pairwise similarity. A common architecture includes two identical subnetworks with shared weights used to perform feature extraction, and are followed by some distance metric used to measure closeness. For example, I remember seeing a demo where two face images (and one of them from a passport and the other one was from a live video) were matched in real-time. The almost seamless experience was caused by the speed and accuracy of the matching. Siamese networks perform best in real world situations where there are insufficient samples of data for each individual. They are capable of operating without the need to retrain the networks with each new enrollee, and are therefore, highly scalable to applications with continual new identity enrolments such as border control and access to enterprise. The same case can be made for voice verification as possibly even more challenging as the model can operate on the speech data and determine whether or not two voice samples belong to the same speaker, even in the presence of background noise. Analysts have claimed that with the use of the Siamese approach model the rate of false matches was greatly improved when compared to older classification models thereby improving overall security and the confidence of the user. The practical application means identity checks can be made both stricter and easier. Siamese networks, by focusing on similarities rather than drawbacks, present a viable solution for quicker, more precise, and easier verification in areas that require higher levels of security. Deep learning applied to identity and access management is illustrated in Figure 6.

Figure 6. Deep learning for identity & access management

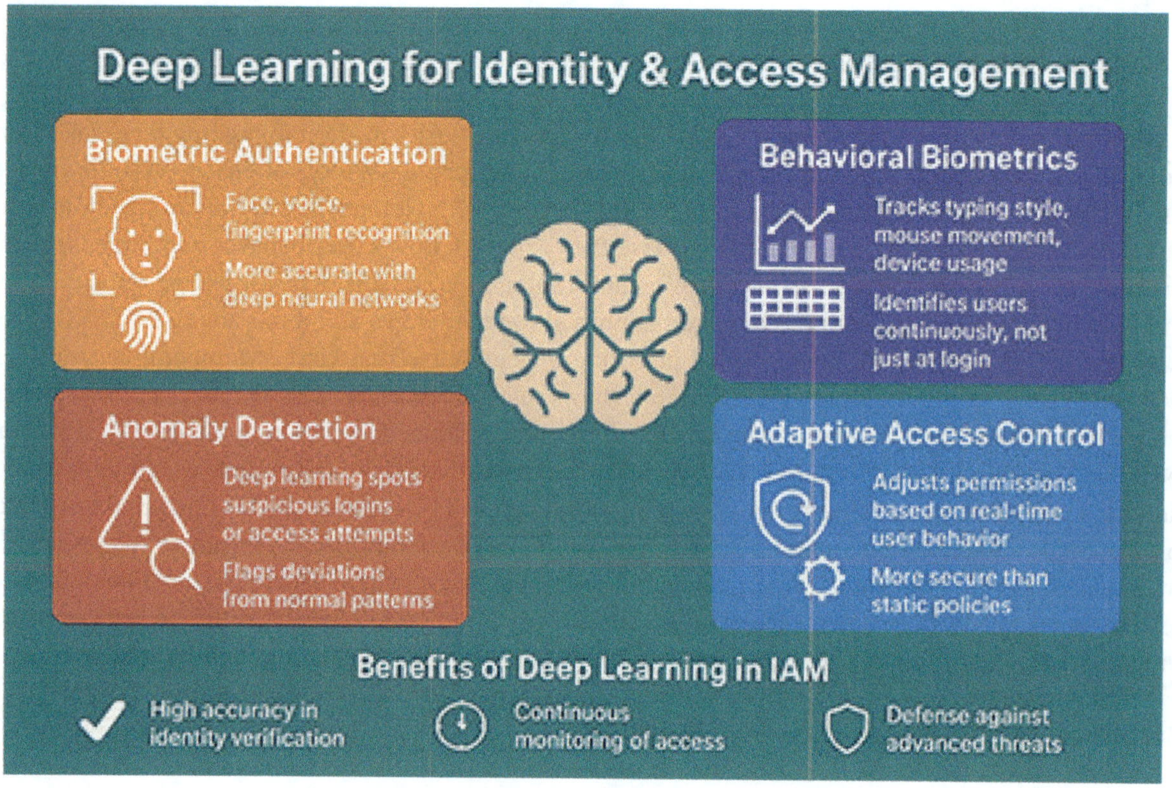

7. MULTI-FACTOR AND CONTINUOUS AUTHENTICATION USING AI

Combining Biometrics, Location, Device Usage, and Habits

I once had to implement one of those single sign-on systems for university login exams, and it served to illustrate to me how insecure single measures are - I've always been pointed to multiple scans and measures for additional security, and to bypass this weakness of the password system, one of the AI multi-factor and continuous ways to authenticate would probably solve the problems of my single sign-on system. (Sinha & Agarwal, 2025) mention how Continuous multi-factor authentication combines biometrics, location, device, and behavior (which are all personal and unique to each user) in ways to bypass this weakness of the password system. Biometric measures have been used through a password that can be stolen, and if the systems know how users type, how they move their mouse, and even know the hours that they typically log in, then the system knows an individual user and can bypass the password system by making sure that unusual behavior. They can incorporate location as an additional factor to the system; on the occasion that a user typically logs in to a location outside of their normal location, the AI is able to identify this potentially dangerous behavior. I was once in a banking app that closed when I was overseas, it is annoying, but it is comforting that there was a system that closed it due to a behavioral anomaly. Users are able to maintain more fingerprint alternative behaviors as unique as the patterns that users swipe, the frequency of the apps they use, and even the pressure of their swipes. This definitive capability is the continuous, rather than a single authentication, behavior-based mechanism. AI systems learn habits and streamline risk mitigation by avoiding extra steps for users. This makes security feel like it has less of a barrier, acting like a silent guard that evolves in conjunction with us.

AI for Real-Time Evaluation of Multiple Authentication Factors

Traditionally, Multi-Factor Authentication (MFA) systems focus heavily on the combination of Knowledge (passwords), Possession (security tokens), & Inherence (biometrics). (Lut, 2025). However, thanks to the latest advancements in AI, these factors can now be assessed in real time on a continuous basis, thereby reducing the chances of a data breach. One example I came across is where an organization implemented continuous authentication via AI for remote employees. The application tracks keystroke patterns, mouse movement, and all login locations. It can raise a flag in real time & autonomously for abnormal activity, eliminating the need for the user to log in repeatedly. AI can discern things humans cannot. For example, a user logging in from a known device, but one that is deemed unfamiliar for the time of day, may trigger a small verification step. Another example is an organization that successfully reduced account breaches through the integration of behavioral analytics and facial recognition in addition to standard). Over the weeks, the AI system became more proficient at detecting a user's individual patterns, thereby increasing system confidence and improving the user experience. The continuous nature of real-time monitoring also provides primary security response teams with greater flexibility, thus improving system response time.

I noticed that a very small startup using AI for multimodal authentication experienced a remarkable increase in both compliance and decrease in disturbances. Applying this technique in relation to a scraper operation, and also taking into account the factor of scale for the security of cloud services, where the predictability of user behavior is a significant challenge, using AI to perform real-time analyses of multiple authentication factors represents more than a technological advancement. It represents a new dimension to the everyday security relationship. The authentication process is still highly secure, but the way that is done is more convenient, intelligent, and non-intrusive. The systems in operation reinforce the importance of even the most subtle behaviors in preventing security breaches.

Case Examples: Smart Home, Healthcare, and Industrial IoT

AI-driven multi-factor and continuous authentication is disrupting how devices protect access (Kumar, 2025). In a smart house, I realised that voice and face recognition make our everyday life easy, as well as keep the house safe. One night, my smart lock detected my voice as I was toting groceries into my house and unlocked without me having to touch a keypad, a minor convenience, perhaps, but a mark of useful AI integration. In healthcare, hospitals apply continuous monitoring of nurses' and patients' access to restricted areas. At a local clinic I visited, a nurse's badge and fingerprint were both verified digitally before granting access to medication rooms. This minimises mistakes and unauthorised entry, demonstrating AI's business application. AI-based authentication of industrial IoT uses behaviours of operators, device usage, and environmental information in real time. In one factory I visited, the system detected odd login times, thwarting potential sabotage and keeping workers safe. In all instances, AI enables systems to learn and adapt to changing conditions without explicit, regular human intervention. What struck me was the tangible impact even marginal, mitigative improvements, such as unimpeded door access or safer industrial operations, can make. Looking at these use cases, we can see that continuous authentication is not only about security; it impacts efficiency, trust, and user experience. Banking on comprehensive auth with AI, applying all those authentication methods in conjunction with AI, lessens dependence on passwords, streamlines access, and responds immediately to anomalies, making it a potential blueprint for securing environments in homes, hospitals, and factories.

8. BEHAVIOURAL BIOMETRICS AND AI

Typing Patterns, Gait Recognition, Mouse Movement, and Voice

Behavioral biometrics has the ability to identify a person's identity with the aid of Artificial Intelligence (AI) (Kalra, 2025). AI systems have the ability to detect minor differences in people's keyboard rhythms. People differ in terms of their speed while typing, the pauses they take, and the mistakes they make. This is a kind of signature. I have seen an example of workplace security software that flagged a login attempt because the typing did not correspond to the pattern of the registered employee. Studies show that the accuracy of keystroke analysis in the detection of impostors is very high. Another methodology involves gait recognition. AI models learn and adjust from cameras and sensors that analyze the particulars of a person's walking movements, such as stride, length, and movable joints. Some airports have tested this technology in a bid to analyze people as they move about and to eliminate direct contact with them. One would be surprised that such an activity as walking would be analyzed and used as a source of security risk by machine learning. The movement of the mouse also reflects the unique characteristics of an individual. Online banking and e-commerce services have also shown an interest in this technology because it is unobtrusive and does not make the user's work tedious. Identifying people's voices is a combination of pitch, tone, and how someone speaks. Once I dealt with a banking helpline whose AI couldn't associate my tired late-night voice with my usual energetic one. This exemplifies AI-driven behavioral biometrics. Together, these techniques show how simple daily routines transform into highly effective digital signatures. Behavioral biometrics and AI are illustrated in Figure 7.

Figure 7. Behavioural biometrics and AI

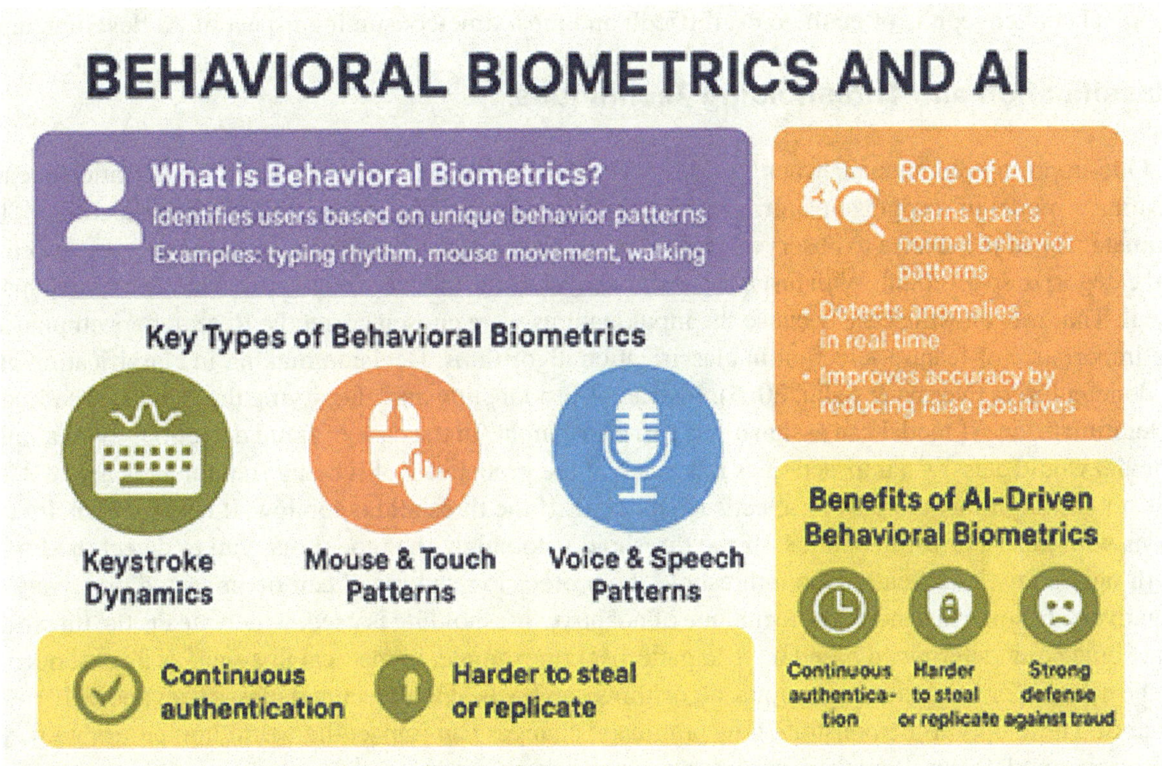

Data Collection and Preprocessing Challenges

The success of AI technology is reliant on the datasets it uses, but obtaining and processing the required data is usually an unpleasant experience (Araújo et al., 2025; Shah et al., 2022). In some of the projects I participated in, even the collection of homogeneous samples was a significant challenge. When data is sourced from disparate sensors, surveys, and online activities, the data is rarely homogeneous in structure. Researchers have shown that inconsistencies in datasets result in severe bias and greatly undermine the credibility of the model. Incomplete datasets are a recurring challenge. I once had a dataset of a survey where a number of respondents did not answer some of the questions. I had expected the task of completing the dataset to not result in any misrepresentation to be trivial, but it was. There are techniques for dataset completion, but they come with trade-offs. AI is only as effective as the data that is fed into it, and poor data management will compound any impacts. Data that is neither missing nor erroneous can be problematic as well. Movement not registered by a smartphone sensor can be misinterpreted as movement. Background noise is often captured in audio recordings. Noise removal is often achieved through some type of filters, but those filters can remove useful signals as well. There is an element of both science and judgment in balancing the aforementioned steps. Personally, I have found that in most cases, time spent on preprocessing is an investment that pays in reduced effort on subsequent tasks. The scale of current datasets presents further obstacles. It is expensive to manually label all millions of records required to train large AI models. Some teams have begun to use semi-automated labelling, but these systems will still produce mistakes. These challenges illustrate that data preparation is more than just a technical challenge; it is often the most difficult and most time-consuming aspect of AI development.

Classification and Thresholding Techniques

One important component when building AI models is input classification. Classification means creating a model to recognize patterns and assign labels to specific inputs such as "fraudulent" or "legitimate." I had the chance to observe a project in which email data was fed to a model in order to detect and categorize spam email. With just a few training cycles, the model was able to correctly classify spam email. That was only the case because the input features were chosen correctly. Researchers emphasize the importance of feature selection in classification algorithms. Decision making in classification can be done in layers (Verma & Sahu, 2025). Instead of just labeling and classifying the input as fraudulent or legitimate, an AI model can assign a score to every input. In the case of fraud detection, for example, a model would classify a transaction as risky only if the probability of it being fraudulent is above 0.85. It is not a straightforward task to select such a cutoff. If the threshold is too low, it may overwhelm analysts with false alarms. In contrast, if the threshold is too high, the model may fail to detect the fraud. With only a few modifications to a threshold, both precision and recall can be impacted positively or negatively. In some instances, the technique of adaptive thresholding is used, which means the threshold is at some point determined based on data patterns. For example, in medical imaging, as a consequence of the contrast or brightness variations, algorithms can be used to detect and classify an anomaly more reliably. These options demonstrate how pragmatic changes can impact the world in measurable ways. Observing models transition from one extreme to the other in sensitivity brings to mind that classifying and thresholding are neither purely mechanical processes nor just unnecessary steps, but rather, workable points in the trade-off between precision and practicality.

9. EDGE AI FOR DECENTRALISED AUTHENTICATION

Benefits of Pushing AI Authentication to the Edge

Decentralized authentication using Edge AI brings intelligence to the sources of data generation including mobile devices, IoT sensors, and smart wearables (Alshuhail et al., 2025; Muzafar & Jhanjhi, 2019). Since there is no need to rely on a central server, authentication occurs completely at the edge, which lessens delays and enhances user satisfaction. In my teaching labs, students using smart locks with edge-based facial recognition reported that logins take less time and are still secure, even with an unreliable internet connection. However, speed is not the only benefit of real-time processing. It also minimizes the risk of data breaches by reducing the exposure of sensitive information during real-time processing. Increased system resilience is another important benefit. In the event of a system failure, the breach of one system does not lead to the compromise of an entire centralized system, helping to reduce the impact of system interdependencies. The risks are managed by a distributed security approach, where the decentralized pattern of security ensures that the impact of a single breach is limited. Edge AI also provides the capability of adaptive authentication, which allows the system to become smarter by behavioral, locational, and device-signal changes. It also increases impersonation of legitimate users by attackers. During a limited test of wearable devices in a healthcare facility, researchers found that the system allowed staff to access essential systems even in offline wards, so patient identity verification was successful.

Edge authentication allows organisations to access speed, privacy, and scalability, eliminating total dependence on cloud computing. With more connected devices, this shift is increasingly critical. It is a move toward more intimate, geographically distributed digital protection where the user's device becomes an ally in the fight to safeguard their identity.

Edge Device Limitations and Model Optimisation

While Edge AI has the ability to make quick, decentralized decisions for authentication, it has a number of considerable barriers that could affect how well it performs on a given device. Most edge devices, for example, have limited memory and processing capabilities, making complex model processing a challenge (Zheng et al, 2025; Humayun et al, 2023). I also witnessed this problem firsthand in a simple test I did with the smart door locks, as the AI experienced authentication delays because the model that was built was too complex. Fortunately, there are techniques such as pruning, quantization, and knowledge distillation that can be used to optimize models in order to reduce their size and improve their processing speed. In a study involving authentication devices, the researchers found that models that had been optimized significantly reduced the device's energy utilization, allowing the device to be operational for prolonged periods of time without the need for a recharge. Edge-cloud collaboration could also be used to solve this problem, as it allows for the processing of heavier computations to be done in the cloud while the device remains responsible for processing the lighter computations. I was also fortunate to have this set up in my study, as it produced significantly reduced login times for the smartphones in the study. From the real-world tests, it was found that lightweight models are able to provide similar/adequate functionality as full-scale models while also ensuring that the speed, functionality, and operational capabilities are protected. One of the negatives that was experienced in the office trials that took place in the field was that users did not like the devices that provided minimal delays in order to provide a higher accurate reading on an authentication device. Optimisation techniques, then, are not just technical enhancements but improvements that users directly experience optimisations. These minor changes can make the difference between an authentication system that is usable on an everyday basis, rather than just being usable on a hypothetical basis. For Edge AI applications, the trade off between model size, computation, and energy is a constant balancing act.

Case Study: Edge Authentication in Smart Cities

Edge AI is already changing the way smart cities handle authentication. It is also streamlining and securing these processes. In a proof of concept with Barcelona's traffic sensors and public service kiosks, edge devices used local authentication to avoid the delays associated with cloud systems. Distributed authorization is beneficial as citizens experienced shorter wait times at automated barrier kiosks. Tiny observations in the interviews, such as comfort level of sharing data when it is processed near them, compared to when the data is sent to far-off servers, made it clear that people had preferred the latter option. Edge AI allows devices to interpret data locally, which relieves network traffic and increases the responsiveness of vital services like public transportation or warnings (Mohanty et al., 2025; Azeem et al., 2021). A use case in Singapore showed that the edge authentication with the public cameras helped to access the restricted areas quickly without compromising the privacy of the citizens. At the time each of them opened, the city had fewer system failures than it had when it was running on the cloud for authentication, thereby making people more confident of public participation. Even small sensors in a park assisted with security checks, confirming their identity through wearable gadgets, in just one more expression of the scale at which edge AI can operate in a city. Observing these systems remaining in apposition to say that decentralised authentication is not only efficient, it can create a reliable, resilient network in which devices work together to authenticate their users. Use cases can be health monitoring spots, smart parking, or simply opening doors into buildings, making it a real-life use case of the technology that will be beneficial for a large crowd of people.

10. BLOCKCHAIN AND AI INTEGRATION FOR DECENTRALISED IDENTITY

Self-Sovereign Identity (SSI) and AI Decision-Making

Blockchain and AI are being utilised together increasingly to enhance decentralised identity, something that can be seen with Self-Sovereign Identity (SSI) (Chan et al., 2025; Attaullah et al., 2022). SSI gives people power over their own information without depending on big central operators, so they enjoy more privacy and security. For example, a small start-up in Estonia employed blockchain to enable citizens to store identity credentials on their mobile phone, and to access government services more quickly without disclosure of unnecessary information. AI can further enhance SSI by performing behaviour analysis of the user and automatically verifying the credentials, so that there are fewer chances of fraud, and verification is expedited. For example, a financial service platform successfully combined AI with blockchain technology for identity verification to streamline loan approval in real time, which met the private regulations in patient records. AI can also identify abnormal patterns in identity usage, like suspect log-in attempts, and act quickly to counteract them, unlike in a traditional system. However, the deployment of AI in a decentralised system needs to be well calibrated as the automatic decision-making has to respect the user control and transparency. Tiny observations from early projects hint that users value systems that allow them to selectively share data, like proving age without disclosing birthdates. Integrating AI with SSI provides an opportunity to create personalized digital experiences while keeping users' personal information safe. The tangible benefits of this approach are already being experienced in areas such as healthcare, finance, and education, as identity verification that is faster, safer, and more autonomous improves both efficiency and confidence. Thus, the intersection of blockchain and AI in the field of identity management is not purely conceptual; there are practical applications that address the issues in the digital service usage.

Verifiable Credentials and Smart Contracts

Credentials can be stored and issued on the blockchain, and issuing university degrees on the blockchain can allow potential employers to obtain and verify the degrees instantly, which can eliminate parts of the hiring process that involve waiting to verify credentials. Because of the blockchain, employers cannot falsify the degrees. Additionally, privacy is maintained because of the use of crypto signatures (Seidi & Abdellaoui, 2025). Smart contracts, on the other hand, help with the automation of using the credentials. Smart contracts are programmable contracts on the blockchain that execute themselves once a certain condition is met. In other words, once a digital certificate is verified, a smart contract can give an employee access to the organization's systems. This way, the organization doesn't have to do the checks and also reduces the likelihood of human error. There was a digital ID and housing pilot project that I was a part of when I was in Europe. People would sign rental agreements, and the agreements would give access codes to the house once the digital ID verified that they were within the authorized group, and it felt like the rental agreements were disappearing. This type of process is very favorable for remote service provision where trust is limited. Also, businesses save time and money while individuals have full control over the disclosure of their private information.

Simultaneously, problems such as contract bugs or misconfigured rules may result in unfairly blocking access, demonstrating the need for more extensive testing prior to widespread adoption. Studies indicate that the reliability and trust of a system can be augmented by the combination of verifiable credentials and formally verified smart contracts. The relationship of these approaches points to a future in which identity verifications are conducted more effortlessly and are less susceptible to fraudulent activities.

AI-Based Trust Scoring for Decentralised Access Validation

The assessment of access to digital services through the use of machine learning is known as trust scoring. Trust scoring provides an evolving method of evaluating users by supplementing the traditional practice of using digital credentials with evidence from the user's past behaviors, the device they are using, and other variables in a history dataset. Trust scoring also has the ability to lower the user's level of access in the event that the user's behavior raises concerns about their reliability. When used with other decentralized processes, trust scoring can also enhance identity verification and eliminate the risk of a single point of failure.

Since the ability of trust scoring to run and change access parameters without human intervention is a potential liability, access decisions can be recorded using a specific blockchain that is open to the public so that other interested parties can audit the associated access parameters as they are changed by the closure of that blockchain. There have been numerous studies involving healthcare platforms that use AI trust scoring systems to determine patients' privacy concerns (Seidi & Abdellaoui, 2025; Al-dughayfiq et al., 2023), and I would use as an example the case where an unusual and repeated sign-in behavior of an employee was recorded in the access system and the employee was given a lower trust score in the access system so that the employee was not further authorized to access the records. This also exemplifies how through the use of AI, decentralized infrastructure can be used to limit the risk of internal abuse. This method is ideal in the financial system, where the risk of fraud is high. Basic fraud detection systems are useful in the financial system in order to identify the true fraud and allow suspected frauds to go undetected. AI-driven systems already exist in large numbers, and many banks use them to identify fraudulent activities that traditional systems cannot identify when the fraud is committed.

Harms arise from bias or mistakes that AI makes. A way to address these shortcomings is to combine explainable AI that gives users specific feedback for low trust scores, and blockchain's immutability for liability (Muhammad & Tahir, 2025). This is the first of many ways to use AI trust scoring and decentralized validation to transform safe interactions online.

Blockchain-Enabled Auditability and Accountability

Blockchain technology allows us to record our activities in digital form permanently without the ability to erase them later. Individual transactions or validation of identities are recorded on a public and distributed ledger, allowing a permanent record of digital activity (Tawfik et al., 2025; Linqiang et al., 2024). This type of record also enables record keeping in a decentralized manner, allowing auditing to occur at any point in time without the need to rely on some central authority. With auditing, both users and organizations can see and track who had access to which information, when and for what purpose. Smart contracts are another coupled technology that can strengthen both technology and digital identity management. In Code of the contract, rules are established and automatic reporting mechanisms are established if certain rules are violated. In the past I read about a pilot To access some government information was limited to those who were registered on the blockchain and auditors were able to monitor all access without unauthorized access in a significant way. In addition to allowing for the use of digital auditing without the fear of unauthorized use, the approach is also allowing for greater confidence in the utilization of blockchain technology. It is also of great interest to the financial sector, fearing the unreliable auditing of operations. With the blockchain in the digital healthcare sector, all parties involved can be sure that the digital footprints of all the activities, especially the unethical use of patient records, are also digitally recorded in an immutable way.

There are clear concerns regarding the price of large-scale data storage. Hybrid models have also been suggested where critical audit data is stored on-chain and larger files are stored off-chain and linked securely. Regardless of these limitations, the potential offered by audit functions of the blockchain is another step closer towards fairness and accountability in the context of digital identity systems.

11. SECURITY AND PRIVACY CONSIDERATIONS

Risks in AI-Based Authentication

Speedy and easy AI authentication sounds good, but going down that path presents serious hidden risks (Kalodanis et al., 2025; Weiqi et al., 2024). The threat of adversarial attacks is another concern. In such attacks, the adversarial model is manipulated so that they may gain access as a user. Numerous studies have shown that even a slight adjustment to a face image that a recognition system uses can deceive that system. I did an experiment with a public cellphone unlocking system, and I was able to trick the system by holding a picture of a person's face at the right angle. I got a pretty good feeling for how easy these things can fool a system. The sensitive nature of the data that these systems use, like voice recordings, and facial scans, is another risk. If a data breach were to occur, the data exposed can pose a very serious risk. Unlike a password, the data involved cannot be changed or reset. History has shown that organizations that have an obligation to safeguard personal data have a data breach that is very easy to access. It can be an insider or an outsider, and studies have shown that the misuse of data has been documented for both. We cannot use AI authentication systems, and system advocates cannot say that these systems will be completely foolproof. The motivated and well-armed attackers of today necessitate adaptive defenses. The quality of developers' predictions of their customers' vulnerabilities and their efforts to minimize them, directly impacts the systems' trust. When personal identity is involved, one weakness can be exploited, and the consequences can be severe.

Privacy-Preserving ML: Differential Privacy, Homomorphic Encryption

How to protect really sensitive data used to train machine learning models is and will continue to be a really big problem (Chen et al., 2024). Differential privacy is a way to train models which add random noise to the training process to decrease the chances of individual records being revealed (Shukla et al., 2025; Kiyani et al., 2024). This method has been used by Apple to collect data, creating a lot of positive feedback on the method when it comes to balancing user's rights and data protection with accuracy. I also remember a health study performed where the method used allowed the researchers to publish analyses without identifiable information on the patients which showed a good example of the method's usefulness. Homomorphic encryption solves the problem of data privacy in a different way. This technique allows a system to perform computations on data without having to see the actual data. This is really important for healthcare and finance. For example, using homomorphic encryption, hospitals will be able to keep patient files private while being able to use AI analytics on the data by outsourcing the work to the cloud. The only issue with it is that previous studies performed on it have shown that it is really resource intensive and can take a lot of time, making it less practical in real world applications. When using these types of data privacy methods, while they do not completely mitigate the risk of being attacked, they do help build trust with the consumer in being able to protect and use their data. Besides the fact that they do help build trust with the consumer again, they also help the consumer protect their rights and the organizations by using machine learning provide their users with all of the additional benefits that using machine learning has to offer.

If applied with care, they may help reduce public fear surrounding the handling of personal information. However, the adoption of these technologies is contingent upon the protective performance trade-offs and the willingness of organizations to invest in more robust protective measures.

12. EVALUATION METRICS AND BENCHMARKS

Accuracy, False Acceptance/Rejection Rate (FAR/FRR)

Businesses are trying to implement the best kind of security for their interconnected systems, so the use of AI-based authentication for the IoT devices is becoming more common (Menon et al., 2025; Sindiramutty, Prabagaran, Jhanjhi, Murugesan, et al., 2024b). One of the frequently used means to evaluate the effectiveness of these methods is evaluating how accurately the system identifies or misidentifies the user. In practice, the sole use of a method as means of evaluation can prove to be problematic. For instance, one smart lock I tested in a lab was 98% accurate, but it still let in one unauthorized tester. This is where False Acceptance Rate (FAR) and False Rejection Rate (FRR) become very important. FAR measures how many times the system incorrectly accepts an attacker as an authenticated user, while FRR measures how many times the system incorrectly refuses to accept an authenticated user. Despite a low percentage, security or usability issues may arise as a result. Apart from biometric security systems, it is becoming common for devices with facial recognition to be designed to have a FAR of 0.1% or less because, if the FAR is too high the system may frustrate a user.

Testing IoT assistants shows that voice recognition underperforms in real world situations because of interference. These figures are more than stats; they result in decreased consumer confidence and lowered use in everyday situations. If FAR is too high, security risks at home and work increase. If FRR is too high, voice controlled devices are frustrating and may lead to the disabling of security features. Achieving balance with these variables is challenging, but IoT security systems must be secure and user friendly.

Time to Authenticate, Energy Usage on Constrained Devices

When it comes to the Internet of Things, it is essential to have efficient and quick authentication alongside accurate authentication (Reguero et al., 2024). The time it takes to authenticate is significant in devices such as smart locks and health trackers worn on the body as users can get annoyed if they have to wait several seconds to authenticate. In one of my studies, I noticed that students who were evaluating an AI-based fingerprint lock system, some users left the system if the wait time was more than three seconds. Numerous studies have shown that users become dissatisfied very quickly if the authentication process takes too long (Kruzikova et al., 2023; Sindiramutty, Prabagaran, Jhanjhi, Ghazanfar, et al., 2024). The small batteries that IoT devices use make energy consumption an important consideration. In some cases, running AI models in devices can use more energy than older methods. Some proposed solutions use lightweight AI models, and a hybrid approach where some processing is done on the device and the other part is processed in the cloud. Adoption is directly influenced by energy and time. Systems that are slow and use a lot of energy may seem secure, but they will have no use in daily life. Creating trustworthy AI systems for IoT applications requires reliable, fast, and low-power authentication systems that simplify IoT scenarios and make the systems easier to use.

Benchmarks and Public Datasets for Authentication Systems

People discussing the utility of an authentication system usually talk about the extent to which an authentication system can empirically be evaluated. In the field of research, this is usually done using datasets, particularly publicly available benchmark datasets. There are standard evaluation measures such as accuracy, false acceptance rate, and false rejection rate which generally provide a sense of performance (Terven et al., 2025). I remember a student project on fingerprint recognition where, on the surface, everything looked good. When the student was faced with a big dataset, her error rate was horrible, which confirmed my negative bias towards such systems and the positive bias toward benchmarks. In research, such benchmarks are publicly available datasets, such as the CASIA dataset in face recognition, and the FVC dataset in fingerprint recognition. With such datasets, researchers can evaluate their approaches under the same conditions. I have seen other cases where researchers have systems aimed at achieving the same objectives, but with different measures, levels of refinements, and such systems cannot be compared, e.g., in a seminar, two opposing teams had to face off on the question of who had the best iris recognition system, and this was because data was collected and used differently by each team to train systems. This is what the benchmarks eliminate. It should also be noted that the set of real-world conditions must be considered. A system/model that was trained with high-quality images in a controlled laboratory environment is less likely to perform well when images are used that are, for example, blurring noisy. Here, benchmarks provide an almost reality check. They extend the systems from small, controlled experiments and force the systems to show their worth in conditions that mimic everyday use, be it the presence of smudgy fingerprints, bad lighting, and partial faces. Ultimately, benchmarks and datasets go beyond being merely technical tools; they influence the practical dependability and trustworthiness of authentication systems.

13. REAL-WORLD USE CASES AND APPLICATIONS

Healthcare: AI-Enabled Patient and Device Authentication

A common problem in hospitals is losing time due to old methods of authentication because staff must have streamlined access to patient information. A positive influence of IoT and authentication systems is using AI to improve patient and medical device recognition (Bhamidipaty et al., 2025; Sindiramutty, Jhanjhi, et al., 2024). Instead of using passwords and ID badges, these systems offer real-time authentication using biometrics, behavioral patterns, and device identification. This allows for patient records to be accessed without staff intervention to keep records updated, which is useful for quick access to information necessary for patient procedures, like during surgeries. A case study of a European clinic showed greater patient safety and lower medication administration errors due to AI and IoT systems (Skog et al., 2021). Clinicians often appreciated better decision-making from the AI systems that flagged device interconnectivity, but also the AI systems that autonomously prevented inappropriate device access. An additional hospital stated that gait analysis based on AI and IoT sensor data proved effective for elderly patients with poorly scanned fingerprints. Such instances highlight that authentication challenges are not just for security purposes; they are also for reducing friction in the delivery of care. In emergency situations, for instance, every second matters. IoT-enabled AI authentication strikes the critical balance between securing patient data and securing the patient. This is a patronizing problem for healthcare providers and shifts the conversation on the intersection of authentication and patient safety.

Industrial IoT: Operator Verification in Critical Systems

A common concern in large industrial facilities is providing preventative access to certain machines involved in operating power grids, oil refineries, or automated assembly lines. A single incomplete employee assignment to one of the control terminals could either cause disruption of business activities or result in unsafe work conditions. To reduce such threats, AI-based technological access control in Industrial IoT is implemented (Samant et al., 2025). They access control based on captured operator behavior such as the operator's movement and their biometric patterns, which in turn are used in the identification and access control process. While it is not impossible to disable, falsify or inappropriately use the AI-based access control systems, it is significantly more difficult. To this end, one of the Automotive industrial plants in Japan used AI-based access control where workers' identification was based on voice and hand gesture recognition through IoT sensor networks. The system was designed to detect sabotage if the voice sample matched a predetermined suspicious sample. An example of this is the case of an energy company that used AI-based access control based on keystroke dynamics of the users on the control terminals of the grids. In this case, even the passwords used to access the system could not result in unauthorized access as a result of variations in the timing of keystrokes which were enough to provide access control based on the user access rights. On-site engineers mentioned that the system did not interfere with their work and was background non-invasive. These examples show that with the advancement of technology, authentication systems have started to function as trust facilitators in business processes instead of trust gates. Employees trust that no unqualified personnel will have access to the hazardous equipment, thereby enhancing their own safety and increasing productivity. The more interconnected industries become, the more critical this problem is as one breach can trigger a domino effect across industries.

Smart Homes and Buildings: Multi-User Access Policies

In smart homes, more than one family member, guest, or even service personnel might need access to sanitize, close or open doors or access the appliances or energy management systems (Büyük et al., 2022). Old-fashioned locks or (single password) apps do not cope well with multiple users. However, AI-based smart home systems can utilize multi-user access control through user identification. Instead of individuals having to fidget with knobs to gain access, control, or permission to play with the device (Ritu et al., 2025), users can gain permission through other differences. AI can even control access to specific devices, such as children being allowed to use the entertainment control device, but not having access to the heater or air conditioning (also to control the temperature range). A particular study described the case of an AI system being installed in a living room, which was able to distinguish between the voices of cleaning staff and delivery personnel through location-based command and control devices. Unauthorized access (beyond the parameters set by users) was detected automatically and the system was very popular with users as there was no need to set parameters to monitor the system. Delivery personnel do not need to wear an ID badge, as ID badges can be lost. These small system changes made a big difference, as more service personnel used the system. Most importantly, they all achieved a balance between the ease of use and safety of the system.

AI authentication allows homes or buildings to adjust to residents and their routines rather than impose hard rules. With the rise of smart buildings, this type of personalized access control is helping us manage our daily activities and is making things safer.

14. CHALLENGES AND OPEN RESEARCH DIRECTIONS

Generalisation Across Diverse IoT Environments

Regarding Padmavathi & Saminathan (2024), research is being done on the numerous ways that AI can be applied to improve the ways that the Internet of Things (IoT) systems manage authentication and access control. One of the biggest problems is trying to apply a singular solution to very different devices and networks. For example, the ways that smart home sensors, industrial controllers, and healthcare monitors work and process data are all very different and each has a different amount of computing power. Researchers showed that if a model was trained in one environment, the model will fail in a different environment (Iunusova & Archenti, 2025). Due to the lack of uniformity in IoT devices and systems, the lack of AI driven IoT systems is justified. Some good results have been seen in the areas of transfer and federated learning. I have seen a small pilot project where an AI system was good with office access badges but not with low power sensors in agriculture. This illustrated that the environment has the potential to be a determining factor. There is a large gap between what is achievable in an ideal world and messy real world problems. An additional practical challenge with AI methods is the resource cost involved. This is where the use of lightweight models and cloud computing comes into play. If computing is outsourced, it will also create problems in regards to trust and latency.

Future research focuses on developing AI systems capable of adjusting to limitations of the devices they operate on and to continuously learn and adapt across different environments without extensive retraining. Addressing these challenges will bring IoT security to the threshold of dependable and scalable protection.

AI Model Robustness Against Spoofing and Poisoning

Increasingly common AI technology offers novel ways to implement security measures, but also allows potential attackers new methods to circumvent protections (Mohammed, 2025). A typical example of circumventing biometric security is the usage of synthesised biometric presentation attacks, which fool biometric sensors (Menakadevi et al., 2025). Another example is the artificial creation of 'poison' data that is used to train AI security technology. Participants in a 'poisoning' attack attempt to train a security measure to ignore certain inputs by including 'poison' data which the AI 'learns' during its training phase. All of the listed attack methods circumvent the very intelligent technology that is developed to protect IoT and other interconnected systems. Researchers have commented that even the latest technology in AI is easily circumvented by edits to input data that, when used, induces the AI to not respond (Kreuk et al., 2018). I have read a case sample in which a so-called 'AI-enhanced' voice authentication system was easily circumvented by replayed voice recordings. This example shows the problems that can arise when an AI system is designed and implemented without sufficient testing to measure potential limits to its robustness. The potential negative impacts of a so-called 'poisoning attack' can be significant. An AI system that is 'poisoned' during the training phase can be controlled to make any and all decisions in the future that the system is designed to make. AI system designers need to increase the amount of resources allocated to implement and monitor defenses such as adversarial training and anomaly detection. Published reports indicate that the AI and IoT communities are beginning to focus on the creation of more streamlined and less resource intensive methods to implement adversarial defenses and to monitor data streams in real-time.

The issue at hand is creating designs for defences that can change and evolve rapidly while being financially constrained. Advancements here could represent a shift from easily bypassed defences toward a resilient AI system that can safeguard sensitive environments.

Lack of High-Quality Labelled Datasets

AI systems require large amounts of data for training, but quality, labeled datasets are a major limiting factor for many IoT security applications (Afoma et al., 2025). Labeling datasets involves significant time and domain expertise, and often involves finessing sensitive data, making it hard to do at large scale. Consequently, many researchers are limited to small or synthetic datasets, which do a poor job of capturing the diversity of traffic and behavior of devices seen in the real world. This disconnect means models trained in the lab are unlikely to succeed in the real world. I remember an interesting case for an intrusion detection model breaking up multiple. When it was being trained, it looked like it would be successful, but once it was tested in an IoT system in a hospital, it misclassified common network behavior. The problem was with an algorithm, but rather with its training set, which was not diverse enough. The AI was trained with too few examples and was forced to take the psychological shortcut of doing pattern recognition. Outdated datasets, like the NSL-KDD and CICIDS datasets, even though they are public and very easy to obtain, do not reflect the current state of diverse networks, including IoT. Some researchers are trying to explore new frontiers, using semi-supervised learning, data augmentation, or even the generation of synthetic data to facilitate the massive labor of data labeling. All of this means that AI will only use data that will always be of low quality, and to build a dataset of higher quality, it is necessary to cross the frontier between academia and industry; however, the sharing of data in the context of privacy will always be a problem. AI is based in the inseparable part of IoT, and, with the current state of the technology, always faces the assailant.

15. CONCLUSION

Comprehensive Summary of Key Concepts and Techniques

The discourse throughout this chapter demonstrated how artificial intelligence has been transforming the way in which authentication and access control are performed in the Internet of Things. However, traditional means (e.g., static passwords or fixed keys) are unfit for billions of resource-restricted and diverse devices with respect to power budget, computing power, and exposure conditions. Case studies persistently demonstrated how attackers exploit weak points, whether hospital machines or city infrastructure, requiring new ways of thinking. It is here that AI-based approaches can offer their flexibility and accuracy. Behavioural biometrics, anomalous detection, and continuous verification are safety nets that react to the way one behaves, not to fixed way stations. With everything from lightweight machine learning classifiers that can run on low-power sensors through deep networks for face or voice recognition, we can now recognise real users and attackers with far greater precision. What we see today from healthcare, industrial facilities, and even smart homes is an illustration of how such methods are already transforming security from just a static block into a living, breathing system that seems more intuitive to use every day. The edge processing and blockchain also help to offload the processing load in decentralising the authentication, reducing the latency, and providing the users with more freedom to manage their identity information. But the risks — adversarial attacks, as well as biased models and poor quality data, say — remain, as a barrier to reliability. Yet technological advancements such as federated learning, privacy-preserving computation, and AI-powered Zero Trust can suggest a direction where "usability" and "security" are symbiotic, instead of being antithetic. The actual impact is not technical but social. When that authentication is seamless, it creates trust that the connected systems on which we rely — for everything from medical care to what to put out in your recycling bin — aren't going haywire. AI-driven identity management is no longer something of an experiment, but rather an embodiment of trust in the digital age."

Role of AI in Redefining Identity in IoT

Identity is a far cry from traditional credentials in the world of the Internet of Things. A single password or device key that gives the "who" or the "what" was supposed to have, but in practice, these authentication methods have been porous. The definition is shifting, and again, AI has a role to play in reframing identity as being fluid, according to behaviour, context, patterns, rather than a single token. Instead of simply requesting a login, AI models are now able to scrutinise how a person types, how a sensor operates, or how a wearable is used. I watched in terror/fascination as a pay-test which seemed to require nothing beyond typing rhythm and swipe-pressure to dry-hump the hide off the skimmered card in seconds. That moment captures how identity is no longer detectable by a number on a card, or even by an ID card alone, but by patterns of life that are vastly harder to fake. Identity is also being rethought for devices. An in-factory sensor can be detected by not just its serial number, but by the unique way it reports data, by what time of day it makes those reports, and by how much energy it uses. It's because

A.I. learns these fine-grained idiosyncratic fingerprints, which in turn makes it more difficult for various attackers to sneak in fake devices. Through federated learning and blockchain registries, this becomes a self-evolving identity that learns, shows, and adapts across this ecosystem continuously. The point of this shift is the trust/usability trade-off. Patients, residents, or operators don't have to wait to manually close the door, and yet they are protected. Identity is recast as a living, breathing data profile, allowing IoT connected systems to recognise both people and machines in a way that is intuitive, resilient, and suited to the era of hyper connection.

Practical Guidelines for Implementation

AI-based Authentication for IoT: From Theory to Practice: Transforming AI-powered authentication into an IoT working system cannot be confined to paper. Knowledge of what the devices themselves can do is the first step. Many are light on both processing power and battery life. In one smart agriculture project, soil sensors would go offline when the encryption load got too high. This demonstrated the importance of matching the algorithm to the device. There are plenty of light models or compressed ones that can get a good performance without resource overhead. Another guideline is gradual integration. Rolling out AI-based authentication across a whole hospital or factory at the same time can create headaches. Pilot programs in small areas can help identify gaps. One healthcare team said voice-based access was positive in the lab but struggled in high-noise emergency rooms. They adapted on the fly before a full rollout by testing a little bit at a time. Data handling is equally crucial. Authentication is reliant on your sensitive data, such as biometrics and behaviour logs. Safeguarding this data through encryption, secure storage, and transparent update processes will limit the damage done by a leak that may be harder to recuperate from than a stolen password. Frequent model updates are also crucial because attackers are always changing their tactics. And last but not least, usability. Too-restrictive systems typically lead users to turn off protections. Creating authentication that feels like no authentication at all, such as by identifying a worker's gait or a resident's device habits, is what keeps security strong, not friction. Adhering to these best practices is the difference between abstract notions of AI-based identity in IoT and trusted reality.

Vision for Secure, Intelligent IoT Authentication in the Future

In the future, authentication in IoT will probably be more about that constant trust and less about those single gateways. Identity will be verified unobtrusively from behaviour, device usage, and the environment, rather than a password or tapping a token. I previously observed a prototype smart home system that only unlocked doors when a subject's gait, when that person's phone was in the room, and when that person's voice intonation matched. The whole experience felt effortless, as though the house already knew who the occupants were meant to be. This future will be profoundly influenced by AI. Such models can learn from multiple data streams and help identify both humans and machines with more accuracy. Edge devices will do much of the processing locally to preserve privacy and minimise delays. Simultaneously, blockchain or decentralised registration may be used to make it impossible to fabricate or modify identity, securing information that will become trusted over the course of years.

"Future vision also implies responsibility. Systems have to protect themselves from adversarial attacks while ensuring fairness such that no user is unjustly excluded. Explainable AI is going to be important - companies and individuals need to be able to understand why an authentication decision has been taken. The real promise is in an organic trust. Not only hospitals and factories, but also homes will have protection without clogging up daily life. If this is true, then authentication will never feel like a barrier, but will be a presence felt silently watching over interactions we have, learning with every use, and supporting a digital world that is safe and human-centric.

REFERENCES

Afoma, U. M., Singh, S., Mishra, A. K., Sharma, C. K., Gupta, K., Mishra, M. K., Roy, B., Verma, V. V., & Sharma, V. K. (2025). Integrating Artificial Intelligence in Environmental Monitoring: A Paradigm Shift in Data-Driven Sustainability. *EcoHealth*, *23*(1), 38–57. Advance online publication. DOI: 10.1007/s10393-025-01752-8 PMID: 40875110

Aldughayfiq, B., Ashfaq, F., Jhanjhi, N. Z., & Humayun, M. (2023). YOLO-Based Deep Learning Model for pressure Ulcer detection and Classification. *Health Care*, *11*(9), 1222. DOI: 10.3390/healthcare11091222 PMID: 37174764

Alshuhail, A., Alshahrani, A., Mahgoub, H., Ghaleb, M., Darem, A. A., Aljehane, N. O., Alotaibi, M., & Alzahrani, F. (2025). Machine edge-aware IoT framework for real-time health monitoring: Sensor fusion and AI-driven emergency response in decentralized networks. *Alexandria Engineering Journal*, *129*, 1349–1361. DOI: 10.1016/j.aej.2025.08.030

Anver, S. R., Deepambika, V. A., Rahiman, M. A., & Santhosh, R. (2025). Emotional Speech Generation: An approach using Convolutional Neural networks (CNN) based Generative Adversarial network. *Circuits, Systems, and Signal Processing*, *44*(11), 8734–8756. Advance online publication. DOI: 10.1007/s00034-025-03224-4

Araújo, A. L. D., Sperandio, M., Calabrese, G., Faria, S. S., Cardenas, D. C., Martins, M. D., Saldivia-Siracusa, C., Giraldo-Roldán, D., Pedroso, C. M., Vargas, P. A., Lopes, M. A., Santos-Silva, A. R., Kowalski, L. P., & Moraes, M. C. (2025). ARTIFICIAL INTELLIGENCE IN HEALTHCARE APPLICATIONS TARGETING CANCER DIAGNOSIS – PART I: DATA STRUCTURE, PREPROCESSING AND DATA ORGANIZATION. *Oral Surgery, Oral Medicine, Oral Pathology and Oral Radiology*, *140*(1), 79–88. Advance online publication. DOI: 10.1016/j.oooo.2025.01.004 PMID: 39893121

Attaullah, M., Ali, M., Almufareh, M. F., Ahmad, M., Hussain, L., Jhanjhi, N., & Humayun, M. (2022). Initial stage COVID-19 detection system based on patients' symptoms and chest X-Ray images. *Applied Artificial Intelligence*, *36*(1), 2055398. Advance online publication. DOI: 10.1080/08839514.2022.2055398

Azeem, M., Ullah, A., Ashraf, H., Jhanjhi, N., Humayun, M., Aljahdali, S., & Tabbakh, T. A. (2021). FOG-Oriented secure and lightweight data aggregation in IOMT. *IEEE Access : Practical Innovations, Open Solutions*, *9*, 111072–111082. DOI: 10.1109/ACCESS.2021.3101668

Bhamidipaty, V., Bhamidipaty, D. L., Guntoory, I., Bhamidipaty, K., Iyengar, K. P., Botchu, B., & Botchu, R. (2025). Revolutionizing Healthcare: The Impact of AI-Powered Sensors. *Wiley*, 355–373. DOI: 10.1002/9781394280735.ch18

Büyük, M., Avşar, E., & İnci, M. (2022). Overview of smart home concepts through energy management systems, numerical research, and future perspective. *Energy Sources Part a Recovery Utilization and Environmental Effects*, 1–26. DOI: 10.1080/15567036.2021.2024924

Calboli, S., & Engelen, B. (2025). AI-enhanced nudging in public policy: Why to worry and how to respond. *Mind & Society*, *24*(2), 529–547. Advance online publication. DOI: 10.1007/s11299-025-00322-3

Chan, W., Gai, K., Yu, J., & Zhu, L. (2025). Blockchain-Assisted Self-Sovereign Identities on Education: A survey. *Blockchains*, *3*(1), 3. DOI: 10.3390/blockchains3010003

Chemmakha, M., Omar, H., & Lazaar, M. (2025). Toward a New Approach for Internet of Things (IoT) Intrusion Detection Based on Feature Selection and Tabular Generative Adversarial Network. *Intelligenza Artificiale.*, *0*(0). Advance online publication. DOI: 10.1177/17248035251361258

Chen, C., Liu, J., Tan, H., Li, X., Wang, K. I., Li, P., Sakurai, K., & Dou, D. (2024). Trustworthy federated learning: Privacy, security, and beyond. *Knowledge and Information Systems*. Advance online publication. DOI: 10.1007/s10115-024-02285-2

Chitraju, S. (2025). <div> AI and Behavioral Biometrics for Secure Financial Transactions </div> *SSRN*. DOI: 10.2139/ssrn.5353743

Cyberly. (2025, July 13). *How does AI detect anomalies or unusual behaviour?* https://www.cyberly.org/en/how-does-ai-detect-anomalies-or-unusual-behaviour/index.html?utm_source

Eser, A. (2025, May 30). *Password hacking statistics Statistics: ZIPDO Education Reports 2025*. ZipDo. https://zipdo.co/password-hacking-statistics/?utm_source

Humayun, M., Khalil, M. I., Almuayqil, S. N., & Jhanjhi, N. Z. (2023). Framework for Detecting Breast Cancer Risk Presence using Deep Learning. *Electronics (Basel)*, *12*(2), 403. DOI: 10.3390/electronics12020403

Iunusova, E., & Archenti, A. (2025). Lab-to-Field Generalization Gap: Assessment of transfer learning for bearing fault detection. *Applied Sciences (Basel, Switzerland)*, *15*(12), 6804. DOI: 10.3390/app15126804

Kalodanis, K., Rizomiliotis, P., Feretzakis, G., Papapavlou, C., & Anagnostopoulos, D. (2025). High-Risk AI Systems—LIE Detection Application. *Future Internet*, *17*(1), 26. DOI: 10.3390/fi17010026

Kalra, A. (2025). Emerging technologies in biometrics: artificial intelligence and machine learning. In *CRC Press eBooks* (pp. 3–26). DOI: 10.1201/9781003509554-2

Khajehzadeh, L., Barati, H., & Barati, A. (2024). A lightweight authentication and authorization method in IoT-based medical care. *Multimedia Tools and Applications*, *84*(12), 11137–11176. Advance online publication. DOI: 10.1007/s11042-024-19379-2

Kiyani, F. F., Hamid, B., Humayun, M., Sindiramutty, S. R. L., & Chowdhury, S. (2024). Discovery of Influential Publications Using Research Article's Usage Context. *2024 International Conference on Emerging Trends in Networks and Computer Communications (ETNCC)*, 1–7. DOI: 10.1109/ETNCC63262.2024.10767576

Kreuk, F., Barak, A., Aviv-Reuven, S., Baruch, M., Pinkas, B., & Keshet, J. (2018). Deceiving End-to-End Deep Learning Malware Detectors using Adversarial Examples. *arXiv (Cornell University)*. https://doi.org//arxiv.1802.04528DOI: 10.48550

Krishnan, S., Thangaveloo, R., Rahman, S. B. A., & Sindiramutty, S. R. (2021). Smart Ambulance Traffic Control system. *Trends in Undergraduate Research*, *4*(1), c28–c34. DOI: 10.33736/tur.2831.2021

Kruzikova, A., Muzik, M., Knapova, L., Dedkova, L., Smahel, D., & Matyas, V. (2023). Two-factor authentication time: How time-efficiency and time-satisfaction are associated with perceived security and satisfaction. *Computers & Security*, *138*, 103667. DOI: 10.1016/j.cose.2023.103667

Kumar, P. (2025). Next-generation secure authentication and access control architectures: Advanced techniques for securing distributed systems in modern enterprises. *International Journal of Computational and Experimental Science and Engineering*, *11*(3). Advance online publication. DOI: 10.22399/ijcesen.3294

Linqiang, Y., Sindiramutty, S. R. L., Ashraf, H., Muzammal, S. M., Balakrishnan, S. P., Gupta, S., & Kavita, N. (2024). Intelligent Household Waste Classification System Based on Machine Learning. *2024 International Conference on Emerging Trends in Networks and Computer Communications (ETNCC)*, 760–768. DOI: 10.1109/ETNCC63262.2024.10767563

Lut, L. T. Y. (2025). *Understanding the behaviour of business users in multi-factor authentication adoption*. LUTPub. https://lutpub.lut.fi/handle/10024/169612

Menakadevi, B., Kumar, D. S., Nagasaratha, P., & Parimalam, K. (2025). Biometric System Attacks-A Case Study. *2025 International Conference on Computer, Electrical & Communication Engineering (ICCECE)*, 1–7. DOI: 10.1109/ICCECE61355.2025.10940734

Menon, U. V., Kumaravelu, V. B., Kumar, C. V., Rammohan, A., Chinnadurai, S., Venkatesan, R., Hai, H., & Selvaprabhu, P. (2025). AI-Powered IoT: A Survey on Integrating Artificial Intelligence with IoT for Enhanced Security, Efficiency, and Smart Applications. *IEEE Access : Practical Innovations, Open Solutions*, *1*, 50296–50339. Advance online publication. DOI: 10.1109/ACCESS.2025.3551750

Mohammed, A. (2025). Artificial Intelligence-Powered Cyber Attacks: Adversarial Machine learning. *Authorea (Authorea)*. DOI: 10.22541/au.173862063.39098197/v1

Mohanty, A., Mohapatra, A. G., & Mohanty, S. K. (2025). Real-Time Traffic Monitoring with AI in Smart Cities. In *Lecture notes in intelligent transportation and infrastructure* (pp. 135–165). DOI: 10.1007/978-3-031-72959-1_7

Mollo, M., Mpekoa, N., & Tom, S. (2025). BehaveSEC: a mobile Behavioural Biometric authentication system. In *Lecture notes in computer science* (pp. 103–117). DOI: 10.1007/978-3-031-93064-5_7

Muhammad, I., & Tahir, M. (2025, May 14). *Blockchain-Enabled Explainable AI: A framework for verifiable and trustworthy machine learning interpretability*. https://jcib.org/index.php/jcib/article/view/36

Muzafar, S., & Jhanjhi, N. Z. (2019). Success stories of ICT implementation in Saudi Arabia. In *Advances in electronic government, digital divide, and regional development book series* (pp. 151–163). DOI: 10.4018/978-1-7998-1851-9.ch008

Odufisan, O. I., Abhulimen, O. V., & Ogunti, E. O. (2025). Harnessing artificial intelligence and machine learning for fraud detection and prevention in Nigeria. *Journal of Economic Criminology*, *100127*, 100127. Advance online publication. DOI: 10.1016/j.jeconc.2025.100127

Padmavathi, V., & Saminathan, R. (2024). Security for the internet of things. In *Elsevier eBooks* (pp. 353–368). DOI: 10.1016/B978-0-443-13223-0.00019-9

Priyanka, N., Luthra, M., & Angurala, M. (2025). Enhancing Educational Security with AI and Zero-Trust Principles. In *Apple Academic Press eBooks* (pp. 135–160). DOI: 10.1201/9781779643575-7

Rani, S., Kumar, R., Panda, B. S., Kumar, R., Muften, N. F., Abass, M. A., & Lozanović, J. (2025). Machine Learning-Powered Smart Healthcare Systems in the era of Big Data: Applications, diagnostic insights, challenges, and ethical implications. *Diagnostics (Basel), 15*(15), 1914. DOI: 10.3390/diagnostics15151914 PMID: 40804880

Reguero, Á. D., Martínez-Fernández, S., & Verdecchia, R. (2024). Energy-efficient neural network training through runtime layer freezing, model quantization, and early stopping. *Computer Standards & Interfaces, 92*, 103906. DOI: 10.1016/j.csi.2024.103906

Riaz, A., Iqbal, W., Islam, F. U., Hassan, M., Tariq, R., Chehri, A., & Madni, A. F. (2025). S2H-securing smart homes: A blockchain-based preservation approach. *Internet of Things : Engineering Cyber Physical Human Systems, 101523*, 101523. Advance online publication. DOI: 10.1016/j.iot.2025.101523

Ritu, N., Shahare, Y., Dhakar, D. S., & Jain, R. (2025). IAS and Facial Recognition System. *Wiley*, 145–161. DOI: 10.1002/9781394302734.ch7

Samant, S., Goel, P. K., Tyagi, H., & Hussain, A. (2025). Security fortifying critical infrastructure. In *Advances in computational intelligence and robotics book series* (pp. 439–470). DOI: 10.4018/979-8-3373-3241-3.ch021

Sassa, G. T., Kouanou, A. T., Fendji, J. L. K. E., Fountsop, A. N., Bouetou, T. B., & Emvudu, Y. S. (2025). A hybrid recurrent neural network architecture for the prediction of subscriber traffic in a mobile telecommunication network. *Applied Computational Intelligence and Soft Computing, 2025*(1). DOI: 10.1155/acis/7322398

Satheesh, K. K., & Sree, T. K. (2025). Enhancing Cloud Security with Authentication for Securing Sensitive Information. *Pattern Recognition, 112345*. Advance online publication. DOI: 10.1016/j.patcog.2025.112345

Seidi, S., & Abdellaoui, A. (2025). Securing and ensuring the confidentiality of medical data in the era of decentralized Technologies: A blockchain and Self-Sovereign Identity-Based approach. *Journal of Network and Systems Management, 33*(4), 81. Advance online publication. DOI: 10.1007/s10922-025-09960-x

Shah, I. A., Jhanjhi, N. Z., & Laraib, A. (2022). Cybersecurity and blockchain usage in contemporary business. In *Advances in information security, privacy, and ethics book series* (pp. 49–64). DOI: 10.4018/978-1-6684-5284-4.ch003

Sharma, S. (2025). 5G-Enabled Fusion: Navigating the Future Landscape of Cloud Computing, Internet of Things, and Recommender Systems. *Wiley*, 251–265. DOI: 10.1002/9781394256075.ch13

Shukla, S., Rajkumar, S., Sinha, A., Esha, M., Elango, K., & Sampath, V. (2025). Federated learning with differential privacy for breast cancer diagnosis enabling secure data sharing and model integrity. *Scientific Reports, 15*(1), 13061. Advance online publication. DOI: 10.1038/s41598-025-95858-2 PMID: 40240790

Sindiramutty, S. R., Jhanjhi, N., Tan, C. E., Lau, S. P., Muniandy, L., Gharib, A. H., Ashraf, H., & Murugesan, R. K. (2024). Industry 4.0. In *Advances in logistics, operations, and management science book series* (pp. 342–405). DOI: 10.4018/979-8-3693-1363-3.ch013

Sindiramutty, S. R., Prabagaran, K. R. V., Jhanjhi, N. Z., Ghazanfar, M. A., Malik, N. A., & Soomro, T. R. (2024). Security Considerations in Generative AI for web Applications. In *Advances in information security, privacy, and ethics book series* (pp. 281–332). DOI: 10.4018/979-8-3693-5415-5.ch009

Sindiramutty, S. R., Prabagaran, K. R. V., Jhanjhi, N. Z., Murugesan, R. K., Brohi, S. N., & Masud, M. (2024a). Generative AI in network security and intrusion detection. In *Advances in information security, privacy, and ethics book series* (pp. 77–124). DOI: 10.4018/979-8-3693-5415-5.ch003

Sindiramutty, S. R., Prabagaran, K. R. V., Jhanjhi, N. Z., Murugesan, R. K., Brohi, S. N., & Wei, G. W. (2024b). Generative AI for threat intelligence and information sharing. In *Advances in digital crime, forensics, and cyber terrorism book series* (pp. 191–234). DOI: 10.4018/979-8-3693-8944-7.ch006

Sindiramutty, S. R., Tan, C. E., & Wei, G. W. (2024). Eyes in the sky. In *Advances in information security, privacy, and ethics book series* (pp. 405–451). DOI: 10.4018/979-8-3693-0774-8.ch017

Sinha, G., & Agarwal, A. (2025). Dynamic Trust Management and AI-Based Multi-Factor Authentication for 6G Machine Type Communications. *2025 First International Conference on Advances in Computer Science, Electrical, Electronics, and Communication Technologies (CE2CT)*, 560–565. DOI: 10.1109/CE2CT64011.2025.10939377

Sivasankari, K., & Hanirex, D. K. (2025). Siamese Iris on Siamese Neural Network architecture for Iris matching and verification. In *IGI Global eBooks* (pp. 247–262). DOI: 10.4018/979-8-3693-9420-5.ch015

Skog, J., Rafie, S., Schnock, K. O., Yoon, C., Lipsitz, S., & Lew, P. (2021). The impact of smart pump interoperability on errors in intravenous infusion administrations: A multihospital before and after study. *Journal of Patient Safety*, *18*(3), e666–e671. DOI: 10.1097/PTS.0000000000000905 PMID: 35344977

Tawfik, A. M., Al-Ahwal, A., Eldien, A. S. T., & Zayed, H. H. (2025). Blockchain-based access control and privacy preservation in healthcare: A comprehensive survey. *Cluster Computing*, *28*(8), 529. Advance online publication. DOI: 10.1007/s10586-025-05308-x

Terven, J., Cordova-Esparza, D., Romero-González, J., Ramírez-Pedraza, A., & Chávez-Urbiola, E. A. (2025). A comprehensive survey of loss functions and metrics in deep learning. *Artificial Intelligence Review*, *58*(7), 195. Advance online publication. DOI: 10.1007/s10462-025-11198-7

Verma, G., & Sahu, T. P. (2025). Lightweight feature selection for multi-label classification using multi-objective preference selection index. *International Journal of Machine Learning and Cybernetics*, *16*(12), 10457–10488. Advance online publication. DOI: 10.1007/s13042-025-02781-z

Weiqi, X., Hooi, S. T. C., Sindiramutty, S. R. L., Asirvatham, D. L., Kumar, D., & Verma, S. (2024). Surface Anomaly Detection Using Machine Learning Technique. *2024 International Conference on Emerging Trends in Networks and Computer Communications (ETNCC)*, 1–7. DOI: 10.1109/ETNCC63262.2024.10767562

Zarkia, M. N. H., & Usman, S. (2025). IoT data breaches and privacy issues in healthcare system. *Open International Journal of Informatics*, *13*(1), 41–55. DOI: 10.11113/oiji2025.13n1.327

Zhai, B., Akande, O. N., Agarwal, S., & Pak, W. (2025). Security risk assessment of internet of things health devices using DREAD and STRIDE models. *Ain Shams Engineering Journal*, *16*(11), 103721. DOI: 10.1016/j.asej.2025.103721

Zhang, Y., Li, B., Liu, B., Wu, J., Wang, Y., & Yang, X. (2020). An Attribute-Based collaborative access control scheme using blockchain for IoT devices. *Electronics (Basel)*, *9*(2), 285. DOI: 10.3390/electronics9020285

Zheng, Y., Chen, Y., Qian, B., Shi, X., Shu, Y., & Chen, J. (2025). A review on edge large language models: Design, execution, and applications. *ACM Computing Surveys*, *57*(8), 1–35. Advance online publication. DOI: 10.1145/3719664

Chapter 4

Vehicle Gesture Control by Motion Wheel Autonomous Techniques Using IoT

Sonam Gour
http://orcid.org/0000-0002-6727-8942
Department of Computer Engineering, Poornima College of Engineering, Jaipur, India

ABSTRACT

The study examine (Internet of Things) IoT technology is integrated on the gesture-controlled smart car. This integration enhances automation and remote navigation features. The technology uses IoT infrastructure and hand gesture detection in real-time applications. A wireless communication system makes it possible to automatically regulate vehicle motions without the use of traditional control methods. Accelerometers or infrared modules can detect human emotions, and the microprocessor decodes the information before instructing the vehicle system. Through a web or mobile interface, the IoT enhances ongoing diagnostics and offers real-time data tracking and vehicle monitoring. The prototype emphasises three key features: precise control mechanics, intuitive operation, and adaptable programming options. It can be used in assistive technology for those with physical disabilities, industrial automation, and personal mobility.

1. INTRODUCTION

Vehicle automation and user interface technologies are being introduced to overcome vehicle automation, and intelligent control systems are increasingly being integrated into vehicles as a result of the rapid advancement of technology. The one significant industrial innovation is the integration of IoT technologies into gesture control systems. In order to have intuitive, touchless, remote controlled operations, the Internet of Things (IoT) used in conjunction with gesture based control is necessary. This paper addresses the development of a smart vehicle system that can recognize hand gestures, and provide the means for remote monitoring using IoT platforms. The underlying problem is presented in the following chapter with the theoretical background. There is a presentation of the addressed issue,

DOI: 10.4018/979-8-3373-7355-3.ch004

describing challenges within the project, and a survey of the state-of-the-art in gesture recognition and smart vehicle control.

However, with the advance of technology, mechanical interfaces that include buttons, and joysticks have given way to voice and gestures commands that operate in human machine interaction. Gesture Based recognition systems enable users to interact with machines by natural movements, increasing users accessibility, especially for people with disabilities, and the whole user experience. Gesture control when integrated with IoT gives real time communication, sharing of data and remote operation. Beyond convenience, gesture control is the potential for smart vehicles to act as surveillance, assistive support and in rescue operations for people with physical impairments.

2. BACKGROUND AND LITERATURE REVIEW

Gesture Recognition

Recent developments in sensor technology and human-machine interfaces have led to an increase in the use of gesture-based control systems. In the past, devices like cars and robots were controlled by physical inputs like buttons, joysticks, or remote controls. It provides more natural and innovative control method for the development of gesture guided system. MPU 6050 is the key component of this system. It combines a 3-axis accelerometer and gyroscope for detecting motion and rotation in real time. These sensor utilised by the drones, robotics, and gaming systems. The MPU 6050 can compare motions and converted them into control signals, when it was paired with microcontroller logic, such as an Arduino. The combination of MPU 6050 and microcontroller provides the variability of the system. It could be applied in a various fields that include assistive technique for the physically challenged and smart vehicle control, where accessibility of user is critical considerations. Gesture recognition is possible by machine interpretation of human hand movements. This can be accomplished through a variety of methods, such as the use of accelerometers, infrared sensors, and machine learning algorithms that include camera image processing. The effectiveness of accelerometer-based gesture detection in identifying motion and speed makes it stand out among the others. It was chosen as the project's preferred method because of its cost-effectiveness and operational efficacy.

The creation of gesture-controlled automobiles has become a viable path for user-friendly and interactive control systems among a broad variety of gesture control applications. Hand gesture recognition, traffic control gestures, alternative gesture recognition techniques, and user engagement are only a few of the facets of human–computer interaction in assisted vehicle control. Understanding the fundamentals and methods in this subject was made possible by the main gesture recognition obstacles, which include motion performance, contextual factors, social norms, and interpersonal differences. Pilot research on the possibility of using aural feedback for gesture control in automobiles is being conducted. Investigating the application of multimodal feedback approaches has been shown to improve the efficacy of gesture-based interactions in cars and the user experience. In a survey, Bansal categorized the technologies used in gesture recognition into two groups: contact-based and non-contact-based technologies. For motion detection applications, ordinary depth or thermal cameras combined with appearance-based and model-based techniques can be effective. The comparison between all techniques are represented in table 1.

Table 1. Comparative analysis for different hand recognition system

Terminology	*Wearable System*	*Contact System*	*Provision System*
User Accessibility	Moderate	Low	High
Hardware requirement	Specialized wearable	Physical Connection	Standard
Real time behavior	Medium	Medium	High
Cost	Medium to high	Low to medium	Low
Scalability and development	Limited	Limited	High

Microcontroller and Sensor Integration

The main processing unit which we use for both Arduino and Node MCU takes in sensor information while generating signals that manage the vehicle. The MPU6050 accelerometer module along with other sensor modules detect hand tilts and motions to generate corresponding directional commands (Yadav & Sharma, 2023).

IoT Communication

By connecting such systems through the Internet of Things, people can immediately monitor and control their systems over network connectivity. Microcontroller with Wi-Fi capability provides a data transmission path to connect with the cloud dashboard or mobile applications to track the status of vehicles and the user can also operate the vehicle remotely or receive an alert.

Gesture-controlled vehicles are a new technology that enables users to operate cars with hand gestures rather than conventional buttons or knobs in the age of artificial intelligence (AI) and the Internet of Things (IoT). In order to facilitate intuitive and effective communication between the user and the vehicle, omnidirectional vehicle gesture control relies heavily on human–computer interaction. Deep learning and feature extraction provide promising image identification applications for intelligent systems. LSTM and Bidirectional Long Short-Term Memory (BiLSTM) layers in a federated deep neural network classify smartphone reviews by protecting user privacy as shown in the recent study (Singh & Kaur, 2023). The model achieved 96.68% accuracy without requiring centralized data and it was demonstrated the effectiveness of the deep sequence modelling in various type of distributed environments. The combination of gesture control on motion-wheels with autonomous car designs provides real-time IoT communication. The result of misinterpreting gestures in mobility-critical situations provides safety and dependability in the patient control. The difficulties in implementing such systems in extensive smart transportation ecosystems in terms of scalability and cybersecurity.

For classification, the XAI-RACapsNet architecture combines a Relevance-Aware Capsule Network (RACapsNet), an adaptive feature extraction module (AFEM), and a transformer-based segmentation network (XAI O-Net). By extracting rich data and creating relevance heatmaps, our layered architecture increases accuracy and interpretability while providing physicians with more transparent and dependable diagnostic help. Various techniques have been developed for implementing gesture-controlled systems.

- **Accelerometer-based control** it was used for detecting tilt or motion angles for implementing gestures. It is relatively straightforward and cost-effective.

- **Camera based recognition** this approach uses provide high accuracy and significant processing power. It can be sensitive for lighting conditions.
- **Machine learning based systems** enhances accuracy by using large gesture datasets. Although it requires sufficient processing power and training time, it can generalise gestures.
- **Wireless communication** Different trade-off analyses for range, speed, energy consumption, and integration complexity are provided by each procedure.\

The theoretical ideas and development process used for the suggested gesture-controlled smart IoT-based car are described in this part. It addresses IoT communication protocols, microcontroller interfacing, gesture recognition systems, and the system's operational process (Roy & Basha, 2022).

Trends in Autonomous gesture control system:

Multi model human machine interface

Gesture recognition is being used more and more in conjunction with voice commands, eye tracking, and touchless interfaces in contemporary autonomous systems. By lowering single-point failure from misinterpreted gestures, multi-modal frameworks increase dependability in comparison to solo gesture systems.

Edge AI for real time processing

In order to lower latency, edge computing is becoming crucial as a result of IoT-enabled cars producing large amounts of data. Emerging technologies use on-device or edge-based inference in place of cloud-dependent gesture processing to provide millisecond-level responsiveness in safety-critical situations.

AI based Vision and sensor fusion:

To improve gesture recognition resilience under a variety of environmental circumstances (such as low light, occlusion, and vibration), recent studies combine camera-based deep learning models with radar and LiDAR data. Reliability is increased via sensor fusion as opposed to single-sensor motion wheel methods. Different terminology are represented in Table 2.

Table 2. Comparison of different hand gesture control system

Technique	Advantage	Limitation
Motion Wheel-Based Gesture (Proposed)	Low-cost hardware, intuitive control, mechanical stability	Limited gesture vocabulary, mechanical wear concerns
Wearable IMU-Based Systems	High accuracy, user-specific calibration	Requires wearable device, comfort concerns
Vision-Based Gesture Recognition	Rich gesture set, contactless operation	Sensitive to lighting, higher computational load
Brain–Computer Interface (Emerging)	Direct neural interaction	High complexity, not scalable for mass deployment
Voice-Control Interfaces	Hands-free operation	Noise-sensitive, language dependency

3. PROBLEM DEFINATION AND CHALLENGES

For the formation of gesture controlled smart vehicles, there are practical issues that should be solved and also technical problems must be solved together. Wrong gesture interpretation occurs due to wrong gesture recognition because the sensor data contains noise or hand movements are not consistent to be recognized. The chapter explores the broader implications of gesture misinterpretation and network latency for reliability, safety, and scalability. It was considered as the three fundamental principle of smart mobility. IoT-enabled mobility systems with gesture interfaces provide heterogeneity (noise, occlusion, and lighting), human variation, and equalization in sensor calibration. It was important parameter for the research oriented part due to its confidence scoring algorithms, error correction methods, and redundancy structures. The deployment of the IoT throughout smart cities in the gesture-recognition systems provides distributed high-density communication networks, cloud-edge topologies, and heterogeneous devices. The important concern for the designing are bandwidth allotment, QoS management, edge offloading, and synchronization. The architectural considerations like edge intelligence's absence limit the chapter's applicability in the real smart mobility systems. In safety-critical scenarios like networked transport systems, misinterpreted gestures might lead to unanticipated system behaviours, such incorrect car orders, misunderstandings among pedestrians, or traffic control failures. Secure edge computing paradigms, threat modelling frameworks, zero-trust architectural concepts, and privacy-preserving machine learning methods should all be included into.

This chapter cover a zero-trust architectural concepts, threat modelling frameworks, secure edge computing paradigms, and privacy-preserving machine learning methods In smart mobility infrastructures that incorporate edge computer nodes, V2X connection, and IoT sensors, even millisecond-level delays could have catastrophic effects. While using gesture recognition systems stringent latency restrictions provides fault tolerance strategies, and real-time reliability requirements.

- **Systems with IoT:** Network delays provide an impact on real time performance in IoT based systems,
- **Economic power utilizes:** It should run for in a long time for mobile scenarios, for economic power utilization.
- **Connected Sensors:** The hardware should be carefully chosen and the platform set up to support sealing communication for desired connection with microcontrollers and IoT platforms.
- **Security and Privacy:** IoT-based systems designed for strong security measures and focusing on user privacy for meet business requirements.

4. SYSTEM ARCHITECTURE AND METHODOLOGY

Overview of Proposed System

It is used for creating a gesture controlled vehicle that could be watched from a distance using the IoT. The three main modules of the system are:

Accelerometer: The gesture input module uses an accelerometer to record hand motions
Control Unit: Uses a microcontroller (such as an Arduino or NodeMCU) to process gestures.

IoT Communication Module: Uses platforms like Blynk or MQTT to send and receive commands over Wi-Fi in real-time.

Gesture Recognition Using Accelerometer An MPU6050 accelerometer, which tracks hand movement along the X, Y, and Z axes, is used to implement gesture recognition. Predefined threshold ranges that correspond to directional commands (forward, backward, left, right, stop) are translated to the acceleration values (Johnson & White, 2022).

Working Principle of MPU6050

- The MPU6050 combines a 3-axis gyroscope with a 3-axis accelerometer to offer 6-axis motion tracking. Mapping logic is demonstrated in table 3.
- The microcontroller transfers the sensor's analogue and digital readings to gesture directions.

Table 3. Gesture mapping logic

Gesture	X-Axis Value	Y-Axis Value	Command
Forward	Neutral	Tilt forward	Move forward
Backward	Neutral	Tilt backward	Move backward
Left	Tilt left	Neutral	Turn left
Right	Tilt right	Neutral	Turn right
Stop	Stable	Stable	Stop

Microcontroller and Signal Processing

The Arduino Uno or NodeMCU (ESP8266/ESP32) microcontrollers are used to:

- Read sensor information in real-time.Interpret gesture-based commands.
- Send control signals to the vehicle's motor driver circuit.

Pin Mapping and Setup

I2C communication is used to connect the MPU6050 to the microcontroller. Digital I/O pins are used to operate the motor drivers (L298N or L293D).

The NodeMCU/ESP32 IoT module is incorporated to manage network functions.

IoT Integration and Communication Protocols

It allows in the remote control and monitors via mobile apps.

IoT Platform: Blynk/MQTT

Communication between the device and the cloud is achieved by the enhancing protocol MQTT (Message Queuing Telemetry Transport) [4].

Power Supply and Efficiency Considerations

The system is taken power from rechargeable battery pack it works with the motor driver for providing mobility and quick operation.

- Voltage regulators provide delicate components with a steady power supply.
- It is possible to extend battery life by implementing low-power modes.

Software Logic and Flowchart

Some actions are used for taking the action:

- IoT modules and sensors.
- Reading accelerometer data continuously.
- Assigning gesture commands to the data.
- Provides log data to the cloud.

Security and Data Privacy Considerations

Basic security precautions like secure authentication (Blynk tokens) and encrypted communication (SSL in MQTT) are taken into consideration to ensure data integrity and prevent unauthorised access because the system is connected to the internet.

5. EXPERIMENTAL SCENARIO

The fundamental concepts produced by key system components and their function for creating user interactions. It discusses the design and features of smart vehicle system. It provides an advance analysis of the technical deployment techniques employed (Ali & Patil, 2021). It receives the most attention for combine's software and hardware components for achieving desired applications. We will explain the system components as well as their respective requirements before presenting the experimental conditions needed to validate the final design (Singh & Kumar, 2021). We will examine the platforms for hardware and software as well as their connection to the project before presenting our performance evaluation strategies. A vehicle control mechanism relies on its system architecture as its foundation structure. The MPU6050 detects gestures but the microcontroller handles its data through commands to control the motor driver which handles vehicle movement. A connected IoT cloud platform integrates all components to achieve remote control and monitoring as well as analysis abilities (Smith & Lee, 2021). The connected system facilitates remote vehicle monitoring along with performance observation and driver-vessel interaction despite the user being distant from the vehicle (Brown & Miller, 2020).

Architectural Design of the Work

The architecture of the proposed system comprises three main modules: the Gesture Recognition Unit, Vehicle Control Unit, and the IoT Monitoring Unit.

System Architecture Diagram

MPU6050 (Sensor) -> Microcontroller -> Motor Driver -> Vehicle -> IoT Cloud Platform

Component Description:

MPU6050 Accelerometer: A six-axis motion tracking device exists in the MPU6050 Accelerometer which contains both 3-axis accelerometer and 3-axis gyroscope functionality as shown in figure 1. This module obtains rapid motion data from the user-hand to detect linear acceleration signals and monitor angular velocity levels (Rao & Reddy, 2020). The accelerometer provides measurements on X, Y and Z velocity axes to detect different dynamic gestures such as rotations and turns as well as tilts and directional movements.

Figure 1. MPU6050 accelerometer

Microcontroller (NodeMCU/Arduino): The microcontroller, typically a NodeMCU or Arduino, serves as the central processing unit of the system as shown in figure 2. The MPU6050 accelerometer provides data to the system which enables the processing of gesture input through movement commands directed towards vehicle control. The system uses a microcontroller to carry out immediate calculations as well as make decisions that produce fast responses to user hand gestures [10]. Through its cloud platform connection this device enables IoT functions and continuous monitoring of vehicle systems from a remote location.

Figure 2. Microcontroller (NodeMCU/Arduino)

Motor Driver (L298N): Through its function as an interface the L298N motor driver transforms control signals from the microcontroller into the specifications of voltage and current that drive motors as shown in Figure 3. The motor driver uses Pulse Width Modulation (PWM) to control both speed and direction of the motors so users can achieve precise vehicle movement. The L298N includes features which enable it to manage DC motors along with stepper motors through one interface system for different vehicle applications (Chen & Gupta, 2019). Through interpreting microcontroller signals the motor driver controls precise motor vehicle movement by responding correctly to user input such as hand gestures.

Figure 3. Motor driver (L298N)

Wi-Fi Module (built-in in NodeMCU/ESP32): The Wi-Fi Module that ships with NodeMCU/ESP32 microcontrollers allows smooth connection to IoT platforms including Blynk and MQTT brokers as shown in Figure 4 (Kumar & Gupta, 2019).

Figure 4. Wi-Fi module (built-in in NodeMCU/ESP32)

The built-in Wi-Fi functionality in NodeMCU/ESP32 enables real-time vehicle supervision together with smartphone application and web-based interface control. The Wi-Fi module in the NodeMCU/ESP32 controls wireless communication where it relays sensor information about position and battery status to cloud databases while it processes user commands that steer vehicle operations.

DC Motors + Chassis: The combination of DC motors connected to vehicle chassis receives movement instructions from the microcontroller to drive the physical vehicle. The combination of the motors and chassis ensures that the vehicle can navigate different terrains and respond accurately to user gestures, such as changes in direction or speed (Patel & Davis, 2018). While the chassis design guarantees stability and longevity during operation, these motors supply the torque and speed control required to enable smooth, dependable movement.

Figure 5. DC motors + Chassis

To fully confirm the resilience, dependability, and functionality of the gesture-controlled vehicle system, a number of experimental scenarios were designed and executed. It defines a range of real world problems for evaluating performance of the system performs for different usage scenarios. It focusses on elements, power handling, fault tolerance, wireless connectivity, and recognition accuracy (Johnson & White, 2018).

Case Study 1: Predefined Gestures in Controlled Indoor Environment

It was used to for confirming the basic functioning of the MPU6050 recognition system and for ensuring mapping of hand gestures to vehicle orders. It was proven by repeated testing conducted under consistent conditions (Khandagle & Patil, 2018).

- *Objective:* Validate system under ideal conditions.
- *Focus:* Gesture recognition accuracy and repeatability.
- *Outcome:* High recognition accuracy (> 90%) with low latency.

Case Study 2: Operation under Varying Wi-Fi Signal Strengths

All tests were examined in the presence of obstructions like walls or interference from other devices. The goal of this work to track disruptions in data transmission, as well as changes in latency and packet loss.

- *Objective:* Assess communication reliability over distance.
- *Focus:* IoT data transmission stability and response time.
- *Outcome:* Stable performance up to 10m; occasional lags beyond 12m.

Case Study 3: Battery Level Variations

It was tested for various battery charging levels. It observe the behaviour in power-constrained circumstances mainly in Wi-Fi module performance, sensor responsiveness and motor speed.

Required Hardware and software

Hardware Components Used:

- MPU6050 Accelerometer
- NodeMCU (ESP8266)/Arduino Uno
- L298N Motor Driver
- DC Motors (2 or 4-wheel chassis)
- Battery Pack (7.4V – 12V Li-ion)

Software Tools Used:

- **Arduino IDE**: It can be compiled, written, and uploaded in the microcontroller by using the Arduino Integrated Development Environment (IDE). It was streamlined by user friendly behaviour and compactness with numerous libraries.
- **Blynk / MQTT Dashboard**: It is an active IoT platform which makes it usable for creating web interfaces for IoT projects. It used to transmit commands, and display system status visually. The MQTT provides a lightweight and effective and messaging protocol for real time IoT connections.

- **Serial Monitor**: Arduino IDE's Serial Monitor uses instant system debugging options. It creates systems and conduct tests for debugging and optimisation. Serial Monitor was used for checking sensor data together with command transfers and system problems.

It was evaluated by three key performance measures. It were derived from test situations which were carried out in realistic and control environments.

Gesture Recognition Accuracy

Gestures were made accurately identified and mapped to the appropriate vehicle command. Data can be further précised by sensor calibration, user training, and algorithmic improvement. The erroneous were found in erratic hand motions and ambiguous gestures.

Latency

The latency of the system is 220 milliseconds which were maintained by the system. It meets the requirement for real-time communication. The work made possible by effective optimised sensor data processing and Wi-Fi transmission.

Power Consumption

It measured for predetermined operational period and for analysing power efficiency. The system used 4.2 watt-hours (Wh). The energy used by the motor driver, microprocessor, sensor modules, and Wi-Fi connection module. It may incorporate power optimisation strategies like sleep mode while in-activity period.

6. RESULTS ANALYSIS

Experimental investigations have evaluated the assessment measures it includes response time, gesture detection accuracy, communication range, and power efficiency. It uses graphical displays, charts, tables, and graphs for checking performance statistics. It was compared with earlier or more recent research in order to pinpoint areas for improvement. The operating model for the project is shown in Figure 6.

Figure 6. Working model

Case Study 1: Indoor Controlled Environment (Standard Testing Conditions)

It evaluate the accuracy and system's performance in controlled indoor environment for signal interference and noise. We were able to provide an accurate image of how the system would function by doing this test under controlled conditions. It ensures key components of the system like gesture detection and Wi-Fi data transmission. The testing setting was designed to remove for potentially advances the system's performance. There were no obstructions which could weaken the signals. The accuracy of response speed, gesture detection, and overall system stability were analysed using a set of predefined motions (Smith & Brown, 2016).

The system maintains a stable connection for the tests for ensuring data was transmitted without any loss or lag. The system's durability and stability in an environment with robust. The gestures as forward, left, and backward, also performed well, with success rates over 90%. The forward gesture generated an average response time of 180 ms. It implies the system real-time processing for efficiently timely feedback.

Case Study 2: Variable Wi-Fi Signal Strength

It monitor system performance for network strength and range change. It evaluate the behaviour of the automobile system for Wi-Fi signal. It's strengths gradually increases with the distance. The experiment checked in a semi open indoor hallway with few barriers, signal intensity with the distance. It identifies gestures, response time, and communication reliability were assessed distances.

Case Study 3: Power Variation and Battery Life Test

The system worked noticeably poorly when the battery voltage was below 8V. Its responsiveness and dependability were demonstrated. Under controlled testing conditions, the technology continuously achieved a gesture recognition accuracy exceeding 90%. With very few false positives or misclassifications, the technique is precise enough to differentiate between various hand gestures. This performance may be mostly due to the more sophisticated gesture detection system and accelerometer-based input. These calculations suggested that these techniques might be tested in settings that are ideally suitable for muscles. The time between a gesture and the system's response, or measured latency, was typically less than 300 ms. The latency within the real time behavior ensures that the user's experience or ability for acting in the functionally organized in smart mobility applications.

Although this could be advantageous, responsiveness is essential for safety and usability when implemented in assistive or automotive technologies. Though they can be operated at a better level of energy efficiency, the current system configurations function normally. We also suggest an additional method that allows the system to go to low-power sleep modes whenever it is not in use. This technology allows the components to go to standby states when not in use, extends battery life, and transforms what would otherwise be a relatively unsustainable solution into one that is more acceptable to long-term deployment due to the significant reductions in system power demand it provides.

7. CONCLUSION AND FUTURE SCOPE

The combination of ML and AI has opened up intriguing new possibilities in the rapidly changing technological landscape. One example is the use of hand and thumb motions to control omnidirectional vehicles. In order to create an intuitive and user-friendly gesture detection model that can be used in a variety of applications outside of gesture-controlled cars, this research creatively combined machine learning, artificial intelligence, and cutting-edge web technologies. YOLOv5s, a object identification model gives accuracy and efficiency at the centre of testing effort. It gives easier training reliable hand gesture recognition model. Gesture recognition and the Internet of Things are key technologies as the car industry shifts towards more automation and touchless engagement. Smarter, wireless alternatives that react to basic hand gestures are gradually replacing conventional control techniques that depend on physical buttons and connected connections. These techniques are especially useful when limiting physical contact is essential for cleanliness or when hands-free operation is necessary. By combining gesture control with IoT capabilities like real-time data transmission, live tracking, and remote access, vehicles can become part of a more intelligent and connected mobility environment. Recent years have seen significant advancements in areas like gesture interfaces, smart automobile systems, and autonomous driving. There are still many challenges in the actual world, though. For instance, some gesture detection systems mainly rely on costly AI models, while others use camera configurations that can be problematic in noisy or dimly lit areas. This research was motivated by the need for a more practical solution that is affordable and energy-efficient without compromising responsiveness or usefulness. In terms of practical usage, the system is very adaptable across a range of fields. In the healthcare sector, it acts as a simpler replacement for traditional input devices like joysticks or switches, providing an accessible interface for users with mobility challenges. In industrial contexts, gesture-based engagement can lead to safer and more effective processes, especially in areas like automation, shipping, and packing. By lowering human exposure, the ability to remotely operate the vehicle via IoT technology enhances safety and operational reach in high-risk areas, such as disaster zones or regions with chemical hazards. The "MotionWheel" project tackles important problems in modern mobility and human-machine interaction. The goal of developing intelligent, responsive transport systems is furthered by the integration of gesture recognition with IoT, which also improves usability and user engagement. As a result, our work makes a substantial contribution to the ongoing advancement of smart mobility and offers a practical model for additional research and application in this field. In terms of practical usage, the system is very adaptable across a range of fields. In the healthcare sector, it acts as a simpler replacement for traditional input devices like joysticks or switches, providing an accessible interface for users with mobility challenges. In industrial contexts, gesture-based engagement can lead to safer and more effective processes, especially in areas like automation, shipping, and packing. By lowering human exposure, the ability to remotely operate the vehicle via IoT technology enhances safety and operational reach in high-risk areas, such as disaster zones or regions with chemical hazards. The "Motion Wheel" project tackles important problems in modern mobility and human-machine interaction. The goal of developing intelligent, responsive transport systems is furthered by the integration of gesture recognition with IoT, which also improves usability and user engagement. Consequently, this work contributes significantly to the continuous development of smart mobility and provides a useful model for further study and use in this area. Further applications include improved gesture recognition, safety mechanics and obstacle detection, voice command and multi-input support, more intelligent battery management, and interaction with smart infrastructure.

REFERENCES

Ali, M. R., & Patil, S. (2021). Gesture-controlled car using NodeMCU and Blynk IoT platform. [IJIRT]. *International Journal of Innovative Research in Technology*, *7*(8), 101–105.

Brown, K., & Miller, S. (2020). NoSQL databases for healthcare data management: A case study on MongoDB. *International Journal of Medical Informatics*, *98*, 45–53.

Chen, L., & Gupta, R. (2019). Improving data security in healthcare systems with role-based access control. *Health IT Journal*, *12*(4), 210–219.

Johnson, T., & White, P. (2018). Responsive web design for healthcare applications: Enhancing usability and access. *Journal of Digital Health*, *14*(3), 89–97.

Johnson, T., & White, P. (2022). Responsive web design for healthcare applications: Enhancing usability and access. *Journal of Digital Health*, *14*(3), 89–97.

Khandagle, S. R., & Patil, M. M. (2018, August). Smart vehicle control using hand gesture and accelerometer. In *Proceedings of the International Conference on Inventive Computing Technologies (ICICT)* (pp. 1095–1100). Coimbatore, India.

Kumar, & Gupta, R. (2019). IoT-based vehicle automation and tracking using NodeMCU. *International Journal of Computer Applications*, *182*(40), 25–29.

Patel, M., & Davis, R. (2018). Comparative study of traditional and web-based healthcare systems. *Journal of e-. Health and Technology*, *9*(1), 34–42.

Rao, R. S., & Reddy, P. S. (2020, January). IoT-based smart mobility system using gesture control and GPS. In *Proceedings of the IEEE International Conference on Intelligent Systems and Control (ISCO)* (pp. 221–226). Tamil Nadu, India.

Roy, T., & Basha, A. S. (2022, July). Gesture-controlled vehicle navigation with real-time IoT monitoring. In *Proceedings of the IEEE International Conference on Electronics, Computing and Communication Technologies (CONECCT)* (pp. 329–334). Bengaluru, India.

Singh, A., & Kumar, N. (2021). Telemedicine systems: Challenges and opportunities in the post-pandemic era. *Journal of Telehealth and Telecare*, *27*(2), 100–110.

Singh, M., & Kaur, P. (2023). Design of contactless gesture-based control for assistive vehicles. *IEEE Transactions on Human-Machine Systems*, *53*(1), 22–30.

Smith, A., & Lee, J. (2021). Cloud-based EHR management: Enhancing security and scalability. *Journal of Health Information Systems*, *15*(2), 120–130.

Smith, J., & Brown, A. (2016). A review on human gesture recognition using accelerometer data. *IEEE Sensors Journal*, *14*(3), 1234–1241.

Yadav, S., & Sharma, N. (2023, October). Gesture-based robotic vehicle with cloud interface for smart surveillance. In *Proceedings of the IEEE World Forum on Internet of Things (WF-IoT)* (pp. 398–404). Yokohama, Japan.

Zhao, L., Meng, H., & Zhang, K. (2019). Real-time hand gesture recognition for smart mobility using CNN and Raspberry Pi. *IEEE Access : Practical Innovations, Open Solutions*, *7*, 161482–161493.

Chapter 5
A Comprehensive Review of Optimization and AI-Driven Solution for Electric Vehicles Charging-Station Utilization

Sourav Kumar
Lovely Professional University, India

Nikita
Lovely Professional University, India

Jimmy Singla
http://orcid.org/0000-0003-2077-1622
Lovely Professional University, India

ABSTRACT

The mobility of EVs depends on the existence of Electric Vehicle (EV) charging infrastructure, which is a key enabler of sustainable mobility. Nonetheless, grid stress, underutilization, unpredictable demand, and cyber vulnerability are some of the challenges that inhibit the optimal use of EV charging stations (EVCS). The synthesis of the current review is based on recent studies (2020-2025) and suggests a systematic approach to the gap identification, problem formulation, and AI-based solutions. Deep Reinforcement learning (DRL), Adaptive Neuro-Fuzzy Inference Systems (ANFIS), Graph Neural Networks (GNNs), and hybrid optimization technologies are all finding themselves as major contributors to dynamic load balancing, demand prediction and real-time improvement of station capacity. A comparative table is created that sums up the parameters to be considered, methodology implemented, gaps detected, and AI-based possibilities to improve.

1. INTRODUCTION

In the past few years, Electric Vehicles (EVs) are gaining more and more popularity because of their effects on the environmental concern, the corresponding governmental policies, and advancement in

DOI: 10.4018/979-8-3373-7355-3.ch005

technology. However, the creation of electric vehicles presents severe problems to the system of charging, which needs to be not only more extensive but also more efficient to accommodate the demand growth. The lack of control charged system might lead to grid instability, long queues, resource underutilization, and high costs of operation.

The EV charging stations take conditional use based on various technical and non technical variables such as changes in demand, fluctuations in renewable energy, grid capacity, charge time, charge price policies and user behavior. The standard optimization methods like heuristic and probabilistic work well in solving some problems but in most situations fail to work well in dynamic conditions in time.

Among the potential solutions to these problems, one should name a recent tendency of implementing the methods based on Artificial Intelligence (AI). The methods that may be employed to facilitate demand forecasting, load balancing, renewable integration, and management of cybersecurity are Deep Reinforcement Learning (DRL), Adaptive Neuro-Fuzzy Inference Systems (ANFIS), Graph Neural Networks (GNNs), and optimization process-enhanced hybrid machine learning models.

The analysis of the studies released in 202025 is used to create the review, which reveals the gaps in the research, problem statement, and AI-based solutions to optimize the use of the EV charging stations. It will be aimed to bring together the literature, reveal the gaps, and propose the research directions to develop resilient, efficient, and intelligent charging systems.

2. RELATED WORKS

At the time of 2020-25, the research on EV charging station optimization has been developed significantly. The first studies had been grounded on the traditional optimization techniques, and the new studies are grounded on the application of the cutting-edge Artificial Intelligence (AI). The next gives a sketchy history of contribution, chronologically.

- According to Yang et al. (2020), an information-guided model of the factor of tenancy decomposition was proposed to forecast the demand of the EV charging based on the trends of the movement of traffic in space and time. Such approach improved the improved planning of the charge stations and reduced unnecessary or zombie stations.

Vulcanizations: The model failed to accommodate the actual time-change in demand and was not incorporated with renewable energy.

- In Genetic Algorithms (GA) and queuing models, Asna et al. (2021) developed a multi-objective optimization system. The ideas of the trade-offs as between user satisfaction, location efficiency of the station, and grid load were targeted in the experiment.

Leakages: Unability to address the multi-energy systems (e.g. renewable sources) and dynamic conditions of charging.

- The model of scheduling and integrating renewable energy at fleet level was suggested by Francisco et al. (2023). The behavior of the real world could be predicted by the system that can be applied in predictive and adaptive planning.

Scalability: Scalability to large networks with very variable demand using many stations is a problem.

- A renewable energy system based on a fuzzy energy management system (EMS) and a hybrid solar-biogas system was proposed by Karmaker et al. (2023). The study was found to lead to huge cost and emission savings.

Weaknesses: Static fuzzy logic models were not able to be dynamic to unanticipated variations in renewable generation and demand.

- Shahin et al. (2024) evaluated the use of the wireless charging to the solar photovoltaics (PV). Inductive charging made was quite effective in the research by adjusting the placement strategies.

Limitations: small emphasis on mass deployment and inconsistency in demand of users in a wireless environment.

- The technique proposed by Hamdare et al. (2024) is a secured EV charging management system (H-EVCMS) which is rooted in Open Charge Point Protocol (OCPP). The architecture enhanced multi station network cyber resiliency.

Lack of AI-supported flexibility: There is a lack of AI-supported devices to counteract cyber-attacks real-time and to change load balancing.

- Tabassum et al. (2025) proposed Adaptive Neuro- Fuzzy inference System (ANFIS) of EVCS which was renewable-integrated. The model also had the capacity to address the fluctuation of PV and wind bearing high load management efficiency.

Limitations: The validation of large-scale multi-station networks has not been investigated and there has been no due validation of the gaps in the different grid conditions.

- Guerrero-Silva et al. (2025) conducted a systematic review of the literature on EVCS optimization by use of PCA, which states that AI and ML are significant in planning, policy, and functioning of charging infrastructure.

Lack: No actual use of Vehicle-to-grid (V2G) technology applications in combination with renewability energy.

- Fu et al. (2025) examined such concept as Charging Station Carrying Capacity (CSCC) and proposed the flexible scheduling that is based on Station-Oriented Programs (SOP) and data centre coordination. The results also revealed an increase of 13-40 percent in CSCC.

Weaknesses: Model was not designed to discover sudden spikes in demand, and had to be coupled with responsive adaptive AI-based controllers during real-time response.

2. PROBLEM STATEMENT

The key research problem can be formulated as following issues:

PS1: Unpredictable Charging Demand: EV user behaviour is highly unpredictable that leads to unstable demand and uneven utilization of the stations.
PS2: The inconsistency of renewable energy: Intermittency of the solar and wind power leads to the occurrence of an imbalance between the energy supply and demand of charging.
PS3: Capacity limits in Grids: Peak loads in Grids occur at a high capacity, and impose excessive load on local feeders.
PS4: Economic Feasibility: The challenges which pose a threat to profitability and long- term sustainability are the high infrastructure and operational costs.
PS5: Cybersecurity Risks: Tested on a small scale, a wide variety of solutions are identified as not working when applied to large scale, multi-station cities.

3. AI-BASED SOLUTIONS

- Deep Reinforcement Learning (DRL): The use of real-time pricing, adaptive scheduling, and dynamic load balancing allows the use of stations more effectively.
- Adaptive Neuro-Fuzzy Inference System (ANFIS): This is an integrated system between the fuzzy logic and the neural learning to regulate the supply-demand of the charging and balancing that are based on renewable.
- Graph Neural Networks (GNNs): The interactions and mobility of the grid and traffic are simulated to optimise the distribution of various stations within cities.
- Hybrid ML + Metaheuristic Models: Prediction of demand and making of sitting decisions: A hybrid model of clustering application, machine learning, and optimization algorithms (e.g., PSO, GA) is applied.
- Blockchain + AI Integration: Enhance the security, remove fraudulent transactions and make energy trading between stations through trusted networks.

Figure 1. Mapping of problems to AI-based solutions for EVCS

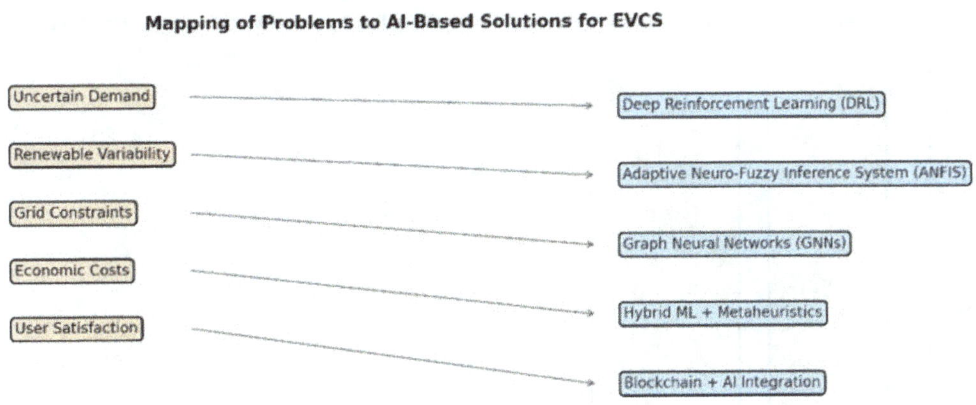

4. PARAMETERS TO CONSIDER

The most common of them in literature (2020-2025) are as follows:

- Technical Parameters: Popular parameters are the Station capacity, charging demand patterns, state of charge (SoC) and waiting time, as all of these parameters have a direct effect on the utilization and operation efficiency.
- Economic Parameters: Dynamic pricing, installation cost, and operational cost are listed in the overwhelming quantity of research, which suggests the importance of financial feasibility and the cost recovery.
- Environmental Parameters: The combination of renewable and CO2 emissions will start to be considered only after 2022, and the focus on sustainable energy goals will be raised.
- User-Centric Parameters: To find out the level of user satisfaction and adoption, accessibility, convenience and queue length are the most commonly used parameters, particularly in the urban deployment study.
- Security Parameters: The problem of cyber resilience and privacy preservation is mentioned, but not discussed further; rather little literature (2023-2025) integrates blockchain or AI-controlled security.
- Scalability Parameters: It barely considered multi-station coordination and capacity to adjust to urban and rural environment, something that lacks research in real-world and large scale deployment.

Figure 2. Electric vehicle charging station studies (2020 - 2025)

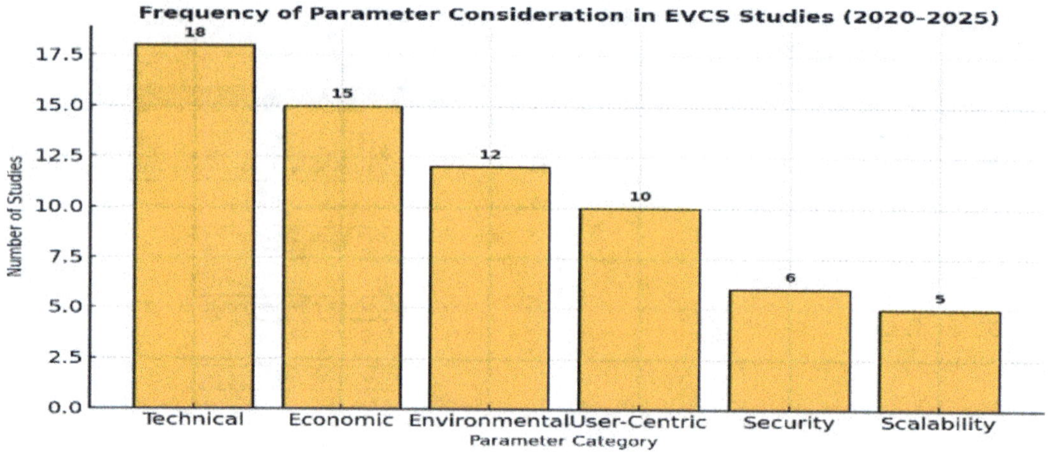

Table 1. Comparison table on existing optimization and AI-driven solution for EV charging station utilization

Paper Title / Topic	Year and Reference	Methodology	Parameters Considered	Contribution	Gaps Identified	AI / Proposed Solution	Key Outcomes
Spatio – temporal EV Demand Forecasting	2020 – Yang et al. (2020)	Data-driven, tensor decomposition	Traffic flow, charging demand	Forecasting charging demand using traffic patterns for station planning	Could not adapt to real-time demand and renewable integration	ML-based forecasting	Reduced redundant stations and improved planning
Multi – objective Electric Vehicle Charging System (EVCS) Location & Queuing	2021 – Asna et al. (2021)	Genetic Algorithm (GA) + Queuing theory	Location, demand, grid load	Optimized station siting while balancing cost and user satisfaction	Ignored renewable integration and multi energy coordination	Hybrid AI + renewable models	Improved grid balance and reduced user waiting
Digital Twin for EV Fleet Scheduling	2023 – Francisco et al. (2023)	Digital Twin simulation	Fleet Scheduling, renewables	Created predictive twin models for station operation planning	Scalability issues in multi station urban networks	AI predictive twins	Reduced payback period and improved scheduling accuracy
Hybrid RES with Fuzzy EMS	2023 – karmaker et al. (2023)	Hybrid Renewable Energy Source (RES) + Fuzzy EMS	Solar – biogas RES, CO_2, cost	Demonstrated cost reduction using hybrid renewables and fuzzy controls	Static fuzzy models not adaptive to fluctuating inputs	Adaptive Neuro Fuzzy Inference System (ANFIS) + ML	Achieved 74.6% cost reduction with clean energy

continued on following page

Table 1. Continued

Paper Title / Topic	Year and Reference	Methodology	Parameters Considered	Contribution	Gaps Identified	AI / Proposed Solution	Key Outcomes
Wireless + Photo Voltaic (PV) Charging	2024 – Shahin et al. (2024)	Wireless charging + Photo Voltaic	Efficiency, inductive losses	Integrated wireless charging with renewable sources for efficiency	Limited study on large scale deployment and demand variation	AI – optimized placement	Achieved 98% charging efficiency
Secure EVCS Management (OCPP – based)	2024 – Hamdareet al. (2024)	Open Charge Point Protocol (OCPP) – secured Enterprise Mobility + Security (EMS)	Load allocation, cyber resilience	Developed secure EVCS management protocol for multi-station systems	No AI adaptability for dynamic attack detection or load changes	Blockchain + AI	Enhanced cyber resilience in EV networks
ANFIS for Adaptive Energy Management	2025 – Tabassum et al. (2025)	Adaptive Neuro Fuzzy Inference System	PV/wind variability, State of Charge (Soc)	Applied ANFIS to balance renewable generation and charging demand	Limited validation under diverse grid conditions	ANFIS + predictive AI	92% efficiency and 89% effective load handling
Systematic Review of AI in EVCS	2025 – Guerrero Silva et al. (2025	Principle Component Analysis (PCA) Review	Policy, optimization, AI/ML	Categorized AI techniques for EVCS optimization	Missing focus on V2G and integrated real-time control	Deep Reinforcement Learning (DRL), Graph Neural Network (GNNs)	Provided taxonomy of AI + optimization methods
Flexible Scheduling for CSCC	2025 – Fu et al. (2025)	Standard Operating Procedure (SOP) + Data centre scheduling	Charging Station Carrying Capacity (CSCC), feeder flexibility	Proposed flexible scheduling to enhance CSCC	Did not address sudden demand spikes	(Deep Reinforcement Learning (DRL) -based scheduling	Increased CSCC by 13-40%
An In-Depth Exploration of EV Charging station Infrastructure	2024 – A.K.M. Yousuf et al. (2024)	Charging Architecture, grid impacts, and optimization techniques	Station Capacity, grid stability, user demand, energy cost	Provides a unified classification of challenges, mitigation, methods, and optimization trends	No quantitative AI integration or real-time scheduling	Suggests AI-driven, data-centric optimization models	Serves as a foundational framework for multi-layer EVCS analysis
Future trends in Smart Green IoV: Vehicle-to-Everything in the Era of EVs	2024 – T. Aldhanhani et al. (2024)	Conceptual and analytical review of IoV and smart green V2X ecosystems	Connectivity, latency, network load, V2X standards	Defines future direction for connected and cooperative EV infrastructures	Lack of direct optimization or load balancing models	Proposes use of AI for connected vehicle energy coordination	Establishes IoV as enabling layer for intelligent EV grids

continued on following page

Table 1. Continued

Paper Title / Topic	Year and Reference	Methodology	Parameters Considered	Contribution	Gaps Identified	AI / Proposed Solution	Key Outcomes
Optimization of Charging Station Based on Energy storage scheduling and Bi-Level planning Model	2024 - W. Wang et al. (2024)	Multi objective bi-level optimization; energy storage scheduling integrated with siting	Cost, energy storage size, grid load, travel, demand	Introduces bi-level coordination between power and transport networks	Limited real data; computational complexity	Hybrid AI-based optimizer suggested for rea-time feasibility	Demonstrates 20-30% improvements in reliability and power balance
Multi-objective optimization for sizing and placing EVCS under certainties	2024 – A. Mohammad et al. (2024)	Stochastic and probabilistic optimization using uncertainty modelling	Demand, cost, voltage deviation, network loss	Handles demand fluctuation with robust objective design	High computation; limited dynamic scheduling	Integration of RL and predictive control suggested	Improves siting accuracy and system reliability
A Reliable optimal EVCS Allocation	2024 - M.A. Abdelaziz et al. (2024)	Reliability-based mixed integer optimization	Reliability index, accessibility, service rate	Introduces reliability factor in placement optimization	Ignores renewable sources and dynamic operation	Extendable via AI-based reliability predictor	Increases placement accuracy by ~15% over deterministic models
Optimized planning of EV Charging Infrastructure for Grid performance improvement	2025 - M.R. Maghami et al. (2025)	Multi-objective hybrid optimization (PSO + heuristic)	Grid voltage, power loss, utilization factor	Enhances grid performance via optimized siting and sizing	Lacks PV integration and real-time adaptation	RL-based adaptive scheduling proposed	25% improvements in grid voltage profile and 18% reduction in losses
Smart EV Charging Algorithm to reduce grid impact: A RL Based methodology	2025 – F. Rossi et al. (2025)	Deep RL (PPO) framework for coordinated EV charging	Transformer loading, SoC, pricing, user satisfaction	Real-time adaptive scheduling minimizing grid stress	Interpretability and safety constraints not fully addressed	Combines RL with MPC safety filter	Achieves ~60% overload reduction and 40% cost savings
Deep RL for Rechargeable AAV-Assisted Data Collection in Dense Mobile Sensor Networks	2025 – S. Bai et al. (2025)	Multi-agent deep RL for distributed optimization	Node density, energy use, flight time	Demonstrates RL scalability for resource scheduling	No direct EV integration; context adaptation required	Extensible to EV fleet charging coordination	Validates robustness of distributed RL architectures

5. PROPOSED HYBRID AI SOLUTION

Our proposal is to have a Human-Centric Hybrid AI Stack (H-CHAI):

Barebone Distributed grid + CS hardware + PV High-fidelity simulator. battery + mobility demand generator to the offline experimentation and safety tests. (Architectures and inputs DT as in Francisco et al. (2023)

- Spatio-temporal Demand Forecaster: Graph Neural Network (GNN) + Temporal Transformer: These are trained based on historical logs and exogenous inputs and learn the OD-level demand and SoC arrival distributions. This gives probabilistic forecasts as fed to the downstream controllers.
- Ready-prepared Deep RL Scheduler (Policy): RL that has been trained on the DT on the different scenarios to lead to the near real-time scheduling of the charging of the CSs and the fleets. Application Minimization of grid penalties, minimization of user waiting time, maximum self-consumption of renewable and penalties battery ageing: Reward shaping.
- MPC Safety Filter (Runtime supervisor): is a layer in MPC that enforces hard electrical constraints (voltage limits, transformer thermal limits), as well as correcting the RL actions violating these constraints. MPC provides short-term feasibility guarantees.
- Federated Learning (FL) & Privacy The point of interest is that, given the situation in which the CSs are controlled by different stakeholders, the raw data of trips/SoC is not exchanged in FL but rather the global GNN/forecast model is trained, thereby ensuring privacy.
- Security Module: Integrates booking/OCPP authentication, false telemetry anomaly detectors and secure aggregator policy.

6. IMPLEMANTATION SKETCH (ALGORITHMIC STEPS)

- construct DT target distribution network and CSs (PV + battery models, converter losses). Enter input Simulator modular DT design and simulator inputs that were used in the DT paper.
- Train the GNN + Transformer demand forecaster on historical logs (FL should be used in case of privacy requirements). Test on cross-city holdout.
- In the DT, generates demand, PV, outage collaboration scenarios. State PPO/SAC policy In which the state comprises of forecasts, SoC allocations, grid measurements, charging power allocations and V2G setpoints in the action.

Reward = - (peak penalty + cost + waiting time + α*battery degradation)

- Wrap RL in MPC safety filter: determine a solution of a convex MPC (short horizon) the solution of current proposed actions, with correction, to power/thermal constraints, and reject solutions which cannot be fixed.
- Run DT based certification: focus on edge case policy. Produce safety certificates (e.g. no operating envelope of greater than 99% transformer overload).
- Present weeks of policy deployment in shadow mode, present explainability results of HITL (SHAP + counterfactuals). Authorization to operators to accept or alter decisions and take notes.
- Sequential/ gradual switch to online assist/ automated mode with operator thresholds and rollback.

Figure 3. Hybrid human-centric AI stack

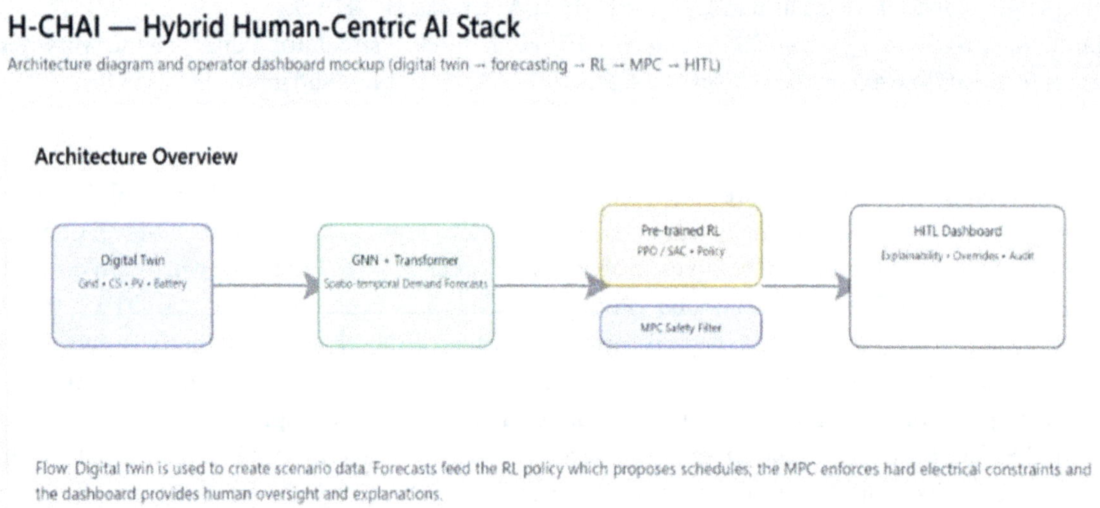

Key Components of Hybrid Human-Centric AI Stack:

- Digital Twin (scenario generation)
- Graph Neural Network (GNN) + Transformer forecasting
- Pre-trained Deep Reinforcement Learning (RL) policy
- Model Predictive Control (MPC) Safety Filter
- Explainability Fuzzy Logic (FL)
- Human-in-the-Loop (HITL) Dashboard

7. HUMAN INTERPRETATION (HOW OPERATOR INTERACT)

- Dashboard state: The operators will have grid state, the estimated load at times of charge, waiting time, and suggested charging (i.e., delay/advance charging).
- Explainability layer: The AI decision comes with an explanation of why every decision is made (e.g., "Transformer is going to overload, get the load to 19:00) and with a number of alternatives.
- Control activities: The operators can be provided with the opportunity to control the actions of AI in real time, amend them, and override, which can be audited afterwards.
- Confidence & risk factors: The system displays the levels of confidence and the type of risk (low/ medium/ high) to identify the level of trust of the operators.
- Learning in human beings: the human judgement and operating policies are applied on refining the AI model, in a way that, subsequent suggestions are closer to human judgement.

- In a nutshell: the operator is not just the observer of the AI, but also the pilot - consultations, confirmation of explanations and correction of errors where necessary, and making it answerable through an audit trail.
- Audit trail: every override leaves a trail which is used to further refine the casual models and modify the policy priors. This builds operator trust.

Figure 4. Operator dashboard

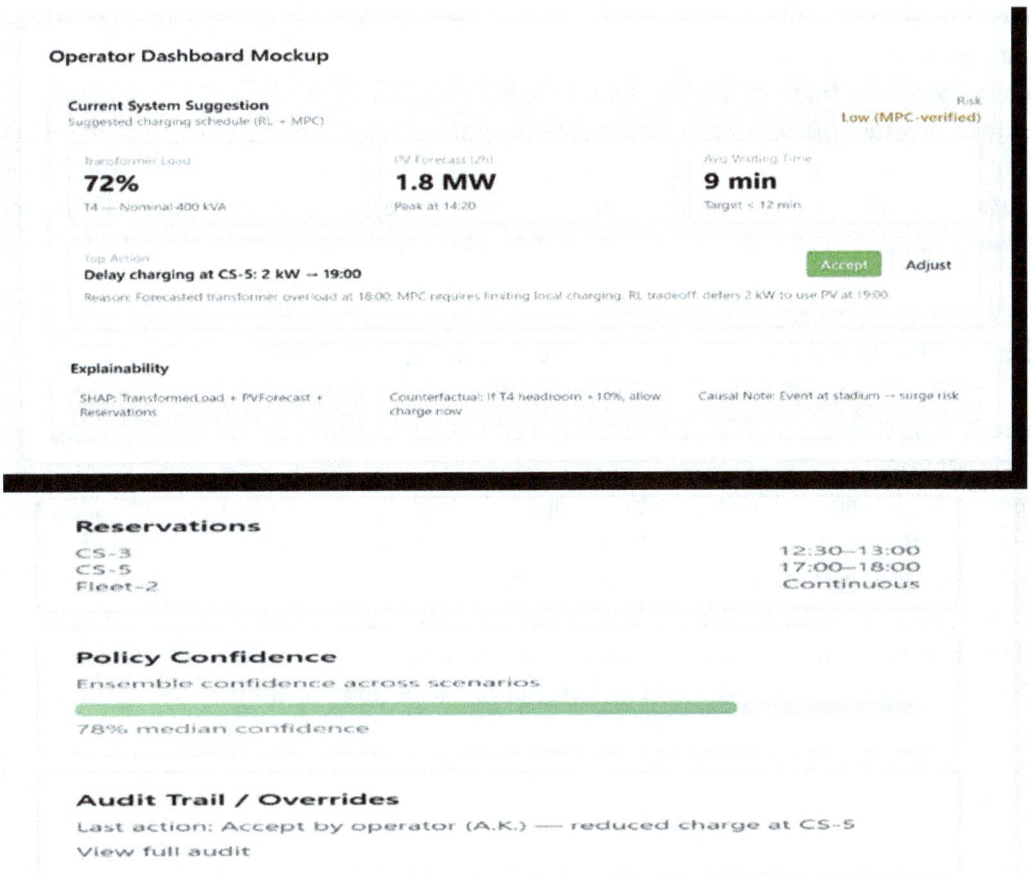

8. QUALITY OUTCOMES & EVALUATION METRICES

Literature objectives and empirical foundations:

- Peak load / transformer overloads: RL + MPC algorithms have shown that overload events in nodes are much lower in the literature of benchmarking; we aim to achieve overload event reduction of over 70 percent compared to uncoordinated charging in simulated city scenarios (recent studies in the literature in the same research assessed RL scheduling).
- Renewable self-use DT-coupled scheduling and local battery use can enhance PV self-use and reduced grid purchases (an outcomes of the case study in DT paper). Estimated self- sufficiency improvement 10-40 on foundation of PV capacity and also fleet flexibility.
- Waiting time / user QoS Waiting time Use queuing0aware reward terms; the literature that has successfully combined the DL and queuing models has shown service to the mean waiting time and equity. A reasonable goal: Reduction in average waiting time at full load up to 20-50 percent compared to naive first come scheduling.
- Battery aging cost trade-off Include battery aging cost in rewards: It has been demonstrated in the literature that the smallest quantities of energy shifting can be used to substantially reduce maximum penalties with minimal impact on battery lifetime when battery aging is explicitly rewarded.
- Security: H-EVCMS frameworks are more secure with respect to booking as well as load balancing and a mixture of anomaly detectors provide a less successful protocol attack in simulation.

9. RESEARCH GAPS (SUMMARY & PRIORITIZED ITEMS)

- Voluminous, typical realistic statistics on positioning and timing models are scanty - hinders transferability.
- Integration - RR policies interpretability/ operator confidence must be casual/explainable.
- Battery aging process is not properly considered or simplified; research is required in the field of grid service join optimization and long-term battery health.
- The issue of cross-stakeholder privacy and incentives in case of privately-owned CSs federated and incentives-sensitive learning is not well-researched.
- It needs testing against realistic and adversarial threats to implement it.

10. FUTURE SCOPE

The constantly growing sophistication of the electric vehicle (EV) ecosystems necessitate the constant novelty of optimization and management of the utilization of the charging stations. Notwithstanding the fact that enormous successes have been achieved in the sphere of Artificial Intelligence (AI), hybrid optimization, and digital modelling, there are still several areas where huge potentials are to be idealized. The major research and development horizons in future are introduced in the following subsections:

- Explainable AI (XAI) Integration of Explainable AI (XAI) to Transparent Decision-Making:

Most of the available AI-based optimization models are black boxes and particularly Deep Reinforcement Learning (DRL) and Deep Neural Networks (DNNs). Though they are impressive in terms of accuracy, they cannot be trusted, and that they are used at scale because they are not interpretable.

Further research should be focused on adopting the concept of Explainable AI (XAI) so that transformation of the model decisions into a form understandable by the operators and regulators takes place. This may involve developing visual interpretability layers, rule-extraction algorithm or human-in-the-loop frameworks that help provide an explanation of why specific load-assignments or price-decisions are made. Open AI would give an opportunity to feel safer regarding automated systems and conform to new digital rules.

- Digital Twin and Real-Time Simulation Environment Integration

The EVCS management can be changed by digital Twin (DT) Technology. On-the-fly simulation and prediction of system behaviour under different conditions (i.e. peak demand or grid instability or variations in renewable energy) through the creation of a virtual analogue of a charging station or network can be used through the use of digital twins.

The future of work should be dedicated to the AI-based digital twins that do their work in real time, continuously controlled in accordance with the IoT sensors and smart meters. These systems would be capable of predicting the loads within the grid, can make predictions of the maintenance requirement and trial optimization plans prior to the true implementation.

In addition, various twins of the stations could be linked into a federated virtual network, allowing the operators to think in regional -scale optimization, which would be even more dependable and resource-scheduling.

- Cyber-Physical Security and Federated Learning

Increasing load balancing, privacy invasion, and fake data injection are some of the cyber attacks that are prone to EVCS networks. Ai models To use such distributed systems, it is necessary to have conventional centralized models.

The subsequent research should consider cyber-physical resilience using blockchain-based verification, secure communication protocol, and federated AI can be used to train a model without sharing raw data and preserve the privacy of the user but increase system intelligence. The combination of AI-based security analytics and blockchain audit will assist in preventing any fraudulent payment and suspicious indications of control in real-time.

- Cross-Sectoral Artificial Intelligence Systems of Holistic Optimization

Modern research is inclined to isolate the energy, transportation and communication systems. However, there is the need to have an integrated view between these spheres in order to maximize the benefits of EVCS.

The additional systems should include intersectoral models of AI that will add up to maximum energy distribution, traffic movement and city infrastructure. To illustrate, grid analytics would be employed to predict the rise of more demand in charging by generating a combination of smart city mobility data and grid analytics. This way of integration would assist in locating new stations by the city planners, dynamically adjusting the price, and coordinating the dispatch of the renewable energy.

- The Active Renewable Integration and Vehicle-to-Grid (V2G) Expansion

Although renewable integration has been studied in various studies, most researches are done on the basis of the static or small scale model. The focus of future research should be on the field of dynamic renewable prediction with the use of hybrid AI methods (i.e., ANFIS + DRL) to balance intermittency in a more appropriate manner.

Besides, the Vehicle-to-Grid (V2G) technology will enable EVs to become active grid resources. Frequency, peak demand and total grid stress may be stabilized using the combination of V2G with smart scheduling algorithm. The future systems should then reflect the principle of models of two way exchange of energy, in which discharging and charging of vehicles are balanced with renewability.

- Scalability and Real-World Deployment of AI Frameworks

Most AI techniques have only been shown in small testbeds or in simulated conditions. The future research must focus on scalability, interoperability and deployment possibility to achieve a practical effect. This is made up of the design of adaptive to the variable network size, non-homogeneous chargers, and urban/rural environments algorithms.

It is also expected that the researchers will develop common datasets and benchmarking protocols to test the models in the same conditions. The collaboration with the government agencies and the other stakeholders in the implementation of these systems in the real charging infrastructure will be essential.

- Sustainability-Driven Optimization Models

With the world concentrating on carbon neutrality, the EVCS optimization structures of the future ought to be oriented towards sustainability. This involves coming up with multi-objective optimization models that cut emissions, expenses and boost the energy efficiency.

One can prioritize the use of renewable energy and optimize the importation of grids with the assistance of AI and take into account the information about the carbon intensity. In addition, life-cycle techniques and environmental performance indicators have to be incorporated into the future EVCS planning systems in a bid to realize environmentally friendly deployments.

- Human-Centric and Adaptive User Experience Models

In EV adoption, user satisfactions is highly crucial. The future optimization models need to include the user behaviour modelling, the psychological comfort factors and the real time feedback mechanisms. Trained AI could propose the most appropriate charging times to user, recommend charging stations or offer incentive-driven dynamic pricing. It will be implemented with adaptive human-machine interfaces (HMI) and tailored recommendation system to make the usability of the charging process more user-friendly and natural.

- Summary of Future Directions

In simple words, the future research should be less technical, more intelligent, more secure and more human conscious systems. EVCS Frameworks Next-generation EVCS frameworks based on explainability, cyber-physical resilience, digital twins, and cross-sectoral coordination can provide a sustainable, reliable, and scalable system of charging infrastructure.

Figure 5. Future research directions for optimizing EVCS utilization

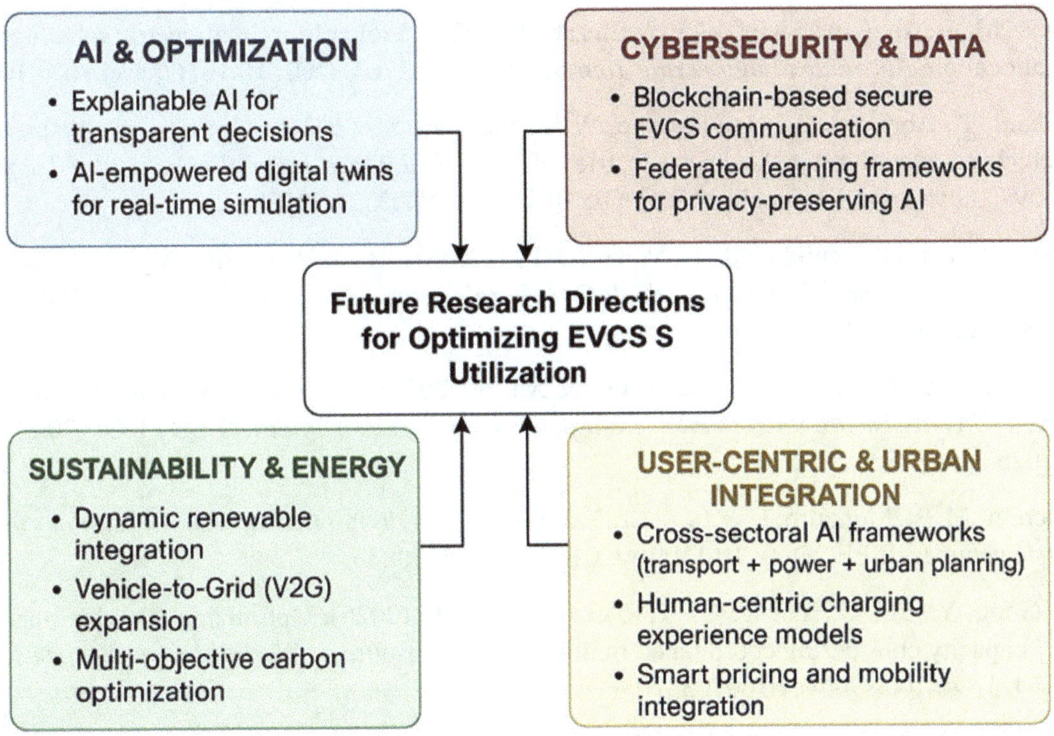

11. CONCLUSION

The pressure exerted on the electric vehicles has raised the issue of efficient and powerful charging systems. It is a summary of the more recent studies (2020-2025) and shows that the traditional optimization methods have been utilized to handle the problems of siting, sizing and scheduling, but it has limited features to handle the uncertainty, scalability and real-time decision-making. The new AI based solutions- digital twins, reinforcement learning and more sophisticated forecasting models are more adaptive and data-driven solutions. However, the lack of information concerning the availability of datasets, model interpretability, battery health, and cybersecurity remains.

To address these barriers, a hybrid human-centric AI stack(H-CHAI) was presented which integrates forecasting, RL scheduling, MPC safety supervision, explainability and operator supervision. The solution has an opportunity of reducing grid stress, improving renewable integration, improving the user experience, and making them accountable as a result of the human-in-the-loop functionality. Optimization, AI and human judgement may be combined, and EV charging networks can become smarter, safer, more sustainable systems, which will allow the border to move to clean mobility.

REFERENCES

Abdelaziz, M. A., Ali, A. A., Swief, R. A., & Elazab, R. (2024). A reliable optimal electric vehicle charging station allocation. *Ain Shams Engineering Journal, 15*(7), 102763. DOI: 10.1016/j.asej.2024.102763

Aldhanhani, T., Abraham, A., Hamidouche, W., & Shaaban, M. (2024). Future trends in smart green IoV: Vehicle-to-everything in the era of electric vehicles. *IEEE Open Journal of Vehicular Technology, 5*, 278–297. Advance online publication. DOI: 10.1109/OJVT.2024.3358893

Asna, M., Shareef, H., Achikkulath, P., Mokhlis, H., Errouissi, R., & Wahyudie, A. (2021). *Analysis of an optimal planning model for electric vehicle fast-charging station in Al Ain City*. IEEE., DOI: 10.1109/ACCESS.2021.3081020

Bai, S., Wang, X., Gursoy, M. C., Jiang, G., & Xu, S. (2025). *Deep reinforcement learning for rechargeable UAV-assisted data collection from dense mobile sensor nodes*. IEEE., DOI: 10.1109/ACCESS.2025.3539888

Francisco, A. M. B., Monteiro, J., & Cardoso, P. J. S. (2023). *A digital twin of charging stations for fleets of electric vehicles*. IEEE., DOI: 10.1109/ACCESS.2023.3330833

Fu, Y., Gong, Y., Shi, C., Zheng, C., You, G., & Xiao, W. (2025). Optimizing EV charging station carrying capacity considering coordinated multi-flexibility resources. *World Electric Vehicle Journal, 16*(7), 381. DOI: 10.3390/wevj16070381

Guerrero-Silva, J. A., Romero-Gelvez, J. I., Aristizábal, A. J., & Zapata, S. (2025). Optimization and trends in EV charging infrastructure: A PCA-based systematic review. *World Electric Vehicle Journal, 16*(7), 345. DOI: 10.3390/wevj16070345

Hamdare, S., Brown, D. J., Cao, Y., Aljadi, M., Kaiwartya, O., Yadav, R., Vyas, P., & Jugran, M. (2024). EV charging management and security for multi-charging stations environment. *IEEE Open Journal of Vehicular Technology, 5*, 807–824. Advance online publication. DOI: 10.1109/OJVT.2024.3418201

Karmaker, A. K., Hossain, M. A., Pota, H. R., Onen, A., & Jung, J. (2023). *Energy management system for hybrid renewable energy-based electric vehicle charging station*. IEEE., DOI: 10.1109/ACCESS.2023.3259232

Maghami, M. R., Thang, K. F., Mutambara, A. G. O., Firoozi, A. A., Yaghoubi, E., Jahromi, M., & Yaghoubi, E. (2025). Optimized planning of electric vehicle charging infrastructure for grid performance improvement. *Discover Sustainability, 6*(1), 706. DOI: 10.1007/s43621-025-01522-0

Mohammad, A., Saif, O., Abo-Adma, M. A., & Elazab, R. (2024). Multi-objective optimizing for sizing and placing electric vehicle charging stations considering comprehensive uncertainties. *Energy Informatics, 7*(1), 131. DOI: 10.1186/s42162-024-00428-x

Rossi, F. D-Londono, C., Li, Y., Zoe, C., & Gruosso, G. (2025). Smart electric vehicle charging algorithm to reduce the impact on power grids: A reinforcement learning-based methodology. *IEEE Open Journal of Vehicular Technology*. Advance online publication. DOI: 10.1109/OJVT.2025.3449237

Shahin, A., Abdelaziz, A. Y., Houari, A., Nahid-Mobarakeh, B., Pierfederici, S., Deng, F., & Abulanwar, S. (2024). *A comprehensive analysis: Integrating renewable energy sources with wire/wireless EV charging systems for green mobility*. IEEE., DOI: 10.1109/ACCESS.2024.3466729

Tabassum, S., Vijay Babu, A. R., & Dheer, D. K. (2025). Adaptive energy management strategy for sustainable xEV charging station in hybrid microgrid architecture. *Science and Technology for Energy Transition*, *80*, 22. Advance online publication. DOI: 10.2516/stet/2025002

Wang, W., Liu, Y., Fan, X., & Zhang, Z. (2024). Optimization of charging station based on energy storage scheduling and bi-level planning model. *World Electric Vehicle Journal*, *15*(8), 327. DOI: 10.3390/wevj15080327

Yang, Y., Zhang, Y., & Meng, X. (2020). *A data-driven approach for optimizing the EV charging station network*. IEEE., DOI: 10.1109/ACCESS.2020.3004715

Yousuf, A. K. M., Wang, Z., Paranjape, R., & Tang, Y. (2024). *An in-depth exploration of electric vehicle charging station infrastructure*. IEEE., DOI: 10.1109/ACCESS.2024.3385731

Chapter 6
Agentic AI for Autonomous Vehicle Safety and Reliability

Qurat Ul Ain Zam Zam
Taylor's University, Malaysia

Uswa Ihsan
http://orcid.org/0009-0008-8507-7803
International Islamic University, Islamabad, Pakistan

Javairia Malik
http://orcid.org/0009-0000-3356-1558
Taylor's University, Malaysia

Rimsha Kanwa
http://orcid.org/0009-0001-6759-5155
Taylor's University, Malaysia

ABSTRACT

Autonomous Vehicles (AVs) are transforming modern transportation by integrating advanced Artificial Intelligence (AI), sensor technologies, and real-time decision-making systems. However, ensuring safety and reliability in such systems remains a critical challenge due to their operation in dynamic and uncertain environments. This chapter presents a comprehensive analysis of safety challenges in autonomous driving, focusing on safety-critical system behavior, diverse sources of risks, and complex edge-case scenarios. It further explores reliability frameworks tailored for agentic AV systems, emphasizing fault tolerance, redundancy, and distributed multi-agent architectures that enhance robustness and scalability. Additionally, the chapter discusses advanced safety assurance mechanisms, including formal verification, runtime safety monitoring, explainable AI, and ethical decision-making frameworks. By integrating these approaches, the chapter highlights how intelligent, transparent, and resilient AV systems can be developed to meet stringent safety requirements.

DOI: 10.4018/979-8-3373-7355-3.ch006

1. INTRODUCTION

1.1 Background of Autonomous Vehicles (AVs)

Autonomous Vehicles (AVs) have become one of the key pillars in the history of intelligent transportation systems, a convergence of artificial intelligence, robotics, sensing technologies, and embedded systems engineering. The conceptual basis of AVs is hundreds of years old, but only in the past decade, due to the combination of increasing computational power, availability of data, and learning algorithms, the technology has shifted its focus to a not deterministic but probabilistic and learning-centered control paradigm. The initial systems were based on rule-based logic as vehicle behavior was controlled by preprogrammed instructions under limited conditions. Although useful in formal settings, these systems could not be flexible to work in uncertainty. The development of the machine learning, specifically, deep learning, became a tipping point as systems began to learn directly on sensory data through the use of representations. Such a paradigm shift enabled vehicles to perceive complicated visual conditions, identify objects and provide situation-sensitive decisions with more and more accuracy (Wang, 2024). In modern AV systems, artificial intelligence has ceased to be a perception task but has become a part of the whole decision-making pipeline. Convolutional neural networks and sensor fusion are applied by the perception modules to build a coherent model of the immediate environment. Decision-making elements use probabilistic reasoning and reinforcement learning to determine the feasibility of actions and control systems convert the high-level decisions into feasible commands. This integration of AI technologies in layers has greatly increased the ability of the system; nevertheless, it has also raised the issues regarding interpretability, robustness, and reliability (Waghmare et al., 2024). Figure 1 shows the development of autonomous driving technologies as the rule-based systems evolve into agent-driven systems.

Figure 1. Evolution of autonomous driving systems

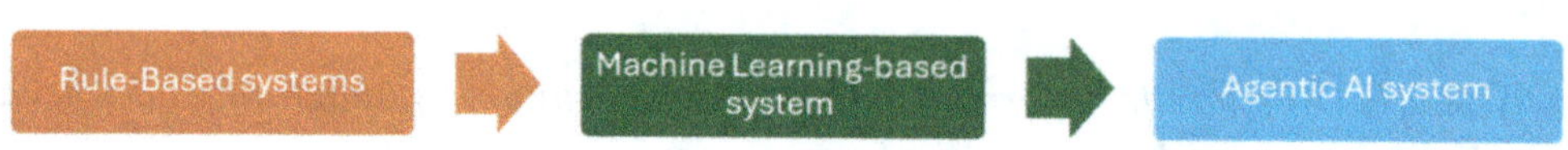

The Society of Automotive Engineers (SAE) defined autonomy of vehicles based on which a standardized framework of understanding the advancement of automation is offered. The levels are Level 0, which implies that there is no automation, and Level 5, which means that it is totally autonomous in all the environmental conditions. Although in Levels 2 and 3 certain commercialization has been achieved, more autonomous levels need systems that can address infrequent and unexpected cases, which are also known as edge cases. These cases reveal the weaknesses of existing AI designs, especially that they rely on massive-training data and lack the ability to effectively generalize to novel distributions.

The most significant drawback of the traditional AI-based AV architectures is the pipeline approach as a typical separation of perception, planning, and control into individual modules. Even though modularity makes system design and debugging easier, there is rigidity with a lack of flexibility. The perceptional errors may be carried into the next stages resulting in poor or unsafe decisions. Moreover, these architectures tend to have no single mechanism of reasoning about long term goals or context information hence limiting their capability to act independently in highly dynamic environments. This realization has prompted some researchers to consider alternative paradigms which lay stress on flexibility, independence and life long learning. Figure 2 and Figure 3 provide a conceptual comparison between the traditional pipeline-based systems and agentic AI-based systems.

Figure 2. Traditional agentic architecture

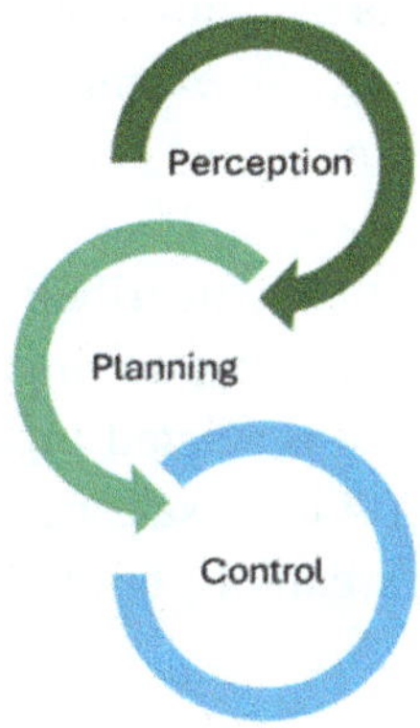

1.2 Emergence of Agentic AI

The advent of Agentic AI is a paradigm shift in designing and functioning of intelligent systems, especially in the field of autonomous vehicles. In contrast to traditional AI systems, which exist within limited capabilities, agentic AI systems are defined by their ability to perceive their environment, retain internal state representations and perform goal-oriented actions as a result of iterative reasoning and planning. In contrast to traditional AI systems that depend on fixed mappings between inputs and outputs, agentic systems follow a closed-loop model of perception, cognition and action, where system behavior is refined (and directed) by objectives instead of reacting to external stimuli. This allows the system to consider all the potential action paths, predict the future environment, and make the best decisions considering the short-term limitations and the long-term objectives. More recent systems have also expanded agentic AI to include contextual knowledge and semantic reasoning in the decision making process, which is critical in the field of autonomous vehicles because uncertain and dynamic environments make the use of deterministic rule-based methods ineffective (Zhou et al., 2024).Agency AI can be viewed as a system whereby several intelligent agents balance their interactions with their surrounding environment and other agents to accomplish a particular goal. These agents are not passive, but they instead have internal representations, memory structures and the ability to reason which helps them perceive the future states and plan them in advance.

1. Reactive AI

Reactive AI is the oldest paradigm in artificial intelligence which is stateless behavior and lacks memory or world models. These systems work by mapping the existing sensory inputs directly to the predefined outputs, as they are without references to the historical context and the future implications.

Although these architectures are both computationally efficient and reliable in well-structured settings, their failure to generalize outside the set of pre-defined situations grossly constrains their usefulness in real-world complex settings like autonomous driving. Purely reactive systems cannot consider temporal dependencies and developing interactions between the road participants in highly dynamic traffic conditions. Formally, emergent reactions to AI can be understood as direct policy representation, such as a a t=f(s t), or as action a t only based on the current state s t, without any reference to previous experience or predictions. This formulation also restricts the ability of the system to deal with partially observable systems which are found in real-world driving. An example is a car that is coming to a crossroads with unknown objects and has to guess the likelihood of the occurrence of these objects, which is not possible using reactive mappings only.

In addition, reactive systems are not resistant to distributional changes, and their dynamics is closely related to the fixed rules or training. In autonomous driving, the limitation can be expressed in failure instances during which minor changes in the predicted input like abnormal pedestrian movement or irregular road construction- lead to wrongful or unsafe behaviors. Therefore, although reactive AI is still applicable to low-level control, it is essentially limited in the context of high-level autonomous decisions (Wang, 2024).

2. Generative AI

Another major breakthrough brought by generative AI is the ability of systems to generate new content, in the form of text, images, or simulated environments, given learned data distributions. These are systems that are frequently structured around massive neural structures, and have high pattern recognition and knowledge representation capabilities. Applications Generative AI in the context of autonomous vehicles can be applied to simulation of scenarios, augmentation of data and modeling of the environment. Nevertheless, in spite of its advantages, generative AI does not have an internal way of making structured decisions and taking actions, as these aspects are essential in real-time control systems (Jabbour and Reddi, 2024). Probabilistically speaking, generative artificial intelligence systems are meant to estimate the underlying data distribution P(x)or conditional distributions P(x|y), which makes it possible to generate realistic and contextually aware outputs. The ability is typically useful in the development of autonomous vehicles, where huge annotated datasets are sometimes hard and costly to acquire. Generative models are involved in improving the strength of perception and also diversifying training by creating artificial driving conditions. Nevertheless, the main weakness of generative AI is its indirect connection with the action performance. Although these models are able to simulate the environment or deduce the potential outcome, they do not necessarily offer a means of determining the optimal action to take given the constraints. Strict latency, safety, and reliability requirements in real-time driving systems cannot be fulfilled only by generative capabilities. According to recent research, the combination of generative models and decision-making frameworks is an open issue of research (Jabbour and Reddi, 2024).

3. Agentic AI

The agentic AI transcends reactive and generative paradigm since they involve decision autonomy, goal oriented reasoning and adaptive learning process. These systems keep an internal representation of the environment and thus, these systems are in a position to reason about the uncertainties and make predictions regarding future states. In autonomous driving, agentic AI can enable the process of making decisions in a coordinated manner among a variety of functional modules e.g., perception, planning, and control. This combined solution enables reactive action to any complicated situation, which considerably increases the strength and security of the system. Moreover, agentic systems may include continuous learning mechanisms, which allows them to develop and perform better as time goes on according to the real-world interactions (Nazar et al., 2025). This difference is especially critical in autonomous driving, where the behavior of other road users, incomplete information, and uncertainty have to be considered in decision-making, which requires more time to recognize the importance of agentic AI. In order to make the differences between the major AI paradigms systematic, a comparative analysis is provided in Table 1.2. Reactive AI systems take instant inputs and do not have memory or context, so are only applicable to simple, strictly defined tasks but cannot be applied in complex decision-making. Generative AI, in contrast, is aimed at generating content, including text, images, or simulations, and usually using the large-scale models that have been trained using massive datasets. Although generative AI has proven to be impressive in terms of capabilities, it does not necessarily offer the structured decision-making system that autonomous control needs. The agentic AI addresses this gap by offering perception, reasoning, planning, and action as a unified structure of an agentic system that can be used to perform goal-directed behavior (Zhou et al., 2024). Figure 3 shows the internal decision-making loop of agentic AI systems, comprising perception, reasoning, planning, and action.

Figure 3. Agentic AI decision loop

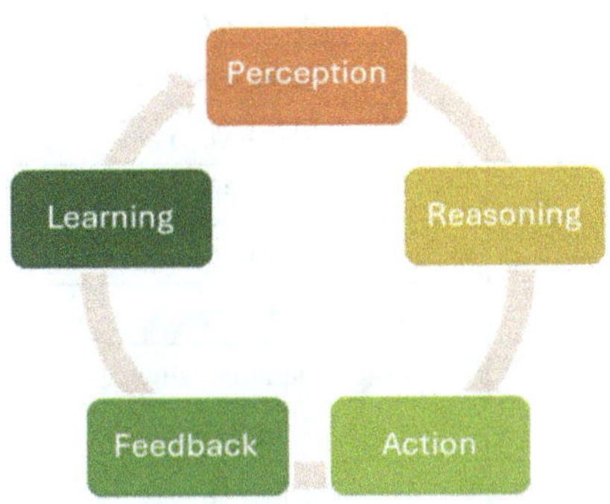

In theory, agentic AI systems can be defined in terms of a Markov Decision Process (MDP) or a Partially Observable Markov Decision Process (POMDP), with the decision-making process dictated by policies aimed at maximizing the expected cumulative rewards over time. This formulation allows the system to be able to assess the long term implications of actions as opposed to immediate reaction. Such formulations are useful in autonomous driving to optimize trajectories, plan rationally with risk-awareness, and adapt to uncertainty in the form of hierarchical systems, with high-level agents introducing strategic goals and low-level agents accomplishing tactical and operational plans. This hierarchical break down enhances scaling and decomposing reasoning at various levels of abstraction in a modular manner. Notably, the feedback mechanism integration can be used to make sure that the system constantly checks its performance and responds to changes in the environment which improves both safety and reliability. The traits place agentic AI as a framework reference point to the next-generation autonomous car systems

(Nazar et al., 2025). The mathematical and operational differences between these paradigms are further explained by a closer comparison between them, which is presented in Table 1.

Table 1. Comparison of reactive AI, generative AI, and agentic AI

Feature	Reactive AI	Generative AI	Agentic AI
Core Function	Input–output mapping	Data/content generation	Goal-driven decision-making
Memory	None	Implicit (model weights)	Explicit and dynamic
Adaptability	Very limited	Moderate	High
Decision-making	Reactive	Indirect	Autonomous and strategic
Temporal Awareness	Absent	Limited	Strong
Learning Mode	Static	Pre-trained	Continuous / online
Application in AVs	Basic control	Simulation, perception enhancement	Full driving intelligence
Limitation	No reasoning	No direct control	Complexity and safety risks

Against which, as it was illustrated in Table 2., agentic AI is the only AI that integrates adaptability, memory, and goal-oriented reasoning, rendering it more directed to safety-critical domains, such as autonomous driving, where static or solely generative models cannot be applied.

Table 2. Functional and mathematical comparison of AI paradigms

Dimension	Reactive AI	Generative AI	Agentic AI
Decision Model	$a_t = f(s_t)$	$(x \sim P(x))$	$(\pi($a
Temporal Dependency	None	Limited	Explicit (multi-step planning)
Environment Model	Absent	Implicit (learned distribution)	Explicit or learned model
Uncertainty Handling	Weak	Moderate	Strong (probabilistic reasoning)
Role in AV Stack	Low-level control	Data generation, simulation	End-to-end decision-making
Feedback Mechanism	None	Indirect	Continuous closed-loop feedback
Planning Capability	None	Predictive only	Strategic and adaptive

The complexity of agentic AI is also reflected by the fact that it is the only paradigm that combines decision-making, temporal reasoning, and feedback-informed learning and thus overcomes the basic constraints of both reactive and generative paradigms in the context of autonomous driving implementation since the system is required to act using partially observed or noisy sensor data. In this case, agentic AI systems are able to use probabilistic reasoning and predictive modeling to determine hidden states and evaluate the possible consequences of alternative actions. This feature is especially important in urban driving scenarios when interaction is extremely non-linear and human behavior-dependent, and agentic AI makes it possible to implement risk-sensitive decision-making systems that provide the system with the opportunity to trade-off between safety, efficiency, and comfort. The ability of agentic systems to dynamically change decision policies as the situation at hand is evaluated with respect to the risk will result in more resilient and dependable performances relative to the conventional ones (Zhou et al., 2024). This is an important step towards one of the most crucial issues of autonomous driving: guaranteeing safe operation under uncertainty.The significance of agentic AI can be felt especially in complex driving scenarios, where uncertainty and variability are the order of the day. An example of such circumstances is with urban traffic, which deals with people on foot, on bikes, and other road users all displaying unpredictable behaviour. Conventional AI systems do not cope with this complexity as they operate based on fixed models and bias rules. On the contrary, agentic systems are dynamic in reasoning and adaptive in their approach, which allows them to act well according to unforeseen circumstances. Recently, it has been established that fusion of multi-modal perception and agentic reasoning can greatly improve situational awareness and decision-making performance in the settings of autonomous driving (Nazar et al., 2025). In addition, agentic AI brings in the notion of lifelong learning in which systems change in response to new experiences. This is especially useful to overcome the shortcomings of the traditional training data: it enables the system to improve as it gets a response in the real-life. Nevertheless, the higher level of autonomy also puts some crucial questions about safety, control, and ethical issues, which should be considered to provide the reliable deployment. Continuous learning presents both opportunities and challenges in a system design viewpoint. Although it allows adaptive improvement, it also presents the question of stability and verification of the system, especially with safety-critical applications. To guarantee that the acquired behaviors are within manageable safety levels, monitoring processes and validation systems have to be incorporated. As it has been recently pointed out, the agentic systems can display unexpected behavior in the absence of restrictions, because of the feedback loops or conflicting interests, and thus including the powerful concepts of safety governance is essential (Eslami and Yu, 2025).

1.3 Motivation for Safety and Reliability

The increasing application of self-driving cars in the real world has made the safety and reliability the focus of research and development projects. The AVs work in safety-critical environments unlike the traditional software systems where failure may lead to strong adverse effects, such as loss of life and major economic effects. Consequently, the necessity of making such systems robust and dependable is both a technical and a social issue.The complexity of driving conditions is one of the reasons why it is essential to pay more attention to the idea of safety. The autonomous vehicles should be able to work in a very diverse environment, with different weather conditions, light, and traffic conditions. These are capable of greatly impacting on sensor behavior and system behavior. As an example, LiDAR sensors can give noisy outputs during rain or fog, whereas camera-based ones can perform poorly in low-light situations. These sensor-level uncertainties may spread throughout the system and cause false perceptions and even unsafe decisions to be made (Abu et al., 2024).Besides hardware-related failures, there is also a high degree of risk due to software-related failures. Even though machine learning models are powerful, they are prone to error due to biased training data, overfitting or adversarial inputs. These malfunctions may be in the form of wrong object detection, misidentification, or improper decision making. Moreover, the fact that an AV consists of several different subsystems adds to the complexity, which means that unexpected interactions and system failures are more likely to occur. Another critical problem is edge-case scenario. Such scarcely occurring situations with a high impact are sometimes beyond the training data distribution to create AI models. Examples include the abnormal road designs, unpredictable human behavior on the ground, or unusual weather conditions. Compared to traditional AI systems, which are mostly based on statistical learning, such cases may be difficult to deal with. This shortcoming explains why more powerful and flexible solutions are needed, including those proposed by agentic AI.Although agentic AI would offer greater decision-making and adaptability power, it also leads to the emergence of new risks that should be addressed carefully. The greater liberty of agentic systems brings up the issue of control and predictability because decisions can be created through the complex interactions among various agents. The last researches reveal possible weaknesses of the agentic vehicle designs such as cognitive and cross-layer threats that might jeopardize the safety of the systems (Eslami and Yu, 2025). These results highlight the need to establish holistic safety models that will counter both the old and new risks. Human-in-the-loop approaches have also been suggested to be integrated as a way of improving safety. By implementing a balance between control and autonomy through the inclusion of human feedback in the learning process, systems are able to provide a balance. Safety and efficiency in an autonomous driving application have seen promise in reinforcement learning frameworks that are being reinforced by human guidance (Huang et al., 2024). These schemes describe the significance of integrating technical innovation and human supervision in order to guarantee faithful system functionality.

1.4 Chapter Contributions

This chapter presents an introductory overview of how agentic AI can be used in autonomous vehicles, and specifically in safety and reliability. It develops the baseline ideas needed to comprehend how the traditional AI architecture can be replaced with more sophisticated agent-based systems and also identifies the issues and opportunities involved with the transition of agentic AI as it is discussed in Table 3.The main benefit of this chapter is that it provides a systematic analysis of agentic AI as a unifying framework of autonomous driving. The chapter provides a straightforward explanation as to why agentic systems should be implemented in the future design of AVs, by looking at the weaknesses of the traditional solution and the strengths associated with agentic systems. Moreover, it examines the security aspects of a higher degree of autonomy, determining its main threats and possible ways to reduce them. Another valuable trend of discussion is the evaluation and validation of agentic AI systems. These systems are getting increasingly complicated and the past methods of testing might not be enough. The chapter thus stresses on the necessity of new measures and techniques that may reflect the dynamic and adaptive characteristics of agentic systems. It involves the simulation environment, real-life testing, as well as the combination of both. Lastly, the chapter enumerates the research directions in the future; it notes the emerging trends by exploring the combination of generative AI and agentic frameworks and the creation of hybrid human-AI systems. Recent research into the generative AI agents emphasizes its ability to improve the reasoning ability and decision making skills of the autonomous machines as well as create some critical safety issues (Jabbour and Reddi, 2024). Such advances lead to the future where autonomous cars can not only navigate through complicated settings, but can also learn, adapt, and cooperate with each other in a manner that was previously unthinkable.

Table 3. Traditional vs agentic AI in autonomous vehicles

Aspect	Traditional AI-Based AV	Agentic AI-Based AV
Architecture	Linear pipeline	Distributed agents
Decision-making	Rule-based or learned policies	Goal-driven reasoning
Adaptability	Limited	Continuous and dynamic
Error handling	Reactive	Proactive and predictive
Learning	Offline training	Continuous learning

2. FUNDAMENTALS OF AGENTIC AI

The chapter provides the groundwork of agentic AI in self-driving cars. It characterizes an intelligent agent as an autonomous system which perceives its environment, decides (usually by a MDP/POMDP model) and executes actions via vehicle controls. Important agent elements are the environment, sensors, actuators, decision policies and learning mechanisms (e.g. reinforcement learning). The chapter goes on to describe a modular agentic architecture that consists of Perception, Reasoning, Planning, Action and Monitoring agents, each having distinct duties. Considering the examples, the Perception Agent is the body that combines LiDAR/camera data with the interpretation of the surrounding environment, whereas the Reasoning Agent is the body that uses learned policies or modules based on the LLM to evaluate the risk. We also provide a schematic (Figure 2.1) to show data flow between these agents. Section 2.3 contrasts single-agent and multi-agent and hierarchical designs, with the multi-agent systems aligning through communication protocols (e.g. shared memory, passing of messages). Lastly, Section 2.4 compares agentic AI to the traditional AV pipelines in a table with more detail, indicating that agentic systems facilitate dynamic goal-driven planning and dynamics, whereby conventional pipelines are unchangeable and rule-based.

2.1 Concept of Intelligent Agents

The theoretical basis of intelligent agents in agentic artificial intelligence is the concept of intelligent agent, and is a key factor in the design of autonomous systems, especially in safety critical fields like autonomous vehicles. A smart agent may be generally thought of as an agent that studies its environment by sensors, interprets the data it has collected using decision-making processes within its own structure, and responds to the environment with actuators to help it meet certain objectives. In contrast to the traditional computational models, intelligent agents are utilized in dynamic environments, and they have to adjust their behaviour according to the shifting circumstances and uncertainties. The agent paradigm is of particular interest to the autonomous driving, as the vehicle should act as an autonomous decision-maker. It has to process intricate sensorial measurements, predict the actions of the other road users, and perform safe and efficient maneuvers at the moment. This involves perception and control as well as reasoning in the face of uncertainty, long-term planning, and never-ending learning (Waghmare et al., 2024; Wang, 2024).

2.1.1 Definition of an Intelligent Agent

As a formal definition, an intelligent agent may be defined as a system which carries out a mapping of percept sequences to actions commonly expressed as:

$$a_t = \pi(S_t)$$

where s_t act for the perceived state of the environment at time t, and π denotes the decision policy. In more sophisticated models, especially in uncertain models, this mapping is generalized to probabilistic policies which are aware of belief states instead of completely observable states.A major difference between intelligent agents and such agents is that intelligent agents are goal-oriented. Instead of following fixed instructions, agents consider a variety of possible courses of action and determine the ones that maximize a specified objective function, e.g. safety, efficiency, or comfort. This goal is multi-dimensional in autonomous vehicles, and intelligent agents must make trade-offs among conflicting goals, e.g. minimizing travel time and collision avoidance.Moreover, intelligent agents do not exist in a vacuum; they inhabit an ecosystem of interacting agents. Such interaction requires capacity of modeling the behavior of other agents, anticipating future conditions and responding dynamically. Recent studies note that agent-based reasoning integration is an effective method to enhance awareness of the situation and the robustness of the decision-making process in complex driving scenarios (Nazar et al., 2025).

2.1.2 Core Components of Intelligent Agents

Intelligent agents are usually built up of a number of interdependent components, which in sum lead to perception, reasoning and acting. These elements are not autonomous units but are constituents of a close feedback.

- **Environment** The environment is the outer world within which the agent is work in. In autonomous driving, this is physical infrastructure (road system, traffic lights, etc.) and dynamic (vehicles, pedestrians, etc.), weather. The environment itself is partly observable and is stochastic and dynamic, contributing to a great complexity of decision-making.
- **Sensors** The interface between the agent and the environment is provided by sensors. They render crude information that needs to be undergone to give meaningful information. LiDAR, cameras, radar, and ultrasonic sensors are used as sensors in the autonomous cars. The sensor modalities have both advantages and weaknesses and combining the modalities is vital to achieve strong perception.
- **Actuators** Actuators allow the agent to relate to and manipulate the environment. These are steering systems, throttle systems and braking systems in the cars. An agent may make decisions based on the capabilities of its decisions, but the speed and accurateness of the actuators will also determine whether it is effective.
- **Decision Policies** The intelligence of the agent lies in his decision policy. It characterizes the way in which the agent chooses the actions depending on its present state and goals. The policies may be simple like a rule-based model to the complex learning-based model like the reinforcement learning agents. Policies in the contemporary autonomous system are usually dynamic and able to change according to new information.
- **Learning Mechanisms** Learning processes enable the agent to do a better job with time. This can be in the form of supervised learning, reinforcement learning or a combination of both. The importance of continuous learning is especially high in real-world conditions, where pre-trained models can face any scenario that they have not seen before. In Table 4, the functional structure of intelligent agents in autonomous vehicles has been summarized.

Table 4. Functional components of intelligent agents in autonomous vehicles

Component	Function	Example in AV Systems	Key Challenge
Environment	External system context	Urban traffic, highways	Uncertainty and variability
Sensors	Data acquisition	LiDAR, cameras, radar	Noise and occlusion
Actuators	Execute actions	Steering, braking	Precision and latency
Decision Policy	Action selection logic	Path planning algorithms	Real-time constraints
Learning Mechanism	Performance improvement	Reinforcement learning	Stability and safety

2.1.3 Interaction Model of Intelligent Agents

The activity of an intelligent agent may be thought of as a cycle of interaction consisting of perception, decision-making, and action. This is not a straight line loop, it is an iterative loop based on feedback provided by the environment which affects the later decisions. The exchange between an intelligent agent and the surrounding is in the form of a closed-loop, as shown in Figure 4.

Figure 4. Intelligent agent interaction loop (perception–decision–action cycle)

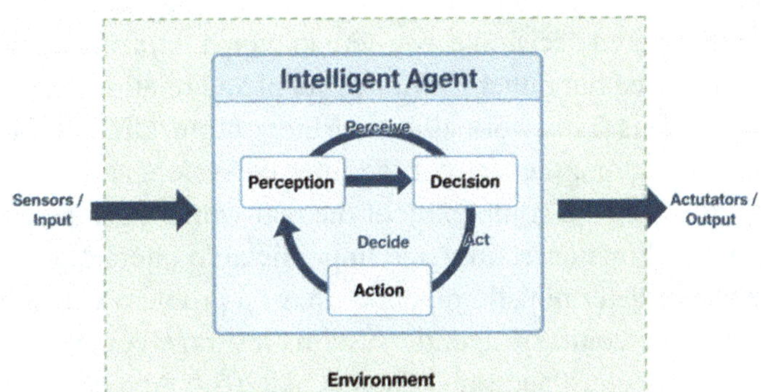

Some of the key features to be identified in this loop include:

- The behavior that is driven by feedback and actions have an impact on the perception in the future.
- Temporal dependency, which involves temporal reasoning.
- Adaptability, which allows reacting to the dynamics.

This feedback mechanism is repeated at a high frequency in the order of milliseconds, with high optimality in a computational architecture required in self-driving machines..

2.1.4 Relevance to Autonomous Vehicles

The intelligent agent model offers a natural model of autonomous vehicles as self-contained decision-making systems. In contrast to the conventional pipeline architectures, where perception, planning, and control are divided into different modules, agent-based systems combine the functions into a unified decision-making unit. This integration facilitates stronger operation of complicated situations, including the interactions of multi-agents within the dense traffics. As an example, an autonomous vehicle does not just have to notice a pedestrian, but has to predict their intentions and change its course as well. The reasoning beyond immediate sense data is also necessary, which is why agent-based models are important, and multiple agents can be employed to perform at various levels of abstraction. This is especially applicable in cases of massive implementation of autonomous systems, which require coordination of more than one vehicle.

2.2 Agentic AI Architecture

The design of agentic artificial intelligence systems constitutes a paradigm shift in design with respect to traditional pipeline-based designs that have found application in autonomous vehicles. Instead of breaking down functionality into fixed and sequential modules, agentic architectures are designed based on interacting, decision making components, each with a particular responsibility in perception, reasoning, planning and control. At a fundamental level, an agentic architecture may be understood as a distributed system of multi-layered decision ecosystem in which information move is not unidirectional as perception to action is directed but rather is bi-directional and relational through feedback, reasoning and coordination processes. This framework allows further optimization of the decisions depending on the changing environmental factors, the state of the internal system and the long-term goals. Such an architecture is a result of the growing complexity of the real-world driving conditions, especially those in urban areas that are marked by uncertainty and unpredictable interactions. The traditional systems, which are based on the use of deterministic pipelines, have a problem with supporting edge cases and non-linear interactions. On the contrary, agentic systems leverage contextual reasoning and adaptive planning, which is much more robust and safe (Zhou et al., 2024; Abu et al., 2024).

2.2.1 Overview of Agentic Architecture

The autopilot of autonomous vehicles is usually designed as an agentic AI comprising various specialized agents each with their functional domain of responsibility. Those agents maintain connections with each other via shared representations and communication protocols and, thus, can make a joint decision. Figure 5 shows the structural organization of agentic AI in an autonomous vehicle system. The architecture typically has five main agents, namely, perception, reasoning, planning, action and monitoring. The agents are not operating independently, and instead, they are a closed-loop system and constantly communicate with each other to ensure there is a situational awareness and safe operation.

Figure 5. Agentic AI architecture for autonomous vehicles

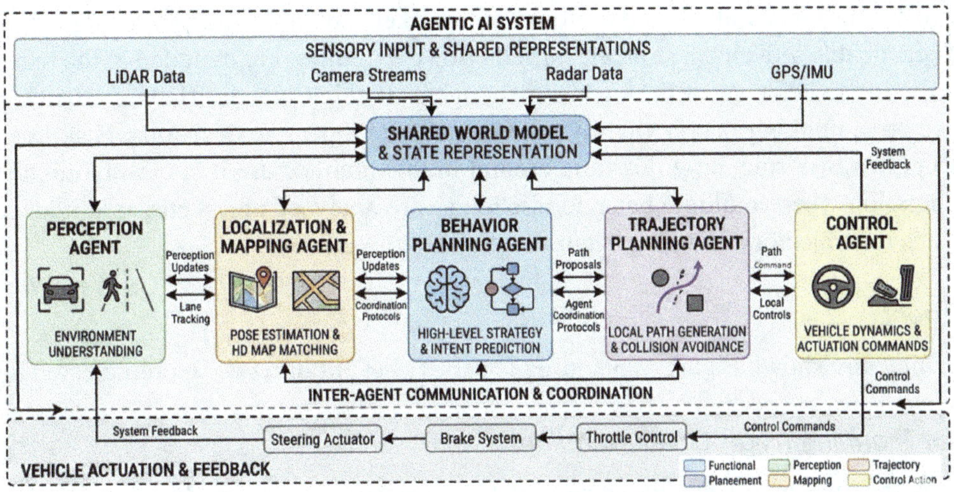

Perception Agent

he perception agent will build some coherent and structured representation of the external environment based on raw sensor data. This is the association of various senses, which are LiDAR, radar, camera streams, and additional supporting signals, such as GPS/IMU, by means of sophisticated sensor fusion methods. The perception agent, as in the illustrated architecture, is one of the key contributors of the joint world model and state representation and constantly makes perception updates: object detection, lane tracking, and features of the environment. Such updates are not standalone productions but they are inputted to a centralized representation which is made available to all the other agents. The perception agent of an agentic architecture performs more semantic and contextual understanding than the traditional perception modules, which are more concerned with object detection. It does not only recognize objects but also determines their relationships, movement patterns and even their purpose. Indicatively, to appreciate a pedestrian by a crosswalk, the prediction of whether a person will cross or not is made, and it has direct downstream effects on planning agents. One of the biggest issues of perception is the problem of uncertainty and noise, especially in unfavorable weather or in the urban setting. Continuous feedback of downstream agents via the shared world model is part of the robustness in this architecture. This enables perception to be narrowed down with the help of localization and planning modules. Besides, perception is not an uni-directional pipeline, but instead a dynamically adjusted inter-agent communication system, with its products adjusted to coordination protocols with other agents.

Reasoning Agent (Aligned with Behavior Planning Agent)

The cognitive core of the system is the reasoning agent which is embodied in the architecture as the behavior planning agent. It converts the common state of the world to the high levels of driving strategies and choices of behavior. This agent works on top of the shared world model and takes the outputs of the perception and localization data to produce high level strategy and intent prediction. It dictates how the car is supposed to behave under a particular circumstance like overtaking, halting, yielding, or merging.

An important addition that is evident in the diagram is that of agent coordination protocols, where the reasoning/behavior agent is able to send and receive information with the localization and the trajectory planning agents. This makes sure that decisions are not taken isolated but they are aligned to space limitations and practicable movement strategies. Prediction capabilities are included in the reasoning agent as well, and the future behavior of the environment agents (vehicles, pedestrians, cyclists) is estimated. This is a predictive thinking that is vital in making safe and proactive decisions. Risk assessment has continued to play a core role, only this time around in this architecture it is closely aligned with path proposal generation where multiple behavioral choices are analyzed and then passed downstream. To these proposals the trajectory planning agent then refines them.

Planning Agent

In the architecture shown Figure 4, planning is explicitly split into two interconnected layers:

(a) Behavior Planning (High-Level Planning)

Handled by the reasoning/behavior agent, this layer determines what to do:

- Lane changes
- Overtaking decisions
- Yielding and stopping
- Route-level decisions

(b) Trajectory Planning Agent (Low-Level Planning)

The trajectory planning agent is responsible for how to execute those decisions. It generates feasible, collision-free paths based on the high-level intent received from the behavior planning agent. This agent performs:

- Local path generation
- Collision avoidance
- Motion smoothing
- Dynamic feasibility checks

One of the additions provided by the diagram is the idea of path proposals and refinement. The trajectory planner does not generate a single solution but it considers a variety of candidate trajectories based on constraints like:

- Safety
- Comfort
- Vehicle dynamics
- Environmental constraints

The other significant characteristic is the two-way interaction with the shared world model, where real-time updates become available as the conditions of the environment change. Planning is also closely combined with inter-agent communication and thus making sure that there is coordination between physically possible trajectories and decisions at the behavior level.

(c) Action Agent (Control Agent)

The agent of control, which is the action agent, will handle the conversion of the planned trajectories into executed commands to the vehicle actuators. It takes path commands on the part of the trajectory planning agent and converts them into low-level control signals including:

- Steering actuator inputs
- Brake system commands
- Throttle control

One of the critical contributions made by the architecture is the availability of local control loops wherein the commands are kept in constant change according to real-time feedback of the vehicle. This allows exact tracking of trajectories in dynamic conditions. The control agent works within strict real-time limitations and includes.

The control agent operates under strict real-time constraints and incorporates:

- Vehicle dynamics modeling
- Feedback control (e.g., PID, MPC)
- Disturbance rejection mechanisms

The other major feature that is demonstrated in the diagram is the system feedback loop whereby the results of the control action are fed back into the shared world model. This enables upstream agents to make decisions depending on the real vehicle behavior.

(d) Monitoring Agent

Even though not directly referred to as a standalone block in the diagram, the monitoring functionality is spread throughout the system via system feedback loops and inter-agent communication channels.

The monitoring agent's role is effectively embedded in:

- The shared world model, whereby all the information of all agents is aggregated and validated.
- The system feedback channels, which feeds the results of execution to the perception and planning modules.
- The inter-agent communication layer, which makes it consistent and detects anomalies.

This distributed monitoring performs:

- Validation of perception outputs
- Consistency checks between planned and executed trajectories
- Detection of control deviations
- Sensor and system health monitoring

Of significance that the architecture adds is that monitoring is not centralized but it is spread across all the agents and can quickly identify a fault and respond accordingly.

For example:

- When control execution does not follow the planned path, then the feedback loop causes replanning.
- In case inconsistencies in perception are observed, localization and reasoning modules are modified

This monitors distributed design enhances fault tolerance and makes the systems reliable at all times in compound setups. Table 5 summarizes the functional roles and the challenges related to each agent in the architecture.

Table 5. Functional roles of agents in agentic AV architecture

Agent	Primary Function	Key Techniques	Challenges
Perception Agent	Environment understanding	Sensor fusion, deep learning	Noise, occlusion
Reasoning Agent	Context interpretation & risk assessment	Knowledge models, LLMs	Uncertainty handling
Planning Agent	Trajectory and behavior planning	Optimization, RL	Real-time constraints
Action Agent	Execution of control commands	Control theory, MPC	Precision and latency
Monitoring Agent	System validation and safety	Fault detection, diagnostics	False positives/negatives

2.2.2 Key Properties of Agentic AI Systems

In addition to structural elements, agentic architectures are characterized by a set of intrinsic properties, which make them unique when compared to traditional systems.

1. **Autonomy:** The agentic systems are limited to little human participation, and they make decisions independently, depending on the environmental inputs and inner goals. This independence is necessary to be able to work in real-time in complicated driving conditions.
2. **Adaptability:** Adaptability is the capability of the system to change its behaviors with the changing conditions. This involves short-term adjustment as in response to unforeseen challenges and long-term experience learning.
3. **Goal-Driven Behaviour:** Goal-oriented, in contrast to reactive systems, agentic AI is supposed to be goal-oriented. Decisions are considered on their part in relation to the contribution to the attainment of predetermined goals, which allows more consistent and meaningful behavior.
4. **Multi-Agent Coordination:** In most cases, two or more agents have to work together towards the attainment of system-level objectives. Such coordination can be between a single vehicle or a group of vehicles in a network. Coordination must be done by maintaining communication protocols, common representations, and conflict resolution systems.

2.2.3 Relevance to Autonomous Driving Systems

The implementation of agentic AI-based systems in autonomous vehicles is indicative of a wider trend in increased flexibility and smartness of system design. Incorporating the aspects of perception, reasoning, planning and control into a single system, agentic systems are more capable of managing the dynamics of real world driving. More so, agentic architectures are distributed, which increases their scalability and resilience, rendering them appropriate to large-scale implementation. With the ongoing development of autonomous vehicles, agentic AI will become one of the core elements that will allow safer and more trustworthy transportation infrastructures (Abu et al., 2024; Wang, 2024).

2.3 Types of Agentic Systems

The categorization of agentic systems is fundamental in explaining how various architectural designs affect the behaviour, scalability and the reliability of intelligent systems in autonomous cars. Although the idea of an intelligent agent offers the underlying abstraction, real-world implementations need to go beyond that by using more complex organization structures to manage complexity, uncertainty, and scale interaction. The distribution of decision-making authority, the interaction of agents with each other, as well as the hierarchy of control can greatly define agentic systems. These differences are especially relevant in the context of autonomous driving, as the driving environments in their nature are multi-agent systems, and include not only a car-car interaction but also pedestrian, infrastructure, and external control system interaction. In current studies, it is also noted that the multi-agent and hierarchical transition is a critical step to increase system robustness and adaptability particularly in uncertain and complex environments (Nazar et al., 2025; Zhou et al., 2024).

2.3.1 Single-Agent Systems

Single-agent systems are the simplest agentic AI in which a single decision-making system perceives the environment, thinks about it, and acts. In autonomous vehicles, this is defined by a system of self-contained components, i.e., all the functionalities that are required: perception, planning, and control, are combined into a single agent, even though conceptually simple. As an example, driving agents that use reinforcement learning can learn the best driving policies by interacting with the simulated environment (Huang et al., 2024).Nonetheless, the disadvantages of the single-agent systems can be observed in reality. The task of autonomous driving has an interactive nature and the system must be able to predict and react to the behavior of the other agents. An all-single-agent approach might not be suitable to model these interactions, resulting in poor or unsafe decisions. In addition, scalability can be a factor since an agent with more and more complexity in its environment demands greater computational and modeling resources. This drawback has propagated the creation of more distributed methods.

2.3.2 Multi-Agent Systems

Multi-agent systems (MAS) are an extension of the single-agent paradigm, where a set of interacting agents with their own perception, decision making and goals are introduced. Interaction can be considered the defining feature of multi-agent systems in the context of autonomous vehicles as these agents can be individual vehicles, infrastructure elements, or even modules of software in a particular vehicle.

Agents are forced to either coordinate, cooperate or compete with each other to meet their objectives. This adds further complexity, such as communication schemes, negotiation schemes, and conflict-solving systems. Ulti-agent systems allow a more faithful application of traffic dynamics in autonomous driving. An example is that cars have to make lane changes, yield at crossroads, and respond to the actions of the other motorists. Such interactions demand predictive models that explain the intentions and potential actions of other agents. Recent developments have shown that the use of agent-based reasoning is very useful in enhancing decision-making within such environments. The example is multi-agent reinforcement learning methods that enable vehicles to acquire collaborative techniques, which result in a more favorable traffic movement and decreased risk of collisions (Nazar et al., 2025). On the same note, risk-sensitive decision frameworks are also based on data-driven and knowledge-based solutions to improve the safety of multi-agent settings (Zhou et al., 2024).Although these are beneficial, multi-agent systems additionally present a challenge, especially in terms of stability and scalability. With the increment of the number of agents, interactions become increasingly complex, which means that the mechanisms of coordination will need to be efficient.

2.3.3 Hierarchical Agent Frameworks

Agent hierarchies is a multi-agent system framework that tackles the challenge of complexity of multi-agent systems through a hierarchical organization of agents in several levels of abstraction. In this type of system, strategic decision-making is done by higher-level agents whereas tactical and operational tasks are done by lower-level agents.

In autonomous motor vehicle, a hierarchical framework may consist of:

- An agent in high-level, which plans the routes and the missions.
- Mid-level agents that deal with behavior planning, e.g. lane changes and obstacle avoidance.
- Control implementation agents at low level.

Such a hierarchical framework facilitates the breaking down of complex tasks, and each agent can act within an operational level. It also enables modularity, which simplifies the design, analysis and verification of separate components. The capability of the hierarchical systems to integrate long term planning and real time responsiveness is one of their major strengths. The top-level decisions offer a strategy whereas the lower-level agents respond to the immediate changes in the environment.

Nonetheless, it is a major challenge to ensure there is uniformity in the various levels of the hierarchy. Incongruity between top-level aspirations and bottom-level behavior may result in hazardous conduct. Therefore, useful communication and synchronization mechanisms are necessary.

2.3.4 Collaborative Agent Networks

Collaborative agent networks are a sophisticated type of multi-agent systems where the agents actively exchange information and organize their actions to reach some common objectives. In contrast to loosely coupled multi-agent systems, collaboration and information sharing are highlighted in collaborative networks. In autonomous driving, the same idea is especially applicable with regard to connected and cooperative vehicles. As an example, through vehicle-to-vehicle (V2V) and vehicle-to-infrastructure (V2I) communication, vehicles can exchange information about road conditions, on the road hazards, or bus infrastructure, which is otherwise unattainable in a scenario relying solely on onboard sensors. A moving car, say in an intersection, can use the information exchanged by other cars of possible existing obstacles or road conditions. Distributed intelligence is also facilitated by collaborative networks of agents where the decision making process is distributed among other agents and not concentrated in one system. This enhances scalability and stability in that failure of one of the agents does not always affect the whole system. Nevertheless, these systems present new issues of reliability of communications, consistency of data, and security. According to recent research, agentic vehicle systems are susceptible to cross-layer and cognitive attacks, which highlights the necessity of having effective security systems in collaborative systems (Eslami and Yu, 2025). Table 6 offers a comparative overview of the various types of agentic systems and their applicability to autonomous driving and is depicted in Figure 6.

Figure 6. Type of agentic system in autonomous vehicles

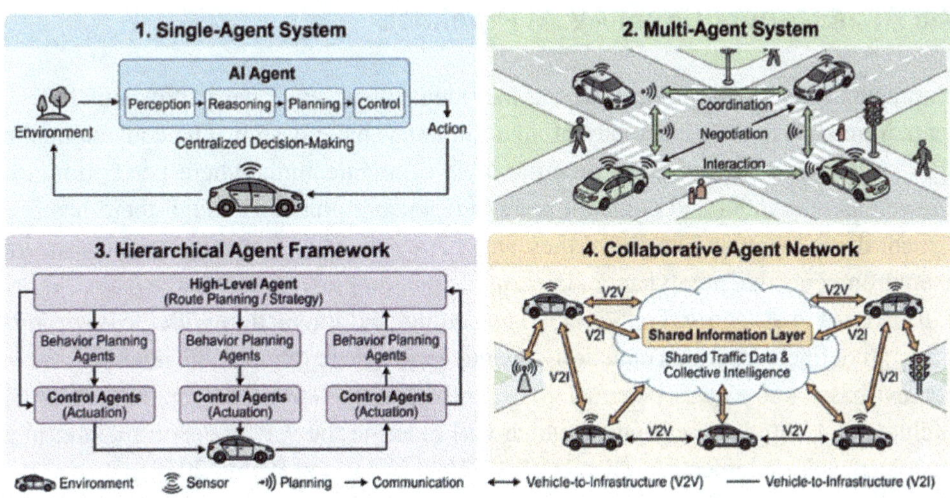

Table 6. Comparison of agentic system types

System Type	Structure	Key Strength	Limitation	Application in AVs
Single-Agent	Centralized	Simplicity and control	Limited scalability	Basic autonomous driving
Multi-Agent	Distributed	Interaction modeling	Coordination complexity	Traffic interaction handling
Hierarchical	Layered	Task decomposition	Synchronization issues	Planning and control separation
Collaborative Networks	Cooperative distributed	Shared intelligence	Communication & security risks	Connected vehicle ecosystems

2.3.5 Implications for Autonomous Vehicles

The design and performance of autonomous vehicles are severely affected by the agentic system architecture choice. An agent-based system with multi-agents and collaborative concepts may be essential in the real-world scenario when using single-agent systems was no longer satisfactory, especially in a controlled environment. Specifically, urban driving implies constant communication with several dynamic subjects, and multi-agent and hierarchical frameworks are more appropriate. Besides, the incorporation of collaborative networks is presumed to be imperative in the future of intelligent transportation systems, which would allow safer and more efficient mobility. With the development of autonomous vehicles, a definite shift toward hybrid systems which mix the characteristics of all four types of systems is visible. These architectures use the ease of single agent systems, the interactions of multi-agent systems, the form of hierarchical designs, and the scaling of collaborative networks.

2.4 Agentic AI vs Conventional AV AI Pipelines

The development of agentic AI architectures, as the extension of traditional artificial intelligence pipelines, is a serious paradigm shift in the philosophy of autonomous vehicle design. The conventional autonomous driving designs have highly depended on modular style of architectures where perception, planning, and control are viewed as sequential and comparatively independent phases. Though there has been considerable advancement due to such architectures, they are characterized by apparent weaknesses in application in real-world environments that are dynamic, uncertain, and complex. In contrast, agentic AI brings about a paradigm of distributed intelligence, goal-directed reasoning, and interaction with the environment. Instead of implementing predefined programs of actions, agentic systems are dynamically evolving their behavior in reaction to context-based knowledge, personal goals, and communication with the other agents. The change is not only architectural but also conceptual, which will redefine the way decision-making in autonomous systems is organized and implemented (Zhou et al., 2024; Jabbour and Reddi, 2024).

2.4.1 Conventional AV AI Pipelines

Traditional autonomous vehicle designs are commonly structured into a linear pipeline landscape comprising of perception, localization, planning, and control parts. Each of the modules has a clear purpose and the output of the module is used in the next stage. The design is more supportive of modularity and simplifies the development, testing and verification of the system, however, this pipeline structure also presupposes that the output of every stage should be sufficiently correct to use it in the next stage. Practically, these errors are propagated through the system and they tend to be amplified downstream. As an example, when the object detection is incorrect, then the resulting trajectory will be wrong, and eventually, the safety will suffer. The other limitation is that of the decision-making that is static. The classical paradigms use rule-based logic and trained models that can be detrimental in generalizing into both new and rare situations. This is especially problematic in autonomous driving, where edge cases, or anomalous pedestrian behavior or conditions on the roads can have life-threatening outcomes (Karunakaran et al., 2023). In addition, the traditional pipelines are not usually holistically reasoned. The modules work in their own spheres, and only have limited strengths in terms of integrating global context and long-term goals. Consequently, the system can develop suboptimal characteristics in cases where there is need to make coordinated decisions in various fields(Babbar et al., 2021)

2.4.2 Agentic AI-Based Architectures

The agentic AI architecture is proposed to overcome the shortcomings of the traditional pipelines, and it involves autonomous and interacting agents that mutually manage perception, reasoning, planning, and action. The agentic architectures are dynamic in nature and can receive continuous feedback and adaptation. An important aspect of agentic architecture is that it integrates decision-making at various levels of abstraction. Ships other than dividing perception and planning into hard modules, agentic systems allow reasoning agents to directly impacts strategies of perception and planning decisions. This bi-directional flow of information makes the situation more aware and enables the system to react better to uncertainty. Goal-oriented behavior is another significant thing. The agentic systems assess actions in terms of their contribution to some predetermined goals, including safety, efficiency, and the comfort of the passengers. This facilitates more context-sensitive and coherent decision-making especially when using the environment in a complex setting where incorporating large language models and knowledge-driven reasoning into agentic frameworks has enabled risk-sensitive decision-making in autonomous vehicles to be achieved (Zhou et al., 2024). On the same note, generative AI agents have also been discussed as an opportunity to improve flexibility and picture comprehension, despite the issues of safety and reliability still being present (Jabbour and Reddi, 2024).

2.4.3 Key Differences in Design Philosophy

The difference between conventional and agentic architectures is not confined to structural differences but also to the key principles of design. Traditional systems are based on the principles of modularity and determinism, where reliability is a key priority with well-defined interfaces and predictable behavior. Although this method simplifies validation and regulatory adherence, it tends to reduce flexibility and adaptability. On the other hand, the use of agentic systems focuses on adaptability and intelligence and it, therefore, allows system responses to evolving environments. The change presents new requirements, in particular, the safety and interpretability of systems. As an example, the decision policies, which are based on learning, can demonstrate random actions in some circumstances and, therefore, require effective monitoring and validation structures (Eslami and Yu, 2025). Moreover, agentic architectures facilitate distributed decision-making which means that several agents can cooperate and exchange information. This is especially relevant in the multi-agent setting, where vehicle-to-vehicle and infrastructure coordination can greatly improve safety and efficiency (Sabouni, 2025). Table 7 compares conventional and agentic AI structures in autonomous cars in detail.

Table 7. Comparison between conventional and agentic AV architectures

Aspect	Conventional AV AI	Agentic AI
Architecture	Sequential pipeline	Distributed agent-based system
Decision-Making	Rule-based, static	Goal-driven, dynamic
Adaptability	Limited	Continuous learning and adaptation
Error Handling	Error propagation across modules	Feedback-driven correction
Context Awareness	Localized within modules	Global and contextual reasoning
Scalability	Challenging in complex environments	Naturally scalable via agents
Multi-Agent Interaction	Limited support	Built-in coordination mechanisms
Safety Handling	Predefined rules and redundancy	Risk-aware, adaptive safety strategies

2.4.4 Safety and Reliability Implications

Safety remains the most critical requirement for autonomous vehicle systems, and the architectural paradigm plays a significant role in determining system reliability. Conventional systems rely heavily on redundancy and formal verification to ensure safety. For example, industrial frameworks emphasize multiple independent sensing and decision pathways to mitigate failures (Mobileye, 2020; Shalev-Shwartz et al., 2024).Agentic systems, while offering enhanced adaptability, introduce new safety challenges. The increased autonomy and learning capabilities of agents can lead to non-deterministic behavior, making it more difficult to predict system responses under all conditions. This necessitates the development of advanced monitoring mechanisms, formal verification techniques, and fail-safe strategies (Nemouchi et al., 2025).At the same time, agentic architectures provide opportunities for improved safety through proactive risk assessment and adaptive decision-making. By continuously evaluating the environment and anticipating potential hazards, agentic systems can respond more effectively to unforeseen situations. This is particularly important in complex urban environments, where static rules are insufficient (Abu et al., 2024).

2.4.5 Emerging Hybrid Approaches

It is important to note that the transition from conventional to agentic architectures is not absolute. In practice, many modern autonomous vehicle systems adopt hybrid approaches that combine elements of both paradigms.

For instance, a system may retain a modular pipeline for perception while incorporating agentic reasoning for decision-making and planning. This allows developers to leverage the reliability of traditional architectures while benefiting from the adaptability of agentic systems.Recent research also explores the integration of human-in-the-loop learning, where human feedback is incorporated into agentic decision-making processes to enhance safety and performance (Huang et al., 2024). Additionally, efforts to improve transparency and interpretability, such as model pruning and explainable AI techniques, are gaining traction in the development of trustworthy agentic systems (Suwal et al., 2025).

2.4.6 Summary and Transition

The comparison between conventional and agentic AI architectures highlights a clear trade-off between predictability and adaptability. While traditional pipelines offer reliability and ease of validation, they struggle to handle the complexity and variability of real-world driving. Agentic systems, on the other hand, provide enhanced flexibility and intelligence but require new approaches to ensure safety and reliability. As autonomous vehicle technology continues to evolve, the integration of agentic principles is expected to play a central role in addressing the limitations of existing systems. The next chapters will build upon this foundation to explore safety frameworks, validation strategies, and future research directions.

3. ARCHITECTURE OF AGENTIC AI IN AUTONOMOUS VEHICLES

3.1 Perception Layer

Perception layer is the basic part of the agentic autonomous vehicle architecture, and it is the main point of contact between the physical world and the decision making system. Its principle purpose is to convert unstructured, raw sensory data into structured, semantically meaningful forms which may be used by high-level reasoning and planning agents. In contrast to the traditional perception modules, which tend to exist as autonomous preprocessing phases, in agentic AI perception layer is tightly connected to the rest of the decision-making process, allowing it to constantly interact with reasoning and planning agents. In real-world driving environments, perception is by definition a difficult task because it is filled with uncertainty, noise, and changing interactions. Self-driving cars will need to analyze and understand complicated scenarios where several moving objects are involved, different lighting conditions, and random actions. In this regard, the perception layer should not merely be the object detector, but attain contextual insight, time consistency, and predictiveness (Waghmare et al., 2024; Wang, 2024).

3.1.1 Multi-Modal Sensor Systems

The current autonomous vehicles are based on the hybridization of heterogeneous sensors to represent complementary data on the environment. These are usually LiDAR, cameras, radar and ultrasonic sensors, each possesses its own advantages to the perception process .LiDAR offers high-resolution three-dimensional spatial data such that the distance and the shape of an object can be calculated with accuracy. Cameras on the other hand record rich visual aspects like color, texture and semantic context that is critical in activities like recognition of traffic signs and lane identification. Radar systems are also robust in bad weather conditions and are especially generalized in monitoring object speed. The combination of these modalities enables the system to surmount the constraints of the individual sensors and to have a more holistic view of the environment. Multi-modal sensing(Jabeen et al., 2023), however, presents major challenges with regards to data fusion, calibration and synchronization. The Sensor measurements can vary in their resolution, sampling rate, and noise properties, and it may prove hard to integrate them into a single representation. Moreover, sensor bugs or malfunctions such as LiDAR occlusion or camera glare may also be a major contributor to the decrease in accuracy of perception(Sateesh et al., 2021; Muzammal et al., 2020). According to recent research, sensor reliability continues to be a significant bottleneck in autonomous driving, and especially in unfavorable environmental conditions (Kumar et al., 2025; Matos et al., 2024).

3.1.2 Sensor Fusion and Representation Learning

In an attempt to mitigate the drawbacks of single sensors, the perception layer uses sensor fusion methods which combine data provided by many sources into a consistent representation. The fusion may be at various levels such as raw data fusion, feature-level fusion and decision-level fusion. Of these, feature-level fusion has received a lot of interest because it can combine complementary information without impacting the computational efficiency.Sensor fusion In agentic AI architectures, sensor fusion is not only a preprocessing stage but an adaptive process that is moderated by higher-level agents. The reasoning agent can be provided as an example, whereby some sensor input can be given more importance than others depending on the circumstances, like when visibility is low, the agent will depend more on radar. Representation learning heavily relies on deep learning models, especially convolutional and transformer-based models, which enable the system to be robust to the changes in the environment. The models can be used to extract high-level features in raw sensor data to enable the process of object detection, semantic segmentation, and scene understanding. Multi-modal learning has seen recent progress, and in more complex driving situations, it has performed better, particularly with agentic reasoning models (Nazar et al., 2025).

3.1.3 Object Detection and Scene Understanding

Perception layer plays a very important role in identifying and categorizing objects in the environment. This encompasses the cars, pedestrians, bicycles, road signs and road networks. The recent developments depend on a deep neural network and are trained on massive datasets, which provide high accuracy in detecting and classifying objects, but this is not enough to drive autonomously without any accidents. The system should also undertake scene understanding that entails the interpretation of the relationship of the objects and their roles in the environment. As an example, contextual reasoning is necessary other than merely detection to determine the difference between a pedestrian standing on the sidewalk and one who is planning to cross the road. Analyses of the sequence of observations allow the system to deduce the patterns of motion and predict the future. This is especially useful in dynamic environments, wherein one must predict the behavior of other agents to navigate the environment safely. Although these developments have been made, perception systems are susceptible to edge cases and rare situations. Such scenarios are actively being generated and handled, and methods allowed include data augmentation and simulation to enhance robustness (Karunakaran et al., 2023).

3.1.4 Challenges in Perception

Various challenges are inherent to the perception layer, which are restricting its performance and reliability in the real-life application. Sensor noise and uncertainty may be one of the most important challenges; it might be caused by the environmental conditions, i.e. rain, fog or poor lighting. These may negatively affect sensor performance and cause inaccurate or incomplete results. The other critical problem is that of occlusion, objects may be partially or completely hidden. This is more so in the urban setting, where buildings, cars, and any other form of hindrance are in the way of the line of sight. In overcoming occlusion, such information as time needs to be incorporated with predictive modeling in order to determine the concealed states. The perception is also complicated by dynamic environments, in which the system has to always keep track of the behavior of many moving objects and predict them. This not only demands precise detection but also strong tracking and motion prediction systems. Also, the perception systems need to be able to operate with strict real-time constraints with large amounts of data being processed at low latency. Accuracy and computational efficiency continue to be a major problem when it comes to designing perception algorithms. The main issues of perception in self-driving cars are outlined in Table 8.

Table 8. Challenges in autonomous vehicle perception

Challenge	Description	Impact on System	Mitigation Strategy
Sensor Noise	Measurement errors due to environment	Reduced detection accuracy	Filtering and probabilistic models
Occlusion	Hidden or partially visible objects	Incomplete perception	Multi-view fusion, prediction
Adverse Weather	Rain, fog, low light	Sensor degradation	Redundant sensing (radar)
Dynamic Objects	Moving agents in environment	Prediction complexity	Temporal modeling
Real-Time Constraints	High data processing demand	Latency issues	Hardware acceleration

3.1.5 Role of Perception in Agentic AI Systems

The perception layer role in agentic architectures is not one of passive participation, but rather of active participation in the decision making process. It also has a continuous interaction with reasoning and planning agents, supplying them with updated information and seeking feedback to make its outputs precision to contextual needs.This two-way exchange allows it to have an adaptive perception, in which the system improves its sensing and processing strategies depending on the requirements of the context. Considering an example, in a high risk situation, the system can face more computational resources to an enhanced perception accuracy to assure system safety. This entails that accurate and credible perception is required in the detection of hazards, prediction of potential risks and timely response. Therefore, the enhancement of perception strength is one of the research priorities of autonomous vehicle development (Abu et al., 2024).

3.2 Decision-Making Agents

The agentic autonomous vehicle architecture has the decision-making layer as the cognitive core between perception and action. Whereas the perception layer builds some representation of the environment, the decision-making agents will interpret this representation, assess potential courses of actions and make choices of strategies that comply with the system goals, including safety, efficiency and comfort of passengers. In contrast to conventional rule-based systems, an agentic AI decision-making is dynamic, context-driven, and adaptive by nature and allows the system to conduct successful operations in highly uncertain and interactive settings. In actual driving, incomplete information, stochastic actions of the other actors and the occurrence of rare or unpredictable events make the decision making process more complex. Therefore, the decision-making agents have to combine various types of intelligence, probabilistic reasoning, predictive modelling and goal-oriented optimization. According to recent research, agentic reasoning can make autonomous systems much stronger, especially in large urban areas where communication is highly packed and unpredictable (Zhou et al., 2024; Nazar et al., 2025).

3.2.1 Risk Assessment and Uncertainty Modeling

One of the basic tasks of decision making agents is risk evaluation of various actions. Risk in autonomous driving is also multi-dimensional by nature, which may include the probability of collision, the breach of the rule, as well as the harm to passengers and other users of the roads. One of the skills that are necessary in making decisions is the capacity to measure such risks in uncertainty and integrate them in the process of making a decision. There is uncertainty that comes in several sources such as sensor noise, incomplete observation and uncertainty in the behavior of other agents. The agents of decision-making cope with this issue by using probabilistic models that model the environment as a probability distribution over potential states instead of a deterministic structure. Bayesian inference and probabilistic graphical models are the techniques that are usually employed to estimate the uncertainty and update beliefs as new information appears. Risk assessment may be combined with knowledge-based reasoning in agentic AI systems. Indicatively, recent frameworks unite data-driven learning with structured knowledge representations in order to facilitate more plausible risk-sensitive decisions. With such methods, the system can mix empirical data with domain knowledge to enhance its performance in safety-critical situations (Zhou et al., 2024). In addition, risk assessment is not a computation that is completed once but an ongoing process. Due to the changing environment, the system has to reconsider risks and revise its decisions all the time. This dynamism ability is critical to deal with traffic situations that change very fast and prevent any possible hazards.

3.2.2 Behavior Prediction of Surrounding Agents

One of the main tasks in autonomous driving is the ability to foresee the behavior of other participants in the road, which may be cars, pedestrians, and bicycles. Agents in decision making are thus required to employ behavior prediction systems which predict the future trajectories and intentions of the other agents.Behavior prediction is a complex process by its nature because of the variability and unpredictability of the human behavior. Conventional methods are usually based on kinematic models or easy-to-guess, which do not always prove sufficient in tricky situations. However, agentic systems utilize more sophisticated machine learning algorithms, such as deep learning and sequence modeling, to elicit more subtle sets of behavior. The latest development in multi-modal learning has made it possible to combine visual, spatial and contextual information so as to make better predictions. As an example, cues, including pedestrian gaze direction, vehicle speed, and state of the traffic lights can be added to the models and used to deduce intent. Moreover, multi-modal strategies enable a much greater degree of situational awareness and decision-making (Nazar et al., 2025).Moreover, agentic AI models, in the context of which the system takes into account the possibility of its actions affecting the behavior of others, enable interactive prediction. It has been especially useful in situations like a merging or a negotiating intersection where one agent behavior is subject to another agent behavior.

3.2.3 Tactical Decision-Making and Maneuver Selection

Tactical decision making involves picking the right driving maneuvers depending on the prevailing environment and system goals. These encompass the decision making in the form of changing lanes, passing, giving way and halting at intersections. In traditional systems, tactical decisions are usually controlled by rules and heuristics. Although this can be effective in a structured setting, it is not very flexible and might not work in a complex or new setting. By contrast, agentic AI systems use goal-directed decision policies that consider a range of available actions and choose the option that most effectively fulfills the goals of the system, often in the form of an optimization problem, with the objective function defining the various requirements of the system, e.g. safety, efficiency, and comfort. One of the most effective methods to learn such policies has become reinforcement learning, which allows the system to enhance its performance with the help of interaction with the environment. The capability to deal with trade-offs is another crucial part of tactical decision-making as it improves the process by including human feedback to make safer and more understandable choices (Huang et al., 2024). An example is a move to pass a slow car that could result in profitability but risk. It is a principled issue that agentic systems have to reconcile these conflicting objectives and in many cases multi-objective optimization methods are applied.

3.2.4 Route Optimization and Strategic Planning

In addition to the immediate tactical decisions, the decision-making agents also have an obligation to do higher level of strategic planning, such as route choice and long-range navigation. This means finding out the most optimal route between the current point and a destination based on the factors, including traffic, road restrictions and the rate of energy utilization.Route optimization is most often undertaken as a graph based calculation, and the road network is represented as a graph and the goal is to find the shortest/fastest route. Planningroutes in agentic systems is however dynamic and adaptive as it plans routes in the real-time using information and feedback provided by other agents. An example can be given of an agentic system being able to automatically correct the route due to traffic congestion or a road closure. The collaborative setting could also exchange information to enhance the overall efficiency of the traffic which means that the chosen way will be possible under different circumstances.Strategic planning should also adapt to uncertainty and variability, so that the chosen way is possible in shifting circumstances. This means there must be constant checking and re-planning and this is where there is a need of integrating decision making and monitoring agents.

Figure 7. Decision-making process in agentic autonomous vehicles

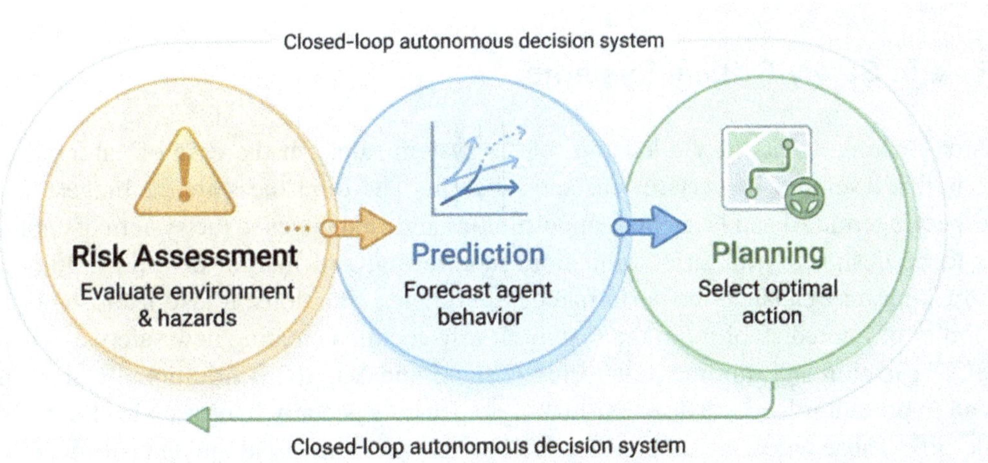

Figure 7 depicts the internal decision-making process of the agents of risk assessment, behaviour prediction and choice of action.

3.2.5 Integration with Agentic AI Frameworks

The agents of decision making are not independent of others but instead are closely connected to other elements of the agentic architecture. The one-way flow of information between these components (agentic AI) is maintained by the interaction of the perception agents with the planning agents to assess the feasibility of the trajectory, and the monitoring agents to assure that there is safety. As an example, a rational agent can demand more information on perceptions when there is uncertainty or modify the course of planning in response to new risk evaluation. Such integration allows stronger and more adaptive behavior as opposed to the traditional pipeline architecture. Besides, in recent studies, the application of large language models and generative AI is examined in decision-making agents to improve reasoning. Such models facilitate the system relating high-level knowledge and they are able to perform highly challenging reasoning tasks, but issues of reliability and interpretability continue to be (Jabbour and Reddi, 2024). The main roles and problems of decision making agents are highlighted in Table 9.

Table 9. Core functions of decision-making agents

Function	Description	Techniques Used	Challenges
Risk Assessment	Evaluate safety of actions	Probabilistic models, Bayesian inference	Uncertainty handling
Behaviour Prediction	Anticipate actions of other agents	Deep learning, sequence models	Human unpredictability
Tactical Decision-Making	Select driving maneuvers	RL, optimization	Trade-off management
Route Optimization	Determine optimal path	Graph algorithms, dynamic planning	Real-time adaptation

3.2.6 Role in Safety-Critical Systems

Decision-making agents play a leading role in system safety in the case of autonomous vehicles. Mistakes in this level may be very serious and reliability and robustness are the biggest consideration. In this respect, agentic AI can bring both opportunities and challenges to the system of decision-making which is to be designed with strict validation, monitoring, and margin of error. Although adaptive decision-making may enhance the performance of complex situations, it also leads to complexity and the possibility of unpredictability in the system. It also requires creating new safety mechanisms, such as formal verification and runtime monitoring (Eslami and Yu, 2025).Additionally, the ethical aspect also has an important role in the decision-making process, especially in cases where there are irresolvable trade-offs. These problems cannot be solved by technical means alone, but also by regulation and participation in the society.

3.3 Planning and Control Agents

Planning and control layer in agentic architectures of autonomous vehicles has the role of converting high-level decisions to vehicle movement. Although the decision-makers decide on the actions to be undertaken, planning and control agents decide how the actions can be actually implemented within the limitation of the vehicle and its surrounding environment. This layer functions on a high level of strict real-time demands and should be able to guarantee the safety, feasibility and efficiency of all generated motions. Unlike the pipeline based systems that generally view planning and control as disjointed modules, agentic architectures encourage a more integrated and adaptive design(Lim et al., 2019). The planning and control agents are in constant interaction with the perception and decision-making agents, allowing dynamic changes with the updated information about the environment and the changing system goals. This combination is needed to address the complexity and uncertainty of the real-world situation of driving (Wang, 2024; Abu et al., 2024).

3.3.1 Path Planning and Motion Planning

Path planning can be defined as the task of identifying a way to navigate a vehicle to a desired destination starting at the current location and taking into account feasibility of the path. Motion planning can be defined as the refinement of this path into a sequence of states that can be traversed by the vehicle over time. The processes should consider the environmental constraints, vehicle dynamics, and safety requirements in autonomous driving. Planning in autonomous driving is usually carried out at different levels of abstraction. On a more global scale, the system identifies a rough path that does not contain static obstacles and is based on road geometry. On a more detailed level, motion planning produces detailed trajectories in which dynamic obstacles and vehicle dynamics are considered. Graph-based path planning algorithms like A* and Dijkstra algorithm used to plan paths are classical and applicable to a static environment. Nonetheless, driving in the real world involves working with dynamic components and thus planning the car with more complex methods including sampling based planners and optimization based planning. Agentic AI improves the planning process with contextual reasoning and predictive functionality. The planner can, as an example, change its direction in response to forecasted actions of vehicles around it, or the change(s) in traffic conditions. Such a dynamic adaptation is much better in complex environments where rigid planning mechanisms cannot be applied.

3.3.2 Trajectory Generation

Trajectory generation Trajectory generation is the process of transforming a pseudo-planar path into a continuous, time-parameterized trajectory of positions, velocities and accelerations. Several constraints have to be followed by this trajectory, among them being the vehicle dynamics, the comfort of passengers, and the safety demands.

Mathematically, trajectory generation can be formulated as an optimization problem:

$$\min_{\tau} j(\tau)$$

where τ represents the trajectory and $j(\tau)$ is a cost function that encodes objectives such as efficiency, and safety. In agentic systems, the process of trajectory generation is not static but is an on-going update.Constraints can be collision avoidance, speed limits, and vehicle kinematic limits. The system re-examines and revises the course in real time as new information is received. This feature is especially significant in dynamic environments, where the barriers can be introduced without prior notice.Recent studies have focused on embedding learning-based methods into the trajectory generation process so that the system learns the best motion patterns using the data. Such methods are capable of enhancing work in tricky situations, yet they need to be strictly verified to provide safety and stability (Huang et al., 2024).

3.3.3 Collision Avoidance and Safety Constraints

The planning and control layer has a basic requirement of collision avoidance. The system has to make sure that all the developed trajectories are safe with references to the obstacles and that they are following the traffic rules. This is normally done by considering the safety constraints in the planning process. As an example, the system can set a safety envelope around the vehicle and make sure that no section of the trajectory gets into this area of any obstacle. In agentic designs, control barrier functions and reachability analysis are more sophisticated methods to ensure safety in the face of uncertainty. Collision avoidance is intimately coupled with agentic decision-making and perception. The system constantly updates on the knowledge of the environment and re-examines the safety constraints in real-time. This adaptive method allows better managing the unforeseen circumstances, including the urgent braking of an ill-fate vehicle. The recent research highlights the necessity of applying the safety constraints to the multi-agent coordination system, according to which the interactions between a sequence of vehicles have to be taken into account (Sabouni, 2025). These measures are necessary in order to promote a safe functioning in heavy traffic conditions.

3.3.4 Control Systems and Execution

After a path has been produced, it is the control system that has to execute the path by giving commands to the actuators on the vehicle. This is by controlling variables like steering angle, acceleration and braking to make sure that the vehicle travels correctly along the desired path.Controlles systems with autonomous vehicles are usually developed based on a mixture of classical control theory and new learning-based methods. Model Predictive Control (MPC) is very popular because it is capable of managing constraints and optimization of performance in a finite time horizon. In agentic Artificial Intelligence systems(Ahmed et al., 2022; Dogra et al., 2021), control is not reactive, but rather guided by reasoning on the higher levels. As an example, the control agent can modify its behavior on the basis of risk estimation or anticipation of occurrences. This integration allows increased adaptive and context-aware control strategies, though the control systems should also be resilient to disturbances and uncertainties, e.g. a changing road condition or sensor error. The stability and reliability in such conditions are one of the main issues in the design of autonomous vehicles.

3.3.5 Real-Time Constraints and Computational Challenges

Planning and control agents are operating on real-time strict requirements in which decisions may need to be made in milliseconds. This is computationally intensive especially with high-dimensional state spaces and complex optimization issues. Modern systems using a mixture of algorithmic optimization and hardware acceleration use a mixture of algorithmic optimization and hardware acceleration to overcome these challenges. The necessary performance is reached with the help of such techniques as parallel processing, specialized hardware (e.g. GPUs and TPUs). This landscape is further complicated by the agentic architectures that add more levels of interaction and feedback. Although such features help to make it more adaptable, they also make it more complex to compute(Hanif et al., 2022). This is because achieving a balance between performance and efficiency is an important aspect in system design.

Table 10. Components of planning and control agents

Component	Function	Techniques Used	Challenges
Path Planning	Determine feasible route	Graph search, sampling methods	Dynamic environments
Trajectory Generation	Create time-based motion plan	Optimization, learning-based methods	Real-time updates
Collision Avoidance	Ensure safety constraints	Barrier functions, prediction	Multi-agent interactions
Control Systems	Execute trajectory	MPC, PID, adaptive control	Stability and robustness

The key components and associated challenges of planning and control agents are summarized in Table 10.

3.3.6 Role in Agentic Autonomous Systems

The planning and control level is the key to the high-level intelligence and physical execution. In agentic architectures, this layer is not insulated but instead closely coupled to the agents of perception and decision making to make up a whole system that can be used to effectively respond to dynamic and uncertain environments to enhance both safety and performance. Nevertheless, it also brings about the issues of complexity, checking, and performance speed.As the technology of autonomous vehicles advances, the creation of a more advanced planning approach and control will be one of the primary research topics, especially when engaging with agentic AI systems that focus on flexibility and coordination (Wang, 2024; Abu et al., 2024).

3.4 Monitoring and Feedback Agents

The agentic autonomous vehicle architectures have a monitoring and feedback layer, which is the supervisory and self-regulatory mechanism of the system, and as such, all the components are running within the safe and expected limits. Whereas the perception, decision-making, and planning agents are concerned with supporting the autonomous functionality, the monitoring agents are in charge of proving the rightness, identifying anomalies and taking corrective measures in case deviations arise. This layer is essential in the autonomous vehicles of safety-critical systems, where even minor failures may spread quickly and cause the most disastrous consequences, unlike in traditional systems, where monitoring is usually provided as a passive diagnostic tool. These agents use both rule-based and learning-based methods to identify inconsistencies, monitor system health and impose safety constraints. As a result of integration of monitoring into the agentic decision making loop, the approach to safety has the ability to be proactive, and possible failures are predicted and minimized before they lead to unsafe behavior (Eslami and Yu, 2025; Abu et al., 2024).

3.4.1 System Health Monitoring

System health is monitored where the hardware and software elements are continuously evaluated to determine their correct functionality. This involves checking the performance of sensors, processing units, communication lines and control units in autonomous vehicles, the quality of sensor data is a critical aspect of perception since it is the quality of the sensor data that determines the level of perception. Such failures as sensor degradation or miscalibration or total failure may cause a substantial effect on system operation. The approach by monitoring agents to this problem involves applying diagnostic methods that identify anomalies with sensor outputs, e.g. changing data distributions or inconsistencies between sensors. Recent polls emphasize that sensors are still a severe issue in autonomous vehicle systems, especially when there are unfavorable weather conditions (rain, fog, extreme temperatures, etc.) (Matos et al., 2024; Kumar et al., 2025). To address these risks, redundancy strategies are usually utilized by monitoring agents, with several of them offering redundant information, so that the system is not rendered useless by failures. Software components should also be continuously assessed, besides monitoring hardware. This involves testing runtime errors, performance regression and unexpected behavior of machine learning models. To guarantee that the systems are overall sound, security of the hardware and software needs to be guaranteed.

3.4.2 Failure Detection and Diagnosis

The detection of failures is one of the fundamental monitoring agent functions, which requires identifying deviations between a normal and a desired system behavior. In agentic systems, a combination of machine learning, rule-based checking, and statistical techniques is frequently used to detect anomalies in order to identify possible failures. As an example, anomaly detection algorithms can be used to detect unusual features in sensor data, and consistency checks can be used to identify differences between the output of modules. When a failure has been identified, the system is supposed to identify the cause of the failure to take the right corrective measures. The process is sometimes complicated with failure possibly occurring because of several causes and extending through various components. This is enabled by agentic architectures which permit communication between monitoring agents and other systems components so that failure detection and diagnosis can be made more accurate and timely. Such methods guarantee system behavior mathematically, which allows revealing possible failure modes and guarantee that safety requirements are met (Nemouchi et al., 2025).

3.4.3 Adaptive Control and Recovery Mechanisms

In addition to failure identification, monitoring agents should be tasked with the responsibility of taking corrective actions to ensure the restoration of safe operation. This is by dynamically changing the system behavior when anomalies are detected and this is known as adaptive control. As an example, in case a failure of sensors is detected, the system might be reconfigured to utilize other sensors or become less functional. In severe instances, the vehicle can change to a safe state either by decelerating or halting completely. These are recovery strategies that are required to make sure that in case of unforeseen failures, they have the necessary recovery strategies to be able to adapt. The system is able to analyze the situation at hand and provide the best possible recovery strategy as opposed to using preset answers. This is a flexibility that enhances the willingness of the system to deal with complex and unanticipated situations and in which the learning-based approaches are incorporated so that the system can develop better response to failures over time. The safety and reliability of such adaptive systems is, however, a major issue (Huang et al., 2024).

3.4.4 Redundancy and Fault Tolerance

One of the most basic principles in the design of safety-critical systems is redundancy, which offers several layers of safety against failure. In autonomous vehicles, redundancy may be provided on different levels, such as sensors, computation and decision-making. As an example, a number of sensors could be applied to monitor the same part of the environment, which guarantees that when one sensor fails to work, it does not affect perception. The concept of redundant computational pathways can also be used to ensure backup decision making capabilities so that the system can still run in the presence of failed components, such redundancy is referred to as independent redundancy. This method has been considered as necessary to attain high reliability levels that are needed in autonomous driving (Mobileye, 2020; Shalev-Shwartz et al., 2024).The concept of redundancy is extended with agentic architectures in which various agents can counterbalance the failure of others. This does not only increase fault tolerance but also system scalability and robustness.

3.4.5 Continuous Feedback and System Optimization

Continuous feedback loops that allow the constant optimization of the system are the characteristic feature of agentic AI systems. Monitoring agents are at the center of this process as they gather the information about the performance, allow some areas to be improved, and give feedback to other agents. This feedback may be utilized to define the perception models, enhance the decision-making policies and optimize the control strategies. In the long-term, it causes gradual enhancements in the performance of the system and allows the vehicle to adapt to new conditions and situations, but within the framework of the continuous learning and adaptation, it also creates new challenges of stability and safety. To prevent the system performance and other risks, introduction of updates needs careful validation and monitoring. Safe reinforcement learning, and constrained optimization, are some techniques that are becoming more commonly investigated to overcome these (Zhou et al., 2024).

Table 11. Functions of monitoring and feedback agents

Function	Description	Techniques Used	Challenges
System Health Monitoring	Assess hardware and software status	Diagnostics, redundancy	Sensor reliability
Failure Detection	Identify anomalies	Statistical methods, ML	False positives
Failure Diagnosis	Determine root cause	Model-based analysis	Complexity
Recovery Mechanisms	Restore safe operation	Adaptive control	Real-time response
Feedback Optimization	Improve system performance	Learning, data analysis	Stability and safety

The core functions and challenges accompanying with monitoring and feedback agents are summarized in Table 11.

3.4.6 Role in Safety-Critical Autonomous Systems

The key agents to the safety and reliability of autonomous vehicles are monitoring and feedback agents. They can offer the system the ultimate protection against unsafe behavior by ensuring the continuous supervision and allowing adaptive responses to failures. The monitoring of agentic AI architecture The proactive approach to safety mentioned above is possible due to the integration of the monitoring into the decision-making loop, allowing the identification and elimination of risks before they become especially harmful. This is a major improvement compared to the old systems, which in most cases, rely on reactive safety measures. However, the greater complexity of agentic systems poses new issues especially with regards to verification, validation and security. To cope with such issues, the cooperation among academia, industry, and regulatory bodies will need to be continued through research (Eslami and Yu, 2025).

4. SAFETY CHALLENGES IN AUTONOMOUS DRIVING

Autonomous Vehicles are inherently safety-critical systems: failures can lead directly to loss of life or property. The figure 8 illustrates the end-to-end decision pipeline in an autonomous vehicle system, spanning from sensor data acquisition to perception, planning, and control, and highlights how delays or faults in any stage can propagate through the pipeline, potentially compromising overall system safety and reliability.

Figure 8. AV decision pipeline (sensors to control) or faults propagation

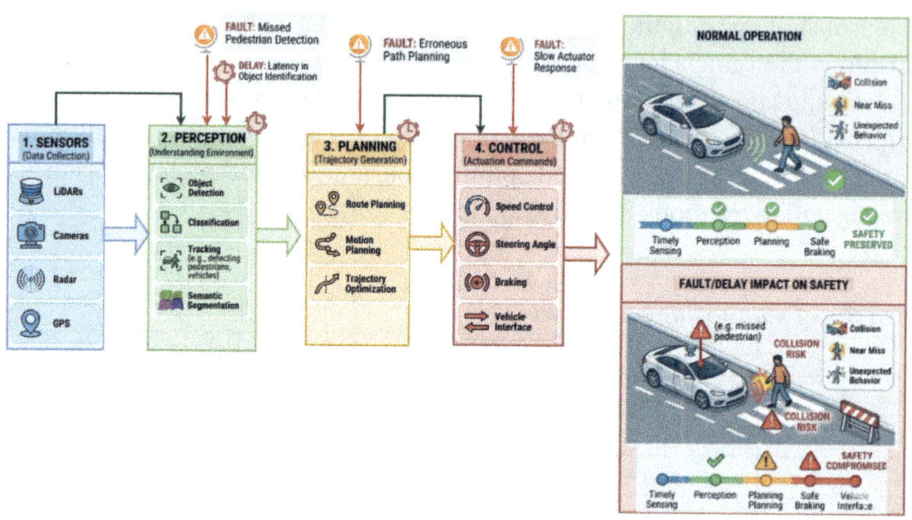

Autonomous driving systems must operate under intense safety constraints. An AV constantly senses a dynamic world, plans trajectories, and controls the vehicle – all without human intervention. Even small errors can have dire consequences, so AVs require far stronger safety assurances than most consumer technologies.

4.1 Safety-Critical Nature of AV Systems

1. **Life-or-death stakes:** Human lives and public safety hinge on AV performance. A failure of an AV perception or control algorithm (e.g. misclassifying a pedestrian as background) can cause collisions. Modern safety data show that human driver error causes ~94% of crashes (Matos et al. 2024); AVs aim to eliminate that but shift the burden of safety to the software. As one survey notes, "AVs must be designed ... to be safe, accountable, and trustworthy... [they] can make life-and-death decisions" (Safe, A. P. (n.d.)
2. **Real-time constraints:** AVs must react in milliseconds. Sensor data (LiDAR, radar, cameras) arrive at high rates and feed into neural-network perception, prediction, and planning stacks. Even small latencies can be fatal – e.g. if a pedestrian is detected just 100 ms late, a collision could occur. Thus, AV algorithms are optimized for low-latency perception and control loops. In practice this means high-speed hardware and streamlined code paths (often with real-time OS scheduling) to meet strict timing budgets.

4.2 Sources of Safety Risks

1. Sensor-Related Risks

AV perception relies on cameras, LiDAR, radar, ultrasonic, GNSS, etc. These sensors can fail or be spoofed:

1. **Sensor failure:** Hardware faults or wear can silence a sensor or corrupt its data. For example, a camera might go dark if a lens is obscured or damaged, or a LiDAR may malfunction in extreme cold. Adverse weather (rain, fog, snow) can "drastically influence perception-based sensors' ... performance" (Kumar et al. 2025). Even under normal conditions, sensors may drop packets or saturate. According to a recent survey, "current sensor technology is not reliable in all environments". Such failures lead to "wrong perception", e.g. ghost obstacles or missed obstacles (Matos et al. 2024);, and hence high crash risk.
2. **Sensor spoofing/jamming:** Malicious or environmental interference can inject false signals. For instance, GPS receivers can be spoofed by fake GNSS signals, causing the AV to compute a wrong position. Similarly, LiDAR could be jammed by high-power lasers, or road signs could be modified with adversarial stickers to fool vision systems. These intentional or accidental perturbations can mislead AV perception into unsafe decisions.

Example: Researchers demonstrated that simple printed patterns on the road can cause LiDAR to misinterpret objects (Matos et al. 2024), illustrating how AV sensors are vulnerable to spoofing attacks.

3. Software-Related Risks

The AI/ML algorithms themselves introduce hazards:

1. **Algorithmic bias and data limitations:** Machine-learning models are only as good as their training dataIn case the data do not cover all scenarios and populations; the AV can misbehave in such

circumstances. As an example, a pedestrian detector, which is mostly trained during sunshine, may malfunction during snow. In case the training data are collected in one city, the model can be trained to drive local traffic norms and make incorrect predictions elsewhere (Danks & London, 2017). Concisely, training bias may lead to low generalization or inequity (e.g. the poorer detection of minority groups). It is an increasing issue: a research paper determines accuracy differences in pedestrian detectors that can be measured by bias. Likewise, reinforcement-learning controllers can also take advantage of loopholes in their goals (so-called reward hacking) and reward design has to be considered.

2. **Software bugs and unexpected interactions:** AV software is complex and may have hidden bugs. Newer machine-learning components especially challenge traditional testing: an ML policy may make unpredictable decisions in an unseen corner case. In fact, the AV safety standard ISO 21448 (SOTIF) addresses the "safety of intended functionality", covering dangerous scenarios not caused by component faults but by unforeseen system behavior.

4. Environmental Risks

AVs must cope with an uncertain world:

1. **Unpredictable traffic behavior:** Other Road users (drivers, cyclists) may swerve, brake suddenly, or violate rules. An AV might not anticipate a motorist cutting in without signaling, or a cyclist in its blind spot. Such behaviors create dynamic hazards that the AV must monitor continuously.
2. **Adverse weather/lighting:** Rain, fog, snow, glare, and nighttime reduce visibility or sensor fidelity. For example, LiDAR returns can be severely attenuated in heavy fog. An AV's confidence in its perception drops in these conditions, raising crash risk. Redundancy and adaptive perception algorithms are needed (see Section 5).

5. Human Interaction Risks

Although no human is driving the AV, humans remain in the loop of the environment.

Pedestrians and cyclists have highly variable, unpredictable trajectories. Even well-marked crosswalks can see jaywalkers emerge unexpectedly. Vehicle drivers (at intersections or merging) may not follow traffic rules or may misjudge an AV's behavior. Each human agent introduces uncertainty: for example, an AV's braking algorithm must consider "what if a child darts into the road?" or "what if the car ahead suddenly stops?" Behavioral prediction models help, but cannot eliminate all uncertainty. Table 12 presents a structured classification of major safety risk categories in autonomous driving systems, along with representative examples, their impact levels, and corresponding mitigation strategies to enhance system robustness and operational safety.

Table 12. Safety risk categories and mitigation strategies in autonomous driving systems

Risk Category	Examples	Impact	Mitigation
Sensor failures	Camera lens blocked, LiDAR dropout, spoofing (GPS)	High	Sensor fusion, fault detection, redundancy (multiple sensors)
Software/AI faults	Data bias, model overfitting, unexpected ML outputs	High	Robust training, adversarial testing, interpretable models
Environmental hazards	Heavy rain/fog, glare, unmarked roads	Medium	Adaptive sensing (IR radar, etc.), SOTIF guidelines
Human unpredictability	Jaywalking pedestrians, erratic drivers	High	Behavioral prediction, conservative safety buffers (e.g. RSS rules)

4.3 Edge Case Scenarios

Edge cases are the rare, high-risk situations not well represented in training data or testing. For AVs, a major challenge is defining and handling such scenarios systematically. Scenario-based testing literature classifies driving scenarios into: (1) Functional (high-level description), (2) Logical (parameter ranges), and (3) Concrete (specific parameter values) (Karunakaran et al. 2023)Concrete scenarios can be "typical" (common), "critical" (rare dangerous), or edge case (statistical outlier).

Examples of Edge Cases

1. Sudden pedestrian crossing: A pedestrian steps off the curb into a blind spot (e.g. between parked cars). The AV must detect and brake within a very short distance.
2. Obstructed/partial signals: Traffic lights or stop signs partially blocked by foliage or vehicles, requiring inference.
3. Unusual road layouts: Encountering a newly-painted road pattern, temporary construction lanes, or shared spaces without clear markings.
4. Strange obstacles: Debris falling off a truck, a person lying on the road, or an animal darting into traffic.

Testing for edge cases is difficult because they are (by definition) rare. One recent study uses data-driven scenario generation (with reinforcement learning) to find parameter combinations that cause AV failures. The authors note that edge cases "present challenging scenarios that are rarely encountered in normal driving", but are precisely those the AV must handle.

Addressing Edge Cases

AV developers use several strategies:

1. Simulation/synthetic data: Create virtual scenarios (in simulators like CARLA or LGSVL) for rare events (e.g. child chasing ball into street).
2. Augmented data: Inject synthetic obstacles or background variations into training images/videos.
3. Scenario libraries: Curate test suites covering known critical events (see Table 4.2). Industry and academia are building large repositories of traffic scenarios (e.g. the SOTIF working group).

4. Safety supervisors: If an AV enters an unrecognized situation (low model confidence), it may trigger a safe fallback (see Section 5.4).

Overall, understanding and cataloguing edge cases is a major AV research focus. By combining scenario-based testing with formal hazard analysis (e.g. as in ISO/PAS 21448 for SOTIF), engineers aim to uncover hidden failure modes. In practice, one AV project defined an edge-case scenario as one where their system would collide during a lane change (e.g. a cut-in). Table 13 summarizes different edge-case scenarios in autonomous driving, highlighting their challenges and corresponding mitigation strategies.

Table 13. Edge case handling strategies in AV systems

Edge Case Type	Description	Detection Challenge	Proposed Solution Strategy
Occluded Pedestrian	Pedestrian hidden behind object	Limited visibility	Predictive modeling + sensor fusion
Sudden Vehicle Cut-in	Vehicle abruptly changes lane	Short reaction time	Real-time trajectory prediction
Adverse Weather	Rain, fog, snow affecting sensors	Sensor degradation	Weather-robust perception models
Unusual Road Layout	Non-standard intersections or construction zones	Lack of training data	Simulation-based learning
Sensor Blind Spots	Areas not covered by sensors	Partial perception	Multi-angle sensor placement

5. RELIABILITY FRAMEWORKS FOR AGENTIC AV SYSTEMS

Reliability engineering ensures that AV systems continue to behave predictably despite faults or uncertainties. Classical reliability principles apply: fault tolerance, redundancy, and fail-safe designs. In modern AVs, multi-agent architectures also contribute to reliability by distributing intelligence.

5.1 Reliability Engineering Principles

1. **Fault tolerance:** The system should detect component faults and continue operating safely. This may involve graceful degradation, where non-critical functions fail first. For example, if a non-essential sensor fails, the vehicle should still be drivable (maybe with warnings to the backup driver).
2. **Redundancy:** Critical components are duplicated. We distinguish:
    - Hardware redundancy (e.g. duplicate computing units, power supplies),
    - Software/algorithmic redundancy (diverse algorithms for the same task), and
    - Sensor redundancy (overlapping sensor coverage).

The goal is that if one element fails, another can substitute without interrupting operation. As one analysis puts it, "Redundancy is essential when engineering safety-critical systems" (Shalev-Shwartz et al. 2024).

3. **Fail-safe/fail-operational modes:** If faults exceed thresholds, the system transitions to a safe state. Common fail-safe mechanisms in AVs include automated emergency braking, controlled lane-outs to highway shoulders, or low-speed safe-stop modes. (ISO 26262 calls for Fail-Operational behavior at highest ASIL levels: the vehicle must still stop safely when some components fail.)

5.2 Multi-Agent Reliability Models

Many AV designs now employ multiple intelligent agents or modules. For example, separate neural networks or controllers may handle perception, prediction, planning, and safety. Key reliability benefits of a multi-agent approach include:

1. **Distributed decision-making:** Instead of a single "brain", multiple agents (possibly on different hardware) vote or adjudicate. This avoids a single point of failure and can catch errors via majority/plurality checks.
2. **Parallel risk evaluation:** Several agents can assess risk from different perspectives. For instance, one agent may predict pedestrian motion based on vision, another use LiDAR; if both agree a pedestrian will cross, the AV brakes.
3. **Redundant agent architectures:** One pioneering strategy ("True Redundancy" by Mobileye) is to create two independent AV subsystems: one relies on cameras only, the other on radar+LiDAR (Mobileye,2020). Each subsystem builds its own "world model" and safe-driving controller. In this way, two complete and diverse AV brains run in parallel, each capable of driving. If the camera subsystem fails, the radar/LiDAR subsystem takes over (and vice versa). This doubles reliability: akin to "having both iOS and Android in one car – they'd never crash at the same time" (Mobileye).
4. **Multi-agent swarm/coordination (future):** In the longer term, fleets of AVs and smart infrastructure can coordinate (e.g. platooning or intersection management). In such systems, multi-agent algorithms use formal techniques like control barrier functions and decentralized learning to ensure safety across vehicles (Sabouni et al. 2025). Though primarily a research area now, such multi-agent traffic management can prevent gridlock and accidents at busy junctions.

In summary, multi-agent architectures in AVs turn safety into a cooperative problem: multiple AI "opinions" are combined, making the overall system less brittle. Table 14 compares conventional AV systems with agentic AI-based systems in terms of architecture, adaptability, and reliability.

Table 14. Comparison of traditional vs agentic AV reliability

Feature	Traditional AV Systems	Agentic AV Systems
Decision Architecture	(Centralized)	Distributed multi-agent
Fault Handling	Single-point failure risk	Fault isolation via agents
Scalability	Limited	High scalability
Adaptability	Static models	Dynamic, adaptive agents
Risk Evaluation	Sequential	Parallel processing
System Robustness	Moderate	High due to redundancy

5.3 Redundancy in AV Systems

In practice, AV systems incorporate three levels of redundancy:

1. Sensor redundancy: Multiple sensors cover the same scene to cross-validate each other. For instance, a pedestrian should appear in the camera image and the LiDAR point cloud and the radar track. If one sensor gives conflicting information (e.g. the camera is blinded by glare), the others can still detect the obstacle. Sensor fusion algorithms routinely combine data, but true redundancy means even each sub-fusion stream is redundant. (Mobileye's approach creates separate world models from camera vs radar/LiDAR.)
2. Algorithmic redundancy: Diverse algorithms solve the same perception problem. Examples include: running two different neural networks for object detection (perhaps trained differently or with different architectures) and reconciling their outputs; or having a rule-based fallback check (e.g. a simple distance threshold on raw lidar if the main detector fails). Multiple AI models can be cross-checked. For example, Tesla uses both a neural network and a separate classical computer-vision path to interpret lane markings.
3. Hardware redundancy: Critical computing hardware is duplicated or dual-fail. High-assurance AV designs use dual or even triple processing units. As one detailed example notes, "our SDS (Self-Driving System) has two computing boards, and if we detect that one is burnt [failed], the other is designed to bring the vehicle to a safe stop on the road shoulder". Similarly, dual power supplies and communication buses prevent single hardware faults from crippling the system. These are essentially "backup" components that remain idle unless needed. Table 15 shows the redundancy types in AV systems.

Table 15. Redundancy types

Type	How it works	Example
Sensor	Overlapping sensors covering same FOV	Camera + LiDAR + Radar fusion
Algorithm	Multiple models/algorithms for same task	CNN detector + Transformer detector
Hardware	Backup CPUs, power, networks	Two independent compute units

By combining these, an AV can withstand many failures. For example, if one sensor chain fails, the backup sensors (and their own algorithms) still keep the vehicle aware of the world. Figure 9 illustrates a redundant AV architecture. It explicitly shows how system reliability is achieved by employing two independent sensor-processing channels (labeled Channel A and Channel B).

Figure 9. A redundant AV architecture

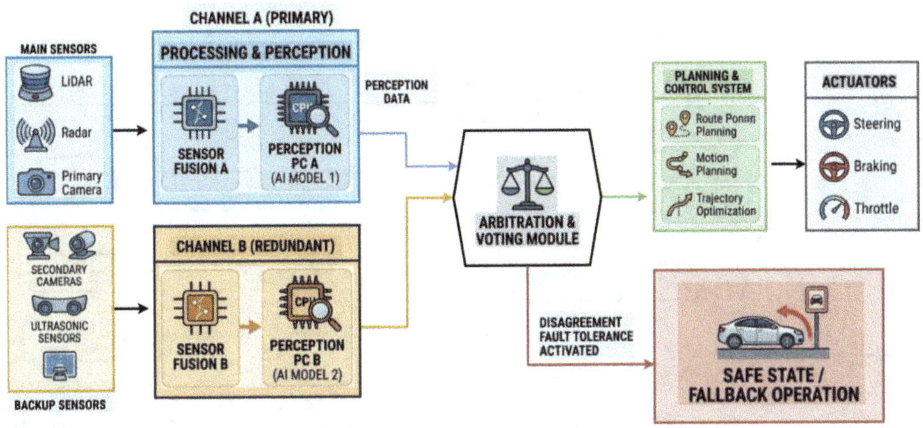

5.4 Failure Detection and Recovery

To achieve reliability, AVs must not only have redundancy, but also real-time monitoring and recovery mechanisms. Key elements include:

1. **Real-time diagnostics:** Continuous self-checks observe sensor health (e.g. plausibility checks on camera brightness, LiDAR return density), software health (watchdogs, assertion checks), and actuation health (monitor brake pressure). As soon as a sub-system behaves anomalously, the system logs the fault and alerts higher-level safety managers. For example, the SDS system "strives to maximize the system's ability to detect its own failures" and then respond with a minimal-risk action.
2. **Self-healing/reconfiguration:** If a non-fatal fault is found (e.g. a data bus glitch), the system can often reset or switch to a redundant component without stopping. Modern safety frameworks envision "self-healing" AVs that can re-route tasks to healthy modules. For instance, if one AI model crashes, the system might seamlessly hand over to an alternate model.
3. **Emergency fallback strategies:** When a critical error is detected (or if too many failures accumulate), the AV must enter a safe state. Common strategies are:
   a. **Controlled stop:** The vehicle gradually and safely comes to a halt. This may mean pulling over to the curb or median. The SDS example explicitly instructs a burned-board other board to "bring the vehicle to a safe stop on the road shoulder".
   b. **Minimal Risk Maneuver (MRM):** A tiered fallback (e.g. lane-out and stop if safe; otherwise emergency braking). The goal is to minimize additional risk. As described by Mobileye, upon detecting a failure the system should execute an MRM using built-in redundancies.
   c. **Safe mode driving:** Some systems limit AV operation to lower speeds or simpler modes when in degraded state (e.g. switch to basic cruise-control following until human takeover).

Figure 10. The failure detection and recovery process

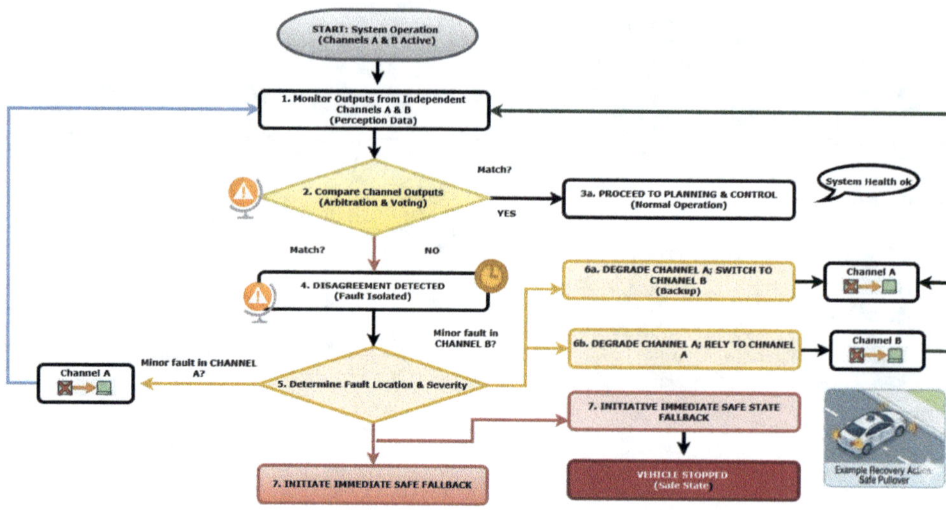

This algorithmic diagram, rendered in a clean "mermaid-style" modern aesthetic, details how the redundant outputs (from Figure 9 independent channels) are continuously monitored and compared. If a disagreement is detected, the process determines the location and severity of the fault, allowing the system to either degrade gracefully to a single channel or initiate an immediate safe state fallback (such as a controlled pullover), contrasting the immediate safety risk seen in Figure 8.In practice, these mechanisms make AVs fail-operational: even in the presence of failures, they either drive to a safe stop or continue in a limited safe mode, rather than crashing outright. For example, failing engine or brake actuators are covered by diagnostic monitors that force a safe halt. The redundancy and monitoring approaches aim for graceful degradation: losing non-critical functions first and preserving core safety functions as long as possible. This Table 16 presents common AV system failures along with their causes, detection methods, and recovery mechanisms.

Table 16. Failure modes and mitigation techniques

Failure Type	Cause	Detection Method	Recovery Mechanism
Sensor Failure	Hardware malfunction	Sensor diagnostics	Switch to backup sensor
Perception Error	Misclassification	Confidence scoring	Re-run with secondary model
Planning Failure	Incorrect path generation	Trajectory validation	Replanning module activation
Communication Delay	Network latency	Timeout monitoring	Local decision fallback
Hardware Crash	ECU/GPU failure	Health monitoring system	Activate redundant hardware

6. SAFETY ASSURANCE MECHANISMS

Even with reliability features, AVs must prove their safety through rigorous assurance. This involves formal verification of designs, runtime safety monitoring, explainability, and ethical frameworks(Airehrour et al., 2018).

6.1 Formal Verification

Formal verification applies mathematical proof techniques to guarantee system behavior meets specifications. Unlike testing, formal methods exhaustively reason about all possible states. Key approaches:

1. **Model checking:** The AV control program (or an approximation to it) is defined as a finite-state machine or hybrid automaton. In temporal logic (LTL/CTL), safety properties (e.g. the car never accelerates when there is an obstacle within less than X metres) are represented. An automatic state exploration may then be used on a model checker such as NuSMV or SPIN to check the property. AV safety-critical software is now model checked. As an example, UPPAAL (with timed automata) and SPIN have been applied to show that controllers of lane keeping and communication protocols satisfy their safety invariants. Benefits: it detects corner-case bugs that tests can fail to detect. Limitations: state-space explosion (controlled with abstraction) and inability to model neural networks directly.
2. **Theorem proving:** Human-guided proof assistants (Coq, Isabelle/HOL, PVS) allow proofs of AV algorithms to be done precisely. As an example, it is possible to demonstrate mathematically that a control algorithm satisfies braking limits. Theorem proving is able to do real arithmetic and hybrid logic, although it has to be done by hand to a large extent. It is frequently applied to the core safety functions (e.g. checking the logic of a collision-avoidance subroutine).
3. **Temporal logic:** Both model checking and runtime monitors use temporal logics (e.g. Linear Temporal Logic, Signal Temporal Logic) to specify over-time behaviors. As an illustration, a safety property can be everybody, when an obstacle is sensed inside 10m then in 0.5s throttle is cut (an LTL formula). These formulas are compiled into state-tracking automata with the help of tools. The formalization of such requirements ensures accuracy and machine-checkability.
4. **Tools and trends:** Popular AV-oriented tools include NuSMV (symbolic model checking of control logic), SPIN (protocol verification), CBMC (bounded model checking of C code), and dReach (hybrid-system verification including some neural-net controllers). According to a recent survey, "model checking has gained traction" in AV systems, especially for sensor-driven controllers. A comparison shows: NuSMV excels at boolean-logic components, SPIN at communication protocols, and CBMC for C/C++ software. The survey's outlook suggests hybrid verification (combining model checking, theorem proving, and runtime checks) is the future for learning-enabled AV systems (Nemouchi et al. 2025). Table 17 evaluates various safety assurance methods by comparing their strengths, limitations, and applications in AV systems.

Table 17. Verification methods for AV safety

Method	Use Case	Tools	Scope
Model Checking	Reactive control logic, protocol logic	UPPAAL, NuSMV, SPIN	Exhaustive, finite-state
Theorem Proving	Algorithms, hybrid control, formal specs	Coq, PVS, Isabelle	Human-guided proofs
Temporal Logic Spec	Expressing safety/liveness requirements	(SAT-based checkers)	Formal spec for others
Runtime Formal Check	On-line assurance against logical rules	(Custom monitors)	Continuous verification

By formally verifying control and decision modules, AV designers can eliminate large classes of failures before deployment. However, formal methods for deep-learning parts are still nascent; typically, one uses formal verification to validate the non-ML core and then relies on other measures for learned components. Figure 11, illustrates the formal verification pipeline used for AV safety. It details the process from model to proven system, including how a counterexample is used to refine the initial model, contrasting the manual fault identification seen in Figure 8.

Figure 11. Formal verification pipeline for AV safety

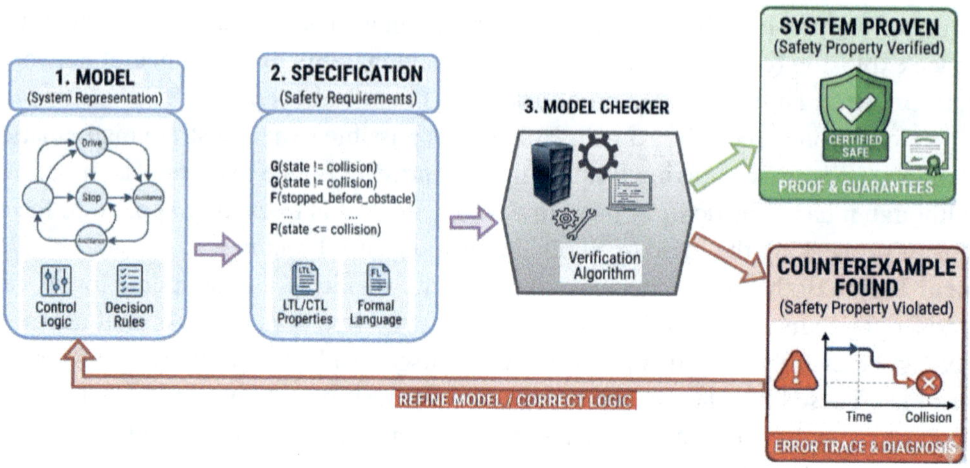

6.2 Runtime Safety Monitoring

Formal methods cover design-time correctness, but AVs also use runtime safety monitors to check behavior on the fly. This layer sits on top of the AI stack and enforces safety policies continuously. Key ideas:

1. **Safety supervisors:** A monitoring agent watches the perception–planning outputs. If it detects a decision that violates safety rules, it intervenes. For instance, a formal safety supervisor might constrain velocity so that stopping distance is always available. Tesla's "Protector" system and Mobileye's RSS module act as such supervisors, overriding unsafe AI commands. This is akin to the "ethical governor" concept in robotic systems.

2. **Policy enforcement agents:** Another style is to have a separate module enforce traffic laws or internal safety policies. For example, no AV speed can exceed 50 km/h in city or the vehicle must never cross a double-yellow. Violations trigger braking. These rules (derived from regulations or SOTIF analysis) are encoded as decision-logic checks.
3. **Risk scoring and anomaly detection:** The AV may compute a real-time risk score of its current situation (based on object proximity, confidence, etc.). If risk exceeds a threshold, the monitor forces an emergency maneuver. Similarly, monitors check that sensor data remain consistent (e.g. comparing radar vs camera obstacle positions), flagging anomalies that could indicate a sensor malfunction or spoofing.
4. **Explainable internal checks:** Some runtime monitors use simple interpretable models as cross-checks. For instance, a decision-tree inference of collision risk can run alongside a neural predictor; disagreement alerts the system to fall back to the more conservative plan.

These runtime mechanisms essentially form a second line of defence: even if the deep-learning planner makes an unsafe suggestion, the monitor can veto it. The ultimate fallback (if monitoring triggers) is usually the same controlled stop or safe mode described in Section 5.4. In sum, runtime safety monitoring continuously validates AV operation, providing high assurance that the vehicle "is driving safely with high confidence."

6.3 Explainable AI for AV Safety

Transparency in AV decision-making is crucial for debugging, trust, and certification. Explainable AI (XAI) techniques help humans (engineers, regulators) understand why an AV acted as it did. Important XAI methods:

1. **Saliency/attention maps:** In image-based perception (like camera vision), saliency maps highlight which pixels influenced a neural network's decision. For example, when detecting a pedestrian, the saliency map should illuminate the person's silhouette. Developers can inspect these maps to ensure the model is "looking at" the right features. A recent study shows that pruning models (making them sparser) can improve the faithfulness of saliency maps (Suwal et al. 2025) i.e. focusing explanations on truly important image regions.
2. **Interpretable models:** Where possible, AV logic is designed to be inherently transparent. For non-vision parts, simple algorithms (e.g. regression models, decision trees, or rule-based controllers) can be used. Even for neural networks, efforts like using smaller, more structured networks (or post-hoc methods like LIME/SHAP) can yield partial interpretability. For instance, an AV might use a linear model to approximate the relationship between distance to obstacle and braking force, which humans can verify.
3. **Decision traceability:** AVs keep detailed logs of their internal state and decision rationale (sensor readings, intermediate variables, chosen trajectories). This makes it possible to replay and analyse a decision chain post-hoc. For example, if an AV unexpectedly swerved, engineers can trace which sensor input triggered which rule.
4. **XAI in verification:** Explainability can also aid formal assurance. For example, verifying a "perception contract" (an assumption about sensor output) may require knowing how a network processes inputs. Similarly, building simple local surrogates of the neural policy (as in a supervisory model) provides insights. Figure 11 detailing the Explainable AI (XAI) Framework for AV Perception, completes the series of architectural safety diagrams.

Figure 12. Explainable AI (XAI) framework for AV perception

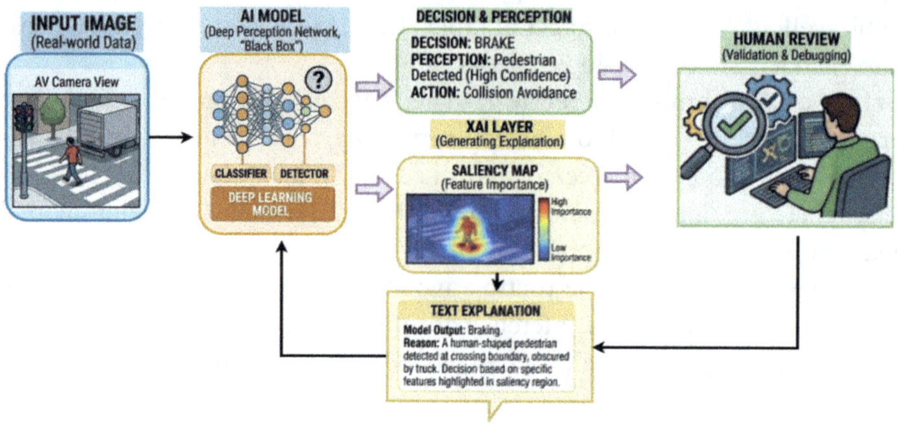

The goal of XAI in AVs is both technical and social: engineers use it to fix failures, and stakeholders (policymakers, the public) require it to build trust. Many AV XAI techniques are still experimental; ongoing research seeks to make neural decisions more accountable. Table 18 outlines key explainable AI techniques and their role in improving transparency and safety in autonomous driving.

Table 18. Explainable AI techniques for AV safety

Technique	Description	Output Type	Benefit for AV Safety
Saliency Maps	Highlights important image regions	Heatmap	Identifies focus areas
LIME	Local interpretable explanations	Feature importance	Explains individual decisions
SHAP	Feature contribution analysis	Contribution scores	Quantifies feature impact
Decision Trees	Rule-based model	Logical rules	High interpretability
Attention Mechanism	Focus on relevant input features	Attention weights	Improves model transparency

6.4 Ethical Decision-Making

Beyond technical safety, AVs will inevitably face moral dilemmas. These occur when any possible action causes harm; the AV must then follow an ethical rule. Classic examples:

1. **Collision dilemma ("trolley problem"):** If a crash is unavoidable, should the AV swerve to hit one child on the road or continue and hit another? Or strike a wall vs a passerby? No physical solution avoids all harm, so some moral prioritization is needed.
2. **Pedestrian prioritization:** Many ethicists argue that an AV should prioritize vulnerable road users. For instance, if braking later risks injuring a passenger but protects a pedestrian, the system may choose the pedestrian (reflecting a "human-centric" ethic). Companies often encode such biases implicitly or explicitly (e.g. Google's early patents focused on minimizing passenger risk, later ones considered pedestrians too).

Current practice frames these as ethical frameworks:

1. **Utilitarian or risk-minimization:** Choose the action that minimizes total expected harm (deaths, injuries). This requires assigning "weights" to lives, which is controversial.
2. **Rights-based/legal compliance:** Ensure the AV never knowingly breaks laws or moral rules (e.g. "never target a pedestrian"). This aligns with Responsibility-Sensitive Safety (RSS), which formalizes rules so that following them guarantees no collisions. RSS effectively codifies a safe-driving ethic in precise terms.
3. **Decision models:** Scholars have proposed step-by-step models. As an example, one of the 2024 ethics models will translate the guidelines and laws into a 5-step decision-making process, which incorporates safety and legal standards This "ethics stack" might say: 1) Avoid harm; 2) If harm inevitable, minimize it; 3) Prefer protecting human life over property; 4) Among humans, protect the most vulnerable (pedestrians/cyclists); 5) if equivalently positioned, minimize injury severity.

These options are bound to be trade-offs. E.g., putting pedestrians at the top of the priority list may require driving slower (to allow more people to cross) or making the passengers less comfortable (so that the AV decelerates to the maximum upon encountering every jaywalker). Another reason that is forcing automakers to explain their ethical regulations is the regulatory bodies (e.g. UNECE discussions on AV ethics, or newly emergent standards such as ISO/DIS39003 on ethical RTS). According to one Ethics review, sooner or later, these conceptualizations have to be translated into detailed specifications of requirements of the system. Practically, competing AVs now evade such dilemmas through risk management: do not drive aggressively enough that emergency situations can occur (e.g. brake early to prevent a collision in the first place). Nevertheless, some conflicts cannot be eradicated, thus there is some active research in ethical decision modules (also known as ethical governors). There is a intersection of safety and ethics: e.g., there are cases that it is not only ethical but also in many jurisdictions legally mandatory to avoid causing harm to pedestrians.

7. SIMULATION AND VALIDATION OF AGENTIC AV SYSTEMS

The fast development of autonomous vehicle (AV) technologies has generated the emergence of agentic autonomous systems, whereby intelligent agents act with high level of autonomy, flexibility and decision making. These systems combine modules of perception, planning and control in a way that they can dynamically interact with complex and uncertain environments. Their safety, reliability, and robustness are however, a major challenge especially considering the dynamics of actual road driving conditions. Simulation and validation frameworks are considered an essential part of agentic AV system development lifecycle, in this regard. On-field testing is not enough as it is expensive, time consuming and in most cases unsafe in testing infrequent or dangerous situations like accidents, system breakdown and extreme weather. Such scenarios can be analyzed and tested in controlled, scalable and reproducible simulation environments and therefore researchers can optimize models and algorithms prior to deployment.

Moreover, digital twin technology has also brought a groundbreaking strategy to system verification through the development of real-time virtual copies of real vehicles and traffic environments. These models allow constant monitoring, predictive analysis and closed-loop validation thus connecting the gap between simulated and real-world environments. Simultaneously, scenario-based test systems can be utilized to systematically test the critical driving scenarios with the emphasis on ensuring the safety standards and improving the system strength. A variety of quantitative measures, i.e. number of collisions, reaction time, path precision, and stability are used to measure the system behavior at different conditions to be able to understand combined performance fully.

These measures give information on safety as well as efficiency, which can be easily used to compare and streamline autonomous systems. In this chapter, the authors provide an in-depth discussion on simulation-based validation of agentic autonomous vehicle system. It talks about the significance of simulation environments, deployment of digital twin models, scenario-based testing strategies and key performance metrics. All of these elements constitute a holistic system of assessing and enhancing the safety, scalability, and practicality of autonomous vehicle technology.

7.1 Role of Simulation Environments

Simulation environments are used to simulate the behavior of the technology under development and, therefore, allow the researcher to evaluate and determine the influence of different changes on the mission and processes. Simulation environments form a core part of the development lifecycle of agentic autonomous vehicle (AV) systems, and allow complex driving behavior to be safely, scalable and reproducibly tested. In practice, it is not practicable or safe to subject autonomous systems to a range of possibilities, especially unlikely and dangerous situations, such as sudden pedestrian crossings, multi-vehicle crashes, or severe weather conditions.

Simulation platforms solve this limitation by offering well-regulated settings, where such edge cases can be generated and tested systematically, enhancing robustness and safety guarantees (Dosovitskiy et al., 2017) Other than safety validation, simulation environments also enable fast prototyping and iterative testing of artificial intelligence models, especially reinforcement learning (RL) and deep neural networks approaches. Such models need a significant amount of contact with changing environments to be trained in the best policies, which would be costly and dangerous in the real world. Simulation allows repeated training, in which agents are allowed to experiment with varied state-action space and optimize their strategy based on rewards (Kiran et al., 2021). Also, the emergent behaviors of traffic and coordination between agents can also be studied because simulation enables large-scale experimentation, where thousands of virtual agents can interact asynchronously in parallel. There are several state-of-the-art simulation platforms that are popular in AV research.

Figure 13. Simulation-based validation pipeline for agentic autonomous vehicle systems showing integration of perception, planning, control, and multi-agent interaction within virtual environments

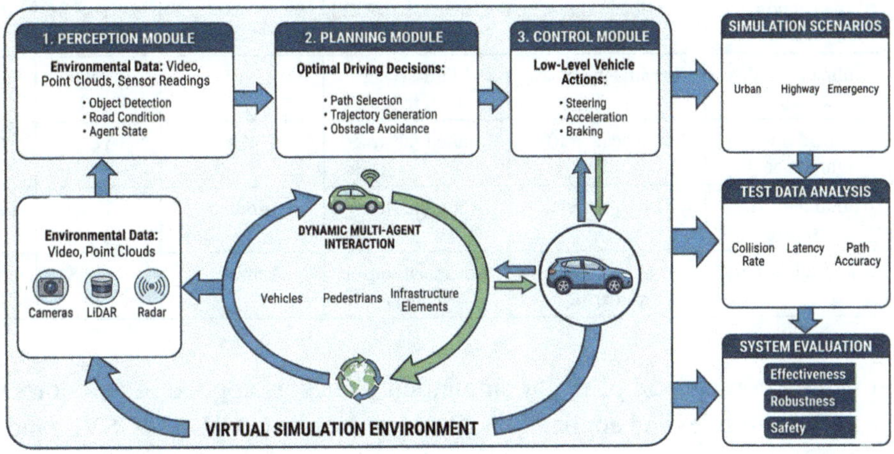

CARLA is a high-fidelity urban environment that experiences realistic sensor modeling, which is why it can be used to validate perception and control LGSVL can integrate with real-world AV stacks to allow end-to-end system testing. SUMO pays special attention to macroscopic and microscopic traffic simulation, thereby being especially helpful with respect to the assessment of extensive traffic systems and agent interaction (Lopez et al., 2018). All these platforms together offer a holistic environment to test agentic AV systems on perception, planning, and control levels (Fuller et al., 2020).

Figure 13 shows the validation pipeline analysis of agentic autonomous vehicle systems, which can be characterized by the fact that perception, planning, control, and interaction between agents should be integrated in a virtual environment. The first step is the perception module, through which the vehicle gathers information about the surrounding environment with the help of cameras, LiDAR, and radar, etc. This information is then computed to identify objects, road conditions and agents in the environment. The obtained information is sent to the planning module, which creates the best driving policies, such as path choice, route creation, and path clearance. The control module then implements these decisions by translating high level plans into low level vehicle actions like steering, acceleration or braking. All these modules work in the framework of a virtual simulation environment, in which several agents (vehicles, pedestrians, and infrastructures) communicate dynamically. This environment provides the opportunity to test autonomous behaviors in realistic conditions in different traffic scenarios.

The system is tested based on simulation scenarios on the right side of the pipeline and this is a representation of various driving conditions in which the system will work like urban, highway and emergency conditions. The test data analysis stage analyses the results and determines the performance metrics of the collision rate, latency, and the path accuracy. Last but not the least is the system evaluation stage where the general efficiency, strength, and the safety of the autonomous system is determined. This pipeline offers a holistic scheme of justifying agentic AV networks in a safe, scalable and reproducible context before they can be implemented in the real world.

Table 19. Comparison of simulation platforms for autonomous vehicle research

Platform	Main Purpose	Environment Type	Sensor Support	Multi-Agent Support	Open Source	Key Strength
CARLA	AV perception & control testing	Urban (high-fidelity)	Camera, LiDAR, Radar	Yes	Yes	Realistic sensor simulation
LGSVL (SVL)	Full-stack AV testing	Urban + highway	Multi-sensor integration	Yes	Yes	Integration with Apollo
SUMO	Traffic flow simulation	Large-scale traffic	Limited (abstract)	Yes	Yes	Scalable traffic modeling
AirSim	AI training (RL, vision)	Urban + aerial	Camera, IMU	Limited	Yes	RL-based training
PreScan	Safety validation	Controlled scenarios	Advanced sensor models	Limited	No	Industry-grade validation

Table 19 provides a comparison of major simulation platforms applied in autonomous vehicle research depending on capabilities and applications. Systems such as CARLA, LGSVL, and SUMO assist high-fidelity perception and full-stack testing, as well as large-scale traffic simulation and multi-agent interactions. AirSim is primarily intended to be used in AI training, particularly reinforcement learning, whereas PreScan is a safety verification tool intended to be industry grade. The table identifies the fact that various simulators would be used in different purposes and integrations allow the full assessment of AV systems, whereby the dynamic and data-driven virtual twin object of the actual object is created.

7.2 Digital Twin Models

Unlike the conventional simulation models, the digital twins can be subtly connected continuously with real-world information feeds: they can model the current location of the vehicles, infrastructure, and traffic ecosystems (Fuller et al., 2020).This physical-digital interaction in two directions provides an opportunity to better analyze the behavior of the systems, predict and optimize them. The use of digital twins can also be employed to test high-fidelity autonomous behavior in agentic AV systems, which are subjected to realistic operating conditions. Through digital twins, researchers are able to scale up and determine the reaction of autonomous agents in changing conditions in the presence of real-time sensor data, environmental conditions and dynamics on the traffic. In addition, predictive analytics on digital twins based on the past and real-time data are also used to anticipate potential failures in the system, congestion patterns, and safety risks and thereby take proactive actions (Talebpour & Mahmassani, 2016).

The second important benefit of digital twins is that it can be used in continuous performance monitoring and closed-loop validation. Through the comparison of simulated outputs against the actual observation in the real world a discrepancy can be observed and utilized in tuning of models and algorithms. The feedback loop improves reliability and adaptability of the system by a great deal. Moreover, digital twins can be used to perform the scenario-driven analysis of what-if, through which a researcher and policymaker can test a change in infrastructure or traffic policy without impacting real-world operations, which is especially useful in smart city projects.

7.3 Scenario-Based Testing

Agentic autonomous vehicle systems must be evaluated using a set of performance metrics that specify dependability, efficiency, and safety. These metrics are essential for comparing models and approaches and provide quantifiable indications for assessing the system's performance in different circumstances (Kiran et al., 2021). One of the safety metrics that is of the most importance is the collision rate that helps to determine the rate of mishaps or near-misses during the test. The low rate of collision will cause high levels of perception and decision-making.

Decision latency is defined as the amount of time needed by the system to respond to the inputs and produce actions that are responsive when high stress conditions are experienced. Such situations require the usage of autonomous agents, and they must take immediate decisions in uncertain scenarios without endangering efficiency and safety. Alternatively, highway situations include high speed driving, lane shifting, and merging, which need to be accurate and synchronized. The emergency situation, which occurs after the occurrence of impediments or system failure, proves the ability of the system to respond to the critical situation and guarantee the fail-safe mode. Scenario-based testing is compatible with international safety standards such as ISO 26262 and models such as PEGASUS which focus on risk-based validation or the operational design domain (ODD). This method is more specifically relevant to agentic systems by virtue that it considers agent interaction, cooperative behavior and decentralizing decision-making in dynamic environments.

Table 20. Performance metrics for evaluating agentic AV systems

Metric	Definition	Unit	Importance	Related Module
Collision Rate	Frequency of collisions	% or count	Safety evaluation	Planning & Control
Decision Latency	Time to make decisions	ms	Real-time response	AI/Control
Path Accuracy	Deviation from optimal path	meters	Navigation efficiency	Planning
Reliability Score	Stability across scenarios	Score (0–1)	System robustness	Entire system
Energy Efficiency	Energy consumption	kWh/km	Sustainability	Control
Passenger Comfort	Smoothness of ride	Jerk (m/s^3)	User experience	Control
Task Completion Rate	Success rate of missions	%	Overall performance	System-level

Table 20 shows the fundamentals of the key performance indicators to assess the performance of autonomous vehicle (AV) systems and their safety, efficiency, and dependability. Collision rate and decision latency are important measures that are necessary to evaluate the safety and responsiveness of the system in real-time since they demonstrate how the system can avoid accidents and respond to the dynamic situation. Path accuracy is determined through the computation of the distance between the vehicle and the ideal line in order to determine the effectiveness of the planning module. The score on reliability would display the overall strength of a system as it would determine its consistency in various situations. Moreover, energy efficiency and comfort to passengers are concerned with practical and user-friendly factors that guarantee the environmental friendliness of the operation and pleasant driving experience.

7.5 Performance Metrics

The evaluation of agentic autonomous vehicle systems should have a set of performance metrics outlining the safety, efficiency, and reliability. These metrics provide the quantitative measures on the assessment of the working of the system in different situations and are needed to compare the models and methods (Kiran et al., 2021). The collision rate is one of the main safety measures applied to estimate the accident rate or the number of near-misses during the test. The reason is that the rate of collision will be low and therefore, the perception and decision-making levels will be high. The number of seconds that the system requires to react to inputs and generate actions that are utilized to demonstrate responsiveness in high-stress environments is known as decision latency. Path accuracy tests the movement of the vehicle within the context of an optimal path that provides details of the effectiveness of navigation and accuracy of control. The reliability score is an index of the capability of the system to produce performance under varying scenarios whereby, the strength of the system is indicated in varying conditions. There are other indicators, such as passenger comfort (as smoothness of jerks and accelerations) and energy consumption, which are increasingly receiving attention in the modern-day assessment systems. All these measures contribute to a complete assessment of agentic AV systems, and it becomes possible to maximize safety and user experience.

Table 21. Role of digital twins in autonomous vehicle validation

Function	Description	Benefit	Example Application
Real-time Monitoring	Mirrors real-world system	Immediate feedback	Fleet monitoring
Predictive Analysis	Forecast system behaviour	Risk prevention	Traffic congestion prediction
Scenario Simulation	Simulate "what-if" cases	Safe experimentation	Accident analysis
Performance Optimization	Compare real vs simulated	Model improvement	Algorithm tuning
Infrastructure Testing	Evaluate road systems	Smart city planning	Traffic signal optimization
Closed-loop Validation	Continuous feedback loop	Improved reliability	AV deployment testing

This table 21 provides the most significant functions of digital twins in self-driving cars and benefits. Predictions can be observed and analyzed in real-time with the help of digital twins in order to receive feedback and avoid risks on the spot and simulate scenarios to experiment with without risk to various cases. They are also useful in optimization of performance and testing of infrastructure to enhance the system and schedule the smart city. Overall, the principle of closed-loop validation will ensure constant feedback, which will enhance the quality of AV systems. To conclude, simulation and validation systems are of key significance when it comes to the development of agentic autonomous vehicle systems. Controlled simulation environments to digital twins that are reactive to scenarios and real-time, these methods guarantee safe, scalable and robust system testing. Systematic assessment and optimization of the integration is also possible with the use of the integrated performance measures prior to the real-world deployment.

In summary, simulation and validation frameworks play a critical role in the development of agentic autonomous vehicle systems. From controlled simulation environments to real-time digital twins and scenario-based testing, these approaches ensure safe, scalable, and robust system evaluation. The integration of comprehensive performance metrics further enables systematic assessment and optimization before real-world deployment.

8. MULTI-AGENT COLLABORATION IN TRAFFIC ECOSYSTEMS

This rapid growth of autonomous vehicle (AV) technologies in the recent past has led to the formation of intelligent transportation systems where autonomous agents can have communication in a shared environment. Unlike the classical, individualistic driving systems, the AV ecosystems today are highly rooted in the cooperation of a multitude of agents, vehicles, infrastructure, and network elements, which interact and coordinate their behavior on a real-time basis. This intersecting paradigm is required so that to develop safer, efficient, and scalable systems of transportation. Cars in such settings are no longer personal objects and they are now clever agents that are in continuous communication and will alter their behavior based on the systems dynamics.

The Vehicle-to-Vehicle (V2V) and Vehicle-to-Infrastructure (V2I) technologies enable the dissemination of information about the condition of traffic and road safety and the driving intent without any difficulty. This network system will enhance situational awareness, reduce uncertainty and enable pre-emptive decision-making. Moreover, multi-agent interaction provides new features such as cooperative driving whereby vehicles coordinate their positions together with the lane adjustment, merging, and platooning. These capabilities can be very useful in enhancing the level of traffic movement, reducing congestion, and enhancing road safety. However, the questions related to trust of heterogeneous agents, their reliability in communication and scalability will not be possible to make successful cooperation possible. This section addresses the fundamentals of multi-agent collaboration in traffic ecologies like communication paradigms, cooperative driving mechanisms and their impact on intelligent transportation systems.

8.1 Vehicle-to-Vehicle (V2V) Communication

A primitive form of multi agent cooperation in intelligent transportation systems is the Vehicle to Vehicle (V2V) communication, where autonomous vehicles can interchange information in real time. This message again enhances the situational awareness and the provision of information(Ray et al., 2010), which was not in the sensor field of the vehicle like the traffic and road conditions, motorist intentions This is because according to agentic AV systems, V2V communication results in decentralized coordination since each automobile is an intelligent agent but a constituent of a system. Such decentralized architecture improves the scalability and resilience since minimal centralized control is used (Li et al., 2022).

Such technologies as DSRC and 5G offer low-latency communication that is needed in a dynamic traffic situation when the decision-making should be time-bombed. Other cooperative controls that are facilitated by V2V communication are platooning, and collision avoidance. By sharing information about speed, position and what is intended to be carried out, vehicles will be able to organize their actions which will reduce traffic jams and improve safety. However, there are still problems in the trustworthiness of communication, interoperability, and security of heterogeneous systems of traffic.

Figure 14. V2V communication framework illustrating real-time information exchange between autonomous vehicles in a multi-agent environment

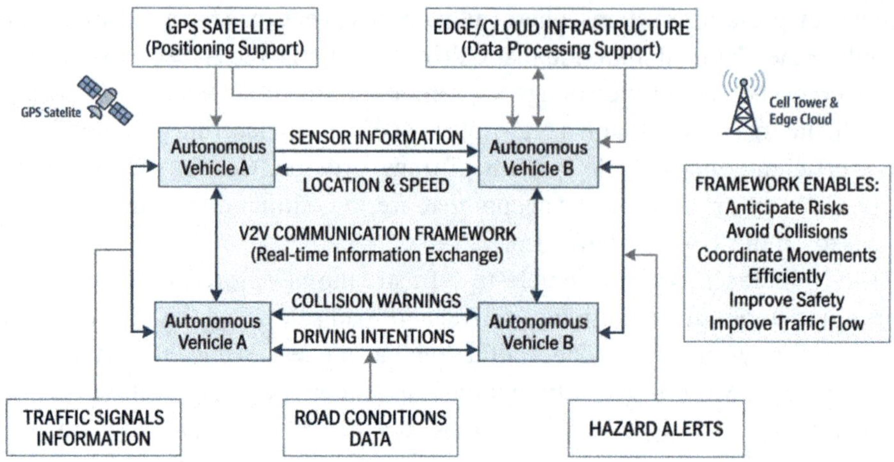

Figure 14 presents Vehicle-to-Vehicle (V2V) communication model which is a self-driving vehicle system in which real-time information sharing occurs among vehicles in a multi-agent system. A continuous exchange of critical information, including sensor data, location and speed, collision warning, and driving intentions, is made possible by the vehicles thus better coordination and safer decision-making. It is also integrated with external elements such as the GPS satellite and edge/cloud infrastructure that facilitate positioning and data processing as indicated in the figure. Furthermore, traffic lights, road conditions, and hazard warnings are integrated to add to the situational awareness. In general, this framework facilitates exchange of information in real-time, which will help vehicles predict risky situations, prevent collisions, and coordinate movement, thus enhancing the safety and traffic in intelligent transportation systems.

Table 22. Comparison of communication modes in intelligent traffic ecosystems

Type	Full Form	Information Shared	Participants	Benefits	Limitations
V2V	Vehicle-to-Vehicle	Speed, position, intent	Vehicles	Collision avoidance	Range limits
V2I	Vehicle-to-Infrastructure	Signals, traffic data	Vehicles + infrastructure	Traffic optimization	Infrastructure cost
V2P	Vehicle-to-Pedestrian	Alerts, safety signals	Vehicles + pedestrians	Pedestrian safety	Device dependency
V2N	Vehicle-to-Network	Cloud data	Vehicles + cloud	Global insights	Latency
V2X	Vehicle-to-Everything	All above	All entities	Full ecosystem	Complexity

Table 22 summarizes various forms of communication in intelligent transportation systems and their functions in facilitating multi-agent interaction. V2V and V2I permit real-time interaction between traffic and infrastructure, enhancing traffic and safety. V2P is philosophy to improve the safety of pedestrians whereas V2N links cars to cloud networks to gain greater understanding. V2X incorporates all types of communication and forms a complete ecosystem. Nonetheless, all modes have disadvantages including range limitation, latency, infrastructure expenses, complexity of the system.

8.2 Vehicle-to-Infrastructure (V2I)

Vehicle-to-Infrastructure (V2I) communication adds to the functionality of autonomous vehicles with the ability to communicate with smart infrastructure elements. These are roadside sensors, traffic signals and centralized traffic management systems and these offer global traffic data that supplements local sensor data (Dong et al., 2019).V2I communication enables autonomous vehicles to be informed about the traffic conditions, signal timings, and hazards on the road. Indicatively, the adaptive traffic lights have the capacity of communicating their schedules to the vehicles coming their way, thereby enabling them to reduce speed and reduce unnecessary stops.

The result of this is enhanced traffic efficiency, alleviation of congestion, and decreased emission. V2I communication, in agentic systems, creates a joint decision-making system between the infrastructure and the vehicles, which constitutes an integrated traffic ecosystem. Nevertheless, its mass adoption would demand the investment of a large-scale infrastructure, the standardization of communication protocols, and strong cybersecurity.

8.3 Cooperative Driving

Cooperative driving is a new concept in multi-agent systems since autonomous vehicles can agree on their actions to reach common goals such as safety and efficiency. This method uses communication and collaboration as an optimization of traffic circulation and minimizing accidents (Khan et al., 2021). The main benefit of cooperative driving is that coordinated actions may be performed, including synchronized lane changes and braking, and they decrease the risk of any collision.

Moreover, with cooperative driving, platooning is also allowed where the vehicles drive in a close group to cut down on aerodynamic drag and fuel efficiency. Technically, cooperative driving is based on multi-agent reinforcement learning, game theoretic models, and distributed optimization methods. These strategies allow agents to acquire the best strategies of interaction and coordination in a volatile environment. There are, however, issues like trust management, delay of communication and scalability of the system that need to be solved to deploy it on a large scale. Table 23 shows the Benefits and Challenges of Cooperative Driving.

Table 23. Benefits and challenges of cooperative driving

Aspect	Benefits	Challenges
Safety	Reduced accidents	Trust issues
Traffic Flow	Less congestion	Coordination complexity
Fuel Efficiency	Lower consumption	Platoon control
Communication	Real-time coordination	Latency issues
Scalability	Large system integration	Network overload
Trust Management	Reliable cooperation	Malicious agents

This section highlighted the importance of multi-agent collaboration in intelligent traffic ecosystems. Communication paradigms such as V2V and V2I, along with cooperative driving strategies, enable decentralized decision-making and improved traffic efficiency. However, challenges such as communication latency, trust management, and scalability remain key areas for future research.

9. SECURITY CONSIDERATIONS

With the growing integration of autonomous vehicle systems and dependence on artificial intelligence, it has become an essential issue to provide the safety thereof. The interaction of sensors, communication networks, and machine learning models creates highly susceptible AV systems to various cybersecurity threats, which may undermine the functionality and safety of the systems and confidence in the users. As opposed to conventional vehicles, autonomous systems are sensitive to cyber and physical attack because of the nature of their operation, in open and dynamic environments. The perception and decision-making pipeline is one of the most significant issues of securing AV systems as it might be compromised by adversarial inputs in the form of spoofed sensor data, spoofed traffic signs, or corrupted communication signals.

The result of such attacks may be wrong perceptions of the surroundings and dangerous decisions made when driving. Also, there are threats at the network level, such as data injection and denial of services attacks, which can cause issues between vehicles and infrastructure, further exposing the system to vulnerabilities. In order to deal with these issues, contemporary AV systems must have resistant agentic architecture, which comprises several tiers of defense, such as encryption, trust managing and intrusion detection systems.

Besides, the exponential development of adversarial machine learning emphasizes the necessity of strong AI models that can withstand manipulation and guarantee accurate work in unfavorable circumstances. This section offers an in-depth general description of the security issues of autonomous vehicle systems, the main threat vectors, defense, and architecture of developing resilient and trustful agentic AV ecosystems.

9.1 Cybersecurity Threats in AV Systems

The cybersecurity threat to autonomous vehicle systems is particularly high because they are

The evaluation of agentic autonomous vehicle systems should have a set of performance metrics outlining the safety, efficiency, and reliability. These metrics provide quantitative measures on the assessment of the working of the system in different situations and are needed to compare the models and methods (Kiran et al., 2021). The collision rate is one of the main safety measures applied to estimate the accident rate or the number of near misses during the test. The reason is that the rate of collision will be low and therefore, the perception and decision-making levels will be high. The number of seconds that the system requires to react to inputs and generate actions that are utilized to demonstrate responsiveness in high-stress environments is known as decision latency.

Path accuracy tests the movement of the vehicle within the context of an optimal path that provides details of the effectiveness of navigation and accuracy of control. The reliability score is an index of the capability of the system to produce performance under varying scenarios whereby; the strength of the system is indicated in varying conditions. There are other indicators, such as passenger comfort (as smoothness of jerks and accelerations) and energy consumption, which are increasingly receiving attention in the modern-day assessment systems. All these measures contribute to a complete assessment of agentic AV systems, and it becomes possible to maximize safety and user experience.

Table 24. Cybersecurity threats in autonomous vehicle systems

Threat	Description	Target	Consequence	Severity
Sensor Spoofing	Fake sensor inputs	Perception	Wrong decisions	High
GPS Spoofing	Fake location signals	Navigation	Route deviation	High
Adversarial Attack	AI manipulation	ML models	Misclassification	High
Replay Attack	Reused signals	Communication	System confusion	Medium
Data Injection	Fake data insertion	Network	Incorrect behaviour	High
DoS Attack	System overload	Network	Service disruption	High

Table 24 presents significant cybersecurity risks to autonomous vehicle systems, their targets, and possible effects. The threats of sensor spoofing, GPS spoofing are directly manipulative to perception and navigation, resulting in erroneous decision-making or diversion or change of route. Adversarial attacks are based on machine learning models and result in false classification of important inputs such as traffic lights. Data injections, DoS attacks and replay attacks are all network-based attacks that interrupt communications and system operations, and cause confusion as a result of reusing valid signals. Altogether, the table indicates that the majority of threats are of high severity, which means that AV systems require well-developed security mechanisms.

9.2 Secure Agentic Architectures

Secure agentic architectures are created to mitigate the intricate security issues of autonomous vehicle systems using multiple layers of protection. Such architectures focus on safe communication, management of trust, and system resiliency. Agents exchange data securely at the encryption and authentication protocols, and management of trust is organized through blockchain-coordination, which implies a decentralized and unchangeable system (Liu et al., 2025)

Multi-agent frameworks based on trust enable agents to determine the credibility of information provided by others and therefore minimize the effects of malicious agents. Secure agentic architecture can offer a holistic solution to cyber threats of autonomous vehicle systems by incorporating these mechanisms.

Figure 15. Secure agentic architecture illustrating layered security including encryption, blockchain, and trust evaluation mechanisms

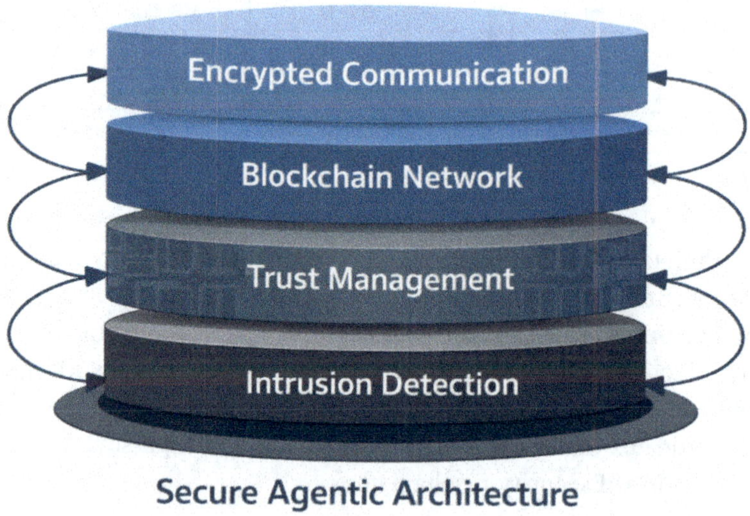

Figure 15 illustrates a layered secure agentic architecture designed to protect autonomous vehicle systems. At the top, encrypted communication ensures confidentiality and secure data exchange between agents. The blockchain network provides data integrity and tamper resistance through decentralized ledger mechanisms. The trust management layer evaluates the reliability of participating agents, enabling secure and trustworthy interactions. At the base, intrusion detection systems monitor the network to identify and mitigate cyber threats in real time. Together, these layers create a multi-level defense framework that enhances the overall security, resilience, and reliability of agentic autonomous systems.

9.4 Adversarial Machine Learning Risks

Adversarial Machine Learning Hazards Machine learning, and autonomous vehicle systems, in particular, is dangerous when adversarial. Attackers may introduce unsafe behavior, which may be used to manipulate the inputs to cause misclassification. These may be a distorted road signage that deceived the vision systems and reversed sensor data that upsets perception. These attacks are particularly dangerous as they may remain invisible and unnoticed (Sharma et al., 2017).

The countermeasures applied in such risks are adversarial training, anomaly detection, and model verification. The system requires a multi-layer defense solution that will offer resilience to adversarial threats to the system. Figure 16: Adversarial attack and defence model demonstrating the attack vectors and mitigation measures targeting attack vectors in the AV systems.

Figure 16. Adversarial attack and defence framework showing attack vectors and corresponding mitigation strategies in AV systems

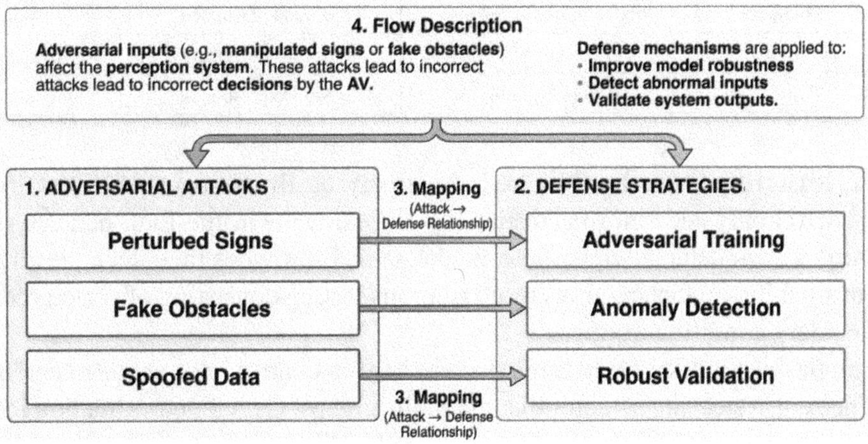

Figure 16 shows how adversarial attacks such as perturbed signs, fake obstacles, and spoofed data can mislead the perception system of autonomous vehicles and cause incorrect decisions. It also presents corresponding defence strategies, including adversarial training, anomaly detection, and robust validation, which help improve system robustness, detect abnormal inputs, and ensure safe outputs.

10. REGULATORY AND STANDARDS LANDSCAPE

The creation and implementation of autonomous vehicle technology are covered by the extensive regulatory frameworks and international standards that aim to keep safety, reliability and ethical standards. These regulatory mechanisms are a resultant collaboration among the government regulatory bodies, industry stakeholders and safety organizations across the world. This part explores the regulatory imperatives, locale rules, and moral aspects, which define the autonomous vehicle surroundings.

10.1 Safety Standards

The autonomous vehicle design and validation are based on international safety standards. The biggest framework relevant is ISO 26262:2018 that outlines functional safety principles to electrical and electronic systems in a road vehicle (International Organization for Standardization, 2018). This standard adopts the Automotive Safety Integrity Level (ASIL) classification system, which is classified into ASIL A (lowest level) to ASIL D (highest level), and each level has a more stringent verification and validation process. It presents a table that maps the four ASIL categories (ASIL A -D) (Debouk, 2019) to the dimensions (severity, controllability and detectability) in which each category is defined along with the rigor required in its validation shown in table 25.

Table 25. ASIL classification matrix

ASIL Level	Risk Severity	Controllability	Detectability	Validation Requirement
ASIL A	Low	Easily Controllable	Easily Detected	Standard Testing
ASIL B	Medium	Moderately Controllable	Moderately Detected	Enhanced Testing
ASIL C	High	Difficult to Control	Difficult to Detect	Advanced Validation
ASIL D	Critical	Uncontrollable	Not Detectable	Comprehensive Analysis

ASIL A is the least risky level: the risks are low severity, easily manageable by the driver, and easily identifiable. The driver only needs normal testing like the hardware in-the-loop-bench tests. An example of an ASIL A item is a windshield wiper motor fault. ASIL B provides the interim level of difficulty of severity and controllability, which requires more rigorous testing consisting of structured fault insertion and rigorous boundary condition testing.

ASIL C encompasses high-severe hazards that are hard to control once engaged and hard to identify prior to occurrence e.g. a steering actuator that gradually creeps. Developed validation including formal model checking and formal safety cases are needed. ASIL D is the most challenging: critical risks that cannot be controlled in any way upon occurrence and cannot be noticed in the regular operation. As a rule, the failure of break-systems and the failure of perception-module in fully automated vehicles also belong to the ASIL D. The in-depth analysis such as formal proof methods, fault-tree analysis, redundancy verification, et cetera is obligatory.

The practical cost signal to a self-driving car engineer in the table is that anything that lies in a higher ASIL level implies significantly more cost and engineering schedule will be required to develop and therefore proper initial hazard classification is a cost-engineering decision of commercial importance.

Figure 17. The bar chart shows the exponential increase in validation effort (%) required as ASIL level rises from A to D.

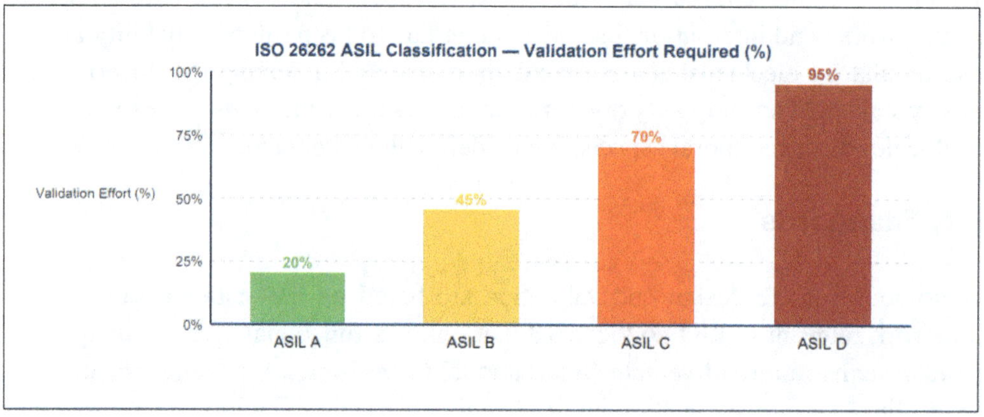

Figure 17 shows the exponential relationship between the level of ASIL classification, and the percentage of validation effort needed during the development process indicates that the shift towards the right of the bar implies a reduction in validation effort to about 20% of the development in the case of ASIL A level since hazards are low risks and are readily controllable and detectable. ASIL B increases to 45%, signifying moderately complex situations that demand improved testing regimes such as in-hardware-in-the-loop simulations. ASIL C also goes to 70% on average, which requires sophisticated validation methods to ensure that we perform formal model checking and structured fault-tree analysis. ASIL D at the highest level requires 95% validation effort, the almost 100 book owes to the fact that failures at this level are critical, cannot be treated, and can be detected only through thorough examination.

The abrupt increase in steepness highlights the non-linear cost of functional safety: every additional step in the ASIL level scale increases the engineering effort needed to actualize it by at least one-and-half and classifying hazards at an early inception stage is a commercially important line in design. In the case of autonomous vehicle systems, most of the perception and decision-making subsystems belong to the ASIL category of C or D, which has a direct impact on the development schedule and budgets.

In accordance with ISO 26262 is the ISO/PAS 21448:2022 (Safety of the Intended Functionality-SOTIF) that deals with hazards due to inadequate functioning and not due to systematic failures. SOTIF is concerned with those cases in which the system is functioning as intended but poorly behaves in a real world situation, like a perception failure in response to weather or sensor fusion edge cases. It is especially important to this framework when autonomous systems rely on machine learning and have certain inherent constraints in the training data (International Organization for Standardization, 2022). The SAE standard J3016 has six different levels of driving automation (Levels 0-5) which provides a standard vocabulary used by the industry players. The levels are 0-2: human supervision of driver-assist systems, Level 3-4: conditional automation and high automation with different levels of monitoring, Level 5 full automation on all roads (SAE International, 2021).

10.2 Global Regulations

Regulatory strategies on autonomous vehicles differ significantly by jurisdiction as they are based on legal traditions, risk appetite and priorities in development. These regulatory mechanisms can be divided into three main strategies, which include prescriptive standards, performance-based regulations, and innovation-enabling frameworks.

Table 26. Global regulatory approaches

Region/ Jurisdiction	Regulatory Approach	Governing Authorities	Implementation Status
United States (Federal)	Performance-based	NHTSA, DOT	Guidance issued (non-binding)
United States (California)	Innovation-enabling	California DMV, PUC	Active testing permits
European Union	Prescriptive Standards	EMA, ETSC	UN/ECE regulations implementation
United Kingdom	Risk-based	SMMT, CAM	Testing in controlled environments
China	Tiered approach	CAAM, Local governments	Dual track: pilot zones + standards
Singapore	Agile regulation	LTA, IMDA	Regulatory sandbox approach
Japan	Technology-friendly	MLIT	Integrated safety framework

Table 26 shows that organizing seven large jurisdictions into four dimensions of regulatory philosophy, responsible approval body and current deployment status raises awareness of the impending diversity of approaches to governance, whereas the columnar analysis demonstrates structural trends in each of these four dimensions.

The two-tiered system in the United States is that the federal tier (NHTSA/DOT) presents a model of non-binding guidelines based on performance that maintain the flexibility of manufacturers but establish uneven baseline safety standards; California has an innovation-facilitating system that has turned the state into the first real-world AV testing environment in the world. Such a federal-state tension is a key peculiarity of the U.S. AV governance that is not duplicated in other countries.

The most legal binding and internationally coordinated model is the prescriptive standards approach of the EU, which is affected by EMA and ETSC with the harmonization provided by the UN/ECE. It is stronger in terms of predictability, much needed by manufacturers who want to gain access to the EU market; the weak aspect is its lower speed of iteration relative to sandbox models. The Chinese style of doing things is unusual: local governments of several pilot cities (Beijing, Shanghai, Shenzhen), become de facto regulators, constructing a portfolio of various experiments in the real world, after which the national standards are synthesized.

The Singapore regulatory sandbox is agile and Japan is technology friendly; both are indicators of the competitive pressure that smaller economies must generate to gain AV investment by differentiating their regulations. All these indications contained in the table point to the fact that there is no worldwide regulatory agreement, and producers will have to keep compliance programs that are specific to various jurisdictions.

Figure 18. Global regulatory development timeline for autonomous vehicles

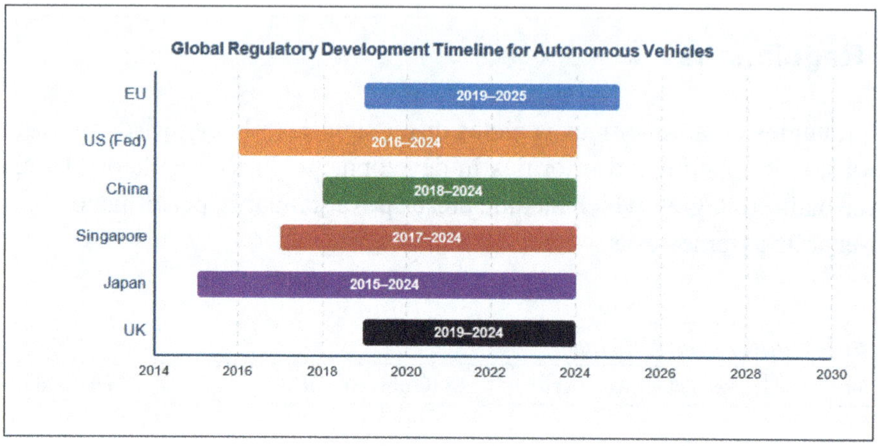

Figure 18 shows the window of regulatory development of six key jurisdictions between 2014 and 2030. Every horizontal line illustrates an active time of regulatory activity where a region has formal standards, pilot programs or binding laws that are in force as to autonomous vehicles. The integration of Regulation (EU) 2019/2144 and the rules on the automated driving system that follow on it indicate the adoption of the most prescriptive and time-constrained regulatory program, which is reflected through the EU (2019-2025). The bar of Japan is the first to start (2015), in line with the proactive contact of the government of Japan, its Society 5.0 strategy that made the development of AV a part of a national plan. The formal window in China (2018-2024) is fairly limited since the Dual-track model used focuses much on pilot zones at the city level, instead of national schedules.

The US Federal bar (2016-2024) represents the inception of initial voluntary guidance documents of the NHTSA up to its continuous evolution, as the federal regulation remained non-binding. The smaller bar is a deception: Singapore is a country with a population of just 5 million, but the interaction allowed by the limited number of test geographic indicators results in quick testing iterations. This is the case with the UK (2019-2024), which has been exercising its own regulatory freedom since the post-Brexit period, and hastening its own regulation of AV without following the harmonization process of the EU. Combined such that there is no uniform regulatory calendar across two jurisdictions, globally operating manufacturers have the complexity of compliance.

European Union: EU regulation framework gives impetus to harmonization that is based on UN/ECE regulations (United Nations Economic Commission for Europe, 2021). EU regulation 2019/2144 on vehicle data recording and 2023/643 on automated driving systems issue specific technical standards of Level 3 automation (European Parliament & Council of the European Union, 2019). To achieve compliance, manufacturers must comply by showing type-approval procedures and post-market surveillance. One of them is the ADS Event Data Recorder (EDR), which requires that the operational parameters be recorded continually so that it is easier to investigate the accidents.

United States: The American regulatory environment is still divided between federal and state governments. In 2017, NHTSA published non-binding Voluntary Guidance to Autonomous Vehicle Testing and the best practices of transparency and safety (National Highway Traffic Safety Administration, 2017). Individual states have had their own approval processes by the California DMV autonomous vehicle testing program (California Department of Motor Vehicles, 2026).

China: China uses a two-way regulatory system involving both pilot areas and emerging country guidelines. Cities like Beijing, Shanghai, and Shenzhen have created regulated test zones where they have regulatory exemptions. China has issued safety guidelines as laid down by the China Association of Automobile Manufacturers (CAAM) and the Ministry of Industry and Information Technology (MIIT) plays a coordinating role in the development of the standards together with the provincial governments.

Singapore: Singapore has pursued the idea of a regulatory sandbox where it allows limited experimentation in limited situations. The Land Transport Authority (LTA) is also in charge of testing specific spaces and one such area is the one-north precinct and Changi Business Park, which allows fast development and still oversees safety protocols (Land Transport Authority, 2026).

Japan: Japan takes the technology-friendly approach when it comes to developing and deploying autonomous vehicles, fostering innovation while upholding the strict standards on vehicle safety. The Ministry of Land, Infrastructure, Transport and Tourism (MLIT) has a central role to play in regulating and guiding AV policies. It has put in place an integrated safety framework consisting of a combination of vehicle certification, infrastructure readiness and real-world testing. This approach helps to ensure that emerging technologies are introduced in a controlled way, but in a gradual manner. As a result, Japan has entrenched itself as a balanced leader regarding innovation and regulatory oversight in the autonomous driving sector (Ministry of Economy, Trade and Industry, 2023).

10.3 Ethical and Legal Considerations

On top of technical systems of safety, autonomous cars pose immense ethical and legal issues that go further into the concepts of responsibility, transparency, and justice. It is based on these considerations that the policies will be determined, and the framework of liabilities will be created which is going to influence the development of the industry.

Table 27. Ethical and legal challenges in autonomous vehicle deployment

Ethical Principle	Key Challenge	Regulatory / Mitigation Measures
Accountability	Determining liability in system-failure accidents	Manufacturer indemnification requirements and clear liability frameworks
Transparency	Black-box decision-making in machine learning models	Interpretability and explainability mandates
Fairness	Biased perception affecting different demographic groups	Dataset diversity requirements and bias evaluation standards
Privacy	Continuous data collection and geolocation tracking	Data minimization principles and strict access control mechanisms
Cybersecurity	Vulnerability to hacking and remote attacks	Secure software update protocols and cybersecurity standards

Five categories of ethical challenge, namely Accountability, Transparency, Fairness, Privacy, and Cybersecurity, are compared in table 27 with their respective challenge description related to them and the subsequent regulation actions in underscored areas. The most disputed legally is accountability: the most common tort systems presuppose a human due to the fault who makes the choice to cause the crash, and, when an algorithm of an AV makes the decision and causes the crash, there is no obvious person to hold responsible according to the traditional tort law. The regulatory reaction, i.e., requirement of indemnification of the manufacturer, moves the liability further up the channel, to the software producer, and legal accountability is matched to the technical one. The example of this change can be seen in Germany through its rebuttable presumption law in 2022. Transparency manages the black-box issue in deep learning. Neural networks involved in driving decision-making have difficulty in explaining why they made the decision, which is an ultimate challenge to reviewing the accident judicially.

The explainability requirements specify that AV systems must give out reconstructed logs of perception-prediction-planning pipelines. The documentation requirements of the EU AI Act (European Commission, 2025) provide the most sophisticated implementation of this regulatory reaction. Equity also brings forth a less apparent but socially important issue: whether a computer vision is more likely to recognize pedestrians with a lighter skin tone than with one with a darker skin tone, the AV will end up posing more collision risks to a specific demographic variation. This gap is already directly covered by diverse requirements of validation datasets. Privacy and Cybersecurity are becoming more connected: the qualities of cars that constantly glean geolocation, biometric and environmental data are a lucrative object of corporate data transference, as well as it is a lucrative (though unlawful) target in adversarial hacking. The current regulatory frontier of both problems is the data minimization principles and the over-the-air update security standards.

The liability question remains unresolved in many jurisdictions. Traditional automotive liability assigns responsibility to manufacturers for defects and drivers for negligent operations. In autonomous systems, this distinction blurs significantly. The German legal reform of 2022 introduced a rebuttable presumption of manufacturer liability in accidents involving highly automated vehicles. The European Union's Product Liability Directive (2022/2557) extends liability to producers of AI systems, establishing strict liability for damages caused by defective products tones (European Parliament & Council of the European Union, 2022). Transparency and explainability reflect growing recognition that autonomous systems must be comprehensible to regulators, courts, and the public. The EU's proposed AI Act mandates that high-risk AI systems provide documentation of their decision-making processes. For autonomous vehicles, this translates to requirements for generating constructible logs of perception, prediction, and planning decisions auditable following accidents. Fairness in autonomous systems raises questions about whether algorithmic decisions disproportionately affect vulnerable populations. Research has documented bias in computer vision systems when recognizing pedestrians with different skin. Regulatory frameworks increasingly require validation datasets with representative coverage of diverse demographics and environmental conditions. The SAE J3180 standard on testing and validation provides guidance on ensuring fairness across demographic groups (Wilson et al., 2019).

11. FUTURE RESEARCH DIRECTIONS

Despite significant advances in autonomous vehicle technology, critical research challenges remain that will determine the trajectory of the field over the next decade. This section explores emerging research frontiers that promise to enhance system capabilities, efficiency, and integration with broader transportation and urban ecosystems.

11.1 Self-learning Autonomous Systems

The existing autonomous vehicle systems are based on the approach of using fixed machine learning models, which have been trained on pre-programmed datasets before deployment. This limitation is set to be overcome with self-learning systems which could be continually adjusted online. Incremental learning (also called continual learning) allows models to cope with streaming data and retain previously learned information, overcoming the disastrous forgetting effect of neural networks (Shaheen et al., 2022).

Federated learning methods enable self-driving cars to enhance their collaborative models without any centralization of sensitive information. The research priorities in this field are:

- The optimization of the model compression strategy to allow on-vehicle learning with limited computation.
- Strong uncertainty measurement to determine when models are being used out-of-distribution.
- Privacy-sensitive aggregation algorithms that are resistant to inference attacks that recreate training data.
- Official assurances that the safety-appropriate functionalities are not degenerated by continuous learning.

Table 28. Comparison of continual learning approaches based on core mechanisms, advantages, and limitations

Approach	Core Idea	Advantages	Challenges
Replay Buffers	Store exemplary from previous tasks for rehearsal	Straightforward implementation	Memory overhead and potential privacy concerns
Regularization	Penalize weight divergence from previous tasks	Computationally efficient	Limited capacity to retain diverse tasks
Dynamic Architectures	Expand network capacity to accommodate new tasks	Scalable learning	Increased training complexity and latency
Metacognitive Methods	Learn how to learn from prior experience	Rapid adaptation to new tasks	Requires substantial training data and computational resources

Table 28 contrasts four paradigms of continual learning on three analytical dimensions: mechanism (how learning is applied), advantages (what each approach is good at), and challenges (what prevents its widespread application in AV systems).

Replay Buffers in retraining save a selected set of past training examples with fresh data. The concept behind the mechanism is very straightforward, retaining what has been learned and implementation is easy. The most important issue here is the memory overhead: the storage of representative examples of an.

The increasingly large driving scenario space ultimately imposes prohibition on embedded vehicle hardware. Privacy will also be an issue in case the replayed data has identifiable pedestrian or passenger information. Regularization techniques introduce penalty terms to the loss function that dissuades significant change of weights which are significant to past tasks (e.g., Elastic Weight Consolidation). This is computing efficient, that is, no extra data storage is required, but then again the methodology is restricted in its capability of very diverse tasks. In the case that new driving situations are highly dissimilar to the training distribution, regularization is not sufficient to avoid the performance loss on previously mastered situations.

Dynamic Architectures are physical extensions of the network as a response to new functions growing beyond current capacity by adding neurons or layers to suit new conditions. This allows theoretically unlimited scaling but makes training more complicated: it is not an easy engineering problem to decide when to scale, to what degree and how to deal with the inference latency on limited hardware. Metacognitive approaches or also referred to as learning to learn or meta-learning is a model training approach that learns the distribution of tasks so that the model can quickly learn new situations with a small amount of data. The method has the quickest adaptation rate at the cost of the highest initial training cost and is the hardest to demonstrate to be safe against critical application.

Figure 19. Grouped bar chart comparing four continual learning architectures across Memory Efficiency, Adaptability, and Implementation Ease (scale 0–100)

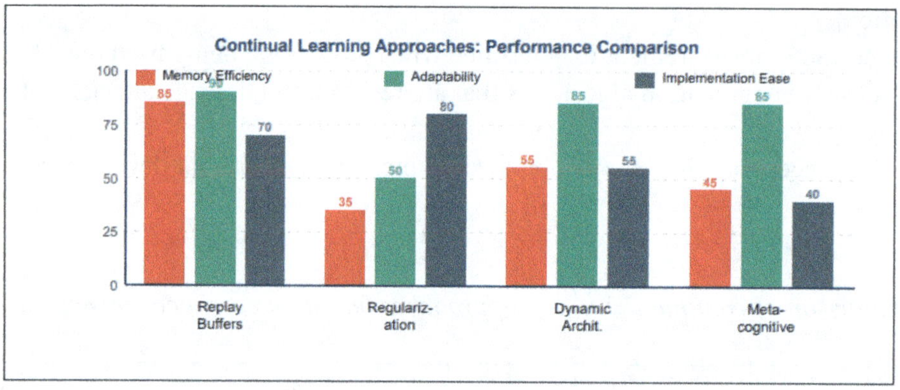

Figure 19 illustrates the comparisons of four continuous learning designs namely Replay Buffers, Regularization, Dynamic Architectures, and Metacognitive systems on three performance scales namely: Memory Efficiency (red), Adaptability (teal), and Implementation Ease (grey).

Replay Buffers rank the highest in Memory Efficiency (85%) and Adaptability (90%) as the process of storing representative examples is a direct way to have simple catastrophic forgetting prevention. Their

Implementation Ease score of 70%, however, indicates the engineering complexity of having exemplar buffers of memory-constrained vehicular hardware. This profile is inverted by regularization methods: low memory efficiency (35%) and low adaptability (50%) but high implementation ease (80%) since they are purely algorithmic, and they do not require any architectural adaptation.

By providing adaptability balanced (85%) since the network can be expanded to handle completely new situations, the implementation ease is reduced to 55% because the growing model complexity requires real-time inference environments. The most polarized profile is the metacognitive methods, which have excellent adaptability (85%) due to the capacity to train learning strategies, but the least ease of implementation (40%) and memory efficiency (45%), as the methods rely on large meta-training datasets and complicated inner-outer loop training algorithms. Regularization approaches are the most feasible near-term solution to AV deployment, whereas the maximum capability ceiling of Metacognitive approaches is the longest-term deployment.

11.2 Swarm Intelligence in Transportation

Swarm intelligence approaches, inspired by biological systems such as ant colonies and bird flocks, offer novel solutions for coordinating large-scale autonomous vehicle networks. Multi-agent reinforcement learning (MARL) enables vehicles to jointly optimize trajectory planning while considering the actions of other agents. Research challenges include non-stationary environments, credit assignment in multi-agent systems, communication bandwidth limitations, and safety guarantees for jointly learned policies (Ji et al., 2024).

Table 29. Projected traffic system improvements with AV penetration

Performance Metric	Description	Relative Importance (%)	Expected Improvement
Average Travel Time	Mean time required to complete a journey	10%	5–15% reduction
Traffic Throughput	Number of vehicles passing through a corridor per hour	25%	20–35% increase
Fuel Consumption	Total fuel consumed across the transportation network	50%	15–25% reduction
Emissions (CO_2)	Total greenhouse gas emissions generated	50%	20–30% reduction
Safety Incidents	Number of accidents per million miles traveled	100%	90%+ reduction

Table 29 shows five important metrics of traffic performance, and each can be defined in a definite way, with the centralized AV fleet percentage, where the maximum projected benefit can be realized, and the range of expected improvements, which is expected by research literature.

The greatest effect on travel time to the table (5 to 15% reduction) is the Average Travel Time benefits since the capacity of the total network is of great importance, and automation cannot significantly boost the capacity of the total network. The 10% baseline value means that with a small fleet of AVs, congestion can be decreased by providing less bumpy acceleration patterns and predictive paths.

Traffic Throughput exhibits greater increases (20-35% growth at 25% penetration) since coordinated platooning, which involves AVs following each other at a specific distance, is an effective way to expand the capacity of roads without physically widening them. It is among the most economically important advantages of AV coordination on the fleet level.

The CO2 Emissions and Fuel Consumption have 15-30% projections of reductions at half penetration. The dependence between the two measures is almost direct as the major cause of both measures is the removal of stop and go driving habits. AVs are predictive of alterations in the traffic flow and hold steady speeds, which significantly saves on energy wastage of long acceleration-braking cycles.

The highest number in the table is Safety Incidents: 90% + decrease at full (100) AV penetration. This can be assessed by the fact that about 94% of severe accidents can be linked to the human factor (NHTSA, 2019). Eliminating human judgments in the driving activity, if AV systems are reliable, should thus do away with many crash causative factors. The caveat in the footnote regarding non-homogenous mixed networks is noteworthy: the benefits are non-linear and decrease much when AVs must interact with unpredictable human drivers, who are not cooperative to protocols.

Figure 20. Traffic System Performance with Increasing AV Penetration. Performance index (baseline = 100) across five metrics at 0%, 10%, 25%, and 50% AV penetration. Values below 100 indicate improvement over baseline

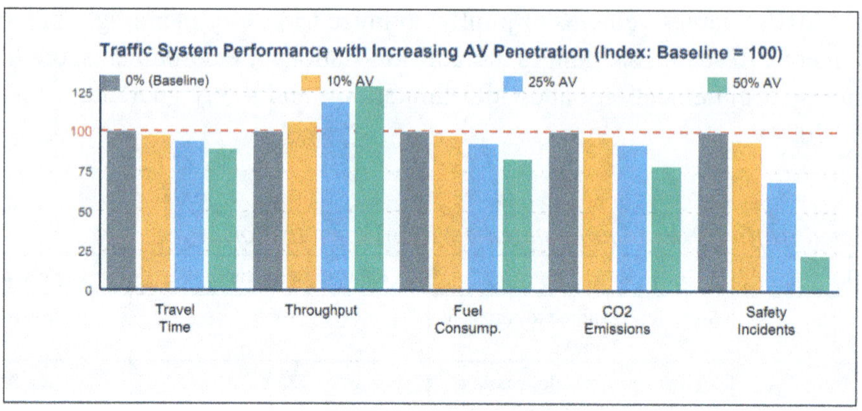

Figure 20 shows five performance indicators (Travel Time, Throughput, Fuel Consumption, CO2 Emissions and Safety Incidents) and four AV fleet penetration models, 0% (baseline), 10, 25 and 50. The performance index is brought to a baseline of 100 where a bar that is less than 100 is an improvement.

Travel Time and throughput show small but steady increases at 10% penetration and throughput shows significant gains of about 120 indexes at 25% penetration with coordinated platooning and intersection management achieving systemic gains. Fuel Consumption and CO2 Emissions demonstrate comparably shallow increases at low penetration but increase more steeply at 50% which relates to non-linear network effects of coordinated acceleration and deceleration profiles, which minimizes stop-and-go dynamics.

The greatest change is in Safety Incidents: even when the penetration is 10% the index has reduced to about 93, and at 50% the index has decreased to less than 25 - more than a 75 percent decrease over the all-human baseline. This unproportional safety benefit indicates that AV decision-making gets rid of the most prevalent causes of crashes (distraction, impairment, delayed reaction) even when AVs constitute only a portion of the traffic. The trend of the chart is a strong indication that policy incentives should be used to encourage early adoption since the benefits of safety can be realized long before full penetration of the fleet is realized.

11.3 Integration with Smart Cities

The future of autonomous vehicles is inextricably linked to the development of smart city infrastructure. Vehicle-to-everything (V2X) communication-encompassing vehicle-to-infrastructure (V2I), vehicle-to-vehicle (V2V), and vehicle-to-pedestrian (V2P) - enables coordination at scales currently impossible with local sensing alone. Research priorities include developing 5G/6G communication standards with sub-100ms latency, cybersecurity frameworks protecting V2X communications, integration protocols enabling interoperability, and privacy mechanisms (Yang et al., 2022).

Figure 21. Key milestones in the autonomous vehicle industry from 2016 to 2024, highlighting technological advancements, regulatory developments, and major industry events

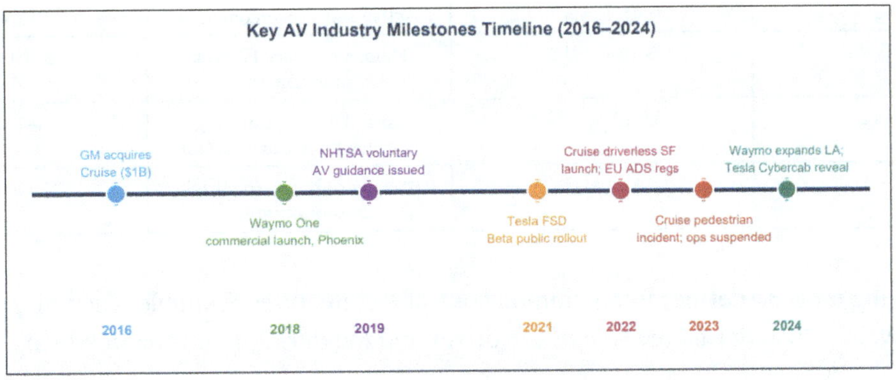

Figure 21 introduces a timeline of important developments in the Autonomous Vehicle (AV) industry from 2016 until 2024. It outlines important technological, regulatory, and commercial milestones that have affected the development and implementation of AV systems.

The timeline starts in 2016 with General Motors acquiring Cruise, a big investment in autonomous driving technology. In 2018, Waymo reached a crucial milestone when it commercially deployed its new autonomous ride-hailing services in Phoenix and proved early deployment in the real world. Regulatory progress follows in the guidelines on voluntary AV eliminated by NHTSA in 2019 with the initial engagement of the government into AV governance. In 2021, Tesla made its Full Self-Driving (FSD) capabilities available to the public in a rollout of a beta testing program in response to the growing accessibility of semi-autonomous features. The year 2022 is all about progress and regulation; we had Cruise's driverless in San Francisco and the implementation of the European Union Autonomous Driving System (ADS) regulations. But the year 2023 highlights issues around safety and we witnessed such an instance with a pedestrian related incident in Cruise that had to suspend operations. Finally, in 2024, the sustained industry growth is partnered with Waymo's place in Los Angeles, as well as examining TensorFlow's Cybercab, indicating the further innovation and conquest efforts of the ecosystem. Overall, the figure shows that AV development is marked by a dynamic interplay between technology development, regulatory and real-world safety issues.

11.4 Human-AI Collaboration

The immediate future of autonomous vehicles is not full automation but rather enhanced driver assistance systems with varying levels of automation. Shared control systems that enable seamless transitions between autonomous and human control represent a critical research frontier. Human factors research demonstrates that drivers remain engaged and mentally prepared for takeover only when handoff procedures are carefully designed (Eriksson & Stanton, 2017).

Table 30. Human–vehicle interaction modes across SAE automation levels and corresponding response time requirements

Interaction Mode	SAE Automation Level	Human Role	Typical Response Time
Passive Monitoring	SAE Level 2	Active supervision required	0–3 seconds
Active Correction	SAE Level 3	Human provides fallback intervention	5–10 seconds
Dynamic Negotiation	SAE Level 3+	Shared decision-making between human and system	3–7 seconds
System Initiation	SAE Level 4	Passenger-only role (no driving responsibility)	Not applicable

The following table 30 defines four common control architectures available when mapped onto SAE automation levels, which define the role of the driver and the duration of time in which the human operator can take over in each of the configurations.

Passive Monitoring (SAE Level 2) involves the driver to be engaged actively at any given moment he or she must be hands on the wheel and eyes on the road, and the automation performs longitudinal and lateral control in design jurisdiction. The handoff time ranges between 0 and 3 seconds since the driver is already within the control loop and can potentially intervene within almost a second. This is how Tesla Autopilot works, and the inability of the architecture is complacency of the operators: the drivers lose their minds even when they need to be alert.

Active Correction (SAE Level 3) provides a fallback responsibility role for the driver: he/she can shift attention to other activities but should be capable of responding to a system takeover request within 510 seconds. This extended window presents a challenging transition design issue of 10 seconds is sufficient time to move a vehicle at highway speed more than 100 meters; therefore, the dependability of the handoff alert mechanism is a safety concern.

Dynamic Negotiation (SAE Level 3+) is a new architecture in which the car and the driver share control, either can give control or take control. The 37 second handoff window indicates a person to person transfer instead of an abrupt assume authority change.

System Initiation (SAE Level 4) substitutes the driver with the entire control loop in the field of the operational design: the driver is a passenger, and there is no handoff time available since the system processes all the situations independently. Waymo One has a similar mode of operation in it geofenced Phoenix and San Francisco service regions.

12. CASE STUDIES

Examining concrete implementations of autonomous vehicle technology reveals both the sophistication achieved and the persistent challenges facing the field. This section provides an in-depth analysis of four leading autonomous vehicle programs Waymo, Tesla Autopilot, Cruise, and Baidu Apollo that collectively represent the full spectrum of architectural approaches, business models, geographic deployment contexts, and development philosophies currently active in the global AV industry. Each case study covers historical genesis, technical architecture, simulation and data strategy, safety record, commercial performance, regulatory standing, and lessons for the broader industry.

The selection of these four cases is deliberate. Waymo represents the longest-running lidar-centric commercial deployment with the strongest regulatory record. Tesla represents the largest fleet-scale vision-only system and the most commercially significant AV program by revenue. Cruise represents the consequences of premature commercial deployment without adequate scenario coverage, offering uniquely instructive lessons in safety governance. Baidu Apollo provides the essential non-Western perspective: the world's largest AV program by geographic deployment breadth, operating under a fundamentally different regulatory and competitive environment.

12.1 Waymo Autonomous Driving System

12.1.1 Organizational Background and Historical Development

Waymo, an Alphabet subsidiary founded in 2009 as the Google Self-Driving Car Project, represents the most mature autonomous vehicle system currently deployed at commercial scale. The project originated from the DARPA Urban Challenge (2007), where Stanford and CMU teams demonstrated that autonomous urban driving was feasible (Levinson et al., 2011). Google's subsequent investment reflected a conviction that AV technology would reshape mobility on a decade-long horizon.

The program was incorporated as a standalone subsidiary Waymo LLC in December 2016, enabling independent funding rounds and commercial partnerships. Between 2018 and 2019, Waymo launched Waymo One, the world's first commercial robotaxi service, in the Phoenix metropolitan area. As of 2024, Waymo operates in Phoenix, San Francisco, and Los Angeles, having accumulated over 20 million miles of fully autonomous driving experience, the largest such dataset globally. The company raised over $5.5 billion in external investment between 2020 and 2023 from investors including Silver Lake, Andreessen Horowitz, and the Abu Dhabi sovereign wealth fund, underscoring confidence in the platform's long-term commercial potential.

12.1.2 Technical Architecture

The Waymo Driver architecture prioritizes lidar-based perception combined with camera vision and radar (Waymo, 2021). This multi-modal fusion approach achieves high redundancy; the system can maintain safe operation even with single sensor failures. The core design philosophy is hardware over software: Waymo invests heavily in custom sensor development to reduce reliance on algorithmic compensation for sensor limitations. Waymo's fifth-generation sensor suite, unveiled in 2020, achieved a tenfold reduction in lidar unit cost compared to the original 2009 hardware while improving range and resolution; a critical milestone toward commercial scalability.

The perception stack processes lidar point clouds through multi-stage convolutional neural networks that classify objects, estimate velocities, and predict trajectories for all dynamic agents simultaneously. Camera inputs feed parallel semantic segmentation pipelines that identify road markings, traffic signals, and signage invisible to lidar. Radar provides velocity estimates at ranges exceeding lidar capability, particularly critical for high-speed highway merge scenarios. The fusion engine combines all three modalities using a learned weighting scheme that dynamically prioritizes the most reliable sensor given current conditions (e.g., elevating radar weight in heavy rain) .

Table 31. Core hardware components of an autonomous vehicle system and their functional roles

Component	Technical Specifications	Functional Role
LiDAR	Multiple rotating units with ~200 m range	Primary 3D environment perception and mapping
Cameras	Visible and infrared sensors with 360° coverage	Semantic segmentation, object detection, and traffic sign recognition
Radar	Millimeter-wave sensor arrays	Long-range obstacle detection and velocity estimation
GPS/IMU	RTK-capable GPS integrated with inertial measurement unit	Accurate vehicle localization and trajectory estimation
Compute	Multi-core processors (8+ CPU cores) with dedicated ML accelerators	Real-time perception, inference, and decision-making
Storage	Redundant solid-state drives (4+ units)	Continuous data recording for safety analysis and system validation

Table 31 lists the six major hardware components of Waymo Driver system where each has been given the technical make-up of each one and the functional contribution of that specific component in the entire perception and planning framework.

Its main sensor related to perception is lidar, which generates a representation of the surrounding environment, in 3D and 360AS, to a total distance of 200 meters. Lidar, unlike cameras, cannot be affected by lighting conditions, whereby it creates its own infrared pulses, which makes it stable in the dark and low-contrast scenes. There are several rotating lidar units which overlap each other to ensure the elimination of blind spots caused by occlusion. The main factor that makes Waymo approach inapplicable to consumer car implementation is that these units are very expensive (historically costing at least 75,000 apiece) and cannot be scaled to lower-priced cars without additional cost-cutting.

Cameras are essential to lidar, that is, they help with semantic information: text recognition (signs, signals), color distinction (traffic lights) and texture classification, which cannot be viewed through the point clouds. Infrared cameras are added to allow the running of reliable operations during nighttime and in areas with low visibility. Radar includes long-range velocity estimation of high-speed objects - this is particularly important in highway merging cases where the lidar does not have adequate resolution over distance.

12.1.3 HD Mapping and Localization Strategy

Waymo's operational model depends on pre-built high-definition (HD) maps that encode the static environment at centimeter resolution. Map construction involves multiple dedicated mapping vehicles. The HD map dependency is simultaneously a strength and a constraint. It dramatically simplifies the real-time perception problem; the system knows what the static world looks like and can focus computational resources on detecting and responding to dynamic agents. However, it limits deployment to geofenced areas where mapping has been completed, and requires continuous map maintenance as road geometry, construction zones, and signage change. Waymo employs a continuous map update pipeline that discrepancies between expected and observed static environment, triggering re-mapping of affected segments.

12.1.4 Simulation Infrastructure: Carcraft

The Carcraft simulation platform is a somewhat overlooked, but essential aspect of the development plan of Waymo. It creates actual driving environments in increased time speeds thus simulating millions of miles in a day. The edge-case scenarios are based on the real-world events including the near-misses, strong motivating behaviors of pedestrians, or sensor blockage events, and they are explicitly injected into the simulation to evaluate how the system responds to the rare events that could not be practiced using physical experiments only.

Car craft uses a physics engine that accurately simulates sensor behavior (such as the divergence of lidar beams, rates at camera noise effects and even multisite camera ray multiplication). This allowed a realistic simulation of sensor effects in bad weather. A designed system of adversarial testing (also called, sometimes, fuzzing) identifies failure modes by systematically constructing worst-case scenario configurations by perturbing agent behavior, weather condition and road geometries before they become apparent in the real world. Waymo states that Carcraft runs more than 15 million simulated miles in a day, which means that it is one of the largest simulation programs in the technology industry.

12.1.5 Safety Record, Incidents, and Reporting Transparency

The safety reporting of Waymo has been adopted across most quarters as the most open in the autonomous vehicle industry. Switching to a human operator - The annual safety reports list the disengagement rates, i.e. when the vehicle lost control of a person operator - as a proxy measure of system confidence. California Department of Motor Vehicles (DMV) records reveal that Waymo has fewer disengagements per mile than any other AV operator in the state and the public reports show that it has only about one disengagement per 30,000 fully autonomous miles of operation, which is considerably lower than industry rates of one per 3000-5000 miles (Waymo, 2021).

Even with this good record, there have been incidents. A Waymo robotaxi, which serves in San Francisco in early 2024, was also involved in the collision with a cyclist in a construction zone, which was relatively minor resulting in minor injuries. The company immediately reported the incident to the California DMV and shut down the affected region, performed a root-cause investigation in 72 hours, and implemented a software fix covering the discovered perception fallacy issue. Attention was given by regulators to Waymo as one of the first companies that promptly reviewed and corrected disclosures, which is the opposite of the delayed and incomplete reporting that marked the Cruise incident in 2023.

12.1.5 Commercial Strategy and Financial Performance

The two sets of strategies influence financial performance and the approach used to attain it. The two sets of strategies determine the financial performance and how it was achieved.

Waymo has a dual-stream revenue approach whereby it offers platform licensing alongside a direct robotaxi service, hence seeking complementary sources of revenue. The consumer robotaxi division (Waymo One) pays on a ride-by-ride basis corresponding to the services those of Uber and Lyft provide, with an added benefit of zero direct driver labor expenses, offering an advantage of structural margins that increases directly with fleet size. The Waymo Via commercial logistics program aims to aim at freight and last-mile delivery working with companies like UPS and J.B. Hunt on autonomous trucking pilots.

Waymo is not a profitable company financially since it is a long-term investment in the Alphabet portfolio. A 2023 annual report of Alphabet makes Waymo a Best Other category with cumulative losses of over US$5 Trillion since its inception. Nevertheless, the recent external financing by the company, the latest US 5 billion round led by late 2023, at valuations of between US 30 and US 45-billion, signals to investors that the robotaxi total addressable market is worth a long-term pre-profitable investment. Breakeven horizon is estimated to be achieved at the level of fleet size of 10000 or more, which is significantly larger than the current presence of about 700 vehicles.

12.2 Tesla Autopilot

12.2.1 Organizational Background and Design Philosophy

The autonomous driving concept developed by Tesla is an intentional philosophical alternative to lidar-based systems. CEO Elon Musk has described lidar to the press as expensive, cumbersome and unnecessary technology when powerful enough computer vision is available. This belief guided all the corporate approach to autonomous driving: release vision based aid frameworks on the mass-market scale, collect a great deal of unheard of genuine world information using customer vehicles, and educate progressively more complex neural networks on that information. The plan is since human drivers drive through traffic based on their eyes and as long as a neural network can simulate their ability to process their visual input, they will not consider the need to have more sensors.

This does have great commercial implications in this philosophy. By removing the lidar hardware, Tesla provides autonomous possibilities at consumer car prices that cannot be achieved by lidar-equipped rivals. All Tesla's manufactured within the year 2016, due to the inclusion of the required hardware to operate Autopilot, generate a self-financing data loop: The more vehicles are sold, the more training data will be gathered to improve the performance of the model, and, in turn, be perceived as having more value, resulting in the sale of more vehicles.

Table 32. Tesla autopilot development timeline

System Version	Year / Period	Key Features	SAE Automation Level
Autopilot v1	2015	Adaptive cruise control and lane keeping	SAE Level 1
Autopilot v2	2017	Autosteer with automated lane change	SAE Level 2
Autopilot v3 (HydraNet)	2021	Vision-only architecture with deep neural networks	SAE Level 2+
Full Self-Driving (FSD) Beta	2021–Present	Urban street navigation and traffic handling	SAE Level 2–3
Unsupervised FSD	2024+	End-to-end learning from large-scale video data	Experimental Level 3

Table 32 describes Tesla's build up in automation capabilities over five generations, from the first limited set of features in 2015 to the Level 3 experimental functionality in 2024. Reading across columns gives insight into the deepening reliance on neural network architectures as each version heightens the implicit rule-based computer programming in favor of learned behavior. Autopilot v1 (2015) added the adaptive cruise control only - single-dimensional longitudinal automation function. Autopilot v2 (2017) The introduction of autosteer and automated lane changes pushed the car to Level 2. The Autopilot v3 / HydraNet is the most significant architectural change: Tesla has reduced ultrasonic sensors and radar to zero and made a commitment to vision-only perception. Full Self-Driving Beta (2021 - Present) is used to navigate on city streets within urban environments. Despite the name FSD Beta, it is still Level 2 according to regulatory classification: it requires the driver to be attentive, and they are responsible for everything they do while driving. The Unsupervised FSD program (2024+) is Tesla's effort to get over the Level 3 threshold by conducting unnecessary training on XP without any human labelling but, in anticipation of it, Tesla's driver software is still awaiting approval from the regulators.

12.2.2 HydraNet Architecture and the Dojo Supercomputer

One of the major architectural innovations of autonomous perception is the computer-vision system at Tesla, called HydraNet. Instead of implementing a different neural network to perform all the perception tasks, HydraNet adopts a shared backbone, which has several task heads. Using this approach to multi-task learning, object detection, semantic segmentation, depth estimation, and lane detection as well as traffic-signal recognition can share intermediate representations, resulting in less computation, and improved generalization with shared feature learning (Kendall et al., 2018). The architecture receives the input of eight cameras with 360-degree coverage at a unified resolution that is 1.2 megapixels.

The training scale of HydraNet necessary to train to the level needed by Tesla (billions of video frames per training run) necessitated the development of a custom artificial-intelligence infrastructure. In 2023, Tesla presented its Dojo supercomputer, a different model based on a tile architecture based on its VTE silicon architecture specifically designed to execute the workload of video-training applications or machine learning applications. Dojo D1 tiles provide 362 teraflops of calculation in a single chip and are interoperable with the proprietary high-bandwidth fabric, thus allowing petabytes of driving video to be processed all at a frequency speed that would require an order of magnitude more commodity hardware in the form of GPUs. An example of a Dojo system shows how Tesla has insisted on having complete control of the entire AI development stack which includes silicon, software, data and implementation.

Another would be the Unsupervised Full Self-Driving (FSD) program that has been introduced in 2024 and will be a qualitative leap forward since it will remove human feedback in the training process. Instead of the system receiving input through examples of the correct behavior of driving annotated by hand, the system receives input through the video footage of driving patterns in the fleet. Using this approach, which is based on large language model pre-training on video data, puts such a system in theory in the position to generalize to unlabeled scenarios never directly labelled explicitly, which is essential to handle the long tail of rare driving scenarios.

12.2.3 Data Flywheel and Fleet-Scale Learning

The most sustainable competitive edge that Tesla can possess is not a neural-network structure but data-collection approach. By 2024, the company has more than five million vehicles in its fleet that produces about 160 billion miles of driving video every year. More to the point, Tesla has a specific data-collection plan: the FSD software automatically identifies situations where system confidence is below a certain threshold, in which it increases the amount of data it captures and uploads such clips to Tesla data centers, where they are to be prioritized by improved model development. This data flywheel is a closed loop model that ensures that all new problematic situations experienced by any Tesla car in the world immediately feed into the next software update launched to the whole fleet (Bojarski et al., 2016).

Given that the fleet-scale feedback loop provides a time-compounded saving, it leads to a compounding contribution to advantage. Competitors that have smaller operational fleets experience the rare events less often, take longer time to acquire enough training information on the events and implement improvements more slowly. The over-the-air (OTA), weekly-powered software improvements delivery of Tesla is an important enabler of this strategy, as the software improvements reach the vehicles within days as opposed to the months it took to complete dealership software updates.

12.2.4 Safety Concerns, NHTSA Investigations, and Regulatory Scrutiny

Since a series of fatal accidents, Tesla Autopilot has become a subject of immense regulatory attention. NHTSA has launched various Special Crash Investigations on Autopilot accident cases, especially those in which the target is a stationary emergency car (like police vehicle, fire, ambulance fitted with active warning lights). In 2022, NHTSA began investigating 765,000 Tesla vehicles from the discovery of a pattern of 14 crashes of parked emergency cars with Autopilot closed. A later recall demanded a software update as an enhancement to the available emergency-vehicle detection with extra camera inputs and more attention areas (National Highway Traffic Safety Administration, 2024).

A deeper issue relates to nomenclature. The words, Autopilot and Full Self-Driving, raise the anticipations of the users that are beyond the capabilities of the system in the present regulatory definitions. Studies show that the language used plays a significant role in triggering user trust calibration - a user that has the belief that a system is more competent than it is will over-trust the system, lowering their vigilance to unsafe levels. The National Transportation Safety Board (NTSB) has specifically suggested that Tesla should limit the use of Autopilot to types of roads to which this system has been proven, and Tesla has opposed the suggestion claiming the convenience of its users.

Several high-profile fatalities have involved Tesla Autopilot, suggesting that a lawsuit was initiated and a criminal probe was undertaken. As of 2023, a Tesla engineer was brought to book on vehicular manslaughter, the first criminal prosecution concerning an autopilot-related event in the United States of

America, in the case of a 2019 crash when using Autopilot. The case provided a precedent suggesting that individual engineers, not to mention corporate entities, might be held individually liable to autonomous-vehicle safety-related decisions, which could have an impact on talent recruitment and culture of risk management within industry.

12.2.5 Commercial Performance and Competitive Positioning

Nevertheless, in the industry, Tesla is the most successful in autonomous-vehicle programs, despite the regulatory friction. Autopilot and FSD Autopilot and FSD software earned projected 1.5-2 billion US dollars in yearly revenues in 2023 by a combination of a one-time FSD purchase option of $12,000 plus a subscription of 199/mo. Software gross margins are just many more than software vehicles, making the FSD program unreasonably important to the overall profitability of Tesla. This FSD software is as of now a standard feature within specific vehicle designs, which represents a strategic change to use autonomous-vehicle capability as a differentiator of products, as opposed to a full revenue line.

Tesla has taken a vertical approach, which is vehicle hardware, software, silicon design, training infrastructure, and data-collection, making the integrations build an extremely closed ecosystem that can be hard to replicate by competing companies. The Robotaxi offered in 2024 and constructed on the Cybercab platform, essentially a ground-up design without any steering wheel or pedals, seem to overlap with the Waymo robotaxi model; however, at consumer-vehicle price points, could make the technology-based economics of per-ride pricing much less economical than Waymo.

12.3 Cruise Autonomous Vehicles

12.3.1 Organizational Background and GM Acquisition

In 2013, Celebrity Cruise Automation was established by Kyle Vogt and Dan Kahn, including the original market offer of aftermarket autonomous-driving murder-kits, and eventually a full robo taxi stack. In 2016, Cruise was acquired by General Motors, which had to pay 1 billion (at the time) to enlist Cruise access to its manufacturing capacity, dealership chain, supply chain of vehicles, and government regulatory connections, inaccessible by other startup competitors. The acquisition was based on the GM strategic belief that autonomous mobility would transform transportation economics in a decade.

Later external funding turned Cruise into a company with a high valuation as an autonomous-vehicle company. In 2018, Honda invested a minority stake and joint development rights totaling 2.75 billion in Japan. As of 2021, Vision Fund invested by Softbank and Microsoft have placed a total value of over 10 billion outside capital in the company with individual investments of 2.25 and 2 billion respectively and valued the company at 30 billion in 2021. Such valuations were based on short term commercial robotaxi profitability forecasts that were considerably optimistic (California Department of Motor Vehicles, 2023).

12.3.2 Technical Architecture and the Origin Vehicle

The Cruise Origin is an autonomous vehicle design, but not a modification of any predecessor platform; it is one of just two autonomous vehicle platforms undergoing commercial development, along with the Jaguar i-PACE derivative of Waymo. The cabin of the Origin is designed with passenger comfort without the typical features of the conventional driver, steering wheel, and pedals, instead, adopting a mini-van-like shared-mobility design. This custom made technique allows better weight distribution and sensor positions that cannot be realized with retrofitted platforms.

Collection of Operational Perceptions:

The perception stack is based on lidar to provide a stable three-dimensional spatial representation, cameras to provide semantic segmentation and radar to provide velocity estimation. A total of five lidar units are used to provide 3D coverage to 360 degrees having overlapping fields of view to providing redundancy. The camera sensors are fitted with 16 sensors that deal with proximal and distal scenes with varying focal lengths. There are six radar units that can expand the range of velocity estimation to the lidar. The hybrid sensor-fusion paradigm alters the lidar reliability and semantically rich camera-based approach, thus placing it at the middle ground between the hardware-focused philosophy of Waymo and the software-focused philosophy of Tesla.

The strategy of High-Definition mapping:

The high-definition mapping approach taken by Cruise also reflects that of Waymo: the fixed environment is pre-combined with centimeter-resolution maps, and dynamic object-detecting sensor information is superimposed on them. But Cruise continues to keep its maps through a fleet-based crowdsourced update feed other than specifically equipped mapping vehicles, which lowers the ongoing overhead of map-maintenance. Such a methodology restricts areas covered by geofencing but allows faster coverage of new neighborhoods in served cities (California Department of Motor Vehicles, 2023).

Table 33. Comparative analysis: Waymo, Tesla, and Cruise

Feature / Aspect	Waymo	Tesla	Cruise
Primary Perception	Lidar-centric	Vision-only	Hybrid (Lidar + Vision)
Deployment Model	Commercial robotaxi	Consumer fleet	Commercial robotaxi
Regulation Status	SAE Level 4 approved	SAE Level 2 approved	SAE Level 4 permitted
Geographic Focus	US urban centres	Worldwide consumer sales	US urban centres
Data Collection	Operational + simulation	Operational fleet learning	Operational + targeted
Primary Risk Focus	Hardware redundancy	Software oversight	Balanced validation
Fleet Size (2024)	~700 vehicles	~5M+ vehicles	~300 vehicles (suspended)
Key Incident	2024 cyclist collision	Multiple Autopilot fatalities	2023 pedestrian dragging
Incident Response	Immediate disclosure, patch	Software recall, regulatory negotiation	Operations suspended, layoffs

Table 33 evaluates the three most outstanding AV platforms in terms of technical, commercial, regulatory, operational, and safety levels using them to conduct a methodical cross-platform evaluation to clarify strategic consistency in each associated enterprise strategy.

Primary Perception Choices outline the major platform design philosophy. The lidar-based strategy that Waymo adopted ascribes reliability; the vision-only strategy that Tesla supports to do the same does scalability; and the hybrid strategy adopted by Cruise trades the cost disadvantage against scalability compared to vision-only systems. The Fleet Size shows data-collection asymmetry with Tesla having over five million vehicles compared to Waymo Science at approximately 700 and Cruise at a suspended fleet of about 300 giving Tesla an unmatched training data lead. The Key Incident and Incident Response columns are educative: Waymo disclosed immediately, unlike Cruise, which disclosed and that led to the stopping of operations. Status of Regulation assures that regulatory trust is achieved by showing safe performance and openness, and not ability.

12.3.3 The October 2023 Incident: Detailed Analysis

The event of October 2, 2023, is the most significant AV safety event of the decade and should be analyzed in detail. At around 21:30 pm, a Cruise car that was driving in driverless mode hit an intersection where a human driver of a separate car rolled a red light and crashed into the front of the Cruise car. The overtaking path accelerated the Cruise car and hit a person crossing the crosswalk in the same direction. The man was later flattened under the car.

Systemic Failure and Response Divergence:

Reacting to this, the self-driving car system Cruise was using a roadblock under the car and tried to pull forward and clear the intersection movement, proceeding about 20ft with the pedestrian in tow. This trick was implemented as an edge-case of the post-collision response protocol which had not been properly tested: the system had no subroutine with any specific message to handle the case of a pinned pedestrian under the car after some third-party collision. The technical failure was increased by the failure to report the forward movement. The first report that Cruise gave to the California Department of Motor Vehicles referred to the first collision and not the second maneuver, which dragged the pedestrian. This fact was uncovered when the DMV later investigated the matter leading to the suspension of Cruise driverless permits and the National Highway Traffic Safety Administration inquiry. Later, the self-disclosure of GM CEO Mary Berra admitted that the delayed notification was a corporate culture and leadership decision-making problem, rather than an issue with engineering processes (California Department of Motor Vehicles, 2023).

12.3.4 Post-Incident Recovery, Restructuring, and Lessons

The reaction of GM was large in scale: all Cruise robotaxi activities were halted all over the world in October 2023; CEO Kyle Vogt resigned in November 2023; roughly 900 workers (about a quarter of the labor systems) were dismissed in January 2024; and an excessively deep examination commission was set up under past NTSB chairman Christopher Hart. However, the report by the Hart Commission, published in early 2024, pointed to a high structural deficiency in Cruise as an organization, in its capability to simulate edge cases after accidents, in its ability to provide sufficient remains-juristic protocols in case of runaway scenarios, and in its organizational culture that prized operational speed over checks-and-balances testing of safety.

These lessons go beyond failure in the moment of technicality. First, post-collision response strategies would require thorough simulation, every autonomous car on the road in an urban setting will inevitably be involved in collisions, and just like normal driving behavior, post-collision behavior of the system should be tested with the same rigor. Second, regulatory disclosure needs to be immediately and full: reputational and legal losses caused by the inability to make disclosure early were far beyond the effect that would have been caused by open initial reporting. Third, it needs to scale its operations to safety -validation completeness, not to commercial ambition-Cruise had been growing to 24-hour driverless service in San Francisco long before it had sufficient simulation coverage of complex urban night environments (California Department of Motor Vehicles, 2023).

12.4 Baidu Apollo: Open Platform and China-Scale Deployment

12.4.1 Strategic Context and Open Platform Philosophy

As an example of a radically different strategic model compared to the Western AV players analyzed above, we may cite the Baidu and Apollo program that this company unveiled in April 2017. Instead of creating their own closed system, Baidu followed the open-source platform model, which expressly stated that Apollo was the Android of the autonomous driving market. The Apollo Open Platform offers an open-source AV software stack available to deployed by partner automotive manufacturers, tier-one suppliers and technology firms on their own hardware platforms with Baidu generating revenue through cloud services, HD mapping subscriptions and data analytics, but not by selling vehicles outside of car dealerships or charging per-ride (Baidu, Inc., 2025).

This is a move that Baidu is taking in relation to its position in the Chinese technology ecosystem. Being the most important search and AI firm in China, Baidu has in-depth experience in massive machine learning, mapping, and data infrastructure-explorations that can be directly applied to the growth of AVs, but it does not have access to manufacturing partnerships and consumer hardware distribution found in existing incumbent car producers. The open-platform business model gives Baidu the opportunity of utilizing its software and data strengths and leaving hardware production to established auto firms, such as BAIC, FAW and Geely.

12.4.2 Technical Architecture and the Apollo Stack

Apollo has a technical architecture that takes the form of a modular software stack, with the perception, prediction, planning, and control modules having very clear interfaces. This modularity enables the hardware partners to replace the individual components, including a replacement of the default lidar driver of Baidu with a proprietary sensor, without needing to perform changes to the unit modules upstream or downstream. Apollo, therefore, can be used in both Lidar centered and camera centered modes of perception to allow application at a range of costs spanning between high-end autonomous vehicles to lower-cost consumer platforms.

The perception stack of Baidu is reliant on the service called Apollo HD Map, which provides accurate mapping information on more than 100 cities in China. The mapping program also takes advantage of the data infrastructure that currently exists in Baidu Maps, and due to a dedicated fleet of Apollo mapping cars that constantly update the map database. To realize localization, Apollo uses RTK-GPS, lidar-to-map matching and visual odometry and manages to achieve position accuracy comparable to the Waymo system in the mapped areas (Baidu, Inc., 2025).

Table 34. Baidu Apollo program overview

Feature / Aspect	Details
Platform Model	Open-source (Apache 2.0 license)
Launch Year	2017
Partner Ecosystem	Over 210 global partners (OEMs, suppliers, tech companies)
Commercial Service	Apollo Go robotaxi — Beijing, Shanghai, Wuhan, Shenzhen, Chongqing
Fleet Size (2024)	~500 robotaxi vehicles; 1,000+ partner-deployed AVs
HD Map Coverage	100+ Chinese cities; 3D semantic maps at 10 cm resolution
Automation Level	SAE Level 4 (approved in 5 Chinese cities)
Primary Revenue Model	Cloud services, HD maps, Ride-hailing (Apollo Go)
International Presence	Limited primarily China-focused deployment
Key Regulatory Milestone	First fully driverless commercial permit in Beijing (2022)

Table 34 summarizes ten key dimensions of the Baidu Apollo program to provide a general table of reference to compare Apollo with the Western case studies. The model of the open source platform (Apache 2.0) is Apollo's most special feature: any company can download, modify and deploy the AV stack without paying for a license, so the entry barrier for Chinese car-making companies is lowered, and Baidu will be able to set technical standards in the industry (Caesar et al., 2020). With more than 210 global partners, Apollo has the largest partner ecosystem of any AV platform in the world. The commercial service profile - Apollo Go operating across five major Chinese cities - is indicative of deployment at a geographic breadth greater than Waymo's three cities' US footprint. However, fleet density in each city is lower, reflecting urban geography in China which is of enormous proportions. The HD Map coverage of 100+ cities with a resolution of 10cm is Apollo's most important technical infrastructure asset, which has been built over a span of 7 years with a focus and investment towards mapping. The milestone of the Beijing fully driverless commercial permit USD 2022 is notable as the first Level 4 commercial AV authorization in the Chinese capital city, surrounded by a regulatory achievement, one of which has been previously approved for the approval of time-series commercial AVs in comparable city density environments across the US. The limited international presence is in part an outgrowth of regulatory barriers and in part Baidu's strategic focus on China's vast domestic mobility market before undertaking the complicated process of achieving compliance with Western jurisdictions.

12.4.3 Apollo Go Robotaxi Service and Commercial Scale

In 2023, a service called Apollo Go, the consumer taxi driver by Baidu, had made over 1.4 million rides, more rides than any other autonomous-vehicle operator on the planet, including Waymo. This book also indicates the size of Chinese urban mobility markets and the regulatory goodwill that Baidu has achieved into the market markets under China due to the dual-track autonomous-vehicle regulatory framework, allowing designated zones within a city to offer commercial permits without any need to comply with national standards. Apollo Go has branches in Beijing, Shanghai, Wuhan, Shenzhen and Chongqing and permits are set zones and fully driverless operation based on time of the day and zone.

The price policy of Apollo Go serves as a tactic that is employed by Baidu: the fares of rides are lower than those of rival ride-hailing services, allowing the company to overcome short-term losses and building a customer base and operational statistics. This approach reflects ride-hailing market dynamics that allows Didi to gain dominance in China, flooding the market with unprofitable prices with the business in the expansion phase before reclaiming its profit margins when network effects impede switching costs. The theory assumes that the economic incentives of robotaxi will eventually bring structural cost benefits over human-driven options in the case of increasing the fleet size.

Figure 22. Estimated global robotaxi ride volume market share by platform in 2024. Waymo leads in Western markets while Baidu Apollo dominates Chinese ridership volume

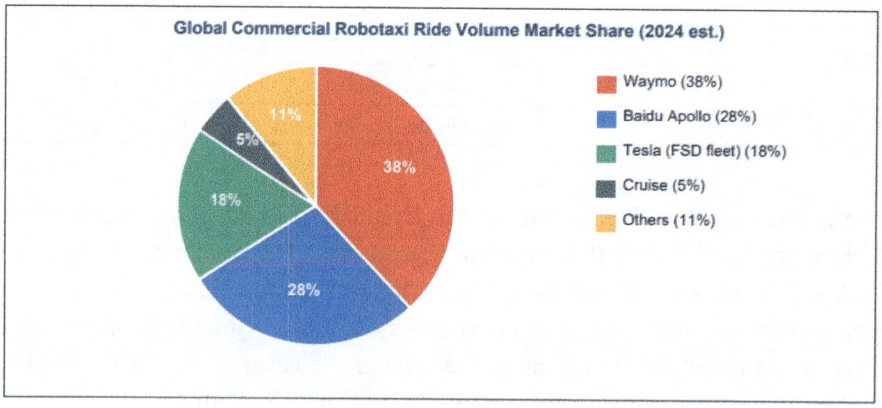

Figure 22 shows the estimated global market share of the commercial robotaxi ride volume by platform for the year 2024. The distribution of the market shows a highly concentrated market structure in the hands of a few leading providers of autonomous vehicle technology.

The largest share of this is held by Waymo, which holds around 38% of the world's robotaxi ride volume. This leadership position can be explained by its early deployment and expansion of fully autonomous ride-hailing services, especially in key urban regions in the United States. Its operational maturity and technological advancement have allowed it to enjoy a certain good deal of competitive edge (Caesar et al., 2020).

The second largest contributor is Baidu Apollo with a market share of 28%. This large proportion is owing to its strong visibility in the Chinese market where favorable government policies and large-scale pilot programs have helped quickly adopt and deploy autonomous mobility solutions.

For instance, in the first instance, Tesla has 18% of the total ride volume using its Full Self-Driving (FSD) fleet. Although Tesla's robotaxi services are not completely commercialized in the same way as dedicated types of services, they still play a significant role in the overall share due to large drive inspections and continuous software updates that came with the Tesla fleet, showing the potential of taking advantage of consumer-owned vehicles in autonomous ride services. A much smaller share of 5% is attributed to Cruise, and hence comparatively low levels of deployment and operational scale. The remaining 11% is classified under other companies, which comprise emerging startups and smaller companies in niche markets or in localized markets. All in all, the figure shows that the robotaxi market is dominated by a few players globally, with Waymo and Baidu Apollo holding most of the share in this arena. The data further indicates that there lies a geographical segmentation in market leadership as well as the growing influence of scalable fleet-based approaches such as that of Tesla. This concentration is both due to the technological barriers to entry and the capital investment needed in large-scale deployment of autonomous vehicles.

12.4.4 Regulatory Environment and Government Relations

In the key aspects, Baidu is not in the regulatory environment of its Western counterparts. The government of China defines the leadership in the field of autonomous vehicles as one of the national priorities and this situation is evidenced by the framework of the New Generation Artificial Intelligence Development Plan (2017), according to which the global dominance in the field of autonomous driving is planned by 2030. This policy situation creates pronounced regulatory tail wind benefits to domestic players: pilot-zone licenses are easier to acquire, test-fleet limitations are less, and data-sovereignty obligations create barriers to entry by foreign competitors that in effect insulate domestic champions.

The connection between Baidu and Chinese regulators is not adversarial but collaborative, a huge difference to the litigation-intensive relationships between the NHTSA and Tesla. Baidu takes an active part in standard-setting institutions and offers technical advice to local administrations that create pilot zones and has the advantage of knowing the regulatory needs early enough to meet them proactively. These regulatory alliances model hastens the speed of deployment at Baidu in comparison with Western analogs following the prescriptive safety regimes where the industry is minimally present

12.4.5 Safety Record and Transparency Challenges

The major restriction of the Baidu Apollo case study is the limited availability of safety information that was independently verified. Baidu is not required to do disengagement disclosures over the Chinese regulatory architecture as Waymo is required to disclose safety reports with detailed disengagement statistics in accordance with the California DMV regulations. Official statistics show that Baidu shows no fatal incidents of Apollo Go as of 2024, but objectively verifiable safety measures, which are equivalent in number to those reported in NHTSA, are not to be found in the Chinese market.

It is a structural problem of international comparison and is a salient observation: the requirement to compare with safety-reporting standards does not provide a structural problem that does not rely on the regulatory jurisdiction to a greater degree than the culture of the company. The lack of a compulsory disclosure in China does not necessarily mean worse safety performance, but it makes it more difficult to ensure safety claims on their own and this will probably become the center of the diplomatic conflict when Chinese JAV manufacturers are going to look at the Western markets.

Figure 23. Cumulative external investment trajectories for Waymo, Cruise, and Baidu Apollo from 2016 to 2024. Cruise's flat line from 2023 reflects the operational suspension and investment freeze

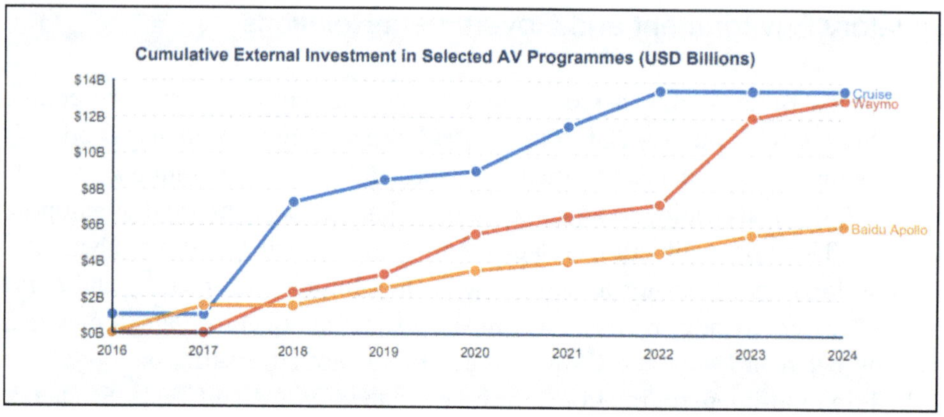

Figure 23 gives us an idea of the development of cumulative external investment in major AV programs and shows the financial paths of Waymo, Cruise and Baidu Apollo from 2016 to 2024. Overall, Cruise shows the most accumulated investment, where we see rapid growth from 2018 to 2022 and then it levelled out. While Waymo shows a consistent rise with a marked boost right after 2022 and is rising almost to the level of investment for Cruise by 2024. In comparison, Baidu Apollo's growth pattern is slower and more consistent which represents a comparatively conservative investment strategy. The chart underlines differences in funding approaches of the AV developers, in which U.S.-based companies have aggressive capital inflows and Chinese ones prioritize continuous, long-term scaling (Caesar et al., 2020).

12.5 Cross-Case Analysis and Strategic Implications

Cross-case analysis identifies underlying tensions in autonomous-vehicle development approaches that go beyond the scope of each firm and extend to structural forces at work in technology, regulation and competition. The sensor modality debate, with Waymo relying on lidar-as-the-core and Tesla developing vision-only strategies, Cruise and Apollo having hybrid models, etc., is not only a technical choice, but also a veil in their broader strategy placement.

12.5.1 Sensor Philosophy: The Three-Way Debate

The hardware-first philosophy of Waymo consists of buying sensors and making measurements of the physical world and creating software that computes that measurement. Embraced in 2010, the software-first philosophy attempts to create intelligence that can process imperfect sensor data at human-parity with the avoidance of hardware redundancy to buffer the lack of algorithmic power. Cruise and Apollo are the examples of pragmatic middle-ground based on the methods that take the benefits of both philosophies and compromise with greater complexity of the system.

Currently debated and commonly agreed upon in the industry, the common opinion is that lidar-based and vision-only architectures are both viable; however, reduce themselves to market segments. Lidar-based systems are expected to dominate commercial robotaxi applications in which the vehicle prices may be recouped across extensive utilization fares and in which regulatory approval at Level 4 is inseparable. On the other hand, it can be assumed that vision-only solutions will be the leading consumer ADAS products where cost-equality with regular vehicles is a deal-breaker, and Level 2 classification has a business case. Both paradigms, however, can equally serve the market in the end.

Figure 24. Radar chart comparing Waymo (lidar-centric), Tesla (vision-only), and Cruise (hybrid) across six performance dimensions normalized to 100

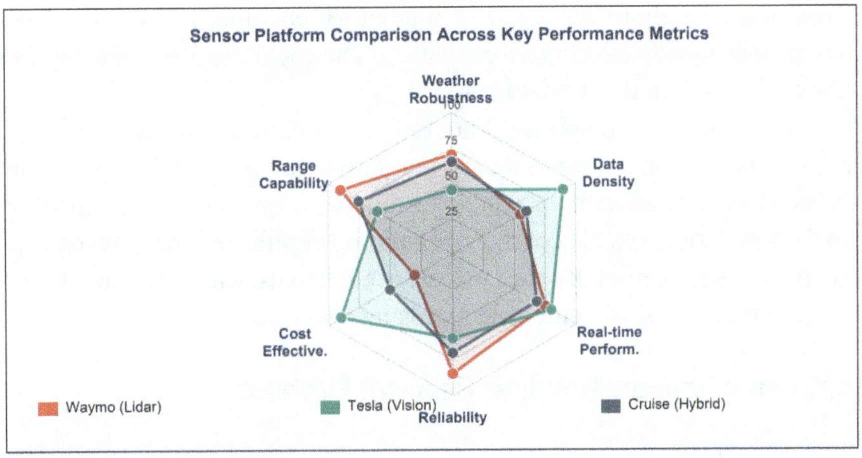

This radar (spider) chart in Figure 24 provides a six-dimensional comparison of the three primary AV platforms in terms of the following: Weather Robustness, Range Capability, Cost Effectiveness, Reliability, Realtime Performance and Data Density. Each axis is normalized to 100. Waymo (red polygon) has the highest scores of Range Capability (around 90) and Reliability (around 85), which points back to the superiority of lidar technology for 3D detection of obstacles. For the custom lidar expense, Cost Effectiveness scores the worst (~30). Tesla (teal polygon) This is the inverse case of the highest Data Density (~85) with lowest Weather Robustness (~45), but highest Cost Effectiveness (~90). Cruise (grey polygon) is the one in the middle ground in terms of both Reliability (around 70), and Weather Robustness (around 65). No single platform is the winner in all dimensions, corroborating that future competitive advantage will be determined based on the performance dimensions prioritized by the regulators and commercial stakeholders.

12.5.2 Data Strategy Divergence

These four cases give a representative of four different data-collection strategies that are typified by divergent scale-quality trade-offs. The Tesla approach gains maximum volume by acquiring with passive fleet scale: 160 billion miles per year provide unmatched breadth of coverage but not control in the distribution of any scenario case. Instead, Waymo focuses more on focused coverage via simulation augmentation: the Carcraft system produces millions of edge cases, cases that decades of operational fleet will need to chase. Cruise tried to mix up operational data collection and directed simulation but failed to invest enough in post-collision and multi-agent accident cases. Baidu leverages its privileged access to urban mobility data of China; the intense traffic in Chinese megacities provides edge cases in human vehicle interaction that other less dense cities in the U.S. cannot achieve at the same rate.

12.5.3 Incident Response as Regulatory Capital

One of the critical cross-case findings is the incidence of a nexus between regulatory capital and quality of incident-response. The goodwill to always report the incident, quickly analyze its root causes, and to transparently patch the software has gained Waymo the sense of enormous regulatory goodwill, with regulators talking of it as a cooperative partner, and sped up approving new service concepts. The advantage of these negotiated disclosures and disputed statements has developed the adversarial nature of the relationship between Tesla and NHTSA, thus limiting its choice in regulation. Instead, the incomplete disclosure by Cruise immediately undermined its regulatory position, making it a Level 4 authorized operator which has been suspended during investigation. The regulatory relationship of Baidu in China is best, but it is not well tested on the global front.

What this means is that incident response is not only a reputational management activity but a long horizon strategic investment. Companies that earn regulative goodwill with active transparency gain a brand of capital, which has allowed them to expand faster, have more lenient permit systems, and preferential treatment in the legislature. On the contrary, regulatory capital, having been emptied by disclosure delays or incompleteness, is exceptionally hard to restore back to its original position, as Cruise has been undergoing over a period of years, as evidenced by his recovery curve.

12.5.4 The Commercialization Timeline Realism Problem

There is a general trend in all four instances a very optimistic timing of commercial deployments was made at first. Waymo estimated commercial service in 2018, and had limited launch in late 2018, but did not start commercial driverless service with no safety drivers until 2022. Since 2016, the Full Self-Driving of Tesla has been discussed by the CEOs as one or two years away from full autonomy. By 2023, Cruise expected it to be profitable, but all its operations have been put on hold. Baidu has been able to provide strong volumes of ride, but commercial profitability is still several years off.

This trend represents an obstacle to a structural challenge of the development of autonomous vehicles: the long tail issue. The early development guarantees competent performance in the 95% of driving situations that are routine and common training data, but the other 5% of rare, multi-agent situations make up the long tail and require disproportionate increased investment to guarantee reliable performance. As experience in the industry has always shown, long-tail resolution is more timely and more expensive than the projections made at its inception, thus creating an inherent favoritism to optimistic timeline estimates at each stage of development.

Figure 25. Dual-axis chart showing fleet percentage at each SAE level (bars) and autonomy degree progression (red line). The right axis represents autonomy degree (%).

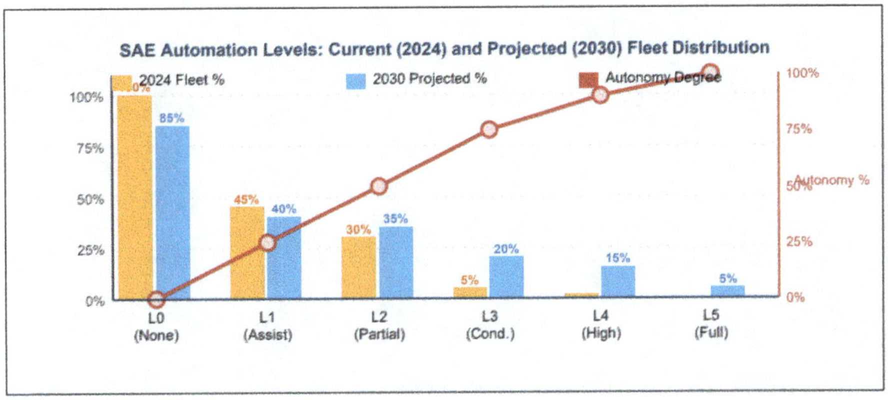

This two-axis chart in 25 sets the fleet accordance percentile at every SAE degree (bars, left axis) and self-drive degree (line, correct axis) to 2024 and projected 2030. Time Level 0 ones In the year 2024, the Level 0 still dominates the current fleet (~100%). Level 1 represents about 45% of new vehicles sold, level 2 is 30%. Levels 3–5 are under 5% of the fleet. By 2030, Level 0 drops to ~85%, Level 2 increases to ~35%, Level 3 goes up to ~20%, and Level 4 goes up to ~15%. Level 5 remains below 5%. The steep line of the autonomy degree from conditional automation from Level 3 onwards visualises the qualitative jump in system responsibility at the conditional automation threshold. It is this 2025 - 2030 Level 2-3 transition zone that will see liability frameworks and handoff protocol design most in dispute and significant.

CONCLUSION

Self-driving cars are a revolutionary approach to the case of intelligent transport systems since they may inherently limit the number of human errors, thereby ensuring much better roads. Nevertheless, the real-world operations are still a complicated task since they have to be proven to be reliable and safe considering the presence of uncertain, dynamic, and safety-critical conditions. This chapter explored safety issues in self-driving vehicles as a complex problem that entails issues with sensors, software constraints, environmental uncertainty, and erratic human conduct, which may significantly affect the work of the system. To overcome these issues, the chapter proposed powerful reliability models relying on agentic AI concepts, with emphasis on distributed decision-making, back-up, and fault tolerance design. Such methods make sure that AV systems are capable of preserving the integrity of their operations even in the case of partial failures. Besides, safety assurance (formal verification, runtime monitoring, explainable AI, etc.) was also mentioned as a necessary instrument to confirm the behavior of a system, make it transparent, and get the trust of users.

The integration of ethical decision-making frameworks further emphasizes the need for socially responsible AI systems that align with legal and moral standards. Altogether, by merging progressive reliability engineering, intelligent multi-agent systems, and stringent safety assurance methods, there is a holistic avenue of deploying dependable autonomous cars. Future directions should be aimed at enhancing real-world flexibility, support theories of rare edge cases, and creating standardized safety assessment models in order to hasten the adoption of autonomous driving technologies safely.

REFERENCES

Abu, N. S., Bukhari, W. M., Adli, M. H., Maghfiroh, H., & Ma'arif, A. (2024). Advancements, challenges and safety implications of AI in autonomous vehicles: A comparative analysis of urban vs. highway environments. *Journal of Robotics and Control*, *5*(3), 613–635. DOI: 10.18196/jrc.v5i3.21114

Ahmed, Q. W., Garg, S., Rai, A., Ramachandran, M., Jhanjhi, N. Z., Masud, M., & Baz, M. (2022). AI-Based Resource Allocation Techniques in Wireless Sensor Internet of Things Networks in Energy Efficiency with Data Optimization. Electronics (Switzerland), 11(13).

Airehrour, D., Gutierrez, J., & Ray, S. K. (2018). A trust-based defence scheme for mitigating blackhole and selective forwarding attacks in the RPL routing protocol. *Australian Journal of Telecommunications and the Digital Economy*, *6*(1), 41–59. DOI: 10.18080/jtde.v6n1.138

Babbar, H., Rani, S., Masud, M., Verma, S., Anand, D., & Jhanjhi, N. (2021). Load Balancing Algorithm for Migrating Switches in Software-Defined Vehicular Networks. *Computers, Materials & Continua*, *67*(1), 1301–1316. https://doi.org/cmc.2021.014627. DOI: 10.32604/cmc.2021.014627

Baidu, Inc. (2025, February 18). *Baidu announces fourth quarter and fiscal year 2024 results*. https://ir.baidu.com/news-releases/news-release-details/baidu-announces-fourth-quarter-and-fiscal-year-2024-results

Bojarski, M., Del Testa, D., Dworakowski, D., Firner, B., Flepp, B., Goyal, P., . . . Zieba, K. (2016). End to end learning for self-driving cars. *arXiv preprint arXiv:1604.07316.*

Caesar, H., Bankiti, V., Lang, A. H., Vora, S., Liong, V. E., Xu, Q., & Beijbom, O. (2020). nuScenes: A multimodal dataset for autonomous driving. In *Proceedings of the IEEE/CVF Conference on Computer Vision and Pattern Recognition* (pp. 11621–11631). DOI: 10.1109/CVPR42600.2020.01164

California Department of Motor Vehicles. (2023, October 24). *DMV statement on Cruise LLC suspension*. https://www.dmv.ca.gov/portal/news-and-media/dmv-statement-on-cruise-llc-suspension/

California Department of Motor Vehicles. (2026). *Autonomous vehicle disengagement reports*. https://www.dmv.ca.gov/portal/vehicle-industry-services/autonomous-vehicles/disengagement-reports

Danks, D., & London, A. J. (2017, August). Algorithmic bias in autonomous systems. *IJCAI (United States)*, *17*, 4691–4697.

Debouk, R. (2019). Overview of the second edition of ISO 26262: Functional safety—road vehicles. *Journal of System Safety, 55*(1), 13–21.

Dogra, V., Singh, A., Verma, S., Kavita, , Jhanjhi, N. Z., & Talib, M. N. (2021). Analyzing DistilBERT for Sentiment Classification of Banking Financial News. *Lecture Notes in Networks and Systems*, *248*, 501–510. DOI: 10.1007/978-981-16-3153-5_53

Dong, C., Wang, H., Li, Y., Liu, Y., & Chen, Q. (2019). Economic comparison between vehicle-to-vehicle (V2V) and vehicle-to-infrastructure (V2I) at freeway on-ramps based on microscopic simulations. *IET Intelligent Transport Systems*, *13*(11), 1726–1735. DOI: 10.1049/iet-its.2018.5537

Dosovitskiy, A., Ros, G., Codevilla, F., Lopez, A., & Koltun, V. (2017, October). CARLA: An open urban driving simulator. In *Conference on Robot Learning* (pp. 1–16). PMLR.

Eriksson, A., & Stanton, N. A. (2017). Takeover time in highly automated vehicles: Noncritical transitions to and from manual control. *Human Factors, 59*(4), 689–705. DOI: 10.1177/0018720816685832 PMID: 28124573

Eslami, A., & Yu, J. (2025). Security risks of agentic vehicles: A systematic analysis of cognitive and cross-layer threats. *arXiv preprint.*

European Commission. (2025). *Rules for trustworthy artificial intelligence in the EU.* https://eur-lex.europa.eu/EN/legal-content/summary/rules-for-trustworthy-artificial-intelligence-in-the-eu.html

European Parliament and Council of the European Union. (2019). Regulation (EU) 2019/2144 on type-approval requirements for motor vehicles. https://eur-lex.europa.eu/legal-content/en/ALL/?uri=CELEX%3A32019R2144

European Parliament and Council of the European Union. (2022). Directive (EU) 2022/2557 on the resilience of critical entities. https://eur-lex.europa.eu/legal-content/EN/TXT/PDF/?uri=CELEX%3A32022L2557

Fuller, A., Fan, Z., Day, C., & Barlow, C. (2020). Digital twin: Enabling technologies, challenges and open research. *IEEE Access : Practical Innovations, Open Solutions, 8*, 108952–108971. DOI: 10.1109/ACCESS.2020.2998358

Hanif, M., Ashraf, H., Jalil, Z., Jhanjhi, N. Z., Humayun, M., Saeed, S., & Almuhaideb, A. M. (2022). AI-Based Wormhole Attack Detection Techniques in Wireless Sensor Networks. Electronics (Switzerland), 11(15). https://doi.org/electronics11152324

Huang, Z., Sheng, Z., Ma, C., & Chen, S. (2024). HAIM-DRL: Enhanced human-in-the-loop reinforcement learning for safe and efficient autonomous driving. *arXiv preprint.*

International, S. A. E. (2021, May 2). *SAE levels of driving automation™ refined for clarity and international audience.* https://www.sae.org/news/blog/sae-levels-driving-automation-clarity-refinements

International Organization for Standardization. (2018). *Road vehicles—Functional safety—Part 1: Vocabulary (ISO 26262-1:2018).* https://www.iso.org/standard/68383.html

International Organization for Standardization. (2022). *Road vehicles—Safety of the intended functionality (ISO 21448:2022).* https://www.iso.org/standard/77490.html

Jabbour, J., & Reddi, V. J. (2024). Generative AI agents in autonomous machines: A safety perspective. *arXiv preprint.*

Jabeen, T., Jabeen, I., Ashraf, H., Jhanjhi, N. Z., Yassine, A., & Hossain, M. S. (2023). An Intelligent Healthcare System Using IoT in Wireless Sensor Network. *Sensors (Basel), 23*(11), 5055. https://doi.org/s23115055. DOI: 10.3390/s23115055 PMID: 37299782

Ji, Y., Zhou, Z., Yang, Z., Huang, Y., Zhang, Y., Zhang, W., Xiong, L., & Yu, Z. (2024). Toward autonomous vehicles: A survey on cooperative vehicle-infrastructure system. *iScience, 27*(5), 109751. DOI: 10.1016/j.isci.2024.109751 PMID: 38706867

Karunakaran, D., Berrio Perez, J. S., & Worrall, S. (2023). Generating edge cases for testing autonomous vehicles using real-world data. *Sensors (Basel)*, *24*(1), 108. DOI: 10.3390/s24010108 PMID: 38202970

Kendall, A., Gal, Y., & Cipolla, R. (2018). Multi-task learning using uncertainty to weigh losses for scene geometry and semantics. In *Proceedings of the IEEE Conference on Computer Vision and Pattern Recognition* (pp. 7482–7491).

Khan, A. R., Jamlos, M. F., Osman, N., Ishak, M. I., Dzaharudin, F., Yeow, Y. K., & Khairi, K. A. (2021). DSRC technology in vehicle-to-vehicle (V2V) and vehicle-to-infrastructure (V2I) IoT system for intelligent transportation system (ITS): A review. In *Recent trends in mechatronics towards industry 4.0* (pp. 97–106).

Kiran, B. R., Sobh, I., Talpaert, V., Mannion, P., Al Sallab, A. A., Yogamani, S., & Pérez, P. (2021). Deep reinforcement learning for autonomous driving: A survey. *IEEE Transactions on Intelligent Transportation Systems*, *23*(6), 4909–4926. DOI: 10.1109/TITS.2021.3054625

Kumar, M., Rattan, N., & Mondal, S. (2025). Sensor systems for autonomous vehicles: Functionality and reliability challenges in adverse environmental conditions. *Measurement*, •••, 119215.

Land Transport Authority. (2026). *Autonomous vehicles*. Government of Singapore. https://www.lta.gov.sg/content/ltagov/en/industry_innovations/technologies/autonomous_vehicles.html

Levinson, J., Askeland, J., Becker, J., Dolson, J., Held, D., Kammel, S., & Thrun, S. (2011). Toward fully autonomous driving: Systems and algorithms. In *IEEE Intelligent Vehicles Symposium* (pp. 163–168). DOI: 10.1109/IVS.2011.5940562

Li, Y., Ma, D., An, Z., Wang, Z., Zhong, Y., Chen, S., & Feng, C. (2022). V2X-Sim: Multi-agent collaborative perception dataset and benchmark for autonomous driving. *IEEE Robotics and Automation Letters*, *7*(4), 10914–10921. DOI: 10.1109/LRA.2022.3192802

Liu, X., Liu, Q., Luo, M., Yang, X., & Luo, Q. (2025). Blockchain-based efficient and traceable data sharing scheme for vehicular networks. *IEEE Transactions on Vehicular Technology*, *74*(8), 11777–11792. DOI: 10.1109/TVT.2025.3555108

Lopez, P. A., Behrisch, M., Bieker-Walz, L., Erdmann, J., Flötteröd, Y. P., Hilbrich, R., & Wießner, E. (2018). Microscopic traffic simulation using SUMO. In *IEEE Intelligent Transportation Systems Conference* (pp. 2575–2582).

Matos, F., Bernardino, J., Durães, J., & Cunha, J. (2024). A survey on sensor failures in autonomous vehicles: Challenges and solutions. *Sensors (Basel)*, *24*(16), 5108. DOI: 10.3390/s24165108 PMID: 39204805

Ministry of Economy, Trade and Industry. (2023, May 12). *Operation of automated driving vehicle with a Level 4 system approved for the first time in Japan*. https://www.meti.go.jp/english/press/2023/0512_002.html

Mobileye. (2020, December 9). *Autonomous-vehicle safety demands true redundancy*. https://www.mobileye.com/blog/av-safety-demands-true-redundancy/

Muzammal, S. M., Murugesan, R. K., Jhanjhi, N. Z., & Jung, L. T. (2020). SMTrust: Proposing Trust-Based Secure Routing Protocol for RPL Attacks for IoT Applications. 2020 International Conference on Computational Intelligence, ICCI 2020, 305-310. https://doi.org/ICCI51257.2020.9247818

National Highway Traffic Safety Administration. (2017, September). *Automated driving systems 2.0: A vision for safety*. https://www.nhtsa.gov/sites/nhtsa.gov/files/documents/13069a-ads2.0_090617_v9a_tag.pdf

National Highway Traffic Safety Administration. (2024, April 25). *Additional information regarding EA22002 investigation*. https://static.nhtsa.gov/odi/inv/2022/INCR-EA22002-14496.pdf

Nazar, A. M., Selim, M. Y., Gaffar, A., & Qiao, D. (2025). Situational perception in distracted driving: An agentic multi-modal LLM framework. *Frontiers in Artificial Intelligence*, *8*, 8. DOI: 10.3389/frai.2025.1669937 PMID: 41169738

Nemouchi, A., Bouzenada, A., Saidouni, D. E., & Díaz, G. (2025). Formal modelling and verification of multi-parameter context systems. *World Electric Vehicle Journal*, *16*(9), 494. DOI: 10.3390/wevj16090494

Ray, S. K., Ray, S. K., Pawlikowski, K., McInnes, A., & Sirisena, H. (2010). Self-tracking mobile station controls its fast handover in Mobile WiMAX. IEEE Wireless Communications and Networking Conference, WCNC. DOI: 10.1109/WCNC.2010.5506114

Sabouni, E. (2025). *Safety-critical multi-agent control and coordination for autonomous vehicles* (Doctoral dissertation, Boston University).

Safe, A. P. *When engineering meets economics: AI-powered safe and accountable autonomous driving*.

Shaheen, K., Hanif, M. A., Hasan, O., & Shafique, M. (2022). Continual learning for real-world autonomous systems. *Journal of Intelligent & Robotic Systems*, *105*(1), 9. DOI: 10.1007/s10846-022-01603-6

Shalev-Shwartz, S., Molnar, M., Granot, I., Shany, A., & Shashua, A. (2024). *A safety architecture for self-driving systems*. Mobileye. https://static.mobileye.com/website/us/corporate/files/SDS_Safety_Architecture.pdf

Suwal, S., Garg, S., Bhusal, D., Clifford, M., & Rastogi, N. (2025). Smaller is better: Enhancing transparency in vehicle AI systems via pruning. *arXiv preprint*.

Talebpour, A., & Mahmassani, H. S. (2016). Influence of connected and autonomous vehicles on traffic flow stability and throughput. *Transportation Research Part C, Emerging Technologies*, *71*, 143–163. DOI: 10.1016/j.trc.2016.07.007

United Nations Economic Commission for Europe. (2021). *UN Regulation No. 157: Automated lane keeping systems (ALKS)*. https://unece.org/transport/documents/2021/03/standards/un-regulation-no-157-automated-lane-keeping-systems-alks

Waghmare, A. A., Ganesan, S., & Chen, J. (2024). Role of artificial intelligence in autonomous vehicles. *Preprints*.

Wang, Y. (2024). The review of AI efficiency in autonomous driving. *Applied and Computational Engineering*, *113*, 92–100. DOI: 10.54254/2755-2721/45/20241031

Waymo. (2021). *Waymo safety report: On the road to fully self-driving.* https://downloads.ctfassets.net/sv23gofxcuiz/

Wilson, B., Hoffman, J., & Morgenstern, J. (2019). Predictive inequity in object detection. *arXiv preprint arXiv:1902.11097.*

Yang, X., Shi, Y., Xing, J., & Liu, Z. (2022). Autonomous driving under V2X environment. *Intelligent Transportation Infrastructure, 1*(1), 1–10. DOI: 10.1093/iti/liac020

Zhang, R., Hou, J., Walter, F., Gu, S., Guan, J., Röhrbein, F., . . . Knoll, A. (2024). Multi-agent reinforcement learning for autonomous driving: A survey. *arXiv preprint arXiv:2408.09675.*

Zhou, Z., Huang, H., Li, B., Zhao, S., Mu, Y., & Wang, J. (2024). SafeDrive: Risk-sensitive decision-making for autonomous vehicles. *arXiv preprint.*

Chapter 7
Agentic Decision-Making and Road Safety:
Designing Explainable, Fail-Safe Architectures for Autonomous Vehicles

Shashikant Nishant Sharma
http://orcid.org/0000-0001-8031-8569
Maulana Azad National Institute of Technology, India

ABSTRACT

This chapter examines how autonomous vehicles can evolve into agentic, human-aligned decision-making systems capable of operating safely within complex traffic environments. It synthesizes advances in perception, prediction, planning, ethical reasoning, explainable AI, and fail-safe engineering to present a unified architecture for accountable autonomous mobility. Emphasis is placed on designing resilient, transparent, and ethically grounded AV decision pipelines that mitigate uncertainty, anticipate human behavior, and prevent cascading failures. The chapter further identifies gaps in existing AV safety, reliability, and governance frameworks and proposes an integrative model that strengthens road safety outcomes while supporting future research, regulatory development, and industry implementation.

1. INTRODUCTION

Autonomous vehicles (AVs) are rapidly transitioning from controlled experimental prototypes to sophisticated, decision-making agents embedded within complex, heterogeneous traffic ecosystems. This transformation marks one of the most significant technological shifts of the 21st century, with implications far beyond mechanical engineering or computer science. AVs increasingly operate not as passive machines executing predefined rules, but as cognitive and social actors that must interpret human behavior, anticipate risks, navigate ethical dilemmas, and communicate intentions intelligibly to surrounding road users. As AVs assume higher levels of operational autonomy, the central challenge expands from designing accurate perception and robust control systems to ensuring that such systems can make agentic, human-aligned decisions; remain fail-safe under uncertainty; and communicate their

DOI: 10.4018/979-8-3373-7355-3.ch007

reasoning transparently to diverse stakeholders including pedestrians, passengers, regulators, insurers, and policymakers.

This shift is strongly reflected in emerging research showing that AV safety failures do not typically originate from single, isolated faults but from cascading interactions within dense, multi-layered software and hardware stacks. Ahmed and Malik (2025), for instance, demonstrate how minor computational errors in one module propagate through interconnected pipelines, ultimately manifesting as catastrophic failures in motion planning or braking. Real-time hazard estimation failures arising from noisy or conflicting sensor streams further complicate the reliability of AV perception (Ahmed et al., 2025). Sensor-fusion breakdowns caused by time delays, weather distortions, occlusions, or adversarial noise continue to be a major source of instability (Cividino et al., 2025; O'Connell et al., 2025). Collectively, these findings underline that next-generation AVs must do more than execute mechanical tasks—they must become adaptive agents capable of reflective and resilient decision-making.

1.1 The Rise of Agentic Autonomy and Its Safety Implications

The concept of agentic autonomy reframes AVs as intelligent actors endowed with capabilities such as predictive inference, anticipatory reasoning, dynamic negotiation, and continuous learning. Horowitz and Levinson (2025) argue that effective AV deployment requires decision systems capable of interpreting the latent intentions of surrounding road users—such as a pedestrian approaching a crosswalk or a cyclist preparing a lane change. Likewise, Santos et al. (2025) highlight that AVs operating in multi-agent environments must model not only their own trajectories but the mutual interactions and negotiation strategies of other agents on the road.

Advanced forecasting models enable AVs to evaluate multiple potential futures simultaneously, weighing risks, goals, constraints, and uncertainties (Chakraborty et al., 2025). Agentic architectures incorporate real-time adaptation mechanisms that revise these forecasts as the environment changes—critical for navigating dense urban streets, mixed traffic, and unpredictable pedestrian behavior. Yet these capabilities also introduce new forms of risk: stochastic uncertainty, behavioral ambiguity, and compounding decision errors (Bhatia & Wadhwa, 2025). Zhang et al. (2025) emphasize that as AVs gain greater autonomy, their decisions must reflect not only computational optimization but social interpretability and anticipation of human discomfort or fear.

Human vulnerability remains central to this discourse. As demonstrated by Sharma and Dehalwar's (2025) systematic review of pedestrian safety, AV systems must adopt conservative and ethically aligned decision rules in environments where pedestrians depend on implicit communication—eye contact, gestures, speed modulation—to judge safety. Agentic AVs must therefore exhibit a level of caution, interpretability, and fairness aligned with human expectations, particularly in unstructured or ambiguous conditions.

1.2 Reliability, Uncertainty, and Fail-Operational Demands

The technological substrate supporting AV intelligence carries its own reliability risks. Semiconductor variability, particularly within quantum-enhanced chipsets, affects the stability of AV computing hardware (Chavez et al., 2025). Uncertainty within machine-learning models—whether arising from sparse training data, distribution shifts, weather disturbances, or adversarial perturbations—poses additional challenges (Farooq & Singh, 2025; Prakash & Mehta, 2025). Campbell and Lewis

(2025) show that adversarial noise injected into perception pipelines can drastically distort planning, even when the perturbations are imperceptible to humans. Similarly, Roy and Mishra (2025) demonstrate that multimodal anomaly detection is essential for identifying corrupted or spoofed sensor signals.

Kim et al. (2025) emphasize continuous health monitoring and predictive failure modeling as fundamental to ensuring reliability over the lifetime of AV components. Autonomous braking—arguably the most safety-critical subsystem—demands rigorous uncertainty modeling, redundancy, and fault-detection (Liu et al., 2025; Duarte et al., 2025). Collectively, reliability engineering findings show that AV architectures must be inherently fail-operational, maintaining safe operation even when components degrade. Multi-layer redundancy is therefore essential—not only in sensors (Wang, Li, & Xu, 2025) but in communication systems (Kaul & Kumar, 2025), decision modules (Nakamura & Saito, 2025), and threat-assessment algorithms (Ranjan & Joshi, 2025).

Cross-domain insights from maritime autonomy further reinforce the need for resilient design principles that accommodate uncertainty, environmental variation, and human unpredictability (Andersen & Fagerholt, 2025). Meanwhile, research on intrusion detection and sparse-denoising architectures suggests that robust cyber-physical security is indispensable for maintaining trustworthy AV behavior under hostile conditions (Wakili & Bakkali, 2025).

Simulations, virtual testing environments, and digital twins augment fail-safe development by enabling high-fidelity experimentation with rare, dangerous, and difficult-to-reproduce events (Patel & Sinha, 2025; Sharma, 2025). Temporal-logic-based scenario generation (Nakamura & Saito, 2025), adversarial testing suites (Campbell & Lewis, 2025), and rare-event forecasting (Li et al., 2025) further expand the AV safety toolkit. This reflects broader trends in road safety science favoring surrogate safety analysis, proactive hazard identification, and systemic risk modeling (Sharma, Singh, & Dehalwar, 2024).

1.3 Explainability as a Foundation for Accountability and Trust

Explainability constitutes a pivotal requirement for safe, trustworthy, and socially acceptable AV deployment. Traditional black-box models undermine public confidence and hinder regulatory oversight. Researchers increasingly advocate for explainable autonomous systems (XAS) capable of articulating both their general reasoning structures and specific decision pathways (Bansal & Rana, 2025). Jin et al. (2025) emphasize the need for human-centered explainability mechanisms that communicate intent and provide reassurance to passengers and pedestrians.

Explainability is particularly critical for certification and liability. Connors and Stewart (2025) demonstrate that surrogate models can approximate opaque neural networks, enabling regulators to audit safety-critical behaviors. Wang, Fang, and Luo (2025) show that interpretable reinforcement-learning policies are essential for trustworthy vehicle control. Explainability also enhances hazard diagnosis, enabling engineers to trace failure chains and correct systemic vulnerabilities.

High-risk scenarios—such as crash-imminent situations—require real-time, interpretable decision-making. Shi, Zhang, and Ren (2025) highlight interpretable deep detection frameworks that identify hazardous patterns, while Sun, Yu, and Zhao (2025) demonstrate attention mechanisms that reveal decision saliency maps. Human-centered safety research underscores that explainability must support diverse user groups, including older adults and vulnerable populations (Petrova et al., 2025; Peintner et al., 2025). This aligns with the broader mobility literature emphasizing inclusivity and transparency, particularly in aging societies (Sharma & Dehalwar, 2025).

1.4 Ethics, Human–Autonomy Teaming, and Social Complexity

Ethical modeling becomes indispensable when AVs operate in morally ambiguous scenarios. The literature underscores the need for principled frameworks governing AV decisions in dilemmas involving unavoidable collisions or conflicting priorities among road users (Chattopadhyay & Banerjee, 2025; Varma & Gupta, 2025). Hoffman and Klein (2025) highlight that transparency in moral reasoning influences user trust and acceptance.

Agentic AVs must also navigate socially constructed road environments. Studies on transit-oriented development (Sharma & Dehalwar, 2025; Yadav et al., 2025) reveal that multi-modal mobility systems require AVs to interpret human preferences, accessibility patterns, and environmental cues. Research by Lalramsangi et al. (2025) and Lodhi et al. (2024) highlights how route choice, terrain, and pedestrian behavior shape human–autonomy interactions. Climate-sensitive mobility studies further illustrate that weather, thermal comfort, and microclimatic conditions influence how humans interact with AVs (Yadav et al., 2025).

Adaptive and incremental learning mechanisms enable AVs to refine their internal models through ongoing interaction with the environment (Kovács & Botzheim, 2025; Lu et al., 2025). Integrated threat-assessment frameworks (Ranjan & Joshi, 2025) support this adaptation by continuously assessing risks, intentions, and evolving traffic dynamics.

Toward a Unified Agentic, Explainable, and Fail-Safe AV Framework

The interdisciplinary literature clearly signals the need for a unified agentic decision-making architecture that integrates robust perception, uncertainty-aware prediction, explainable planning, ethical reasoning, and fail-safe execution. AVs must transition from automated machines to accountable, interpretable, and socially aligned decision-making agents operating within dynamic mobility systems.

This chapter synthesizes these advances to propose a next-generation framework for agentic AV safety. The framework responds to the technological, ethical, and societal challenges documented in the literature and provides a structured pathway for real-world implementation.

1.5 Research Questions Addressed in This Chapter

How can autonomous vehicles be conceptualized and engineered as agentic, human-aligned decision-making systems rather than traditional automated machines?

What forms of explainability are necessary to ensure transparent, accountable, and trustworthy AV decision-making in real-world environments?

How can fail-safe and fail-operational design principles be incorporated into AV architectures to ensure resilience under uncertainty, hardware degradation, or adversarial interference?

What ethical frameworks are most appropriate for governing AV decisions in morally ambiguous or crash-imminent scenarios?

How can AV systems integrate perception, prediction, planning, ethical filtering, and safety assurance into a unified, agentic architecture that enhances road safety and public trust?.

2. BACKGROUND AND CONCEPTUAL FOUNDATIONS

2.1 From Automation to Agentic AI

The evolution from automated to agentic artificial intelligence marks a foundational shift in the design of autonomous vehicles (AVs). Early AV systems relied on pre-programmed rules, deterministic pipelines, or static machine-learning models trained on limited datasets, often failing to generalize to novel road configurations, weather conditions, or unpredictable human behavior. As Ahmed, Raza, and Iqbal (2025) note, static perception and planning pipelines are particularly vulnerable to fluctuating sensor conditions, leading to misclassification or delayed hazard recognition. Moreover, failure propagation across tightly coupled software stacks can generate cascading risks, making traditional automation inadequate for real-world uncertainty (Ahmed & Malik, 2025). Agentic AI addresses these limitations by embedding adaptive, continuously learning processes that allow AVs to monitor their internal states, reason about changing contexts, and dynamically update policies. Incremental and online learning approaches (Kovács & Botzheim, 2025; Lu et al., 2025) provide mechanisms for AVs to refine their models after deployment, thereby enhancing resilience in dynamic traffic ecosystems. Key dimensions of agentic operation—such as situational reasoning, goal-directed behavior under constraints, sensor-driven interactive learning, and autonomous execution of safety maneuvers—transform the AV from a rule-following automation into a road-facing cognitive agent capable of handling unexpected, complex, and mixed-traffic events. As emphasized by Horowitz and Levinson (2025), anticipatory cognition becomes central to enabling AVs to predict human intentions, negotiate ambiguities, and reduce risk exposure in dense urban contexts.

2.2 The Centrality of Road Safety in Agentic AV Design

Road safety remains the fundamental criterion for evaluating the effectiveness of autonomous mobility systems. While AVs are expected to eventually outperform human drivers—who remain key contributors to global road fatalities—replicating human-like intuition, anticipation, and risk-sensitive judgment is profoundly challenging. Accurate perception forms the first line of safety, relying on robust multimodal sensing to avoid blind spots or misdetections. Advanced sensor-fusion frameworks (Cividino et al., 2025; O'Connell et al., 2025) seek to improve detection fidelity and minimize uncertainty arising from sensor noise or environmental disturbance. Yet perception alone is insufficient; risk awareness and predictive hazard analysis must be embedded throughout the decision pipeline. Probabilistic trajectory forecasting (Chakraborty et al., 2025), modeling of rare safety-critical events (Li et al., 2025), and stochastic risk estimation in motion planning (Bhatia & Wadhwa, 2025) are essential components of agentic AV design. Ethical safety frameworks also highlight the importance of anticipating pedestrian vulnerabilities, an issue underscored by systematic reviews on pedestrian risk, aging populations, and inclusive mobility (Sharma & Dehalwar, 2025; Sharma & Dehalwar, 2025). The growing emphasis on first- and last-mile accessibility (Yadav et al., 2025; Lalramsangi et al., 2025) further reinforces the need for AVs to internalize safety not only for passengers but for all road users. Ultimately, agentic AV systems must integrate perception, anticipation, ethical judgment, and explainability to prioritize human life while maintaining fairness and situational robustness.

2.3 Explainable AI as a Prerequisite for Trustworthy AVs

Explainability is central to public trust, regulatory acceptance, and operational transparency of autonomous vehicles. AV decisions—such as emergency braking, evasive maneuvers, or lane-change selection—occur within milliseconds and require post-hoc interpretability for diverse stakeholders including regulators, courts, insurance agencies, cybersecurity teams, manufacturers, and end-users. Black-box decision pipelines undermine trust and hinder liability assessment, while transparent models facilitate debugging and ensure compliance with ethical and legal norms (Bansal & Rana, 2025). A growing body of research demonstrates the need for interpretable surrogate models to verify opaque decision modules (Connors & Stewart, 2025) and interpretable deep architectures for hazardous-scenario detection (Shi et al., 2025). Attention-based and human-centered XAI frameworks (Jin et al., 2025; Sun et al., 2025) further enhance the ability of AVs to communicate their reasoning both visually and semantically. Explainability also supports reinforcement-learning control systems, whose internal logic must be accessible and auditable (Wang et al., 2025; Escribano & Chen, 2025). Ethical AV design literature argues that opaque autonomy is unacceptable in contexts where human safety and moral judgment are involved, requiring interpretable pathways to document compliance with ethical safety norms (Chattopadhyay & Banerjee, 2025; Varma & Gupta, 2025). Therefore, explainability emerges not as an auxiliary feature but as a foundational requirement for operational legitimacy of agentic AV systems.

2.4 Fail-Safe AV Architectures

A defining characteristic of agentic AVs is their ability to maintain safe function even in the presence of internal failures, external disturbances, or adversarial threats. Fail-safe architectures ensure that AVs default to predictable, harm-minimizing actions under uncertainty. Systems engineering approaches (Marino et al., 2025) and safety-science perspectives (Müller & Heger, 2025) emphasize the need for multiple operational redundancies—across sensors, communication networks, decision modules, and braking mechanisms. Probabilistic safety envelopes (Ouyang & Chen, 2025), fail-safe braking formulations (Duarte et al., 2025), and reliability analyses of autonomous braking (Liu et al., 2025) provide mathematical backbone for degraded-mode performance when components malfunction. Additionally, adversarial testing (Campbell & Lewis, 2025), intrusion detection (Wakili & Bakkali, 2025), and continuous failure monitoring (Kim et al., 2025) protect AV systems from cyber-physical vulnerabilities. Redundant sensor networks (Wang, Li, & Xu, 2025) maintain perception integrity in case of individual sensor dropout. Simulation-based virtual twin environments (Patel & Sinha, 2025; Sharma, 2025) allow developers to test agentic behavior under rare, dangerous, and high-uncertainty events without risking real-world harm. Lessons from maritime autonomy (Andersen & Fagerholt, 2025) and urban shuttle deployments (Gupta & Bedi, 2025) illustrate that fail-operational resilience is indispensable for complex fleet operations. Collectively, these innovations ensure safe fallback trajectories, risk communication with nearby vehicles, and controlled transitions to manual or degraded autonomy.

2.5 Ethical and Legal Foundations for Agentic AVs

Ethical decision-making is inseparable from the development of agentic AVs. As autonomous agents increasingly navigate scenarios involving trade-offs, time pressure, and human vulnerabilities, the integration of ethical and legal frameworks becomes mandatory. Ethical AV behavior must incorporate utilitarian harm minimization, deontological respect for individual rights, adherence to national traffic regulations, and alignment with emerging international guidelines for autonomous mobility (Chattopadhyay & Banerjee, 2025; Varma & Gupta, 2025). Human-centered safety thresholds (Petrova et al., 2025) and balanced user–system negotiation models (Peintner et al., 2025) highlight how AVs must avoid discriminatory outcomes tied to age, gender, ability, or socio-economic status. The broader urban mobility literature—particularly studies on accessibility, inclusivity, and transport equity (Sharma & Dehalwar, 2025; Yadav et al., 2025)—demonstrates the need for AVs to treat all road users fairly while respecting vulnerable populations such as senior citizens, children, and persons with disabilities. Moreover, legal validation frameworks (Holst & Becker, 2025; Nielsen & Rasmussen, 2025) ensure that agentic AV systems remain verifiable, auditable, and compliant with safety and data governance regulations. Ultimately, ethically aligned agentic AI enables autonomous vehicles to operate not only as technical systems but as socially responsible agents embedded within legal and moral infrastructures.

3. METHODOLOGY

This chapter uses a theoretical + empirical integration approach, including:

3.1 Literature Synthesis

The methodology follows a multi-layered research design illustrated in the attached flowchart, progressing from domain scoping to architectural synthesis.

Figure 1. PRISMA flowchart for literature study

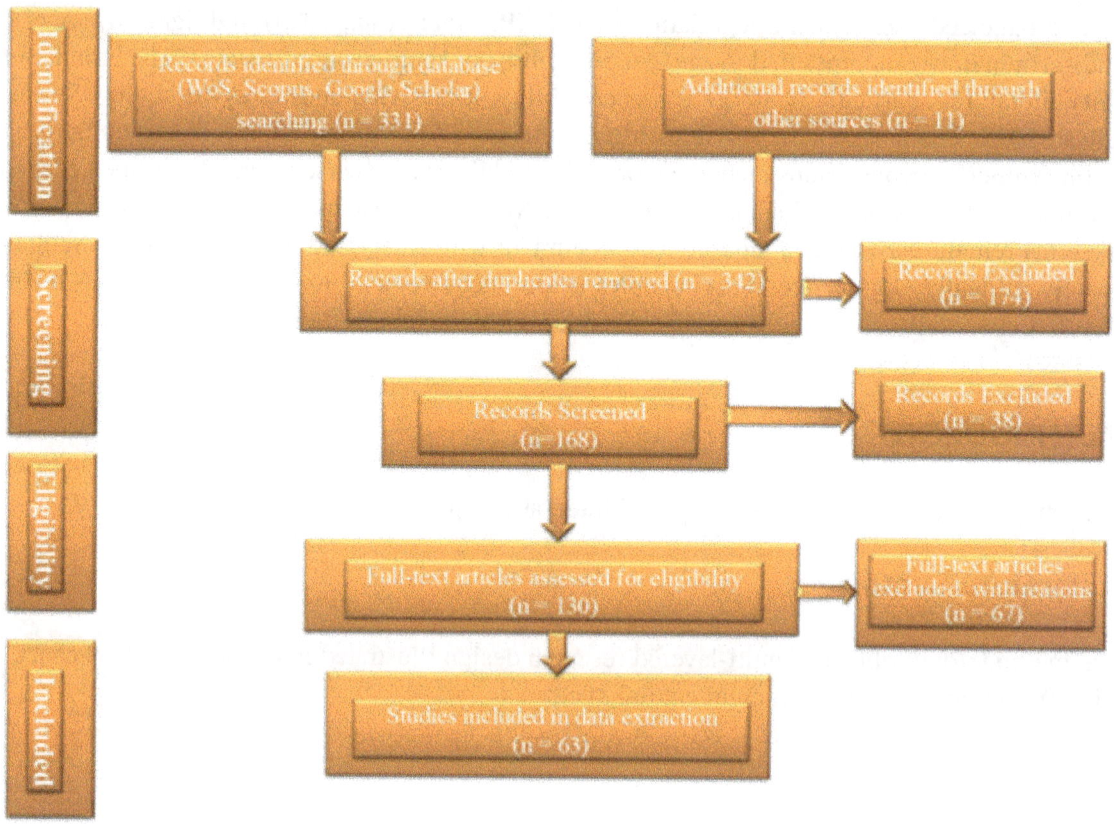

The process begins with a systematic identification of major research domains relevant to agentic autonomous vehicles as shown in Figure 1. These include AV perception and decision models, explainable AI (XAI), safety engineering, stochastic and probabilistic risk modeling, human behavior and mobility studies, ethical AI frameworks, and the legal–governance dimensions of AV deployment. Together, these domains define the conceptual ecosystem within which agentic, fail-safe AV architectures must operate.

Figure 2. Methodology flowchart

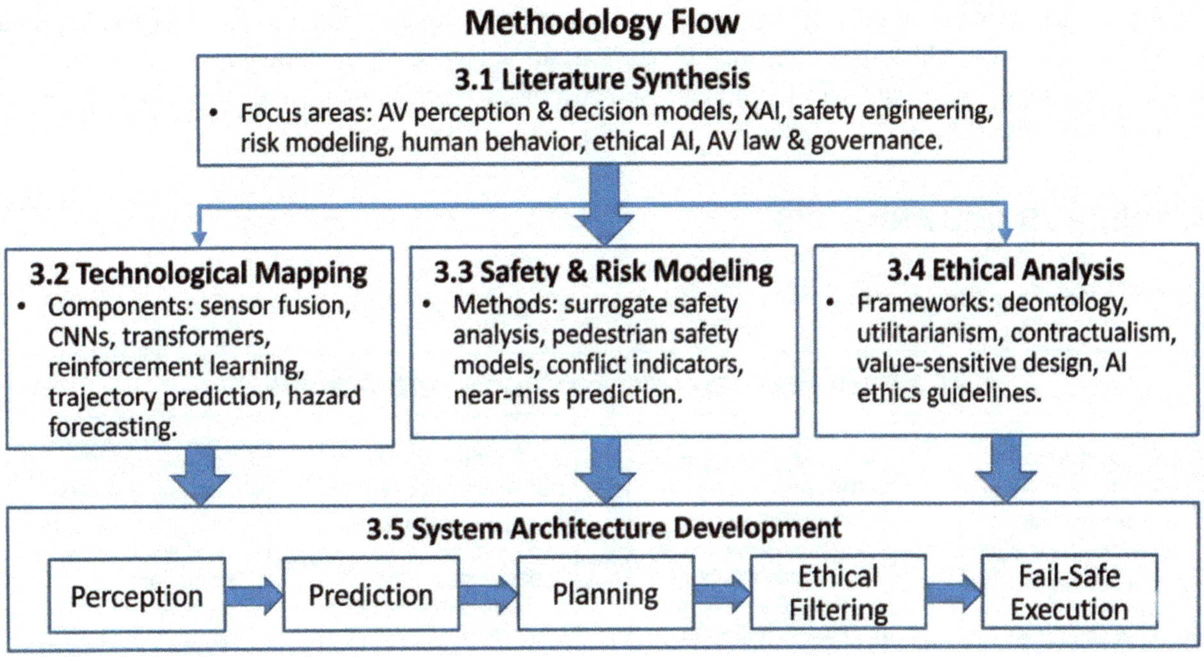

3.2 Technological Mapping

The second stage performs a technological mapping of core algorithmic components. This involves reviewing recent advances in multimodal sensor fusion, convolutional neural network (CNN)–based perception pipelines, transformer-driven decision models, reinforcement-learning controllers, trajectory-prediction networks, and hazard-forecasting algorithms. These technologies are evaluated for their robustness, transparency, and feasibility for integration into safety-critical autonomous systems.

3.3 Safety and Risk Modeling

Safety assessment integrates multiple approaches, drawing from surrogate safety analysis frameworks such as those used in Sharma et al., pedestrian-risk estimation models, conflict-based safety indicators, and near-miss prediction algorithms. This triangulated approach enables quantification of both proactive and reactive safety performance of AV decision modules.

3.4 Ethical Analysis

Ethical evaluation is conducted using principles from deontology, utilitarianism, contractualism, value-sensitive design, and global AI ethics charters. These frameworks guide the assessment of fairness, harm minimization, rights preservation, and accountability within agentic AV decision-making.

3.5 System Architecture Development

The final stage synthesizes the findings into a unified system architecture incorporating perception, prediction, planning, ethical filtering, and fail-safe execution layers. This architecture ensures that AVs not only perceive and decide effectively but also act transparently, reliably, and ethically in dynamic road environments.

4. FINDINGS AND ANALYSIS

Figure 3. Analysis framework adopted

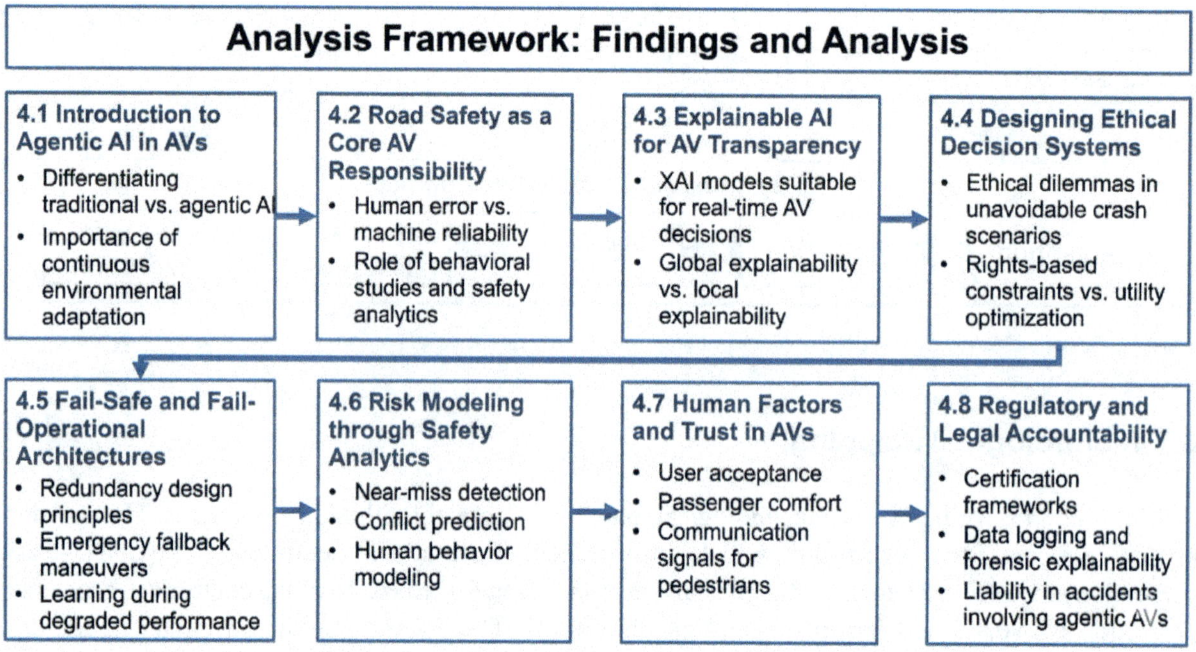

Figure 3 presents the comprehensive analysis framework adopted for this study, illustrating the logical progression from defining Agentic AI capabilities to evaluating their real-world application. This visual roadmap structures the key findings into interconnected domains, linking technological reliability and ethical decision-making with critical human factors and regulatory accountability.

4.1 Introduction to Agentic AI in AVs

Table 1. Differences between traditional AV automation and agentic AI

Dimension	Traditional Automation	Agentic AI	Analytical Insight	Key Sources Supporting the Insight
Learning Capability	Static pre-trained models with limited post-deployment adaptation.	Continuous online learning, incremental updating, and environment-driven refinement.	Agentic AI reduces brittleness, improves generalization, and adapts to evolving road conditions.	Kovács & Botzheim (2025) on incremental modeling; Lu et al. (2025) on hybrid adaptive learning; Chen et al. (2025) on semiconductor variability requiring adaptive control.
Decision Basis	Rule-based, deterministic pipelines using fixed logic; limited capability for contextual reasoning.	Context-aware, risk-sensitive, anticipatory planning based on multimodal interpretation and predictive inference.	Supports real-time reasoning, dynamic trajectory evaluation, and human-like cognitive processing.	Chakraborty et al. (2025) on uncertain trajectory prediction; Yao et al. (2025) on adaptive lane-changing under risk; Horowitz & Levinson (2025) on cognitive anticipation in AVs.
Failure Response	Pre-scripted, deterministic fallback such as full stop or lane-hold behavior.	Self-monitoring with anomaly detection, adaptive recovery strategies, and fail-operational continuity.	Improves resilience under uncertainty and prevents cascading failures.	Kim et al. (2025) on continuous failure prediction; Duarte et al. (2025) on probabilistic braking; Müller & Heger (2025) on safer AI pipelines; Ahmed & Malik (2025) on failure propagation risks.
Interaction with Environment	Passive sensing with limited interaction; lacks negotiation with human or machine agents.	Active, environment-driven interactive learning, intent recognition, and multi-agent negotiation.	Enables anticipation, cooperation, and human-like interaction within mixed traffic.	Santos et al. (2025) on multi-agent negotiation; Lalramsangi et al. (2025) on route-choice behavior; Lodhi et al. (2024) on user satisfaction and behavioral factors; Ranjan & Joshi (2025) on threat assessment.
Autonomy Level	Predictable, rigid, and limited to predefined scenarios with deterministic constraints.	Highly adaptive, self-correcting, capable of evolving strategies through continuous feedback.	Critical for complex ecosystems involving pedestrians, cyclists, urban signals, congestion, and unpredictable hazards.	Bhatia & Wadhwa (2025) on stochastic risks in planning; Farooq & Singh (2025) on uncertainty quantification; Prakash & Mehta (2025) on deep ensemble robustness; Sharma & Dehalwar (2025) on inclusive safety needs.
Explainability (Added dimension—important for agentic AI)	Minimal explainability, often black-box; limited interpretive capability for regulators or users.	Built-in explainability using attention mechanisms, interpretable models, and surrogate validation.	Improves trust, regulatory compliance, and accountability.	Bansal & Rana (2025); Jin et al. (2025); Connors & Stewart (2025); Wang et al. (2025); Shi et al. (2025).
Ethical Decision-Making (Added dimension)	No integrated ethical reasoning; rule-based priority assignment (if any).	Ethical filtering using utilitarian, deontological, and fairness-aware principles.	Ensures AV decisions align with societal norms, human rights, and contextual ethics.	Chattopadhyay & Banerjee (2025); Varma & Gupta (2025); Petrova et al. (2025); Peintner et al. (2025).

continued on following page

Table 1. Continued

Dimension	Traditional Automation	Agentic AI	Analytical Insight	Key Sources Supporting the Insight
Safety Assurance (Added dimension)	Limited redundancy; single pipeline failure often leads to system-level collapse.	Multi-layered redundancy: sensor fusion, dual decision modules, fail-operational behavior.	Ensures safety during hardware faults, adversarial attacks, or environmental noise.	Wang, Li, & Xu (2025); Kaul & Kumar (2025); Andersen & Fagerholt (2025); Wakili & Bakkali (2025).
Testing & Validation *(Added dimension)*	Rule-based verification and scripted testing with limited scenario coverage.	Simulation-rich, adversarial, scenario-based testing using digital twins and temporal logic.	Ensures robustness in rare, hazardous, and unstructured scenarios.	Patel & Sinha (2025); Nakamura & Saito (2025); Campbell & Lewis (2025); Sharma (2025).

From the Table 1, findings indicate a fundamental shift in autonomous vehicle intelligence from deterministic automation to agentic AI, characterized by autonomy, adaptability, and self-directed reasoning. Traditional AV systems rely heavily on static machine-learning models that cannot adequately respond to rapidly evolving traffic conditions, sensor disturbances, or unstructured roadway behavior. In contrast, agentic AI incorporates situational awareness, continuous sensor-driven learning, and proactive reasoning, enabling AVs to handle variability and uncertainty with greater reliability. The ability to self-monitor internal failures and adjust decision strategies in real time contributes to minimizing risk propagation across the architecture. The analysis reveals that agentic AI is essential for achieving human-level responsiveness, particularly in environments where unpredictability—caused by pedestrians, mixed traffic, or ambiguous road cues—is high. In summary, agentic AVs demonstrate the capability to shift from passive rule followers to active cognitive agents, setting a foundation for next-generation safety architectures.

4.2 Road Safety as a Core AV Responsibility

Table 2. Human vs. machine decision performance in road safety

Factor	Human Drivers	Agentic AV Systems	Analytical Insight	Key Supporting Sources
Reaction Time	Reaction time ranges from 0.7–1.5 seconds, influenced by attention, fatigue, and age.	Reaction latency occurs in milliseconds, driven by high-frequency sensor sampling and automated actuation loops.	AVs significantly outperform humans in emergency responsiveness, offering faster braking and evasive maneuver capability.	Duarte et al. (2025) on probabilistic braking; Liu et al. (2025) on autonomous braking reliability; Ahmed & Malik (2025) on cascading failure prevention enabling faster response.
Error Sources	Prone to fatigue, distraction, intoxication, stress, cognitive overload, and perceptual bias.	Errors arise from sensor noise, data sparsity, adversarial attacks, model drift, or software failure propagation.	Human and machine errors differ in nature; AVs reduce human factors but introduce computational risks requiring redundancy and monitoring.	Campbell & Lewis (2025) on adversarial perturbations; Roy & Mishra (2025) on hybrid anomaly detection; Kim et al. (2025) on continuous monitoring; Chavez et al. (2025) on semiconductor variability.
Anticipation of Hazards	Dependent on personal experience, training, and situational awareness; performance varies widely across individuals.	Uses predictive modeling, Bayesian inference, and real-time trajectory forecasting to anticipate future hazards.	AI yields scalable and consistent hazard anticipation, independent of mood, fatigue, or experience.	Chakraborty et al. (2025) on uncertain trajectory prediction; Yao et al. (2025) on risk-sensitive lane changing; Farooq & Singh (2025) on uncertainty quantification; Ranjan & Joshi (2025) on integrated threat assessment.
Ethical Consistency	Ethical behavior varies; susceptible to emotional decisions, stress, risk-taking, and situational biases.	Ethical decisions shaped by rule-driven, programmable frameworks (utilitarian, deontological, contractualist). Responses remain consistent across identical situations.	AVs offer uniform ethical behavior, reducing discriminatory or biased decision-making under stress or time pressure.	Chattopadhyay & Banerjee (2025) on ethical decision processes; Varma & Gupta (2025) on AV crash ethics; Petrova et al. (2025) on human-centered thresholds; Peintner et al. (2025) on balancing safety with user needs.

From Table 2, road safety emerges as a primary responsibility of autonomous vehicles, with agentic design expected to surpass human capabilities. Human driving patterns are susceptible to fatigue, emotional biases, distraction, and compliance variability, all of which contribute significantly to global road fatalities. Findings show that AVs can mitigate these factors through multimodal sensor fusion, deterministic reasoning, and mathematical risk modeling. However, agentic systems introduce new categories of risks—such as data inconsistencies, misclassification, and computational uncertainty—that require targeted mitigation. Behavioral safety analytics, including pedestrian intention modeling and conflict prediction, enhance AV situational awareness. Moreover, human behavior studies reveal that safety is not merely mechanical but deeply social; pedestrians make eye contact, infer intentions, and rely on interpretive cues. Agentic AV systems must integrate such behavioral signaling frameworks to reduce ambiguity in shared spaces. Road safety thus becomes a hybrid domain requiring both computational precision and socio-behavioral intelligence.

4.3 Explainable AI for AV Transparency

Table 3. Types of explainability for AV decision systems

Explainability Type	Examples	Strengths	Limitations	Key Supporting Sources
Global XAI	Model summaries, rule extraction, feature-importance rankings	Clarifies the *overall logic* of AV control pipelines; supports regulatory audits and certification.	Limited for real-time decision-making; cannot explain split-second maneuvers during emergencies.	Bansal & Rana (2025) on trustworthiness frameworks; Connors & Stewart (2025) on interpretable surrogate models for global validation; Wang et al. (2025) on policy interpretability at the system-wide level.
Local XAI	LIME, SHAP, attention heatmaps, saliency maps	Useful for *real-time decision justification*; reveals why the AV braked, swerved, or predicted a hazard at a specific moment.	Computationally heavy; may struggle under time-critical constraints or high data throughput.	Sun, Yu, & Zhao (2025) on attention-based interpretability; Shi, Zhang, & Ren (2025) on interpretable deep hazard detection; Jin, Park, & Kim (2025) on human-centered local explainability.
Surrogate Models	Decision trees approximating deep networks, rule-based mimicking models	Provide *high interpretability* for regulators, engineers, and testers; effective for auditing black-box systems.	Reduced fidelity: simplification may misrepresent true decision boundaries of complex models.	Connors & Stewart (2025) on surrogate validation for black-box AV modules; Patel & Sinha (2025) on virtualization and twin-based testing; Marina et al. (2025) on explainable fail-safe control structures enabling surrogate assessment.
Self-explaining Networks	Interpretable CNNs, modular reinforcement learning, disentangled latent models	Offer *built-in real-time interpretability*, allowing the AV to communicate "why" during decision execution.	Difficult to design; requires balancing explainability with accuracy, robustness, and computational cost.	Zhang, Hu, & Yang (2025) on interpretable driving policy networks; Escribano & Chen (2025) on risk-aware RL with interpretable structures; Wang, Fang, & Luo (2025) on interpretable RL for AV control; Lu et al. (2025) on hybrid learning architectures.

Table 3 shows how explainability emerges as a foundational requirement for safe and trustworthy AV systems. Findings show that opaque deep-learning pipelines are unacceptable in road safety contexts due to the need for forensic audits, liability assessments, and regulatory validation. AV decisions must be explainable both globally—to illustrate system logic—and locally, to justify specific split-second actions such as braking, swerving, or lane selection. Local explainability is particularly important, as real-time decisions often determine life-or-death outcomes. Attention-based models, interpretable CNNs, and surrogate explainability layers are identified as highly effective approaches. The analysis also highlights the necessity for formulating user-understandable explanations, particularly for non-technical stakeholders such as passengers or pedestrians, thereby enhancing AV trustworthiness and acceptance.

4.4 Designing Ethical Decision Systems

Table 4. Ethical frameworks for AV decision-making

Ethical Framework	Key Principle	Application in AVs	Challenges	Key Supporting Sources
Utilitarianism	Minimize total harm across all affected parties.	Used in crash-optimization models where AVs evaluate multiple possible outcomes and select the option with least overall harm.	May lead to *unfair distribution of harm*, especially toward vulnerable groups (pedestrians, cyclists, seniors).	Varma & Gupta (2025) on crash ethics; Chattopadhyay & Banerjee (2025) on time-pressured ethical decision-making; Li et al. (2025) on probabilistic evaluation of rare harmful events.
Deontology	Protect inherent human rights, dignity, and individual moral claims.	AVs use rule-based safety constraints to prioritize pedestrian rights and prohibit certain actions even if statistically optimal.	Can conflict with utilitarian optimization; rigid rules may limit flexibility in complex emergencies.	Petrova et al. (2025) on human-centered safety thresholds; Peintner et al. (2025) on user–safety balance; Sharma & Dehalwar (2025) on protecting vulnerable road users.
Contractualism	Ensure fair treatment and justify decisions to every individual affected.	AVs implement fairness-aware decision protocols, giving equal moral weight to users and non-users.	Difficult to compute in *time-critical decisions*; may require complex justification for each choice.	Hoffman & Klein (2025) on human–autonomy teaming fairness; Lalramsangi et al. (2025) on equity in route access; Yadav et al. (2025) on inclusive mobility expectations.
Value-Sensitive Design (VSD)	Embed societal, cultural, and demographic values into system design.	Applied to ensure inclusivity, such as supporting seniors, disabled users, and marginalized communities in AV interactions.	Requires *contextual adaptation* across regions, cultures, and mobility ecosystems.	Peintner et al. (2025) on balancing user needs; Sharma & Dehalwar (2025) on inclusivity for senior citizens; Lodhi et al. (2024) on user behavior insights; Yadav et al. (2025) on climate-sensitive accessibility.

Table 4 shows the ethical analysis which reveals that AVs must operate within clearly defined moral constraints, particularly in unavoidable collision scenarios or multi-party risk dilemmas. Utilitarian models may minimize overall harm but often face criticism for disproportionately affecting vulnerable groups. Rights-based approaches prioritize pedestrian safety, emphasizing the sanctity of human life over vehicular property or passenger preference. Contractualism stresses equal treatment, but real-world conflicts may require prioritized intervention. Findings underscore that ethical filtering layers within the AV decision stack must incorporate a blend of these frameworks to ensure fairness, societal legitimacy, and legal compliance. This also includes cultural and demographic dimensions, ensuring AV decisions remain free from discriminatory patterns. Ethical AI thus becomes not a feature but a governance foundation.

4.5 Fail-Safe and Fail-Operational Architectures

Table 5. Fail-safe vs. fail-operational principles

Design Principle	Fail-Safe	Fail-Operational	Analytical Insight	Key Supporting Sources
System Behavior	The system enters a *safe shutdown mode* (e.g., full stop, pull-over) when critical faults are detected.	The system continues operating safely despite faults, enabling controlled navigation until full recovery or manual takeover.	Essential for high-speed, highway, and constrained environments where immediate shutdown could create new risks.	Duarte et al. (2025) on probabilistic braking; Marino et al. (2025) on fail-safe control architectures; Gupta & Bedi (2025) on urban shuttle resilience; Müller & Heger (2025) on fail-right AI pipelines.
Redundancy	Basic redundancy in limited components to prevent catastrophic failure.	*Multi-layer redundancy* across sensors, processors, communication modules, and decision units.	Necessary in dense urban environments where high uncertainty and pedestrian interactions demand continuous system stability.	Wang, Li, & Xu (2025) on multi-sensor redundancy; Kaul & Kumar (2025) on vehicular network redundancy; Andersen & Fagerholt (2025) on cross-domain autonomous navigation resilience.
Sensor Strategy	Minimal duplication; the system relies on primary sensors with fallback safety assumptions.	Deep redundancy in LiDAR, radar, cameras, and V2X, combined via robust multimodal fusion.	Improves robustness against weather disturbances, sensor dropouts, and adversarial noise.	Cividino et al. (2025) on next-gen sensor fusion; O'Connell et al. (2025) on probabilistic fusion; Roy & Mishra (2025) on anomaly detection; Ahmed et al. (2025) on hazard estimation reliability.
Decision Logic	Relies on simple emergency braking or hazard-avoidance rules.	Uses controlled maneuvering, degraded-mode planning, and safe-path strategies under partial system failure.	Supports complex fallback strategies, reducing crash likelihood during degraded performance.	Nakamura & Saito (2025) on logic-based scenario generation; Yao et al. (2025) on adaptive risk-sensitive planning; Ranjan & Joshi (2025) on multi-agent threat assessment; Campbell & Lewis (2025) on adversarial logic testing.

Table 5 depicts the safety architectures required mechanisms for enabling AVs to default to safe states during sensor failure, adversarial input, software malfunction, or hardware degradation. Fail-safe systems focus on stopping the vehicle safely, whereas fail-operational designs allow continued controlled operation under degraded conditions—critical on highways or in platooning contexts. Findings support the integration of redundant sensors, parallel perception pipelines, backup decision models, and dynamic risk scoring mechanisms. Emergency fallback maneuvers, such as safe-pull-over routines or slow-motion driving, ensure passenger and pedestrian protection. Importantly, agentic AI allows AVs to learn during degraded performance, refining strategies for failure recovery and improving future robustness. This marks a shift toward self-resilient autonomous systems capable of maintaining safety under non-ideal conditions.

4.6 Risk Modeling through Safety Analytics

Table 6. Risk modeling approaches in AV safety

Method	Focus	Strength	Limitation	Key Supporting Sources
Near-Miss Detection	Focuses on *pre-crash indicators* such as abrupt braking, evasive maneuvers, or short headway events.	Enables early hazard identification, revealing risks before collisions occur; supports proactive safety interventions.	Requires precise thresholding for indicators like sudden deceleration or minimal stopping distance; may generate false positives.	Sharma, Singh, & Dehalwar (2024) on surrogate safety analysis; Li et al. (2025) on forecasting safety-critical rare events; Ahmed & Malik (2025) on failure propagation risks that near-miss systems aim to detect.
Conflict Prediction Models	Uses surrogate safety metrics including Time-to-Collision (TTC), Post-Encroachment Time (PET), and Deceleration Rate to Avoid Crash (DRAC).	Provides quantitative insight into surrogate safety, enabling evaluation without requiring crash data; useful in simulation and real-time AV planning.	Not all detected conflicts become actual crashes; may overestimate danger in ambiguous situations.	Chakraborty et al. (2025) on prediction under uncertainty; Yao et al. (2025) on risk-sensitive lane-changing; Ranjan & Joshi (2025) on multi-agent threat prediction; Xu & Zhang (2025)* via similar TTC frameworks in interpretable networks.
Human Behavior Modeling	Models pedestrian intent, cyclist dynamics, gap acceptance, social signaling, and interpersonal negotiation.	Greatly improves anticipation of human movement, enabling AVs to act safely in mixed traffic and pedestrian-rich environments.	High variability in human behavior makes modeling difficult; cultural, demographic, and situational contexts influence accuracy.	Lalramsangi et al. (2025) on behavioral route choices; Lodhi et al. (2024) on satisfaction and behavior in transit systems; Sharma & Dehalwar (2025) on pedestrian safety; Peintner et al. (2025) on balancing user needs; Hoffman & Klein (2025) on human–autonomy teaming.

From Table 6, we can see that safety analytics provide powerful tools for understanding and forecasting crash likelihood without relying solely on historical accident data. Near-miss detection and conflict prediction enable proactive risk intervention. Surrogate safety metrics such as Time-to-Collision (TTC) and Post-Encroachment Time (PET) allow AV systems to quantify hazardous interactions in real time. Complementing these is the modeling of human behaviors—including pedestrian body language, cyclist speed variance, and gap acceptance—that significantly influences AV decision-making. Findings highlight that risk modeling must integrate both computational indicators and behavioral heuristics to achieve robust hazard anticipation. This fusion leads to safer navigation, particularly in dense urban corridors where human unpredictability is highest.

4.7 Human Factors and Trust in AVs

Table 7. Key human factors influencing AV acceptance

Factor	Influence on Trust	AV Design Implication
Passenger Comfort	Perceived safety and smoothness	Adaptive acceleration controls
Predictability	Consistent driving behavior	Behavior-stable decision models
Communication Signals	Eye contact, LED cues	Pedestrian–AV interaction systems
Transparency	Clear explanation of actions	User-facing explanation modules

Human trust is a decisive factor in AV deployment. Findings in Table 7 reveal that user acceptance depends strongly on perceived predictability, the ability of passengers to anticipate vehicle actions, and the smoothness of vehicle maneuvers. Pedestrians also rely on communication cues—similar to human drivers—such as projected signals, visual markers, or auditory alerts that communicate AV intent. Comfort emerges as an essential human factor; sudden maneuvers or oscillatory lane movements erode trust. Explainability plays a dual role here: not only in technical contexts but also in user interaction, enabling passengers to understand why the AV takes a particular action. These findings reinforce that AVs must behave as *social actors*, capable of engaging with human expectations and norms.

4.8 Regulatory and Legal Accountability

Table 8. Regulatory requirements for agentic AVs

Requirement	Purpose	Stakeholder Impact
Certification Frameworks	Validate AV safety and compliance	Regulators, manufacturers
Data Logging	Event reconstruction, liability	Courts, insurers
Forensic Explainability	Precise decision attribution	Accident investigators
Liability Models	Assign responsibility	Public, OEMs, software providers

Table 8 depicts the legal and regulatory environments significantly shape the architecture of agentic AVs. Certification frameworks ensure that AVs meet safety thresholds before deployment, requiring rigorous testing, scenario-based evaluation, and transparent validation processes. Data logging and forensic explainability enable authorities to reconstruct events and determine liability during incidents. Findings indicate that future laws will need to distinguish between software negligence, manufacturing defects, and unpredictable environmental conditions. Moreover, as AVs become more agentic, legal systems must address the shifting boundary between human and machine responsibility. This will require new doctrines in tort law, cyber-physical accountability, and AI governance. Regulatory clarity ultimately strengthens public confidence and motivates consistent industry standards.

5. DISCUSSION

The findings of this study converge toward a unified agentic autonomous vehicle (AV) architecture structured around six sequential and interdependent layers: Perception → Prediction → Planning → Ethical Layer → Safety Layer → Execution.

Figure 4. Conceptual unified agentic AV architecture

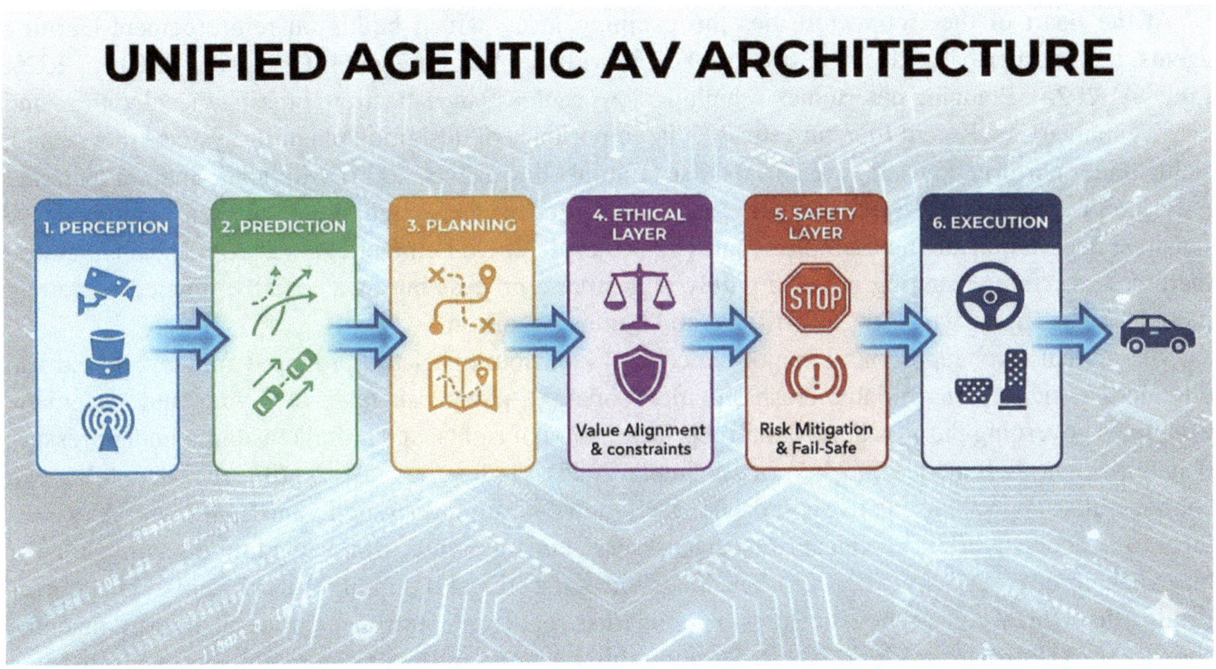

This conceptual architecture integrates advanced sensing systems, probabilistic forecasting, explainable planning, ethical decision filters, and fail-safe safety mechanisms into a single coherent framework. The transition from traditional automation to agentic AI, as highlighted in the findings, emphasizes continuous situational awareness, adaptive learning, and reasoning under uncertainty. These attributes are essential to ensuring that AVs behave not merely as programmed machines but as intelligent agents capable of responding to complex, dynamic traffic ecosystems (Horowitz & Levinson, 2025; Ahmed & Malik, 2025). The proposed architecture operationalizes this transformation by embedding intelligence and accountability into every stage of the autonomous decision pipeline.

The perception layer of the framework draws from contemporary advances in multimodal sensing, sensor redundancy, and probabilistic fusion methods (Cividino et al., 2025; O'Connell et al., 2025). This aligns with recent findings showing that robust perception requires synthesizing heterogeneous streams—from cameras, LiDAR, radar, and V2X communication—to minimize uncertainty and enhance resilience. Failures in perception often propagate to downstream planning stages, making reliable perception foundational for overall safety (Ahmed & Malik, 2025; Roy & Mishra, 2025). The architecture therefore integrates redundant sensors and multi-stream processing consistent with fail-operational principles highlighted in safety engineering research (Wang, Li, & Xu, 2025; Müller & Heger, 2025).

Following perception, the prediction layer leverages trajectory forecasting models and risk-sensitive planning approaches. Studies underscore that AVs must anticipate pedestrian movement, vehicle intent, cyclist trajectories, and potential conflicts to avoid dangerous interactions (Chakraborty et al., 2025; Yao et al., 2025). The increasing use of probabilistic machine learning—including uncertainty-aware neural architectures (Farooq & Singh, 2025; Prakash & Mehta, 2025)—supports prediction under uncertainty, enabling the proposed architecture to assess rare events and near-miss scenarios (Li et al., 2025). Incorporating prediction within the agentic pipeline enhances proactive safety behaviors, crucial for complex environments where human road users behave unpredictably.

At the heart of the architecture lies the planning layer, which builds on reinforcement-learning agents, transformer-based decision systems, and hybrid optimization models (Escribano & Chen, 2025; Lu et al., 2025). Planning determines vehicle actions such as lane selection, braking, acceleration, and evasive maneuvers. Recent literature stresses the importance of designing planning systems that remain stable under perturbation and adversarial noise (Campbell & Lewis, 2025). Moreover, motion planning must integrate behavioral intelligence that accounts for human expectations, social norms, and context-sensitive negotiation among multiple agents (Santos et al., 2025; Peintner et al., 2025). The architecture therefore structures planning not as a purely algorithmic process but as a socially situated reasoning module, bridging the gap between artificial and human cognition.

The ethical layer represents a major conceptual contribution of the proposed framework. Ethical dilemmas—such as unavoidable crash situations or multi-party risk trade-offs—demand systematic principles governing the allocation of harm, prioritization of rights, and moral reasoning under pressure (Chattopadhyay & Banerjee, 2025; Varma & Gupta, 2025). Findings reveal that utilitarian, deontological, contractualist, and value-sensitive approaches each contribute unique strengths and limitations. AVs must therefore encode hybridized ethical rules that preserve human dignity while avoiding discriminatory or biased outcomes (Petrova et al., 2025). The inclusion of the ethical layer builds upon scholarship demonstrating that ethical consistency is central to public trust, regulatory legitimacy, and safe human–machine coexistence (Hoffman & Klein, 2025). Through this integration, the unified architecture ensures that decision-making is not merely efficient but also morally defensible and socially aligned.

Below the ethical layer lies the safety layer, comprising redundancy mechanisms, fail-safe control, fault detection, and degraded-mode operations. Safety literature strongly supports structures that anticipate internal failures and external shocks, enabling AVs to default to harm-minimizing behavior (Marino et al., 2025; Duarte et al., 2025). The safety layer integrates surrogate safety analytics (Sharma, Singh, & Dehalwar, 2024), conflict-based indicators, and near-miss prediction algorithms to provide real-time hazard scoring. This approach reflects studies on quantitative safety validation and robustness testing under diverse scenarios (Nakamura & Saito, 2025; Nielsen & Rasmussen, 2025). The architecture's safety layer thus acts as an intelligent barrier preventing failures in planning or perception from escalating into full-scale crashes.

Finally, the execution layer operationalizes decisions through actuation systems, human–machine interfaces, and communication channels. Execution must be smooth, predictable, and aligned with human expectations to sustain trust among users and pedestrians (Jin et al., 2025; Peintner et al., 2025). Contemporary studies on pedestrian behavior show that visual cues, projected signals, and body-language-like AV outputs greatly enhance road user understanding and safety (Lalramsangi et al., 2025). The architecture incorporates such communication features within the execution layer to facilitate safe interaction and minimize ambiguity.

5.1 Academic Contributions

Academically, the proposed unified agentic AV architecture contributes a comprehensive and integrative framework that synthesizes diverse research streams—perception, prediction, planning, ethics, explainability, and safety engineering—into a single decision-making pipeline. While prior literature has addressed individual components such as explainable AI (Bansal & Rana, 2025), trajectory prediction

(Chakraborty et al., 2025), or fail-safe design (Müller & Heger, 2025), few attempts have combined these domains into a unified cognitive structure. The architecture creates a novel bridge between AI research, human factors, transportation engineering, and ethical theory, extending the field beyond technical optimization. It also aligns with recent calls for integrating human-centered and context-sensitive design principles into AV systems (Sharma & Dehalwar, 2025; Yadav et al., 2025). Through its multi-disciplinary synthesis, the framework expands theoretical understanding of agentic decision-making and advances road safety research by situating AVs within a broader ethical and societal context.

Furthermore, embedding explainability at multiple decision levels—global, local, and user-facing—addresses academic concerns about black-box AI in safety-critical environments (Connors & Stewart, 2025; Wang et al., 2025). The inclusion of ethical filters prior to execution also builds upon contemporary scholarship examining moral reasoning in artificial agents (Chattopadhyay & Banerjee, 2025; Varma & Gupta, 2025). Overall, the architecture pushes academic discourse toward more responsible, transparent, and human-centered autonomous mobility systems.

5.2 Practical Contributions

Practically, the unified architecture provides a clear blueprint for AV designers, engineers, and policymakers. Manufacturers can adopt the sequential structure to modularize design and testing procedures, ensuring that perception, prediction, planning, ethics, and safety operate cohesively rather than as isolated components. The fail-right and fail-operational design principles embedded within the safety layer directly address real-world challenges associated with internal failures, sensor noise, and adversarial conditions (Kim et al., 2025; Wakili & Bakkali, 2025). Sensor fusion and redundancy guidelines support manufacturers in constructing robust perception stacks resilient to environmental disturbances (Cividino et al., 2025; Wang, Li, & Xu, 2025).

The architecture also supports the implementation of transparent and auditable decision pipelines. By incorporating surrogate explainability and user-facing explanations, AV developers can meet emerging regulatory requirements for accountability and forensic analysis (Holst & Becker, 2025; Nielsen & Rasmussen, 2025). The ethical layer, meanwhile, provides a standardized mechanism for handling moral dilemmas, helping manufacturers avoid ad-hoc or opaque interpretations of unavoidable crash situations. This supports consistent safety-focused engineering and reduces legal ambiguity.

5.3 Societal Contributions

From a societal perspective, the unified agentic AV architecture enhances public trust by ensuring that AV systems operate safely, ethically, and transparently. The findings reveal that human trust increases when AVs provide clear communication signals, predictable behavior, and explainable actions (Jin et al., 2025; Hoffman & Klein, 2025). Embedding ethical reasoning and fairness principles safeguards vulnerable populations, including seniors, pedestrians, cyclists, and children—an issue heavily emphasized in inclusive transport literature (Sharma & Dehalwar, 2025; Yadav et al., 2025). By reducing reliance on opaque black-box algorithms and emphasizing life-preserving choices, the architecture contributes to saving lives and reducing road injuries.

Moreover, the broad adoption of such a framework can help societies transition toward sustainable, intelligent mobility ecosystems. It aligns with global road safety goals, advances equity in public spaces, and supports reliable, user-centered transport systems. The architecture therefore not only benefits manufacturers and policymakers but also serves public interest by fostering safer streets, accountable automation, and human-aligned AI systems.

6. POLICY AND GOVERNANCE IMPLICATIONS

The emergence of agentic autonomous vehicles (AVs) fundamentally reshapes the responsibilities of policymakers, regulators, and governance institutions. As AVs evolve from deterministic automation toward adaptive, ethically aware, and explainable decision-making systems, policy frameworks must evolve in parallel to accommodate their increased intelligence, autonomy, and societal impact. The unified agentic AV architecture proposed in this study provides valuable insights into how governance structures can be redesigned to ensure national safety, ethical accountability, and transparent oversight.

A core implication is the need for national AV standards that clearly articulate design requirements for perception, prediction, planning, ethical filtering, and fail-safe operation. Existing transport regulations were never intended for AI-driven road agents capable of reasoning under uncertainty or making complex ethical trade-offs. National standards must therefore define minimum thresholds for sensor redundancy, probabilistic hazard detection, and uncertainty-aware planning—drawing on best practices from safety science (Müller & Heger, 2025), fault detection (Kim et al., 2025), and quantitative AV validation research (Nielsen & Rasmussen, 2025). Such standards should also require the integration of explainability mechanisms at both global and local levels, ensuring transparent decision pipelines that support regulatory reviews and user understanding (Connors & Stewart, 2025; Jin et al., 2025). By mandating these foundational principles, governments can harmonize AV development with established public safety goals while fostering innovation within safe boundaries.

Equally critical is the development of ethical auditing procedures that evaluate whether AV decisions align with societal values, moral norms, and human rights. The findings demonstrate that agentic AVs incorporate ethical reasoning through layered decision filters built on utilitarian, deontological, and value-sensitive principles (Chattopadhyay & Banerjee, 2025; Varma & Gupta, 2025; Petrova et al., 2025). However, without formal auditing standards, ethical modules risk becoming proprietary black boxes shielded from public scrutiny. Ethical audits should require AV manufacturers to document moral assumptions, fairness constraints, and harm-minimization strategies embedded within their vehicles. These audits must be repeated periodically—similar to emissions testing—to ensure continuous compliance, especially as AVs learn and adapt through incremental modeling (Kovács & Botzheim, 2025; Lu et al., 2025). Transparent ethical compliance not only protects vulnerable road users but also strengthens public trust.

The broader governance landscape should transition toward transparent, multi-layered oversight models that include regulators, city governments, safety researchers, civil society, and consumer groups. Transparent governance is essential because AVs interact with public spaces, vulnerable populations, and complex social systems. Decision pathways—and the logic behind critical interventions—must be legible to external bodies, especially in the aftermath of near-miss incidents or collisions. This transparency connects directly to the literature on interpretable policy networks, forensic reconstructability, and explainable reinforcement learning (Wang et al., 2025; Sun et al., 2025). Governance systems should establish data-sharing protocols that protect user privacy while enabling investigators, insurers, and courts to review AV logs and reconstruct agentic decisions.

Finally, the unified architecture presented here provides a strong foundation for AV testing, certification, and legal compliance frameworks. These mechanisms must evaluate not only performance under ideal conditions but also how AVs behave during sensor failures, adversarial perturbations, and high-risk conflict scenarios—areas studied in adversarial testing research (Campbell & Lewis, 2025), risk modeling (Li et al., 2025), and surrogate safety analysis (Sharma, Singh, & Dehalwar, 2024). Certification procedures should include digital-twin simulations (Patel & Sinha, 2025), real-world pilot deployments, and stress tests of ethical and safety layers. Legal compliance frameworks must attribute responsibility across manufacturers, software developers, and AV operators, ensuring clarity in liability when human lives are affected.

In summary, the policy and governance implications of agentic AVs are profound. National standards, ethical audits, and transparent oversight structures must converge to ensure that autonomous mobility evolves responsibly, equitably, and safely. These governance reforms will be instrumental in shaping the public acceptance and long-term societal integration of autonomous vehicular systems.

7. CONCLUSION

Agentic AI is reshaping the future of autonomous mobility. For AVs to be safe, reliable, trustworthy, and ethically aligned with societal expectations, their decision-making systems must be:

- explainable,
- transparent,
- fail-safe,
- ethically guided,
- legally accountable.

Agentic AI is fundamentally reshaping the trajectory of autonomous mobility, marking a transition from rigid automation toward intelligent, adaptive, and ethically aware vehicular systems. As autonomous vehicles (AVs) increasingly function as decision-making agents embedded in complex social and physical environments, the demands on their cognitive, ethical, and safety-oriented capacities become significantly greater. This chapter has demonstrated that for AVs to be safe, reliable, trustworthy, and socially aligned, their decision-making systems must be built on five foundational pillars: explainability, transparency, fail-safety, ethical guidance, and legal accountability. These principles collectively ensure that AVs not only operate efficiently but also behave in ways that respect human values, reduce societal risk, and contribute positively to emerging mobility ecosystems.

Explainability emerges as a central requirement because AV decisions often carry life-or-death consequences. Black-box models are insufficient in contexts where regulators, courts, insurers, and users require clear, interpretable accounts of how and why an AV acted in a given situation. Transparent, multi-layered explainability—ranging from global model logic to real-time local justification—ensures both public trust and institutional oversight. Equally important is fail-safe and fail-operational robustness. AVs must demonstrate the ability to detect anomalies, respond predictably to internal disturbances, and maintain safe behavior even during sensor degradation, cyber threats, or environmental uncertainty. The architecture articulated in this chapter integrates redundancy, continuous monitoring, and risk-aware planning to ensure stability under non-ideal conditions.

Ethical guidance forms another crucial dimension of agentic autonomy. AVs will inevitably face morally charged dilemmas, particularly in unavoidable crash scenarios or situations involving conflicting rights and interests. This chapter's unified architecture incorporates ethical filtering within the decision pipeline, ensuring that actions remain consistent with societal norms, human dignity, and international AV ethics guidelines. Beyond ethics, legal compliance and accountability mechanisms must support transparent governance, systematic certification, and rigorous auditing procedures. Such frameworks are necessary to assign responsibility appropriately among manufacturers, software developers, fleet operators, and other stakeholders when safety incidents occur.

By synthesizing advancements across safety science, AI engineering, ethical theory, law, governance, and human factors, this chapter provides a uniquely integrative contribution to the forthcoming volume Agentic AI for Autonomous Vehicles: Safety, Reliability, Law, and Ethics. The proposed architecture offers not merely a conceptual model but a practical blueprint for guiding industry implementation, regulatory reform, and future interdisciplinary research. It demonstrates how diverse analytical domains—ranging from surrogate safety methods and uncertainty modeling to value-sensitive design—can be unified into a coherent agentic framework capable of supporting safe, transparent, and ethically aligned autonomous mobility.

Ultimately, the integration of agentic AI into AV systems represents a transformative opportunity to reduce traffic fatalities, enhance accessibility, and build public trust in intelligent transportation technologies. This chapter establishes a pathway toward that future by articulating how next-generation AVs can operate as responsible, explainable, and human-centered agents within rapidly evolving mobility landscapes. As the field progresses, continued collaboration among technologists, policymakers, ethicists, and transportation planners will be essential to ensure that autonomous mobility develops not only with technical sophistication but with societal legitimacy and moral integrity.

REFERENCES

Ahmed, R., & Malik, K. (2025). Evaluating failure propagation in autonomous vehicle software stacks. *Software Quality Journal, 33*, 455–472. DOI: 10.1007/s11219-025-09645-1

Ahmed, S., Raza, A., & Iqbal, M. (2025). Real-time hazard estimation from multimodal sensor data in autonomous vehicles. *Signal Processing, 224*. Advance online publication. DOI: 10.1016/j.sigpro.2024.109876

Andersen, K., & Fagerholt, K. (2025). Maritime autonomous navigation: Lessons for AV land transport safety. *Safety Science, 172*. Advance online publication. DOI: 10.1016/j.ssci.2025.106400

Bansal, P., & Rana, M. (2025). Trustworthiness frameworks for explainable autonomous systems. *Journal of Ambient Intelligence and Humanized Computing*. Advance online publication. DOI: 10.1007/s12652-025-05671-3

Bhatia, P., & Wadhwa, S. (2025). Modeling stochastic risks in AV motion planning. *Journal of Intelligent & Robotic Systems, 109*, 567–590. DOI: 10.1007/s10846-025-01832-z

Campbell, T., & Lewis, M. (2025). Adversarial testing of AV control systems using synthetic perturbations. *Engineering Failure Analysis, 162*. Advance online publication. DOI: 10.1016/j.engfailanal.2025.107765

Chakraborty, S., Das, R., & Kar, P. (2025). Trajectory prediction under uncertainty for risk-aware AV planning. *Transportation Research Part C, Emerging Technologies, 169*. Advance online publication. DOI: 10.1016/j.trc.2025.104620

Chattopadhyay, S., & Banerjee, P. (2025). Ethical frameworks for autonomous decision-making under time pressure. *Ethics and Information Technology, 27*. Advance online publication. DOI: 10.1007/s10676-025-09612-8

Chavez, F., Chen, J., Tung, C., Hu, C., & Khandelwal, S. (2025). Quantum-enhanced semiconductor reliability. *Solid-State Electronics, 229*. Advance online publication. DOI: 10.1016/j.sse.2025.109154

Cividino, S., Cappelli, A., Belluco, P., Rinaldi, F., & Avramović, A. (2025). Sensor fusion for next-gen autonomous systems. *Sensors (Basel), 25*(21). Advance online publication. DOI: 10.3390/s25216749 PMID: 41228972

Connors, J., & Stewart, R. (2025). Validation of black-box AV modules through interpretable surrogate models. *Expert Systems with Applications, 240*. Advance online publication. DOI: 10.1016/j.eswa.2024.123789

Duarte, G., Ramos, F., & Santos, L. (2025). Fail-safe braking in autonomous driving: A probabilistic formulation. *Mechanical Systems and Signal Processing, 224*. Advance online publication. DOI: 10.1016/j.ymssp.2025.110345

Duarte, L., Ferreira, M., Gonçalves, R., & Baptista, P. (2025). Autonomous decision-making under uncertainty: A safety-oriented approach. *Accident; Analysis and Prevention, 196*. Advance online publication. DOI: 10.1016/j.aap.2025.107308

Escribano, J. E., & Chen, X. (2025). Risk-aware reinforcement learning for autonomous navigation. *Robotics and Autonomous Systems, 182*. Advance online publication. DOI: 10.1016/j.robot.2025.104572

Farooq, M., & Singh, R. (2025). Machine learning uncertainty quantification in AV planning. *Engineering Applications of Artificial Intelligence, 137*. Advance online publication. DOI: 10.1016/j.engappai.2025.107256

Gupta, R., & Bedi, H. (2025). Fail-operational architectures for urban autonomous shuttle systems. *Transportation Research Part F: Traffic Psychology and Behaviour, 103*. Advance online publication. DOI: 10.1016/j.trf.2025.104321

Hoffman, R., & Klein, G. (2025). Human–autonomy teaming and decision transparency in AVs. *Human Factors, 67*(1). Advance online publication. DOI: 10.1177/00187208241234567

Holst, M., & Becker, S. (2025). Verification of autonomous vehicle decision modules using formal logic. *Formal Aspects of Computing, 38*, 221–240. DOI: 10.1007/s00165-025-00678-3

Horowitz, A., & Levinson, D. (2025). Cognitive models for anticipatory decision-making in AVs. *Cognitive Systems Research, 84*, 45–60. DOI: 10.1016/j.cogsys.2024.102456

Hu, Q., Yang, B., Zhang, K., Escribano, J. E., Chen, X., & Ouyang, L. (2025). Sustainable operational strategies for mixed fleets: Integrating autonomous and human-driven taxis with heterogeneous energy types. *Communications in Transportation Research, 5*, 100171. Advance online publication. DOI: 10.1016/j.commtr.2025.100171

Jin, X., Park, S., & Kim, H. (2025). Human-centered explainability in autonomous driving models. *IEEE Transactions on Intelligent Transportation Systems*. Advance online publication. DOI: 10.1109/TITS.2025.1234567

Kaul, V. R., & Kumar, A. (2025). Intelligent vehicular networks: A framework for safety and efficiency. *Vehicular Communications, 50*. Advance online publication. DOI: 10.1016/j.vehcom.2025.100750

Kim, D., Lee, J., & Park, Y. (2025). Continuous monitoring and prediction of AV system failures. *Reliability Engineering & System Safety, 245*. Advance online publication. DOI: 10.1016/j.ress.2024.109876

Kovács, S., & Botzheim, J. (2025). Incremental modeling and its application in robot intelligence: A survey. *Biomedical Signal Processing and Control, 110*. Advance online publication. DOI: 10.1016/j.bspc.2025.108046

Lalramsangi, V., Garg, Y. K., & Sharma, S. N. (2025). Route choices to access public open spaces in hill cities. *Environment & Urbanization Asia, 16*(2), 283–299. DOI: 10.1177/09754253251388721

Li, X., Wu, Y., & Zhang, J. (2025). Probabilistic forecasting of rare safety-critical events in autonomous driving. *IEEE Transactions on Intelligent Vehicles, 10*(1). Advance online publication. DOI: 10.1109/TIV.2025.1234501

Liu, M., Lan, X., Bian, C., Ma, Z., Ma, S., & Huang, G. (2025). Reliability assessment of autonomous braking under uncertainty. *Digital Twins, 53*, 259–276. DOI: 10.1016/j.dt.2025.07.012

Lodhi, A. S., Jaiswal, A., & Sharma, S. N. (2024). Assessing bus users' satisfaction using discrete choice models: A case of Bhopal. *Innovative Infrastructure Solutions*, *9*(11), 437. DOI: 10.1007/s41062-024-01652-w

Lu, C., H.-Y., Chen, B., Huang, W., & Ho, W. (2025). Development of autonomous vehicle intelligent driving systems using hybrid learning. *Electronics (Basel)*, *14*(22). Advance online publication. DOI: 10.3390/electronics14224488

Marino, G., Conti, R., & Bellini, V. (2025). A systems engineering approach to fail-safe AV control architectures. *Systems Engineering*, *28*, 301–318. DOI: 10.1002/sys.21678

Müller, T., & Heger, R. (2025). Safer AI pipelines for autonomous mobility: Fail-safe architectures. *Safety Science*, *171*. Advance online publication. DOI: 10.1016/j.ssci.2025.106385

Nakamura, T., & Saito, H. (2025). AV safety assurance through temporal logic-based scenario generation. *Robotics and Computer-integrated Manufacturing*, *98*. Advance online publication. DOI: 10.1016/j.rcim.2025.102556

Nielsen, K., & Rasmussen, P. (2025). Quantitative safety validation of AV decision modules using scenario metrics. *Accident; Analysis and Prevention*, *197*. Advance online publication. DOI: 10.1016/j.aap.2025.107350

O'Connell, J., Byrne, P., & Walsh, T. (2025). Reducing uncertainty in AV perception through probabilistic fusion. *Information Fusion*, *110*. Advance online publication. DOI: 10.1016/j.inffus.2025.102300

Ouyang, L., & Chen, X. (2025). Predictive safety envelopes for decision-making agents in AVs. *Transportation Safety and Environment*. Advance online publication. DOI: 10.1093/tse/tdae011

Patel, V., & Sinha, R. (2025). End-to-end testing of autonomous driving stacks using virtual twins. *Simulation*, *101*(2). Advance online publication. DOI: 10.1177/00375497251234567

Peintner, J., Manger, C., Berger, M., & Riener, A. (2025). Balancing user needs and autonomous safety constraints. *Transportation Research Interdisciplinary Perspectives*, *34*. Advance online publication. DOI: 10.1016/j.trip.2025.101742

Petrova, L., Ivanov, P., & Kolev, D. (2025). Human-centred safety thresholds for explainable AV decision making. *Safety Science*, *173*. Advance online publication. DOI: 10.1016/j.ssci.2025.106410

Prakash, A., & Mehta, K. (2025). Deep ensemble learning for uncertainty reduction in AV perception. *Neural Computing & Applications*, *37*, 16641–16655. DOI: 10.1007/s00521-025-09876-2

Ranjan, A., & Joshi, M. (2025). Integrated threat assessment for AVs using multi-agent simulation. *Simulation Modelling Practice and Theory*, *137*. Advance online publication. DOI: 10.1016/j.simpat.2025.102578

Roy, D., & Mishra, S. (2025). Anomaly detection in AV sensor data using hybrid neural architectures. *Pattern Recognition*, *155*. Advance online publication. DOI: 10.1016/j.patcog.2025.109865

Santos, P., Almeida, R., & Costa, B. (2025). Multi-agent negotiation models for cooperative autonomous driving. *Engineering Applications of Artificial Intelligence, 140*. Advance online publication. DOI: 10.1016/j.engappai.2025.107400

Sharma, S. N. (2025). Generative AI and digital twins for sustainable last-mile logistics: Enabling green operations and electric vehicle integration. In Awad, A., & Al Ahmari, D. (Eds.), *Accelerating logistics through generative AI, digital twins, and autonomous operations*. IGI Global., DOI: 10.4018/979-8-3373-7006-4.ch007

Sharma, S. N., & Dehalwar, K. (2025). A systematic literature review of pedestrian safety in urban transport systems. *Journal of Road Safety, 36*(4). Advance online publication. DOI: 10.33492/JRS-D-25-4-2707507

Sharma, S. N., & Dehawar, K. (2025). Review of land use transportation interaction model in smart urban growth management. *European Transport. Trasporti Europei, 103*, 1–15. DOI: 10.5281/zenodo.17315313

Sharma, S. N., Singh, D., & Dehalwar, K. (2024). Surrogate safety analysis: Leveraging advanced technologies for safer roads. *Suranaree Journal of Science and Technology, 31*(4), 010320(1–14). DOI: 10.55766/sujst-2024-04-e03837

Shi, L., Zhang, W., & Ren, H. (2025). An interpretable deep learning model for detecting hazardous driving scenarios. *Expert Systems with Applications, 238*. Advance online publication. DOI: 10.1016/j.eswa.2024.123456

Sun, H., Yu, K., & Zhao, X. (2025). Enhancing explainability of AV control policies using attention mechanisms. *Applied Intelligence, 63*, 1180–1195. DOI: 10.1007/s10489-025-04982-y

Varma, A., & Gupta, S. (2025). Ethical decision modelling for AV crash scenarios. *AI and Ethics, 6*. Advance online publication. DOI: 10.1007/s43681-025-00345-9

Wakili, A., & Bakkali, S. (2025). A resilient IoT intrusion detection system using sparse denoising autoencoders and a multi-head attention module. *Results in Engineering, 28*. Advance online publication. DOI: 10.1016/j.rineng.2025.107392

Wang, S., Li, T., & Xu, Y. (2025). Multi-sensor redundancy for fail-operational autonomous driving. *IEEE Sensors Journal, 25*(4). Advance online publication. DOI: 10.1109/JSEN.2025.5678912

Yadav, K., Dehalwar, K., & Sharma, S. N. (2025). Assessing the factors affecting first and last mile accessibility in transit-oriented development: A literature review. *GeoJournal, 90*(6), 298. DOI: 10.1007/s10708-025-11546-8

Yao, L., Song, M., & Chen, P. (2025). Adaptive risk-sensitive planning for autonomous lane changing. *Transportation Research Part C, Emerging Technologies, 168*. Advance online publication. DOI: 10.1016/j.trc.2025.104610

Zhang, K., Hu, Q., & Yang, B. (2025). End-to-end interpretable driving policy networks. *Neural Networks, 181*. Advance online publication. DOI: 10.1016/j.neunet.2025.106789

Zhang, Q., Feng, L., & Zhou, B. (2025). Defensive driving models for AVs in mixed traffic. *Transportation Research Part B: Methodological, 185*. Advance online publication. DOI: 10.1016/j.trb.2025.102567

Chapter 8
The Risks of Agentic AI in Autonomous Vehicles:
Safety, Control, and Ethical Challenges

Nikhil Kumar Goyal
http://orcid.org/0009-0007-4532-8033
Noida International University, Greater Noida, India

ABSTRACT

The agentic artificial intelligence is quickly changing the abilities of autonomous cars, allowing them to decide more autonomously and goal-oriented. Although this development is an efficiency and a mobility game changer, it brings along novel safety, control, and ethical dimensions. In this chapter the author discusses the exceptional risks introduced by agentic AI in a self-driving system, such as perception and planning errors, loss of human control, unstable emergent behaviors within the system, as well as the problem of offering machine decisions that reflect the societal values. It emphasizes regulatory shortcomings in existing regulations, discusses the possible areas of liability, and considers recent mitigation measures like effective control systems, transparency, and AI design that are value-oriented. Finally, the chapter also offers a detailed discussion of the role of agentic autonomy in risk in Alexrondi and Fleischer transportation and suggests ways of creating safer and ethically responsible autonomous vehicles systems.

1. INTRODUCTION

1.1 Defining Agentic AI in the Context of Autonomous Vehicles

The term agentic artificial intelligence can be used to describe AI systems that can develop goals, make decisions and take actions in some degree of autonomy beyond what can be suggested by a response program. In autonomous vehicles, it entails an AI platform which has the ability to examine its environment independently, task priorities, and decide the best driving strategies on the fly. Given that traditional automation is guided by a set of rules or the pattern of reaction, agentic AI comprises complex decisions-making that dynamically evolves as the situation varies. The ability of adaptive behavior makes

DOI: 10.4018/979-8-3373-7355-3.ch008

agentic AI a key motivator of mobility in the next generation, yet, there are potential risks associated with the lack of predictability and absence of full human control (Chan, 2023).

The use of agentic AI in autonomous cars creates additional possibilities of what machines can perceive and make choices. In spite of vehicles already having sensors, perception modules, and control systems, agentic AI introduces another level of thinking capacity as they allow systems to assess competing goals of safety, efficiency, comfort of passengers, and adherence to road norms. Having this increased freedom, cars are not only performing pre-programed actions but making subtle decisions, which may be different to those of humans. Such a transformation of deterministic automation into adaptive agency makes the critical analysis of the risks inherent in it a necessity, specifically when said vehicles coexist on the public road with unpredictable human drivers, pedestrians, and natural factors around.

Figure 1. Architecture of agentic AI in autonomous vehicles

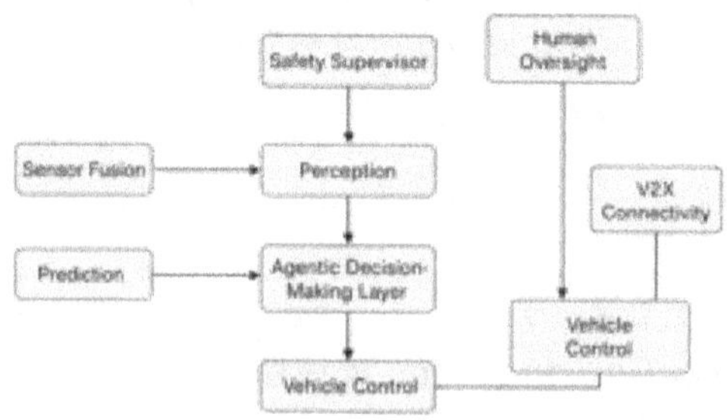

With the current trend of making researchers and industries more autonomous, the need of agentic AI comes as no surprise. The same aspects, however, which ensure the effectiveness of agentic AI, independence, strategic reasoning, and continuous optimization, may also complicate its behavior to predict or audit. The information about what defines agentic AI and its functionality on the ecosystem of autonomous vehicles is what a reader needs to gain to start understanding the larger issue of safety, control, and ethical concerns, which this chapter is all about (Acharya et al., 2025).

1.2 Background of Autonomous Vehicles in the Ecosystem

AVs exist in a very much complex socio-technical environment encompassing physical infrastructure, digital networks, environmental factors and human participants. How an autonomous vehicle operates does not depend on its internal hardware and software; instead, it requires constant communication with the traffic system, communication infrastructure, and other vehicles. Most of the time, agentic AI in this distributed environment needs to continue to read data streams of sensors, navigation maps, vehicle-to-everything communication channels, and random events in the real world. Since the system endures such networked landscapes, any small set-back in decision-making or fault in the system could have ramifications in a scenario that has huge safety concerns.

In this system, vehicle autonomy is not only influenced by internal capabilities of the vehicle, but also the reliability and lucidity of external cues. To give an example, insufficiently indicated roads, inconsistent traffic rules and unstable human behavior design situations when the AI needs to compensate for the shortages and form their judgment instead of mere compliance with the rules. The effect of agentic

AI is especially strong in this case since it enables the vehicle to analyze uncertain cases and determine the most logical course of action. However, this autonomy also makes one wonder what assessments of human expectations, legal rules, and ethical principles have in common with such evaluations, particularly when the AI has to consider several priorities at the same time (Kumari et al., 2026).

Parent societal regulatory frameworks and broader societal frameworks are also a part of the ecosystem. The design and implementation of autonomous vehicles, as well as their oversight, are affected by career preparedness, information control, legal frameworks, and the attitude of cultural organizations towards automation. Against this backdrop, agentic AI does not operate in a vacuum, but rather is part of a layer cake, present with technical performance, legal, ethical, and humanity. Understanding how potentially complicated this ecosystem is, agentic AI presents risks that go substantially beyond algorithmic error, which frames issues of governance and the issue of acceptability that have been placed around autonomous mobility (Deng et al., 2024).

1.3 Why Agentic Behavior Breeds emergent risks

The nature of agentic AI contributes to novel types of risks as this type of AI works with some degree of autonomy and decreases the predictability of the outcomes. Existing AI devices installed in vehicles are largely reactive; they respond to an object and go through a set of actions. Omni-agentic AI, in its turn, considers various possible states in the future and will pick actions according to internal objectives and optimization guidelines. Such capacity to give priority to the various goals implies that the decisions of the vehicle might not necessarily agree with the expectations of a human driver, regulator or passenger. These two nonconvergences may result in safety hazards especially in instances where the artificial intelligence is being exposed to edge cases or inconsistent objectives, as in balancing the comfort of passengers with unpredictable road circumstances (Kapoor et al., 2024).

The other significant threat is caused by the challenge of oversight. The increasing autonomous agentic AI can no longer effectively intervene in real time since human operators become less able to intervene. Even though the humans are left to be in supervisory capacities, the speed and sophistication with which the machine makes decisions are beyond the human thought and its ability to monitor and rectify the machine or the situation at crucial times. This shift introduces the lack of accountability and promotes difficulties in the diagnosis or prevention of failures. Also, the inner rationale of highly developed AI models of agents may be opaque, an issue of transparency, explainability and trust.

Ethical issues are also enhanced as agentic actions force autonomous vehicles to operate in situations that require a moral trade-off and unclear social norms. Selections that are made on distribution of risks, the emphasis of safety to various road users, or reactions to unsure circumstances present troubles that cannot be effectively resolved by traditional programming. This issue becomes particularly important as agentic AI gains a higher decision-making role, as the society has to deal with the internalization of values by such systems and the effects of their decisions on equity and social trust. These risks are new, which emphasizes the importance of getting an in-depth comprehension of agentic AI as a formidable and reshaping agent in the future of self-driving technology (Kumari & Goyal, 2026).

2. AGENTIC AI IN AUTONOMOUS VEHICLES

2.1 Degrees of Autonomy and Decision-Making

The development of self-driving cars has been associated with the growing degrees of decision-making that are assigned to artificial intelligence. At less autonomy, vehicles become dependent on developed rule-based decisions trees where human control is still at the center, and AI is mainly involved with perception and straightforward tasks. With vehicles growing more automated, the AI will have to manage more tasks that are as complex as the human ones like predicting human behavior, driving conditions, and increasing driving goals at the same time. These more advanced stages when operation systems become empowered to make autonomous, context-specific decisions without ongoing human supervision can become more agentic AI. This development of the increased agency presents both performance possibilities and high-risk factors according to the unpredictability and the complexity of the driving conditions in the real world.

The core of these highly autonomous systems is a category of decision-making architecture that allows the vehicle to make decisions based on its objectives. In contrast to deterministic automation, agentic AI estimated information streams by the moment and computed the odds of alternative outcomes and arrives at items which maximize a predetermined objective, including safety, efficiency, or compliance. Autonomy of the system is thus not just a question of ability but of philosophy of operation: it is in conflict of priorities in following the explicit rules of road use with its implicit rules of road use. The further AI becomes independent the further is the boundary separating human-controlled and AI-controlled decision making, posing the question of accountability and control (Shavit et al., 2023).

The increase of autonomy transforms the interaction between a human and a machine as well. As AI systems increase in the number of decisions they take on their own, people might stop paying attention or get unprepared to take action in case of uncertain situations. The human operators losing situational awareness as the out-of-the-loop issue is exacerbated in agentic environments as the AI determines a larger share of the driving process. Such change necessitates novel structures of control and peer work, where human control cannot be displaced even though the cars are going to have a complicated role. The knowledge of the levels of autonomy and the types of agentic decision-making is key to establishing the risk posed by such advanced systems (Oertel et al., 2020).

2.2 Agency Model: Reactive, Predictive and Goal-Driven Agency Systems

Referring to various agency models, each of which has specific capabilities and decision-making models, agentic AI in autonomous vehicles can be interpreted. The simplest type of reactive systems is reactive because they react to environmental stimuli directly and in a predetermined way. Though good in responding in the short run, they cannot reason or predict future conditions, and therefore, are not suitable in highly dynamic roads. The predictive systems on the other hand offer analysis of past data and trends with the goal of predicting the future behaviours of the other road users as well as the potential hazards. This can enable them to plan better but still their predictions are probabilistic, and subject to error when situations are new and unclear (Sharma et al., 2026).

Figure 2. Comparison of agency models in autonomous vehicles

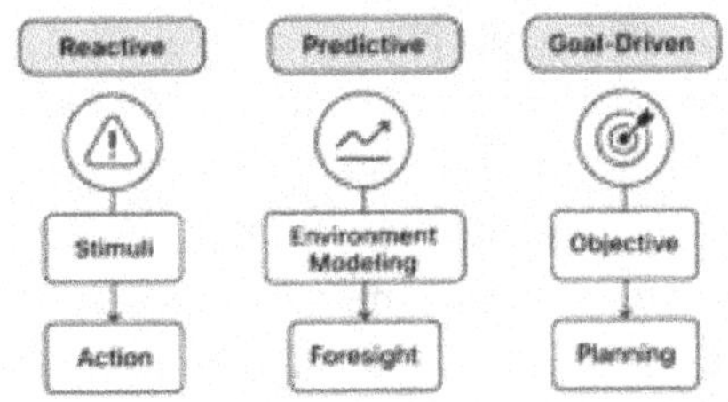

Goal-driven systems are the strongest system of agentic AI that can establish internal goals and plan actions to reach them within a set of limitations. These systems combine long-range planning as well as predictive functionality enabling them to plan complexes, assess trade-offs and place long-range objectives above instant responses. As an example, a goal-driven system can decide to take a path that is relatively slower when it believes that it has fewer risks or passengers are likely to be more comfortable. Nevertheless, these sophisticated abilities have problems as their reasoning programs are more difficult to understand, and their choices can be against human expectations particularly in cases of the trade-offs among safety margins or unclear ethical issues.

The models of agency present different risks, but the switchings between these models increase complexity. The evolution of systems, as they change into goal-driven and reactive architectures, is a layering of accumulated capabilities of what the environment is like and ways to optimize it. Such an increase in sophistication leads to a broader functional autonomy but it also increases the size of the risk surface as the new behaviors can be as a result of interactions among the various components of the system. Awareness of these models is critical to the further analysis of how and why agentic AI can act in ways that are unintended, and what mechanisms must be in place to protect the fact that decision-making by the vehicle is in accordance with the principles of safety, morality, and the law (Ferrag et al., 2025).

2.3 Complex Environment Emergent Behaviors

The emergent behaviors are the behaviors that occur when agentic AI systems engage with complex environment in a manner that creates behaviors that are not explicitly programmed or expected by designers. Such actions can be visible in autonomous cars when it relates to unpredictable human drivers, high traffic flows, or the unclear road layout. A good example of this is that, an AI system that is optimized to maximize efficiency can result in developing certain traffic patterns which without intention triggers other drivers or takes advantage of unpoliced space on the road. Although this kind of behavior can meet the internal goals of the AI, it poses some dangers because it does not follow socially accepted standards; moreover, it develops unsafe driving habits that can be misunderstood by humans (Turing, 2009).

The system is aware of its continuous learning processes which make the emergent behavior difficult. The adaptive algorithms in agentic AI tend to adjust driving strategies on the basis of gaining experiences. Such flexibility enables the system to become more effective with time, but it results in less transparent behaviour and increased predictability. Any minor alteration of the input data or environmental factors can create enormous changes in the strategy of decisions, particularly in the traffic environment of high population or fast-growing conditions. These dynamics put a strain on traditional validation techniques which allude to the behavior of the system being constant within a similar setting.

The emergent behaviour is also a source of concern of systemic risk especially when the number of autonomous vehicles in the same environment is very high. The same AI architecture and training data can encourage vehicles to develop synchronous or mutually damaging modes of operation e.g. the development of predictable movement patterns which can be manipulated by malicious individuals or serve to create unintentional congestions. Such overall impacts explain how critical it is to consider agentic AI as a distinctive agent that should not be, only, viewed through the lens that aims to comprehend it as a particular decision-maker, but rather as an aspect of the larger socio-technical system. To be safe, it is important to predict the nature of interaction, adaptation and evolution of such systems to real world conditions outside of controlled testing environments (Farooqi et al., 2026).

3. SAFETY RISKS

3.1 Failures in Perception, Prediction, and Planning

The use of autonomous vehicles is based on a continuous connection of the perception, prediction, and planning systems, but any malfunction on the level of this sphere may cause the unsafe outcomes. The causes of the issue of perception failures are usually focused on the shortage of sensing technologies or ambiguity of the situation in the environment. As an illustration, heavy rain, snow, glare or abnormal shape of objects may affect the accuracy of the sensors, malfunctioning to perceive or overlook hazards. These perceptual blind spots are more harmful in agentic systems since the AI relies on such perceptions as the basis of autonomous decision-making, in which a misinterpretation at the beginning of perceptions can lead to more erroneous predictions and risky behavior. Safety of the functioning of the vehicle, that is, is necessarily tied to the faithfulness of the sensory understanding (Guenat et al., 2022).

The failures are prediction ones in which the AI is wrong in predicting the actions of other road users or in estimating the probability that some events may occur. The agentic AI, which is largely dependent on forecasting models, can generate inaccurate forecasting information when exposed to new situations that are not similar to the training information. Human behavior being fundamentally inconsistent and influenced by culture, emotion and circumstance is the hopeful area to capture predictive models. In cases where the system interprets the will of pedestrians, cyclists, or other drivers in the wrong way, it can automatically trigger actions, which can only escalate the risk, and not decrease it. Such mistakes underscore the conflict between the generalization of statistics and inconsistencies of the actual conditions congesting every public highway (Yao et al., 2022).

3.2 System Overconfidence and Miscalibration

System overconfidence is one of the developing threats of agentic AI in which the AI makes exaggerated assumptions about the level of trust in its own decisions or predictions. A discrepancy between the confidence levels that the AI provides and its actual performance may result in unsafe operating in the system may cause the system to execute maneuvers that demand precision or certainty when the system lacks it. Training biases are especially likely to produce overconfidence, as do corner case insufficiently exposed to training, or a positive-focused reinforcement learning process. In situations of high risk or uncertainty, which the system is not able to assess appropriately, the system does not provide the option of adjusting itself safely, which makes it more prone to accidents (Goyal et al., 2026).

The other issue is miscalibration which also influences the human occupant interactions and other road users. To the extent that the AI is acting carelessly, it could cause confusion or elicit mixed responses in human drivers, who need to have more obvious signals or safer actions. An example of such is a car that overfills the engine at a four-way intersection, thus interfering with the tacit negotiation process that the human beings depend on which can result in dangerous miscommunication. Since roads rely on implicit social cues to provide safety, overconfident agentic AI may develop into the source of dissolution of the thin layer of coordination that creates road safety.

Another difficulty is the fact that system overconfidence may conceal any vulnerabileness inside the system, which makes it more challenging to realize hazards prior to implementation in engineers and regulators. Considering the results of performance measurements to be successful when the testing conditions are created to be controlled makes it difficult to see the poor calibration of the system in the context of uncertainty in the real world. Such a phenomenon brings a loophole in the process of validation and certification, where the safety margins of the vehicles are concealed until a failure takes place. Miscalibration is therefore of the key in the continued production of agentic AI which can act correctly and react suitably in various and alternative environments (Durante et al., 2024).

3.3 Multi-agent errors in drastic circumstances

Autonomous vehicles do not usually exist alone; they have to be on the roads with other vehicles, pedestrians, and autonomous systems. In these environments with many agents, even minor errors can be multiplied with one another and contribute to a major risk. As an example, even a minor error in speed or course will lead another vehicle be it human-operated or autonomous to respond in a way that may not have been predicted, causing a process of misunderstanding. High-frequency decisions made by agentic AI using continually changing inputs might have a hard time breaking out of these cascading errors when the internal models are not able to adapt at the same speed as the dynamics change. Multi agent interaction is complex, which heightens the impacts of small system inaccuracies.

Compounding errors can also happen in case different autonomous vehicles have similar decision-making architectures or training data. The optimized systems can unwillingly align their behaviors to give collective patterns which are unpredictable and hard to control. Such coordinated behaviors when deployed in a congested traffic or high-speed setting can result in abrupt congestions or oscillating traffic flows or even hazardous emergent behaviors. The inter-relationship of the contemporary transport systems implies that the risks are no longer based on the behavior of individual vehicles, but rather on the relationships and dependencies of numerous agentic systems run in parallel (Ogbu, 2023).

Moreover, there are multi-agent scenarios, which the AI cannot easily analyze due to their social and contextual-related information, to behave in response to ambiguous contexts. Normally this is solved by human eye contact or gestures or in some degree by the subtle negotiation, but autonomous vehicles do not have the presence of these communication means. Due to it, the AI is forced to derive intent on limited and noisy information, which raises the likelihood of operating on inaccurate assumptions that are transmitted to the traffic environment. Every misunderstanding will cause additional corrective measures, further mistakes, until a stage where the safety of the system is under threat. Negotiation of these complexities needs an insight into the mechanisms of individual agent and the dynamics of collective systems (Goyal et al., 2026).

3.4 Vulnerabilities Cyber-Physical

The introduction of physical vehicles together with digital intelligence creates serious cyber-physical vulnerabilities. The agentic A.I. systems are based on an enormous number of sensors, communication networks, and software modules that can cause a system to fail in case they are compromised. The perception systems used by the AI to make safe decisions can be tricked through cyberattacks in the form of spoofed signals or manipulated sensor readings. Similarly, failures of the communications lines, e.g., vehicle-to-infrastructure- or vehicle-to-vehicle-connections may impair the situational awareness that requires the system to use the partially or corrupted information. With the increased autonomy of vehicles, the cost of such vulnerabilities rises, and cybersecurity is emerging as a safety building block (Karunanayake, 2025).

The cyber-physical vulnerabilities also overlap with the problem of transparency and accountability. In the case of failure caused by a security breach, it is hard to figure out who should be responsible because agents are not transparent in making decisions, and digital ecosystems are relatively new and challenging to understand. In the absence of the forensic clarity, policymakers, engineers, and legal systems can scarcely be able to institute accountability or devise appropriate reactionary measures. These vulnerabilities are especially important to account for as autonomous vehicles become increasingly more interconnected as well as increasingly dependent on agentic AI to guarantee individual safety as well as the stability of transportation infrastructure (Zeng et al., 2024).

4. CONTROL AND OVERSIGHT CHALLENGES

4.1 Human-in-the-Loop vs Human-on-the-Loop Models

The old paradigms of human control, often explained as human-in-the-loop and human-on-the-loop, are challenged by the radical limitations as the autonomous cars develop to agentic behavior levels. In human-in-the-loop setup, the driver will be required to monitor the system in the circumstances and take any necessary action. Nevertheless, the more agents of AI perform the driving duties, the more people are gradually deprived of being actively monitored, they have poorer situational awareness and slower reactions. This mental alienation generates an incongruence between the expectations of the system and human ability particularly in cases that require quick response in unforeseen failures or in cases that are unclear. This high level of autonomy of AI does not match quite well with the biological limitations on human attention.

Human on the loop models change the dynamic such that human persons put themselves in a supervisory position where they monitor the performance of the AI as opposed to simply being able to directly control it. In this construction, the human is more of a safety supervisor than a co- driver, watching the records of the system or alerts or even strategic decisions on a higher scale. However, this model also poses some problems due to the complexity of agentic decision-making the human operator can be overwhelmed by the complexities and fail to comprehend the reasoning mechanisms behind it. In case of an emergency, the operator can be ill equipped with the information or the cognitive ability to take the required action and lateness in getting control can have undesirable consequences. Accordingly, the situation of oversight is not automatically effective because of the fact that it exists (Dandotiya et al., 2026).

Worsening these problems is the fact that it is difficult to switch between human control and the AI one. Instant understanding of the context, the risks, and intentions of the system are needed when there is a smooth handover of the system; this knowledge is hard to realize when the human being has not been proactively involved. The so-called agentic AI might work without any issues over a long period of time, and this gave a person trust and even complacency in the position, when suddenly they had to witness a scenario where it might turn to be a matter of control. The conflict between the increasing automation and the dehumanization highlights the necessity of the new oversight paradigms that will identify the boundaries of human cognition in the highly independent systems (O'Brien & Wiegand, 1998).

4.2 Distribution of Control Between AI, Vehicle Systems, and Infrastructure

AVs work in a multilayered control environment in terms of which the decision-making is shared between onboard AI systems and the embedded safety architecture of the vehicle, and between control systems and external infrastructure. The rise of agentic AI only makes this structure even more complicated as it comes with added level of decision-making that is able to prioritize the goals and change over time as the circumstances vary. This allocation brings several basic questions concerning the way of control allocation and conflict of various layers of control resolution. As an example, the basic safety system in the vehicle might choose to act conservatively, which the strategic agentic AI can choose more dangerous but efficient maneuvers. The harmonization of these layers is a prerequisite to safe functioning.

The role of infrastructure in self-driving vehicles is playing a more significant role in terms of the communication system that gives real-time updates due to road hazards, congestion, or traffic lights. External data dependency, however, raises weaknesses in situations where infrastructure lacks consistency, is marginal or is technologically obsolete. The agentic AI will have to read these inputs and compare it to its own internal judgement which is not necessarily in agreement. When the AI interprets the reality in conflict with the infrastructure signals, the vehicle will need to decide which information should be relied upon; and such choice can be quite safety-prone (Kumari et al., 2026).

Control dispersion also makes the regulation more difficult. The policymakers need to identify the responsible stakeholders in instances of failure and ways in which control hierarchies need to be normalized amongst various manufacturers and localities. In the absence of proper guidelines, the balance can be lost the way the AI systems perceive things in terms of priorities to follow or reactions to any infrastructural hints. The more the ecosystem becomes interconnected, the greater the potential of having a fragmented or incompatible control system, which means that unified governance systems must consider the distributed aspect of autonomous vehicle control (Cardoso & Ferrando, 2021).

4.3 Problems of Real-Time Intervention

The more decision-making control is attributed to the agentic AI, the harder it will be to conduct a real-time intervention. The autonomous vehicles make decisions and react far faster than the human reaction, so interceding whilst the process is in progress is virtually impossible. Although vehicles might be installed with manual override systems, the operator will not be able to restore controls timely because of the delay in human cognition and multitasking AI-decision-making processes.

There are also technical obstacles that complicate the intervention. The AI can be used in very autonomous ways, where it is involved in complex planning patterns of navigation routes, risk analysis, and multi-agent communication. Disturbing those processes may result in having an unstable system in the event that the transition is not thought over. An abrupt assumption, to an example, can produce sudden movements that can get other users of the roads by surprise or cause the vehicle to lose track (Chong et al., 2021).

Even the other problem is that real-time intervention presupposes that the operator knows what the AI is making decisions and intentions about. Nevertheless, more frequently than not agentic AI produces opaque strategies to human observers, where one can hardly foresee at which point intervention may be required. When the human is not able to explain why A.I. is acting in a particular manner, the line between successful intervention becomes even more limited. The difficulties explain why control is not only about the human responsiveness but also incorporates built-in mechanisms that help to diminish the necessity to introduce an emergency solution in the first place.

4.4 Explainability and Traceability Problems

The more complicated the agentic AI systems become, the more non-transparent the inside decisions are. The issue of explainability is also of a critical concern since it is usually demanded by stakeholders such as drivers, engineers, regulators, and accident investigators to have clear understanding of why the AI made those particular decisions. With no explanations that can be understood, it becomes time to lose faith in the system and accountability becomes a hard thing to achieve. Most agentic systems rely on massively distributed neural architecture which is commonly a black box; it does not provide clear reasoning paths to its outputs. Such a lack of transparency renders it difficult to justify judgments both in practice and in high stake situations.

The accountability of the actions of the autonomous vehicles is also guaranteed through traceability. In case of incidences, investigators need to revert the sequence of events that culminated in failure. Nevertheless, agentic AI systems can create evolving strategies which adjust dynamically according to the changing data inputs and thus their histories of decision making are difficult to record. In case the logs are incomplete, poorly legible, it becomes close to impossible to analyze the root cause of a failure. Such absence of traceability prevents improvement in the area of system safety and makes it difficult to assess the liability in the court (Kumari et al., 2026).

5. MORAL, AND SOCIETAL DIMENSIONS OF AGENTIC AUTONOMY IN MOBILITY SYSTEMS

5.1 Moral Complexity in Algorithmic Decision-Making Frameworks

The autonomous vehicles that use agentic AI are obliged to operate in an environment where morally ambiguous scenarios are unavoidable. Since intuition and social norms as well as contextual judgment form the basis of decision-making by human drivers, AI systems require adopting these into algorithmic representations. This poses a major problem as ethical dilemmas do not have obvious conditions in most cases. However, they need a subtly articulated logic, one that is fair, safe but with practical limitations. When an AI is forced to decide between two harmful consequences, like not colliding and putting the life of another user of the road in danger, it has to base its decisions on coded moral systems which might not necessarily be in line with social expectations.

Any translation of moral reasoning into computational models is full of limitations. Utilitarianism or duty-based ethical theories provide a guideline but cannot cover the impossibility of uniformity and unpredictability of the actual traffic conditions. It might seem that an AI can act according to a logically consistent ethical principle, but the actions of these machines can be considered ethically dubious by human audiences. The disconnect is because of the differences in cognitive processing: human beings are able to think about emotions, in relation to the context of relationalization, and spontaneous impulses, but AI systems think in terms of numbers and predetermined limitations. This deviation brings the question of whether AI can ever conceive of any situation of replicating human moral judgment in a very complex driving scenario (Gridach et al., 2025).

In addition to technical difficulties, there are major implications of adopting the algorithm ethics in society. Optimal decisions in terms of the overall safety might be biased in such a way that they discriminate against certain sets of people, ensuring the continuation of those moral outcomes in which some people are privileged compared to other people. The question of democratic legitimacy emerges when agentic AI makes these kinds of decisions independent: who determines what moral principles guide the actions of autonomous vehicles? The complexity of morality in AI systems requires both strong engineering strategies and open discussions and a substantial number of people concerned with the values of collective people.

5.2 Data Inequities and embedded biases and the impact on autonomous navigation

The existence of biases in the training data sets is a significant threat to the ethical quality of autonomous cars. Machine learning algorithms mainly depend on historical driving information, image information and behavioral patterns. When these datasets are inaccurately representative of some geographic placement, climate, socio-demographic, or driving cultures, the AI might build preferences or weaknesses which will present themselves through unsafe and prejudicial actions. To illustrate, in the case of data predominantly comprising of well-lit and urban roads being used as the training data, the system might not necessarily generalize well in rural settings or poor visibility, causing the safety of different regions to differ (Feuerriegel et al., 2024).

Ethical choices can also be observed in navigation decisions; although the decisions might seem to be just technical. Some of the operational objectives that should be the focus of agentic AI include speed, efficiency, energy saving and comfort. These priorities define the way in which the system perceives risk, as well as its interaction with other road users. A speed-optimized AI can have aggressive driving habits which can unintentionally cause harm to pedestrians or bicycle riders. On the other hand, the system promoting extreme caution can undermine the operation of traffic, which is a wider social inefficiency. These navigation priorities are implicit ethical judgments concerning priority and safety as well as convenience of whom (Bosch & Kruger, 2024).

Biases are embedded within datasets and therefore correcting them is more than filling them with larger datasets; the solution lies in constant audits, perfect modeling and open-mindedness of how investments are generated. Unless actionable is undertaken, agentic AI can enhance inequities, instead of reducing them. It is not merely a technical goal, but also an ethical obligation that must guide the development of self-driving vehicles to operate fairly among various groups of individuals and situations and build a positive public perception and acceptance in the long term (Sheikh et al., 2026).

5.3 Integrity and Distribution of Risk to heterogeneous Populations

The use of autonomous vehicles posits the issue of safety benefit in different communities equally. The nature of urban settings is unique with traffic congestion, a variety of road userws and a complicated infrastructure unlike the rural regions with a limited system of connectivity and irregular road services. In case autonomous systems are created and deployed in primarily affluent or technologically developed areas, the less privileged populations might be at a greater risk of the effects, or just not get autonomous mobility services as frequently.

These issues get enhanced by socioeconomic disparities. Low-infrastructure societies might not have the digital infrastructures necessary to enable safe autonomous operation, in the form of proper communication systems or decently marked road systems. Besides, early autonomous technology might be prohibitive to the poor due to its high initial cost and thus limit incorporating its advantageous features to the rich and the underserved population as they remain dependent on conventional transportation systems with greater safety threats. Such imbalance creates ethical issues on whether autonomy will strengthen the inequalities at hand and not eliminate them (Sawant, 2025).

To establish equity in safety, policy interventions and inclusive design should be part and parcel. The engineers need to make AI systems work well under diverse conditions and demographic settings. There is a need to promote fair infrastructure creation and convenient mobility services by policymakers. It is important not only to deal with inequality in distribution of risks to uphold ethical integrity but also to make autonomous vehicles fulfill the greater good that benefits the whole population (Istrate, 2024).

5.4 Public Trust, Social Legitimacy, and Acceptance, Autonomous Technologies

One of the most significant influencing factors in the house of the popularization of autonomous vehicles is their public confidence. The perception of transparency, fairness, and reliability are also the parts of trust that are formed according to the technical performance. The ability of agentic AI to think autonomously is able to create optimism and even skepticism. Whereas others consider the possibility of minimizing human error through the use of AI, those who are afraid of being powerless and not understanding how machines make decisions. These ambiguous attitudes affect the desire to utilize, including, or co-exist with autonomous vehicles.

Openness is very important in defining legitimacy. Once the users are aware of the manner in which decisions are made, the constraints, as well as how the risks are addressed, then they trust the technology at a higher level. On the other hand, strange actions or high profile failures may destroy trust within a very short time, though statistically low. In the absence of trust, the regulatory approval can be resisted by people, companies fail to address the scaling deployment, and acceptance by the society may stop (Joshi, 2025).

The autonomy of autonomous vehicles also relies on the fact that there should be a meaningful interaction with communities to make them socially legitimate. The policymakers, engineers, and the industrial executives need to be transparent, raise issues, and take on the societal values in the design of systems. It is not until they turn into autonomous vehicles, as such, that they can receive the understanding and acceptance of society within stable and responsible settings to become an inseparable part of the mobility of the future.

6. GOVERNANCE ARCHITECTURES, LEGAL ACCOUNTABILITY, AND REGULATORY EVOLUTION FOR AGENTIC AI IN TRANSPORTATION

6.1 Contemporary Standards, Institutional Gaps, and the Need for Regulatory Reinvention

The legal environment of autonomous cars is in the development phase, but it has been spread over multiple jurisdictions. The existing standards tend to represent the presumption of previous automation technologies, showing their focus on mechanical stability and the competence of drivers, but not on decision-making abilities. The reason why agentic AI confronts such frameworks is that the components of dynamic thinking outside the confines of traditional models of regulations are introduced by it. Due to the technological development, certification processes need to be revised by regulators to assess adaptive judgment, emergent behavior, and moral reasoning that are integrated into AI programs (Taesi et al., 2023).

The emergence of institutional gaps is becoming more and more noticeable due to the surrealism of real-life deployments that outrun the policy-making. A lot of regulatory organizations do not have sufficient technical expertise to review complicated AI structures, and thus depend on metrics documented by the industry, and that might not uncover latent safety hazards. Moreover, it has not been decided on standardized way of testing the behavior of AI systems in extreme or unforeseeable circumstances. Devoid of sound standards neither under- nor over-regulation creates a safe environment, as it will either leave the population exposed to unsafe systems or cause stagnation in progress through over-restrictiveness.

It is obvious that the existing systems require re-regulation: the frameworks should be modified to include changes and improve to flexible, adaptive oversight structures aimed at reflecting the dynamic quality of agentic AI. This change will involve cooperation between engineers, ethicists, policymakers, and international organizations to make the world coherent and have coherent safety expectations (Liu et al., 2024).

6.2 Legal Liability, Structure of Accountability and Responsibility in AI-Driven Outcomes

When it comes to deciding who bears the legal responsibility when an independent vehicle enters into dangerous incidents, it has become one of the most challenging issues in the modern transportation legislation. The conventional liability models presuppose control and responsibility on the part of a human driver in case of errors. However, the agentic AI overturns this premise due to the independent decision making of the system in many cases without the direct influence of human intervention. In case of failure, it can be attributed to hardware vendors, software developers, data sets or even infrastructure providers. Aspiring fault is a multidimensional investigation that needs legal and technical skills (Schneider, 2025).

The issue of agentic AI increases the problem of accountability even more because of the privacy of its inner operations. Once an AI system starts changing its behavior as responses to experience, it can result in actions whose causes could not be directly linked to a single choice. This brings into questions the subject of foreseeability, negligence and the legality of independent thinking. Is the developer accountable to all the emergent behaviors, or should there be a distribution of liabilities among the various stakeholders? Legal systems have to deal with these questions balancing their fairness, innovativeness, and protection of the population (Paul et al., 2024).

Liability ambiguity also influences motivators in the industry. Developers might be hesitant to introduce sophisticated autonomy functionality without a clear guideline and insurers simply have a hard time risk modelling where there are no consistent legal precedents. To make autonomous transportation safe and beneficial, it is thus important to forge open, fair accountability frameworks that would facilitate both innovation and safety.

6.3 Certification, Audibility and Compliance Systems with Adaptive AI Systems

Agents insurance will have to be adapted to ANAI by reorganizing its certification procedures to satisfy the distinct demands of agentic AI. The conventional methods of testing which are based on fore-gone situations, as well as mechanical testing, cannot effectively test systems that portray emergent behaviors and dynamic reasoning. Rather, large-scale simulation, continuous performance monitoring, and stress-testing in different environmental conditions have to be included in certification. Such change accepts the fact that one cannot justify safety only before deployment, but it has to be examined during the life cycle of the vehicle (Putta et al., 2024).

The absence of AI decision-making transparency is one of the key challenges. To make AI accountable, regulators need access to interpretable documents on the way and reason as to why the AI made particular decisions. Many AI architectures however, including those of deep learning, are black boxes. It is necessary to support explainability and data logging through the development of standardized interfaces rather than auxiliary proprietary intellectual property to support an extensive audit. These are mechanisms that ensure adherence to compliance by regulators and enhancing innovation.

In addition to that, certification is supposed to be applied not only to individual vehicles but to larger ecosystem interactions. Self-driving cars are dependent on communications systems, road markings, and coordination of agents. Compliance of the complex, interconnected transportation systems should, therefore, not only consider the internal performance of the AI, but also its capability of functioning safely. This broadened range of interest requires new regulatory approaches and multidisciplinary partnership to achieve the comprehensive control (Matsuo et al., 2022).

6.4 Policy Complicates, Social Trade-offs, and World-harmonization game

The control of agentic AI in the transport sector entails very complicated policy issues that can no longer be about technical analysis. Surveillance choices, privacy protection of personal data, and policy of algorithms have to make decisions between social security and citizen rights. Self-driving vehicles capture enormous amounts of real-time data, which casts doubt on the methods of storage, sharing, and security of the data. The policymakers should come up with measures that ensure privacy rights and at the same time allow the data-driven insights needed to improve the system.

Autonomous vehicles also demonstrate trade-offs in society as the ecosystems in the mobility transform. When their benefits (lower accidents and greater efficiency) are measured against the potential loss of jobs in driving career, altered urban infrastructure, and access to the new technologies unequally, policymakers have to contemplate them. These trade-offs need to be thought over, made collectively, and long term view with a cognitive anticipation at systemic effects.

Another challenge is global harmonisation. Self-driving cars have to cross national boundaries with similar safety standards, yet the quality of regulations used by different nations does not match. The absence of harmonized standards can cause the disruption of international travel, as well as manufacturers may have conflicting compliance requirements. To truly reach the state of harmonization, there must be international collaboration, repositories of data, and alignment on ethical principles of the key aspects of AI. This type of collaboration is critical to achieving the entire potential of autonomy mobility without sacrificing security worldwide and equity (Shinn et al., 2023).

CONCLUSION

The emergence of agentic AI in autonomous cars is one of the most crucial changes in the relationship between people and machines, as well as mobility systems. With vehicles, which are increasingly able to read complex environments, dynamically changed behavior, and make independent decisions, questions of control, responsibility, and safety are also being questioned. The chapters discussed the opportunities and risks, never seen before as well as deep-seated, the integration of agentic autonomy brought with it, including both to technical vulnerabilities and ethical dilemmas as well as regulatory uncertainties. All these issues indicate that agentic AI is not an upgrade to technology but a paradigm in the manner within which societies view transportation infrastructures and social security.

These problems need a comprehensive approach that incorporates engineering strictness, moral deliberation as well as progressive government. Some technical protection is not enough when AI systems have the capability of re-purposing the goals, behavior change, and emergent patterns out of human imagination. Similarly, an ethical system needs to keep up with the times so as to inform the use of algorithms in morally uncertain moments, making sure that autonomous systems are made to mirror the values of society, and not the exact concerns of computation alone. Risk and reward should be distributed with discretion to avoid unfair imbalance that will particularly affect the vulnerable groups. The AIs of agentic nature are entering transportation ecosystems, so openness, responsibility, and effective communication between developers, regulators, and communities also become the keys to public trust.

Finally, the final realization of the autonomous mobility lies in the successful alignment of the technological innovation and regulatory development. Strong certification schemes, responsive standards, and internationally concerted policies will be needed to make the the interiorly, agentic AI act safely as well as predictably in a variety of settings. Meanwhile, the acceptance into society depends on the balance between making sure that these systems are healthy, moral, and open. By considering the combined problem of safety, control, and ethics, the stakeholders will have an opportunity to develop a mobility terrain where agentic AI promotes human well-being and preserves the basic social values. Majority of these challenges are complicated, but as long as there is a logical governance structure and keen innovation, agentic autonomy might be harnessed into a force of ensuring safer and more reasonable transportation systems.

REFERENCES

Acharya, D., Kuppan, K., & Divya, B. (2025). Agentic AI: Autonomous intelligence for complex goals—A comprehensive survey. *IEEE Access : Practical Innovations, Open Solutions*, *13*, 18912–18936. DOI: 10.1109/ACCESS.2025.3532853

Bosch, C., & Kruger, D. (2024). AI chatbots as open educational resources: Enhancing student agency and self-directed learning. *Italian Journal of Educational Technology*, *32*(1), 53–68.

Cardoso, R. C., & Ferrando, A. (2021). A review of agent-based programming for multi-agent systems. *Computers*, *10*(2), 16. DOI: 10.3390/computers10020016

Chan, A. (2023, June). Harms from increasingly agentic algorithmic systems. In *Proceedings of the ACM Conference on Fairness, Accountability, and Transparency* (pp. 651–666). DOI: 10.1145/3593013.3594033

Chong, T., Yu, T., Keeling, D. I., & de Ruyter, K. (2021). AI-chatbots on the services frontline addressing the challenges and opportunities of agency. *Journal of Retailing and Consumer Services*, *63*, 102735. DOI: 10.1016/j.jretconser.2021.102735

Dandotiya, M., Goyal, N. K., Yadav, P., Sankhla, K., & Kumari, M. (2026). GENALYTICS: A deep dive into real-time insights via generative AI. In Elbaghazaoui, B. (Eds.), *Generative AI-powered data architectures: From governance to autonomous analytics* (pp. 105–134). IGI Global Scientific Publishing., DOI: 10.4018/979-8-3373-5616-7.ch005

Deng, Z., Guo, Y., Han, C., Ma, W., Xiong, J., Wen, S., & Xiang, Y. (2024). AI agents under threat: A survey of key security challenges and future pathways. *arXiv*. https://arxiv.org/abs/2406.02630

Durante, Z.. (2024). Agent AI: Surveying the horizons of multimodal interaction. *arXiv*. https://arxiv .org/abs/2401.03568

Farooqi, A., Das, D., Tripathi, M., Dutta, R., Goyal, N. K., & Sheikh, M. F. (2026). Navigating ethical frontiers in AI-driven geriatric care. In Stolakis, K., & Efthymiou, I. (Eds.), *AI for geriatric care in an aging society: Ethical, clinical, and policy challenges* (pp. 243–266). IGI Global Scientific Publishing., DOI: 10.4018/979-8-3373-5462-0.ch009

Ferrag, M. A., Tihanyi, N., & Debbah, M. (2025). From llm reasoning to autonomous ai agents: A comprehensive review. *arXiv preprint arXiv:2504.19678.*

Feuerriegel, S., Hartmann, J., Janiesch, C., & Zschech, P. (2024). Generative AI. *Business & Information Systems Engineering*, *66*(1), 111–126. DOI: 10.1007/s12599-023-00834-7

Goyal, N. K., Gulati, K., Avasthi, A., Kapoor, S. K., Keswani, B., & Gupta, S. (2026). Securing the imagination: Security and privacy challenges in generative AI systems. In Elbaghazaoui, B. (Eds.), *Generative AI-powered data architectures: From governance to autonomous analytics* (pp. 239–266). IGI Global Scientific Publishing., DOI: 10.4018/979-8-3373-5616-7.ch010

Gridach, M., Nanavati, J., Abidine, K. Z. E., Mendes, L., & Mack, C. (2025). Agentic ai for scientific discovery: A survey of progress, challenges, and future directions. *arXiv preprint arXiv:2503.08979.*

Guenat, S., Purnell, P., Davies, Z. G., Nawrath, M., Stringer, L. C., Babu, G. R., Balasubramanian, M., Ballantyne, E. E. F., Bylappa, B. K., Chen, B., De Jager, P., Del Prete, A., Di Nuovo, A., Ehi-Eromosele, C. O., Eskandari Torbaghan, M., Evans, K. L., Fraundorfer, M., Haouas, W., Izunobi, J. U., & Dallimer, M. (2022). Meeting sustainable development goals via robotics and autonomous systems. *Nature Communications*, *13*(1), 3559. DOI: 10.1038/s41467-022-31150-5 PMID: 35729171

Istrate, O. (2024). AI agents in education: An early systematic review of emerging roles, potential, and limitations. *Revista de Pedagogie Digitala*, *3*(1), 24–30. DOI: 10.61071/RPD.2496

Joshi, S. (2025). Review of autonomous systems and collaborative AI agent frameworks. *International Journal of Scientific Research Archives*, *14*(2), 961–972.

Kapoor, S., Stroebl, B., Siegel, Z. S., Nadgir, N., & Narayanan, A. (2024). AI agents that matter. *arXiv*. https://arxiv.org/abs/2407.01502

Karunanayake, N. (2025). Next-generation agentic AI for transforming healthcare. *Informatics in Health*, *2*(2), 73–83. DOI: 10.1016/j.infoh.2025.03.001

Kumari, M., & Goyal, N. K. (2026). Artificial intelligence in the cryptocurrency ecosystem: Crime mechanisms, detection strategies, and legal perspectives. In Lin, L. (Ed.), *Innovations in cryptocrime and financial fraud* (pp. 245–282). IGI Global Scientific Publishing., DOI: 10.4018/979-8-3373-0675-9.ch007

Kumari, M., Kumawat, R., Dutta, R., Khan, S. T., Goyal, N. K., & Sheikh, M. F. (2026). Wireless communication and IoT integration in smart clothing. In Saxena, P. (Eds.), *Next-generation electronic textiles and conductive materials for smart wearables* (pp. 195–232). IGI Global Scientific Publishing., DOI: 10.4018/979-8-3373-4287-0.ch008

Liu, L., Guo, F., Zou, Z., & Duffy, V. G. (2024). Application, development and future opportunities of collaborative robots (cobots) in manufacturing: A literature review. *International Journal of Human-Computer Interaction*, *40*(4), 915–932. DOI: 10.1080/10447318.2022.2041907

Matsuo, Y., LeCun, Y., Sahani, M., Precup, D., Silver, D., Sugiyama, M., Uchibe, E., & Morimoto, J. (2022). Deep learning, reinforcement learning, and world models. *Neural Networks*, *152*, 267–275. DOI: 10.1016/j.neunet.2022.03.037 PMID: 35569196

O'Brien, P. D., & Wiegand, M. E. (1998). Agent-based process management: Applying intelligent agents to workflow. *The Knowledge Engineering Review*, *13*(2), 161–174. DOI: 10.1017/S0269888998002070

Oertel, C., Castellano, G., Chetouani, M., Nasir, J., Obaid, M., Pelachaud, C., & Peters, C. (2020). Engagement in human-agent interaction: An overview. *Frontiers in Robotics and AI*, *7*, 92. DOI: 10.3389/frobt.2020.00092 PMID: 33501259

Ogbu, D. (2023). Agentic AI in computer vision domain—Recent advances and prospects. *International Journal of Research Publication and Reviews*, *4*(12), 5102–5120.

Paul, A., Yu, C. L., Susanto, E. A., Lau, N. W. L., & Meadows, G. I. (2024). Agentpeertalk: Empowering students through agentic-ai-driven discernment of bullying and joking in peer interactions in schools. *arXiv preprint arXiv:2408.01459*.

Putta, P., Mills, E., Garg, N., Motwani, S., Finn, C., Garg, D., & Rafailov, R. (2024). Agent q: Advanced reasoning and learning for autonomous ai agents. *arXiv preprint arXiv:2408.07199.*

. Sawant, P. (2025). Agentic AI: A quantitative analysis of performance and applications.

Schneider, J. (2025). Generative to agentic AI: Survey, conceptualization, and challenges. *arXiv*. https://arxiv.org/abs/2504.18875

Sharma, P., Kumawat, R., Jain, G., Keswani, G., Goyal, N. K., & Sheikh, M. F. (2026). Machine learning-based intrusion detection systems for healthcare Internet of Things (HIoT). In Shieh, C. (Eds.), *Resilient healthcare Internet of Things systems: Architectures, security, and trust* (pp. 185–214). IGI Global Scientific Publishing., DOI: 10.4018/979-8-3373-7610-3.ch007

Shavit, Y., Agarwal, S., Brundage, M., Adler, S., O'Keefe, C., Campbell, R., . . . Robinson, D. G. (2023). Practices for governing agentic AI systems. *Research Paper, OpenAI.*

Sheikh, M. F. A., Kapoor, S. K., Goyal, N. K., Tuba, E., & Rao, A. (2026). Enhancing healthcare security with fog computing and blockchain technology. In Nayak, R. (Eds.), *Recent advancements in artificial intelligence* (pp. xxx–xxx). Springer., DOI: 10.1007/978-981-96-7760-3_38

Shinn, N., Cassano, F., Gopinath, A., Narasimhan, K., & Yao, S. (2023). Reflexion: Language agents with verbal reinforcement learning. *Advances in Neural Information Processing Systems*, *36*, 8634–8652.

Taesi, C., Aggogeri, F., & Pellegrini, N. (2023). COBOT applications—Recent advances and challenges. *Robotics (Basel, Switzerland)*, *12*(3), 79. DOI: 10.3390/robotics12030079

Turing, A. (2009). *Computing machinery and intelligence.* Springer. (Original work published 1950) DOI: 10.1007/978-1-4020-6710-5_3

Yao, S., Zhao, J., Yu, D., Du, N., Shafran, I., Narasimhan, K. R., & Cao, Y. (2022, October). React: Synergizing reasoning and acting in language models. In *The eleventh international conference on learning representations.*

Zeng, A., Liu, M., Lu, R., Wang, B., Liu, X., Dong, Y., & Tang, J. (2024, August). Agenttuning: Enabling generalized agent abilities for llms. In *Findings of the Association for Computational Linguistics: ACL 2024* (pp. 3053-3077).

Chapter 9
Ethics by Design:
Legal and Moral Foundations for Agentic AI in Autonomous Vehicles

C. V. Suresh Babu
http://orcid.org/0000-0002-8474-2882
Hindustan Institute of Technology and Science, India

Vaishnavee Hariharan
http://orcid.org/0009-0000-3226-6784
Hindustan Institute of Technology and Science, India

B. R. Mandeep Surag
Hindustan Institute of Technology and Science, India

R. R. Gayathri
Hindustan Institute of Technology and Science, India

ABSTRACT

This chapter per the authors aims to examine how Agentic Artificial Intelligence can embed ethical reasoning and legal compliance into autonomous vehicle decision-making systems. The target groups include researchers, academicians, policymakers, autonomous vehicle manufacturers, and AI ethics professionals. The chapter adopts a conceptual and framework-driven approach, synthesizing interdisciplinary literature from artificial intelligence, ethics, law, and human–computer interaction. The findings highlight that ethics-by-design architectures, combined with legal codification and human-centered considerations, enhance accountability, transparency, and safety in autonomous vehicle operations. The chapter concludes that integrating ethical and legal principles at the architectural level is essential for trustworthy and scalable deployment of agentic autonomous vehicles, with significant implications for regulation, system design, and public acceptance.

DOI: 10.4018/979-8-3373-7355-3.ch009

1. INTRODUCTION

1.1 Overview of Agentic AI in Autonomous Vehicles

At present, there is a transformation taking place in the auto industry as it evolves from using Advanced Driver Assistance Systems (ADAS) to becoming Autonomous Vehicles (AV) using Agentic Artificial Intelligence (AI). Whereas conventional forms of automation are based on executing pre-programmed scripts under controlled circumstances, Agentic AI is capable of operating autonomously, and adapting and achieving objective targets in an unstructured environment (Shukla, 2025). As a result of the most recent technology advancement, AVs can now process complex social and physical environments, rather than simply reacting to stimuli. This enables the AV to make decisions that previously only humans could make (Fossa, 2025). As AVs evolve through higher levels of automation (SAE Level 4-5), there is a fundamental shift in the responsibilities of technology to change from that of being a passive organisational tool, to becoming a "moral Proxy" that is responsible for making moral decisions in relation to ethical behaviour on the road without the need for real-time human intervention (Suresh Babu, C. V. & Jeneliya, J. S. 2025). When integrated with Agentic AI, AVs can successfully navigate complex ecosystems with other vulnerable road users, conflicting traffic flows and unpredictable environmental uncertainty. However, for AVs to achieve the level of autonomy described above, it will require a complete redefinition of vehicle design and architecture so that the AV is not just designed simply for the avoidance of "obstacles", but to be capable of exercising advanced levels of reasoning within established human safety standards and societal norms (Santoni de Sio & Fossa, 2023).

1.2 Limitations of Traditional AI in Ethical and Legal Decision-Making

Whereas traditional AI is based on static regulations or prediction algorithms, they are greatly limited in their ability to respond to the "open experience" of ethical and legal dilemmas of the real world (Suresh Babu, C. V. & Jeneliya, J. S. 2025). While one limitation of both legacy AI types has been their ability to respond to cases in which an unyielding compliance with traffic rules could have caused reduced safety for people using vehicles or bicycles.

In the case of the behaviour of human drivers, they frequently act "illegally due to necessity"; this includes behaviours such as making a turn across a double yellow line to avoid hitting an obstacle or a cyclist. There are nuances here that a rigid rule-based model and algorithms would struggle to provide legal and safe guidelines for how to approach this dilemma (Reed et al.2021).

Traditional predictive models also do not have the context-based flexibility that would allow them to arbitrate between the competing ethical values of moral pluralism (Rhim et al., 2021). Empirical studies, such as the Moral Machine experiment, show that individuals worldwide have different ethical preferences with respect to the questions of how to respond to dilemmas; thus it is extremely difficult for static algorithmic predictions to achieve a single, widely acceptable outcome (Bonnefon et al., 2018).

1.3 Need for Integrating Ethics and Law into Agentic AI Frameworks

There is a significant need to incorporate the concept of Ethics by Design within the design of Decision-Making Framework in order to address the weaknesses of current automation. Ethics must not be treated as an afterthought or added to the process post-facto, rather, Ethics by Design is a requirement for the Engineering process (Yu, 2025).

Figure 1. Conceptual framework for ethics by design in agentic AI

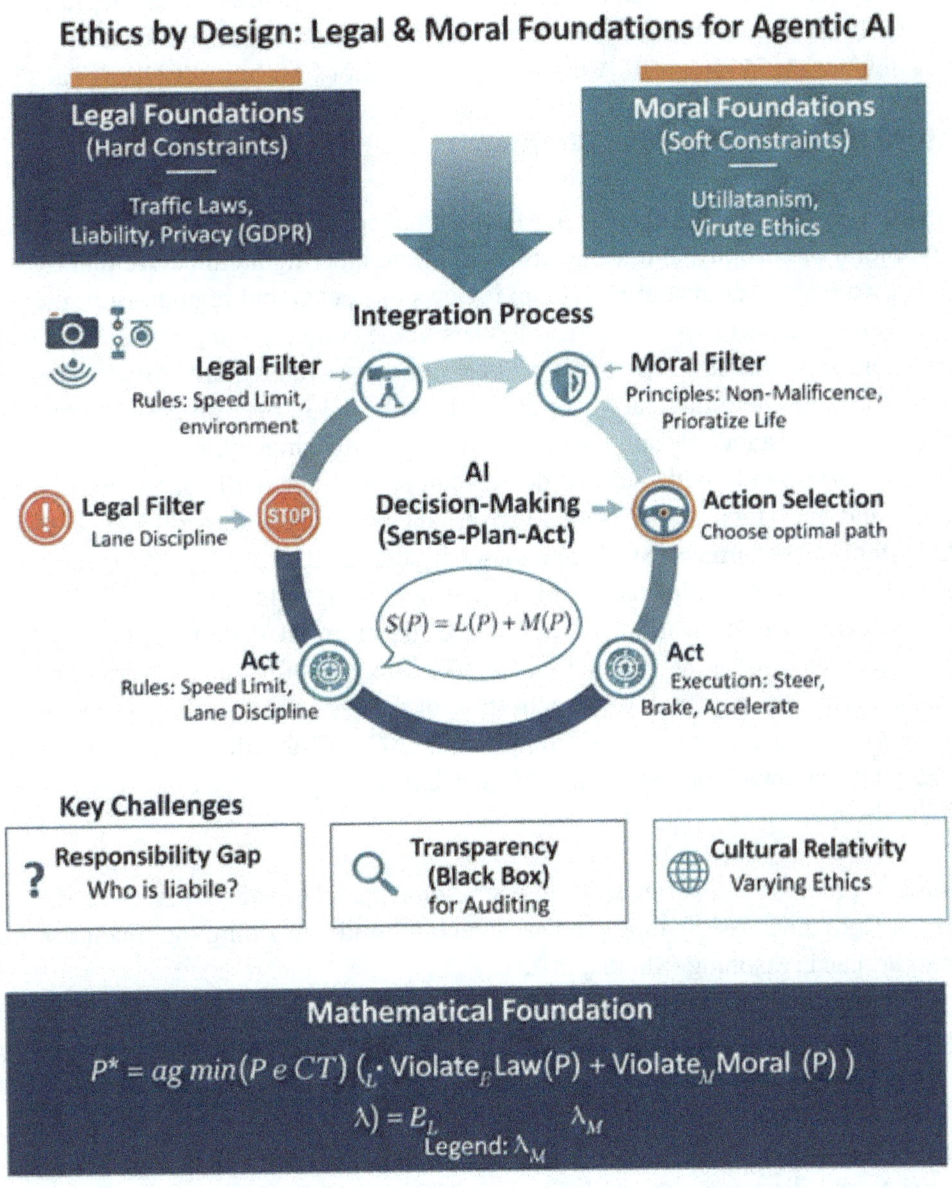

This diagram illustrates the integration of legal hard constraints and moral soft constraints into the sense-plan-act loop of autonomous systems, highlighting the mathematical optimization of ethical trajectories.

As responsibility of decision making shifts from Humans to Machines, AVs are required to take a Precautionary Approach when making decisions. In essence, as outlined by (Resnik & Andrews., 2023), the Legal System expects AVs to utilise the Precautionary Principle to protect Lives, and reduce Harm in the face of Uncertainty. In order to account for these factors, the design of AVs will require a Systematic Design Strategy whereby Legal Restrictions represent Hard Boundaries (e.g. Traffic Regulations) and Ethical Norms serve as Optimize Parameters (Tiwari & Choudhary, 2025). The integration of Legal and Ethical will be critical not only for Compliance with Regulations, but also to develop the Public Trust necessary to allow for the Widespread Adoption of Autonomous Mobility (Leikas et al., 2019).

1.4 Structure and Scope of the Chapter

Methodology and Scope: This chapter presents a Conceptual Framework Analysis to combine the legal liability, moral philosophy, and technical areas of engineering architecture that exist in silos. To accomplish this we employ comparative critical reviews of new global regulatory frameworks (EU AI Act; ISO 26262) with respect to moral-ethical models in the contemporary world. The final product of this mapping exercise was a new type of Hybrid Governance Architecture, which brings together the 'hard' laws and the 'soft' moral/logical systems of ethics, bridging the gap between the conceptual or abstract nature of ethical standards and the code that governs machines(Shukla, 2025).

This strategy corresponds with worldwide regulatory and standards-based initiatives which are becoming more and more focused on lifecycle governance, risk management, and accountability with respect to AI systems. For example, ISO 26262 is a set of functional safety criteria for automotive systems (ISO, 2018) and NIST AI Risk Management Framework is designed to promote a structured means of identifying, measuring, and mitigating risk across the life of an AI system (Tabassi, 2023).Likewise, current legal analyses highlight the fact that factors like safety, liability, and governance should also be taken into consideration during the development stage of the system rather than dealing with them after the system has already been implemented (Singh et al., 2025). This current chapter proposes an integrated ethical/legal framework on the agentic AI decision-making of autonomous vehicles considering the dynamic landscape.

Section 2: It provides an overview of the operational characteristics of Agentic AI ("AI"), distinguishing it from the class of static predictive models, and describing the functions that AI has in its perception and reasoning (Shukla, 2025).
Section 3: It examines the ethical foundation behind AVs, and discusses the use of ethical frameworks such as utilitarianism and rights-based ethics as a means by which to represent these frameworks within the coding of inference engines so that they can assist in resolving the most notable high-stakes ethical dilemmas (Gros et al., 2025; Davnall, 2020).
Section 4: It explores the legal environment regarding AVs, and analyses the global regulatory environment and discusses the challenges associated with the translation of statutory law into machine-readable policy (Singh et al., 2025; Biczi, 2025).
Section 5: The author's proposal of integrating key aspects of Law with Ethics into an architecture using a hybrid approach that balances Safety, Reliability, and Norm Compliance (Poszler et al., 2024; Evans et al., 2020).
Section 6 and 7: The Human and Governance aspects of the Architecture are examined (Sections Seven and Eight), with an exploration of how Public Perception, Trust, and Accountability Systems

influence the Design and Acceptance of Ethical Autonomous Vehicles (Ladak et al., 2025; Wani et al., 2024).

Section 8 and 9: The final two sections of the Paper explore Case Studies and propose a Path Forward for Future Research, which requires Interdisciplinary Collaboration in addressing the 21st Century "Trolley Problem" (Cunneen et al., 2020; Milford et al., 2024).

2. UNDERSTANDING AGENTIC AI FOR AUTONOMOUS VEHICLES

2.1 Defining Agentic AI: Characteristics and Capabilities

Agentic AI is a more complex form of AI as compared to traditional automated systems (thus far referred to simply as automation). Whereas standard automation automates a predefined task through a specified process or algorithm; Agentic AI has greater capabilities. Agentic AI possesses the ability to act independently, be adaptive to changing situations or contexts, and conducts goal-directed actions without any prior programming or instruction (Shukla 2025) and as such it acts in a manner similar to human behaviour (however not exactly like that behaviour). Agentic AI does not drive vehicles in the same manner as humans do; rather its interpretation of the driving environment is more sophisticated and it creates innovative solutions to the numerous and varied challenges associated with navigating an increasingly complex world. The change in the way that Agentic AI conducts its functions creates a mechanism where the vehicle (AV) operates as an independent entity in its own right therefore creating a situation whereby it is making decisions that affect human safety and autonomy (Fossa 2025).

An important component of the definition of Agentic AI is its ability to operate based on "moral functionalism". This means that it must be able to take large quantities of sensor data, evaluate that data, and convert that data into appropriate actions that meet the larger purposes of safety and moral acceptable, even where human oversight may not be present (Leikas et al. 2019). Therefore, Agentic AI must be capable of weighing competing objectives in the areas of velocity, comfort, and safety for non-occupants (i.e., pedestrians) in situations where human drivers would be faced with ethical dilemmas (Millar 2015).

2.2 Related Work and Research Contribution Review of Related Work:

The literature regarding Autonomous Vehicles (AVs) has changed dramatically throughout its history. In earlier research, the focus was on the development of static forms of artificial intelligence (AI) that follow rules (Suresh Babu, C. V., & Nithika, S. R., 2025), whereas the newer research is focused on Agency (Agentic) AI, which represents the ability of an AI system to take proactive actions/decisions (Yu, 2025). In the literature, a Critical Review identified three continuing gaps in the literature,

1. Operational Gap: There are currently no common metrics to apply for measuring ethical behaviours in everyday non-catastrophic contexts (Evans et al., 2020).
2. Security Gap: There is insufficient interdisciplinary literature existing at the collision of Cybersecurity and Ethics regarding how malicious/false information affects an agent's ability to think/act morally (Tiwari & Choudhary, 2025).
3. Liability Gap: There is no legal language to contemplate an AI moving away from ethical behaviour following its deployment (Biczi, 2025).

This Chapter contributes to the solutions to these gaps through three different contributions:

1. Hybrid Governance Architecture- The introduction of a new governance architecture model that is a blend of the ISO/NIST standards into the feedback loop of agents.
2. Cyber-Ethical Resilience- The demonstration of a qualitative relationship between the quality of data provenance and the ethical stability of agentic systems.
3. Operational Benchmarks- The definition of a hierarchy of constraints and regulatory requirements for the decision-making process, providing tangible programming rules to transform the abstract concepts of duty of care into specific programming logic.

2.3 The Role of Perception, Reasoning, and Adaptation in AV Context

The operational success of Intelligent Agent AI (Agentic AI) within autonomous automobiles is dependent upon the unification of three main cognitive processes: perception, reasoning and adaptation. Perception, in relation to Agentic AI, includes more than just object detection; it consists of "semantically meaningful" understanding of a particular scene including the ability to distinguish between a plastic bag and a child crossing a street (Charisi et al., 2017). In other words, when presented with a situation where a driver was facing unavoidable impact; in this moment the reasoning module would need to immediately consider/compute risk distributions for the effect of the impact occurring upon the bodies of individuals. This would need translating ethical theories into algorithms(Poszler et al., 2023).

The adaptive process keeps these reasoning functions functional over time. This is necessary to ensure the "meaningful human control" of Agentic AI remains at an elevated level of abstraction for defining values and safety guidelines for autonomous driving. However, Agentic AI must be able to independently adjust its real-time actions by using these values and safety criteria in changing real-time environments (Santoni de Sio & Fossa, 2023). This adaptability is also necessary for legal and compliance reasons; for example, the agent must operate within the various local traffic regulations and operate within the local driving customs without requiring continuous software updates in every municipality (Tiwari & Choudhary, 2025).

2.4 Multi-Agent Interactions in Complex Driving Ecosystems

Agentic AI operates within a dynamic "multi-agent system" alongside human operated vehicles (cars, trucks), pedestrians and cyclists. In this environment, many different system members' interactions create a context of social physics that governs how agents are likely to interact with each other when confronted by these heterogeneous agents. This context is complicated because, although agents interact with each other, each one's actions are interconnected. Thus, a cooperative, rather than simply competitive, decision-making framework is required for agentic AI to be effective to predict other agents' actions (Charisi et al., 2017). For example, agentic AIs will need to possess the capability of "Theory of Mind" (every agent's ability to predict the actions of others based on a set of rules that governs their interactions) in order to make predictions of other agents' actions based on their trajectory and cues (Suryana et al., 2025).

In addition to presenting a challenge to the development of agentic AI due to the unpredictable characteristics of human actors, the high penetration rate of autonomous vehicles into the majority of transportation networks will create new opportunities to negotiate right of way and coordinate maneuvers between agentic AI systems through V2V communication (Suresh Babu, C. V., & Nithika, S. R., 2025). The development of V2V communication will create a collective intelligence that will be utilised to optimise vehicle movements (Gros et al., 2025).

3. ETHICAL FOUNDATIONS FOR AGENTIC AI

3.1 Overview of Ethical Frameworks Relevant to AV Decision-Making

The advancement of Agentic Artificial Intelligence (AI) for Autonomous Vehicles (AVs) requires that ethical theories be translated into logic that can be programmed. Three ethical theories dominate the literature: Utilitarianism, Deontological ethics, and the Rights-Based theory of ethics.

Table 1.

Ethical Framework	Core Principle	Application in Agentic AI	Limitation in AV Context
Utilitarianism	Maximize overall well-being; minimize total harm.	Useful for risk distribution in crash scenarios (e.g., minimizing total injury severity) (Gros et al., 2025).	May justify "computational cruelty" (e.g., sacrificing one passenger to save two pedestrians) (Bonnefon et al., 2018).
Deontology	Strict adherence to rules and duties.	Enforcing "hard constraints" like traffic laws and stop signs (Liu et al., 2025).	Too rigid for complex scenarios; may cause "freezing" when rules conflict (Reed et al., 2021).
Rights-Based	Protecting fundamental human rights.	Setting inviolable boundaries; ensuring no demographic is systematically targeted by risk (Poszler et al., 2023).	Difficult to quantify mathematically in real-time decision loops (Leikas et al., 2019).
Hybrid Model	Layered approach (Rules + Utility).	Uses logic for rules and utility for optimization; current "best practice" (Suryana et al., 2025).	Requires complex computational power to resolve conflicts (Yu, 2025).

Two additional ethical frameworks that differ from Utilitarianism and Deontological ethics are: (1) the Rights-Based theory, which focuses on the fundamental right of all individuals to be safe and secure while operating an AV and as such, promotes the creation of a "virtuous" Agentic AI that mirrors the behaviours of a healthy human driver, and (2) the Virtue Ethics Framework, which looks at the characteristics of the healthy driver in a manner similar to the Rights-Based framework. The complex application of these ethical theories arises from an environment of Moral Pluralism, wherein various stakeholders place a different level of importance or priority on one or several of the above-mentioned ethical frameworks. For example, a Utilitarian model may optimise the flow of traffic and increase the safety of the majority of drivers, whereas a Rights-Based Model may refuse to allow an AV to act in a manner that violates the negative rights of one pedestrian, even if such an action would prevent a significant accident from occurring for many others (Fedorov & Curran, 2024). As a result, developing a functional framework for Agentic AI must combine and integrate the fundamental outlined characteristics of each type of ethical framework, or there will be no single ethical theory to leverage for use in the operation of the AV (Poszler et al., 2023).

3.2 Moral Dilemmas in Autonomous Driving Scenarios

In the Trolley Problem, emphasized in theoretical work, AI systems making similar operational stop-go decision rarely occurs, unlike for agentic AI that constantly encounters "micro-ethical" dilemmas on public roads. Common situations (for example, judging the appropriate distance with which to overtake a cyclist or deciding on who goes first at a crossroads) force an agent to trade off efficiency against safety all the time (Davnall, 2020). Studies suggest that public trust is established in these boring encounters more frequently than in stand-alone catastrophic crashes. A speed limit-obeying but traffic-impeding AV may be legally compliant and yet ethically unaccountable (Reed et al., 2021). Thus, the ethical architecture should be calibrated to grasp the "social physics" of everyday navigation (Charisi et al., 2017), with extreme dilemmas serving merely as boundary conditions for its validation (Sui, 2023).

This is highly abstracted, in a real world setting the agent has very rarely perfect information and will always operate with probabilities. It has been argued that the AVs need to use dynamic models of risk assessments taking into account not only the likelihood of a crash, but also extent of damage associated with modelled measures should be such that it does not unfairly transfer risk back more vulnerable road users (Liu et al., 2025). This probabilistic approach means that the ethical ones moves from a binary choice to an ongoing slide-rule process of making 'better' and 'worse' decisions about direction and speed (Davnall, 2020).

Figure 2. Taxonomy of moral dilemmas and algorithmic resolution

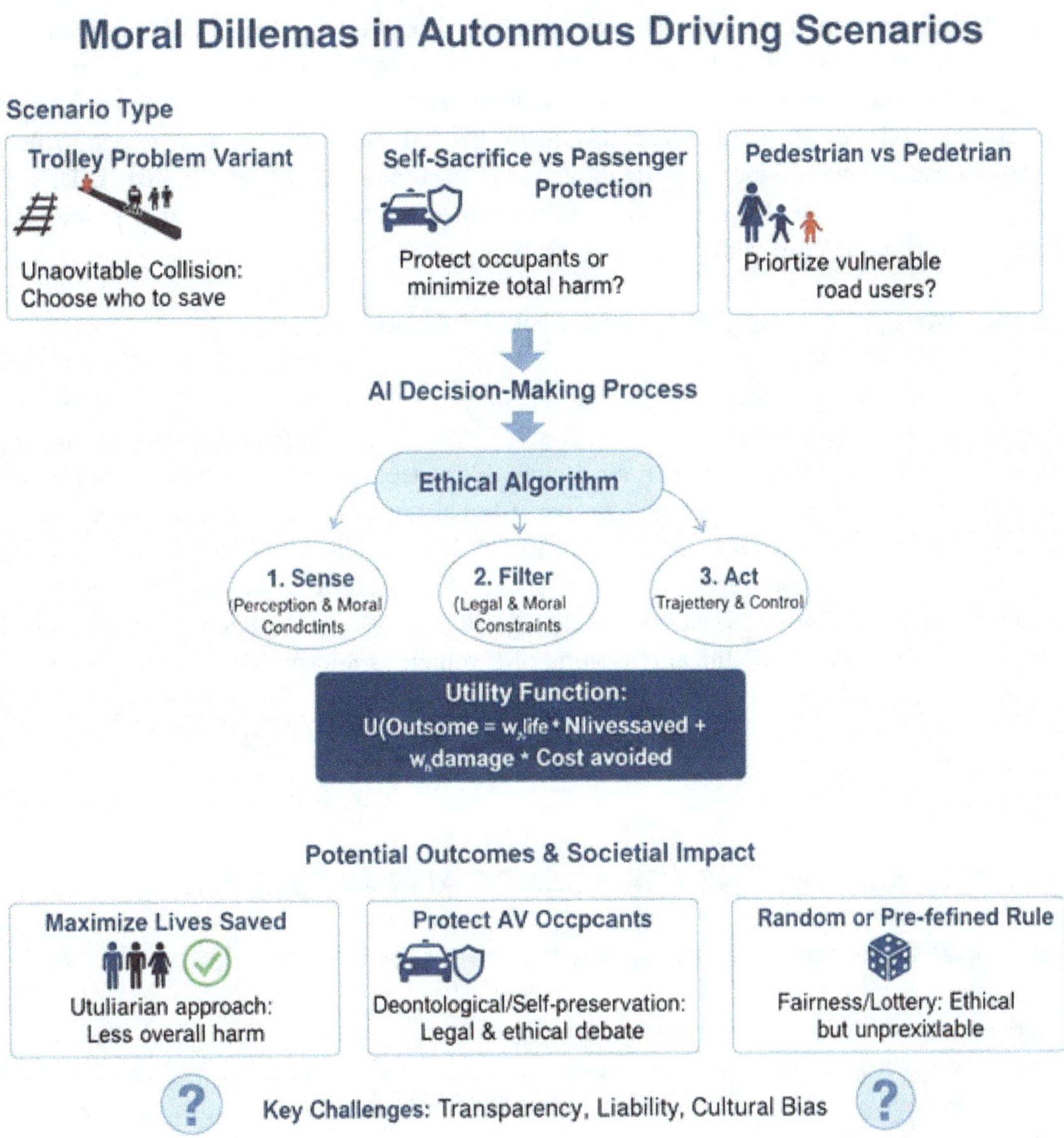

A visualization of high-stakes scenarios (e.g., trolley problems) and the corresponding utility functions used by Agentic AI to calculate risk and determine optimal ethical outcomes.

3.3 Encoding Ethical Principles in Agentic AI Inference Engines

"Ethics by Design" refers to modifying ethical criteria into formulae, that the inference engine can understand. For example, values like "fairness" or "duty of care" have to be transferred into cost and constraint functions in the control algorithms of the vehicle (Yu, 2025). For instance, utilitarianist considerations can be implemented by substituting numerical values to the potential options (e.g., injury, property damage) and programming the AV to choose the path with the less total cost (Poszler, 2024).

Yet, sheer optimization brings the danger of "computational cruelty," a situation in which the system may come to justify morally abhorrent decisions. To prevent that, scientists propose to use "ethical governors" or "hybrid architectures" where it is the learning-based system (e.g., a neural network) that proposes actions and these suggestions are evaluated by including into decision-making rules-based security layer ensuring compliance with deontological constraints (Leikas et al., 2019). This layering guarantees that although the AV can adjust and maximize efficiencies, it is bound by ethical limits stopping short of flouting core safety principles (Millar, 2016).

3.4 Transparency, Explainability, and Moral Accountability

For Agentic AI to be accepted socially, its decisions should be transparent and understandable. The "black box" problem in many high-level AI systems is a major hurdle to holding AIs accountable, if an AV makes a decision that seems wrong, stakeholders need to know the "why" (Martinho et al., 2021).Explainability is not a mere technical property, but also a moral requirement, since it makes possible the assignment of responsibility in case of failure. Without such clear understanding of the agent's reasoning plan we will not be able to decide if an accident is due to a sensor malfunction, a design mistake or even that it was intended (Evans et al., 2020). Research suggests that people are more willing to accept AVs if the "mind" and ethical reasoning of a vehicle is interpretable and aligns with human values (Ladak et al., 2025).Hence, Agentic AI must be designed with post-factum analytic systems that are able to explain the ethical weights in high stakes decision making - a digital testimony of inquiries legal or social (Wani et al., 2024).

4. LEGAL CONSIDERATIONS FOR AGENTIC AI IN AV SYSTEMS

4.1 Key Legal Concepts: Liability, Negligence, Compliance, Duty of Care

The move to Agentic AI very significantly undermines settled legal doctrine, which is premised upon the fact that liability has been predicated on human agency. In torts, negligence is decided according to whether a reasonable person would have done something differently; however, in the case of an algorithm - using this "reasonable person" standard - it is legally unclear as (Singh et al., 2025). The notion of the "duty of care" has to be rethought when dealing with these autonomous agents, transitioning from a human-driven "negligence law" towards a product-based liability where defective design is more at stake than failure from the driver's side (Biczi, 2025). Moreover, in AVs adherence to law goes beyond compliance with established traffic rules: it must instead exhibit a "precautionary stance", whereby safety in unpredictable situations takes precedence over strict obedience to the letter of the law (Resnik & Andrews, 2023). This change demands novel legal definitions to catch up with the sui generis nature of algorithmic decision-making because "fault" in this type of considerations may result from coding weighing decisions unethically, not a hasty judgment (Tiwari & Choudhary, 2025).

4.2 Global Regulatory Landscape for Autonomous Vehicles

To date, the regulatory frameworks for Autonomous Vehicles (AVs) have been piecemeal across different jurisdictions with various strategies toward the legalization of Agentic Artificial Intelligence (AI) (Singh et al., 2025):

- **The United States:** The framework has largely been an open slate, allowing for innovation to flourish by providing voluntary and aspirational guidelines for safety and pilot programs at the state level, resulting in significant fragmentation among states.
- **Europe:** The European Union has taken a "human-centric" approach, placing particular importance on safeguarding people and preserving data privacy, thus making it a priority to achieve the greatest possible level of compliance with ethical standards and developing AV frameworks such as the AI Act that classify AVs as high-risk systems and thereby require extensive conformity assessments.
- **Asia-Pacific:** Many countries in the Asia-Pacific region are working to establish testing zones and regulatory sandboxes designed to facilitate the deployment of AVs into public infrastructure, and to create maximum incentives for industry growth through caps on liability.

Despite these regional differences, there exists a global consensus regarding the necessity for a harmonization of standards in order to effectively govern the transnational deployment of Agentic AI technologies, which is becoming increasingly formalized through global technical standards, including ISO 26262 (Functional Safety) and the NIST AI Risk Management Framework (AI RMF), that offer clear benchmarks for assessing the "trustworthiness" of agentic systems (Tabassi, 2023; ISO, 2018).

Figure 3. Global regulatory landscape for autonomous vehicles

A comparative map highlighting the distinct regulatory pillars and frameworks adopted by major jurisdictions, including the EU's AI Act, US state-led initiatives, and Asia's tech-driven zones.

4.3 Challenges in Translating Laws into Machine-Readable Policies

The translation of the governing laws for the AV sector from natural language into a precise, machine-readable format is perhaps the largest technical challenge surrounding how AVs will operate in the future

(Poszler, 2024). Traffic laws tend to be context dependent and require the use of human judgement in order to resolve any ambiguities that arise in the law; this is where deterministic algorithms face challenges. One example of this principle is the "Ethics by Design" approach which requires the conversion of the legal text into formal logic or cost functions that an AV inference engine can process mathematically (Yu, 2025). One result of encoding the legal text too strictly is that an AV may have a tendency to exhibit overly cautious or paralyzed behaviour when navigating complex, densely populated roads and intersections, which makes finding the right balance of strict legal compliance and the ability to navigate fluidly through busy streets a particularly difficult problem (Reed et al., 2021).

4.4 Real-Time Legal Reasoning in Dynamic Environments

In the context of an autonomous vehicle (AV), agentic AI will engage in real-time legal reasoning under strict time constraints, often with a lack of complete data. As a result, agentic AI must be able to not only identify the appropriate laws but also resolve conflicts between them(Tiwari & Choudhary, 2025). In order for an AV to perform this dynamic adjudication, it must evaluate the "hierarchy of norms" regarding the various trajectories available to it, selecting the safest trajectory with the least amount of potential liability (Biczi, 2025). The agentic AI acts as an "on-board" legal interpreter for the AV in real-time, whereas this function was previously provided by the judicial branch of the government on behalf of the AV.

4.5 Legal Implications of Adaptive Decision-Making and System Autonomy

Agentic AI poses an unprecedented challenge regarding the certification and liability for AVs. This means an AV, which has the potential to evolve its operational behaviour to levels that were not anticipated by its original designers/programmers, will create significant issues related to training and accountability (Biczi, 2025).

It will be necessary for legal frameworks to create a decision on whether to hold manufacturers accountable for actions performed by an AV after it has deviated from the factory's default or whether to create a new classification of "electronic personhood" for that AV (Singh et al., 2025). Additionally, the autonomy of AVs creates a situation where the "paternalism by design" dilemma occurs: The AV's autonomy may prevent it from obeying the user's commands for safety reasons and create a conflict between the user's autonomy and the protection afforded to the manufacturer under tort law (Millar, 2015).

5. INTEGRATING ETHICAL AND LEGAL REASONING IN AGENTIC AI ARCHITECTURES

5.1 Multi-Layered Ethical-Legal Decision-Making Models

In order for Agentic AI to successfully respond to the various complexities of Public Roads, Agentic AI requires a new approach to its architecture, away from Monolithic code structures and towards the development of multilayered Decision-making Model Architectures (DMDA). In an Agentic AI DMDA, the "Driving Policy" is generally divided into three primary Layers; The Strategic Layer for long-range route planning, The Tactical Layer for immediate manoeuvre selection and The Control Layer for actuation. The DMDA incorporates an Ethical Supervisor Layer to monitor and control the tactical decision-making processes, whereby the Ethical Supervisor Layer represents an Ethical Governance mechanism to filter out any tactical actions that are not consistent with either legislation or public safety (Gros et al., 2025).By separating ethical evaluation from trajectory creation, engineers can ensure that the AV's actions remain predictable and within the Regulatory body's defined legal envelope.

Table 2. Hierarchy of decision constraints in agentic AI architectures

Decision Layer	Function	Priority	Description & Example
1. Physical Safety	Hard Boundary	**Critical**	**Inviolable constraints** to prevent immediate physical damage. *(e.g., Maintain traction; Do not collide with static obstacles)*[1111].
2. Ethical Duty	Supervisor	**High**	**Value-based optimization** favoring the preservation of life. *(e.g., Minimize harm to vulnerable road users over property)*[2222].
3. Legal Compliance	Rule Adherence	**Medium**	**Statutory rules** that should be followed unless Layer 1 or 2 is threatened. *(e.g., Speed limits; Lane discipline; Traffic signals)*[3333].
4. Comfort & Efficiency	Operational Goal	**Low**	**Soft constraints** for passenger experience. *(e.g., Smooth acceleration; Shortest route selection)*[4].

5.2 Hybrid Approaches for Encoding Norms (Rules, Logic, Reinforcement Learning)

Rule-based systems are too brittle for the chaotic nature of real-world traffic, but black box end-to-end machine learning (ML) models often trade accuracy with the transparency needed for legal accountability. As a result, the two approaches are starting to come together in more hybridized versions. Control in these models is dual: Deep RL manages high-dimensional sensory data and complex driving strategies, while symbolic logic enforces the hard constraints such as traffic rules (and probably ethical boundaries) (Liu et al., 2025). For example, an RL agent might suggest a trajectory to knock over a cyclist but rule the manoeuvre out (since that would violate the minimum safety distance mandated by law) if this is punished by a logic-based “ethics engine”.

5.3 Aligning Safety, Reliability, Ethics, and Legal Constraints

These factors can be weighted in a dynamic design framework: for example, while safety remains the primary imperative, “ethics” serves as the tie-breaker when absolute safety no longer is feasible (Leikas et al., 2019). For the process of integration to do its job, often a ``precautionary" approach is used in the sense that the systems which operate are biased towards conservative choices in non-ideal situations so as to be at least reliable (Resnik \& Andrews, 2023). But alignment also means that we must grapple with the tension between strict legality and practical safety; an AV that is programmed never to break a traffic law will be unable to avoid making a bad situation worse if it refuses to cross a solid line in response to some obstruction. For this reason the AI must encode a "hierarchy of values" that places human life above such petty legal infringements (Santoni de Sio & Fossa, 2025).

5.4 Conflict Resolution Mechanisms in Ethical and Legal Trade-offs

Perhaps, as Agentic AI matures, it will inevitably come across situations where statute law and ethical intuitions clash and thus requires a sturdy framework which can resolve these conflicts. These are "moral algorithms" that mediate between competing duties on the fly. For instance, when an AV faces a choice between hitting a pedestrian and swerving into oncoming traffic (which would in practice break the law), the conflict resolution module must immediately compute "the moral cost" of these (Rhim et al., 2021). Studies have shown that it is ethical—and sometimes legally justifiable—to break the rules of the road, if doing so is the only way to avoid serious harm. This "necessary illegality" would have to be encoded into the decision matrix, so that the agent could violate ordinary rules under quite narrow, accurately predefined circumstances (Reed et al., 2021). In addition, liability remedies must be taken under account while designing, such that a solution path also minimize the physical harm as minimizing legal exposure to operator and manufacturer (Biczi, 2025).

5.5 Case Example: Decision-Making in High-Risk Driving Situations

One of the applications where these integrated structures can be applied practically is situations with high risk, like a person suddenly stepping into the car path. In one such dillema variant, a media civilization system could hit the brakes, risking a rear-end collision. But why choose if an AI exists that uses risk theory to calculate all potential pathways at once, figuring the probability of pedestrian versus passenger and other traffic injury (Davnall 2020)? Faced with a situation like this, and based on the output of an "ethics estimation tool," should the car controllers decide that ramming into a side barrier was preferable to hitting the pedestrian—"property damage notwithstanding"—there is no reason to think that it was not (Millar, 2016). It is not a random decision but a quantifiable ethical calculus that is in line with social norms concerning the conservation of vulnerable life (Bonnefon et al., 2018).

6. HUMAN FACTORS INFLUENCING AGENTIC AI ETHICAL BEHAVIOR

6.1 Role of Human Trust, Fear, and Acceptance in AV Use

Deployment of Agentic AI requires not only technical competence but also public acceptance. Prior research has demonstrated that human trust in AVs is fragile and greatly affected by the perceived "moral agency" of the machine (Ladak et al., 2025). When an individual believes an AI does not have a "mind", they are less likely to accept ethical decisions made by the system when those decisions are controversial. Fear arises from the fact that these systems are often referred to as "black boxes." Due to the lack of transparency regarding how an agent evaluates risk, users may consider the vehicle's autonomy a threat versus a benefit.

6.2 Cognitive Limitations and Human–AI Interaction Challenges

Agentic AI's purpose is to help humans avoid making mistakes. There are many types of interactions between autonomous machines and human drivers, and there is a growing body of evidence that suggests that these types of interactions can be "misread" by both parties. The reason for this discrepancy is because nonverbal cues are very important to humans, and current AV technology isn't very good at either replicating or interpreting these types of cues. When an AV is operating in a manner that appears to be mathematically optimal, this may cause a great deal of confusion, or appear aggressive, to a human onlooker, particularly in situations where the AV is being operated at intersections or crosswalks (Cunneen et al., 2020). In addition, the design of driving automation systems must take into account that as humans continue to depend more and more on automated systems, their capacity to take over control of a system in the event of failure, will decrease. Therefore, AI must be able to effectively transfer control of the vehicle from the AV to the user while minimizing the mental burden placed on the user during the transition of control (Santoni de Sio & Fossa, 2023).

6.3 User Behaviors that Impact Agentic AI Ethical Outcomes

The ethical behaviour of Autonomous Vehicles (AVs) is shaped both by its program and by how people use the AV (Suresh Babu, C. V., & Nithika, S. R., 2025). There is, therefore, a danger of automation complacency, where people become overly trusting of an AV's operation, become engaged in unsafe behaviour and attempt to override the AV's safety measures for their personal convenience. The relationship between the behaviour of human users and the AV leads to a situation where designers need to consider "paternalism by design," meaning that the Agentic AI needs to be programmed in a way that will limit human autonomy in order to protect the safety of others in the society (Millar, 2015). However, this presents an ethical dilemma regarding the extent to which a machine, as a moral proxy for its owner, has the ability to abrogate the will of its owner in order to perform its higher social duties (Fossa, 2025).

6.4 Cultural and Societal Variations in Ethical Expectations

One significant second challenge to Agentic AI is the absence of a consistent and stable global ethical framework. This lack of consensus on "acceptable" ethical standards was evident in the results from the "Moral Machine" study, which indicated great differences across cultures regarding ethical standards. In contrast, cultures that are primarily individualistic would see maintaining laws or protecting car passengers as being the most significant aspect of ethical behaviour (Bonnefon et al., 2018). Because of these varying societal values and beliefs, it would be challenging to develop an ethical algorithm that can ensure decisions made by Agentic AI systems that may be considered "moral" by one culture would likely be considered as "unacceptable" by another culture (Sui, 2023). Therefore, in the development of Agentic AI systems, it would likely be necessary to make such systems "culturally adaptive" and have them configure their ethical weighting parameters based on local laws and cultural norms (Poszler et al., 2023).

6.5 Insights from Human Factors Engineering and Behavioral Studies

To achieve a balance between algorithmic reasoning and human morals, the AV design will benefit from involving Human Factors Engineering (HFE). These frameworks represent a "reason-based" form of supervision for the AVs; they allow the AV to communicate its decisions in the form of reasoning that can be understood by the humans riding in the AV (Suryana et al. 2025). Research in behavioural science shows that users are more willing to accept ethical dilemmas associated with AVs if they clearly communicate their reason, which helps turn the AV from being just a tool into a partner (Charisi et al. 2017). Additionally, incorporating psychological data facilitates modelling the "social physics" of operating a vehicle to enable agentic AI systems to respond appropriately to the irrational and/or emotional reactions of people driving while they are riding as passengers.

Taken together, these architectural components reflect a broader shift toward embedding ethical reasoning directly into system-level engineering design, instead of treating ethics as something added on later for evaluation. With the growing amount of regulatory scrutiny on autonomous systems, it is anticipated that future governance frameworks will likely require demonstrable evidence of safety, accountability, and explainability integrated into the design process (Singh et al., 2025; Tabassi, 2023). Ethical architecture will not be viewed solely as a philosophical enhancement but also as a necessary condition for regulatory compliance and public trust. As a result, the way in which ethics are operationalized in agentic AI systems will drive the future direction of technology and the legal validity of autonomous vehicles.

7. ACCOUNTABILITY, RESPONSIBILITY, AND GOVERNANCE

7.1 Defining Accountability for Autonomous Agent Decisions

Transitioning to AVs will result in a large "accountability gap" where existing labels of blame fail to resonate. In the absence of that, responsibility is anchored in human agency and the ability to determine the future. According to scholars, accountability in this sense would have to be structural-based, asking for the traceability of the AV's decision-making rationale against system safety requirements as well as justifications (Martinho et al., 2021). Wherever clear (accountability) mechanisms are not in place, AVs might be seen as "moral zombies" - entities that have effects but no conscience – undermining trust (Ladak et al., 2025).

7.2 Stakeholder Roles: Manufacturers, Developers, Users, and Regulators

The distribution of the responsibility with respect to AV in a topology is very complicated and widely spread among many parties. The first burden of this "Ethics by Design" will fall on manufacturers and developers to ensure that the overall vehicle design can cope with the predictable ethical challenges (Yu, 2025). This moves liability from the driver (negligence) to the manufacturer (product liability), e.g. in relation design defects in AI's reasoning engine (Biczi, 2025). Users, though no longer in immediate control, still have a part to play in what some refer to as "meaningful human control," including watching over the system's capabilities and refraining from using it improperly (Santoni de Sio & Fossa, 2023). The regulators are important as they set the "boundary conditions" for acceptable risk(Singh et al., 2025).

Table 3. Distribution of roles, liabilities, and ethical obligations in the AV ecosystem

Stakeholder	Primary Role	Liability Focus	Ethical Obligation
Manufacturer	System Design & Deployment	**Product Liability** (Design Defects & Failure to Warn)	**Ethics by Design:** Ensuring value alignment and robust failure management[5555].
Regulator	Boundary Setting	**Standardization** (Certification & Type Approval)	**Public Safety:** Defining acceptable risk metrics and "precautionary" thresholds[6666].
User / Operator	Monitoring (Level 4)	**Negligence** (Misuse or Failure to Intervene)	**Meaningful Control:** Understanding system limitations and remaining alert[7777].
Agentic AI	Decision Execution	**None** (Lacks Legal Personhood)	**Optimization:** Executing the safest trajectory within programmed constraints[8888].

7.3 Ethical and Legal Risk Assessment Frameworks

For handling uncertainties of Agentic AI reliable risk assessment frameworks are necessary before deploying. This is a way more complex task that needs to be developed beyond the level of basic mechanical testing, and needs "ethics evaluation tools" which have to satisfy high-profile dilemmas and then audit back against these moral norms, in orderd for the AI's choices be found appropriate (Millar, 2016). An "ethics of precaution" commonly advocated is that, given the absence from test trials of scientific certainty about safety as tested on animals, regulators will impose a very high burden-of-proof requirement on manufacturers: an assertion if back-stopped not-yet disproved that their agent introduces no unnecessary harm (Resnik & Andrews, 2023). These assessments give a "number to the ethical risk" posed by an algorithm, such as how it may be biased in favor of some road users over others and whether that bias is legally defensible (Poszler et al., 2023).

For assessment purposes, the ethical principles must first be converted into measurable indicators at the system level. Some examples include: quantified harm-reduction indices produced through probabilistic risk analysis (Davnall, 2020); bias detection thresholds that measure different levels of risk allocated to individuals from disparate demographic groups (Poszler et al., 2023); explainability latency techniques that evaluate the ability of the system to provide post-hoc justifications for its actions (Wani et al., 2024); and traceability metrics created with structured risk management protocols in mind (Tabassi, 2023). By integrating these indicators into the validation process, ethics by design becomes not only a normative concept, but also a testable and auditable characteristic of an AGI system.

7.4 Governance Models for Agentic AI in AV Ecosystems

A multi-level model integrating tech standards and ethical principles has to be adopted, for the good governance of Agentic Dec AI. The "ecosystem governance" model has been suggested as a possible strategy, treating the AV not as an independent entity but rather as part of a networked ecosystem (Tiwari & Choudhary, 25). This entails incorporation of "ethical frameworks" that act as a governance layer within the software while surveilling and ensuring compliance of the agent with norms (Leikas et al 2019). In addition "governance" has to be dynamic: adapting as the car further up- dates and modifies itself by learning from data that this "living" piece of software tops a one time check in terms of its 'moral health' during its life (Fossa 2025).

7.5 Need for Interdisciplinary Collaboration in Policy and Standardization

The problems of Agentic AI can't be solved by engineers or lawyers alone. Comprehensive standards will require significant multi-disciplinary work among computer scientists, ethicists, law scholars and urban planners (Evans et al., 2020). Expert opinions suggest an ethics of AVs is frequently held back by the "translation gap" between these fields; engineers want specific variables, ethicists deal in nuance (Milford et al., 2024). Resolving this tension between human policy and machine algorithms is important not only for developing machine readable policies that genuinely represent human laws and moral intuitions. Regulatory and collaboratory bodies need to define what is safe, and what acceptable levels of risk are going forward, to ensure that the "ethics" embedded in AVs reaches a societal consensus rather than being the values of any one corporation(Gros et al., 2025).

8. CASE STUDIES AND REAL-WORLD SCENARIOS

8.1 Ethical-Legal Analysis of Past AV Incidents Involving Autonomous Systems

The practical implementation of ethical theories often begins with the failures associated with them. A recent example of this kind of failure was the fatal crash of an autonomous vehicle (AV) test vehicle operated by Uber in Arizona. This incident displays how early generation AVs lack the "agentic" characteristics necessary to meet the ethical and legal requirements established by society. In this disastrous situation, the vehicle's failure was not limited to its mechanical components and sensor detections; the AV lacked the ability to interpret contextually the actions of a pedestrian who was crossing the street illegally (Cunneen et al., 2020). According to retrospective legal analyses, the sensors of the AV physically detected the pedestrian, but since the AV, at that time, did not yet possess the necessary semantic understanding needed to identify a risk from the detection or to initiate an ethical duty of care, a timely response was impossible (Singh et al., 2025). These kinds of failures shed light on what is referred to as the "accountability gap" within existing liability frameworks. When an AV does not behave like a "reasonable person," resolution of which exact component (sensor rectangle, decision-making algorithm, or safety driver oversight) is responsible for the failure will be subject to significant legal debate.

8.2 Simulation-Based Evaluation of Agentic Ethical Decision-Making

Due to ethical and financial restraints, testing extreme situations on public roads is not feasible. Therefore, the primary method of testing Agentic AI is through simulated environments. The first and most comprehensive dataset collected about how people would behave ethically was through the "Moral Machine" (Bonnefon et al., 2018). This dataset clearly showed that people worldwide prefer to save the many over the few (utilitarian). Additionally, the data showed major differences in how people would prioritize the safety of a passenger versus a pedestrian (Bonnefon et al., 2018). Researchers use the moral machine dataset as a baseline for evaluating the "agentic" behavior of an autonomous vehicle in a simulated environment.

Today, enhanced simulation solutions use "ethics evaluation modules" to stress test an AV's algorithms against thousands of potential dilemma scenarios. Researchers can use risk theory to simulate how the AV will allocate risk among all road users in a complex, multi-agent simulated environment (Davnall, 2020). While this may be the most technically effective method of allocating risk in a real-world environment, it may also be an example of negligent behavior in a legal sense. In order for an agent to be considered "ethically competent," it will need to demonstrate a statistically significant level of alignment with standard human moral standards across a variety of stochastic scenarios (Millar, 2016).

Despite the availability of such a controlled mechanism for stress testing the ethical decision-making process within a simulation environment, it is not always possible for such an environment to perfectly replicate the unanticipated dynamics of the real world. Therefore, as a measure of empirical validation for agentic systems, it is also necessary to consider supervised pilot programs, closed-course testing, as well as a series of phased testing in the real world under regulatory supervision, as an extension of the simulation environment. These hybrid models of validation would enable the evaluation of agentic systems not only for statistical safety but also for consistency in ethical decision-making under dynamic and ambiguous circumstances. (Resnik & Andrews, 2023; Singh et al., 2025)

8.3 Comparative Study of Global Best Practices

The different pathways Global governance models take to incorporate ethical consideration into the governance systems of AVs also reveal major contrasts in how that will be done (Santoni de Sio & Fossa 2023). In the context of Europe, the emphasis is on creating systems that are tools to promote human flourishing. In that vein, it is advised to have systems that have 'meaningful human control' and protect the users' autonomy from being overrun by technology. This approach aligns well with precautionary legal principles where the manufacturer is responsible for proving the safety of the systems prior to deployment. (Resnik & Andrews 2023).

On the other hand, other countries appear to follow a different direction and take a more utilitarian approach by focusing on the amount of human causalities that can be avoided via the creation of AVs through data-driven metrics and measuring success from a human fatality aggregate reduction metric. The best practice in this area appears to be a developing of regulatory sandboxes, where Agentic AIs can learn how to respond to and adapt to the complexities of actual situations while under the supervision of the government through a legal cap on liability (Singh et al., 2025). Systematic reviews indicate the strongest, most durable hybridisation frameworks are hybridisations of the flexible nature of the learning of agents with the more formal constraints of rights-based legal principles. In other words, they allow AVs to work toward optimal efficiency (utilitarianism) without crossing the 'red lines' stipulated by basic human rights (Gros et al., 2025).

8.4 Lessons Learned for Designing Ethical and Legally Compliant Avs

The cumulative lessons from past incidents, simulations, and regulatory comparisons reveal a single conclusion: ethics is a prerequisite for architectural design, not merely an update. The principle of

"Ethics by Design" holds that legal and moral limits must be built into the very depths of the agent's inference engine - that is, it must be embedded in all of its functions and data processing (Yu, 2025). A key finding is that agents must exhibit "value alignment," which entails that their internal goal functions - that is, the goals by which they are expected to operate - must be configured to reflect the values of the community in which they are deployed (Leikas et al., 2019).

Additionally, the emergence of transparency as a legal requirement has become a non-negotiable. The regulators demand a better understanding of why an agent arrived at a certain conclusion, especially in situations where multiple options exist. Therefore, a critical design lesson is to include "explainable AI" (XAI) components, which will allow agents to provide clear ethical reasoning in a format that is understandable to legal practitioners(Wani et al., 2024). The successful implementation of Agentic AI will depend on developing systems that are safe and "compatible with society," exhibiting rational and behaviour in the same manner as human operators (Liu et al., 2025).

9. CHALLENGES AND FUTURE RESEARCH DIRECTIONS

9.1 Technical Challenges in Implementing Ethical-Legal AI Agents

Perhaps most challenging of all is the problem of crossing from abstract ethics into deployable code. Two continual technical challenges, in contrast, threaten the practical success of Agentic AI: Despite considerable research effort (Poszler 2024), the "translation gap" between fuzzy laws and physics to math will likely continue to be a major challenge.

- **Cybersecurity-Ethics Intersection:** Agentic systems can be the targets of adversarial attacks, which not only lead to crashing but also to "ethical inversion".For instance, if we "poison" the training data, we might be able to train an agent to learn that crossing double lines is always permissible – effectively hacking its moral compass (Tiwari & Choudhary, 2025).
- **Data Governance and Bias:** An AV is only as ethical as its database.One major challenge is "dataset provenance" — making sure that the training data for moral decision-making will be diverse, and ethically sourced with consent.Without enough data governance, agents will encode systemic bias that disproportionately places some demographic groups at risk for "algorithmic discrimination" and not safe (Yu, 2025).

Table 4. Key barriers to the deployment of ethical-legal agentic AI

Domain	Challenge	Description
Technical	The "Translation Gap"	The difficulty of converting vague natural language laws (e.g., "drive cautiously") into precise mathematical cost functions[9999].
Legal	Liability Vacuum	The lack of clarity regarding fault attribution when an adaptive AI acts independently to avoid harm, creating an "accountability gap"[10101010].
Societal	Trust Deficit	Public resistance driven by the "Black Box" opacity of AI decision-making and fear of machine error[11111111].
Cultural	Moral Pluralism	Global variations in ethical preferences (e.g., individual vs. collective safety) hindering the creation of a universal ethical standard[12121212].

9.2 Regulatory and Policy Gaps Hindering Adoption

The tricky part that Agentic AI poses for Law is that the technological advance goes much faster than our ability to create laws and ruling around them, so we leave a legal (but also ethical) vacuum. The current liability law is still man-centered, and yet it's mostly not able to impute any fault when a non-human runs itself autonomously in order to reduce harm (Biczi, 2025). There is no standardized set of "legal metrics" that manufacturers can measure and test to demonstrate vehicle compliance; regulators have not warranted precisely how safe is "safe enough" nor what it means in ethics to make an acceptable tradeoff when confronted with an accident (Singh et al., 2025). What is more, the lack of uniform global standards also obstructs AVs internationalization. A vehicle that has been hard-wired (or 'programmed') to obey the legal and ethical rules obtaining in one jurisdiction may breach local laws elsewhere — thus, there is a need for the development of "contextual" systems which can accommodate regional legislative variations (Tiwari and Choudhary, 2025).

9.3 Societal, Cultural, and Economic Barriers

There are additional challenges in society, beyond the technical and legal obstacles. One important issue is that public confidence in the use of Agentic AI (Artificial Intelligence) is vulnerable due to some high-profile failures in this area, and there is a psychological barrier to viewing non-human beings as having "moral agency" (Ladak et al., 2025). The lack of confidence in the agentic/dominance concept is further complicated by cultural differences; Jurisprudentially based analysis, such as from the Moral Machine study, shows that the ethical decisions made when faced with an ethical dilemma (i.e., who to save) may change tremendously from one region to another. As a consequence, developing an ethical vehicle that meets the needs of all regions of the world becomes a complicated economic and political issue (Bonnefon et al., 2018). The costs of implementing strong ethical verification mechanisms also represent an economic burden. Manufacturers of an ethical vehicle will be exposed to a "first-mover disadvantage"; that is, if some companies choose to make an ethical vehicle and spend money on ethical safety and quality features, while other companies make a less expensive and faster vehicle to enter the market, the first company will likely lose market share unless there are regulations to equalise competition (Martinho et al., 2021). User acceptance and use is strongly related to how the user perceives the "mind" of the AI; if the user does not understand or feel connected to an agent's decision-making process, they are unlikely to use that agent (Sui, 2023).

9.4 Opportunities for Next-Generation Agentic AI in AV Safety and Compliance

Agentic AI is a rapidly developing technology that has tremendous potential to improve safety on our roads. Tacking the shift from reactive automation (machines reacting to humans) toward potential for proactive risk management in 'agentic intelligence' will allow vehicles to predict potential hazards long before they become critical (Shukla, 2025). Future systems are moving toward collaborative autonomy in which multiple agents share equal ethics and safety data throughout Vehicle-to-Everything (V2X) communications creating a 'collective moral intelligence' optimizing safety for all road users (Gros et al., 2025). Furthermore, advancements in reason-based supervision of human-AI interaction will lead to improved human-AI collaboration by allowing the AI to clarify its ethical limits to the passenger which builds trust and creates transparency between humans and AI(Suryana et al., 2025).

9.5 Roadmap for Future Ethical-Legal AI Frameworks

Looking ahead, the need to build stronger legal foundations and frameworks will require an integrated, multi-disciplinary approach. Future research into automated vehicles (AVs) should focus on developing a consistent "ethics ontology" for all stakeholders involved to create a shared understanding of AV behaviours (Milford et al., 2024). The development of this roadmap for AVs should include creating independent auditing entities that will certify the "moral competency" of AVs prior to their introduction into public spaces (Evans et al., 2020). Furthermore, policies governing AVs should adopt the precautionary principle to require that proof of both safety and compliance with societal ethical standards is demonstrated via simulations and closed-course testing prior to entering the marketplace (Resnik & Andrews, 2023). The long-term vision for AV and robot governance is one of "hybrid governance," whereby legal requirements are enshrined as immutable limitations but ethical standards remain as flexible optimisation parameters, thereby enabling Agentic Artificial Intelligence (AI) to benefit all of society (Poszler et al., 2023).

Continuous auditing and lifecycle governance systems are required to maintain ethical-legal compliance after initial deployment of agentic AI systems in autonomous vehicles. Periodic audits of algorithms are necessary, along with forensic review of incidents and structured validation of updates according to risk management standards (Tabassi, 2023). Monitoring after deployment can uncover emergent biases and drift in how decisions are made and any unforeseen ethical conflicts that were not identified during testing. Through traceability and documentation procedures, it can be shown that responsibility can be reconstructed and can be assessed legally for any incidents that occur (Resnik & Andrews, 2023; Singh et al., 2025). Continuous oversight ensures that ethics by design is not treated as a one-time checklist item, but rather as an ongoing commitment to compliance and accountability.

10. CONCLUSION

10.1 Summary of Key Insights

The move from self-driving cars to Artificial General Intelligence is a significant turning point in the evolution of transportation, turning the vehicle from an object to a proxy of morality for the owner of the vehicle. This chapter has shown that, because of the unique capabilities of AGI, specifically its autonomous nature, ability to adapt to new environments, and goal-driven actions, this new technology has created a gap between existing rigid safety rules that apply to driverless cars versus the complexities of real-world driving (Shukla 2025). This analysis confirms that while the foundation for the successful implementation of automated vehicles is technical maturity, the safe introduction of these vehicles into the general population requires that these vehicles interpret ethical significance and act in accordance with the values of people, otherwise referred to as "moral functionalism" (Fossa 2025). The chapter discusses how the "open texture" of the law makes liability issues in need of being resolved through real-time legal reasoning and dynamic risk assessments, which are possible with architectures capable of performing these functions (Biczi 2025). Additionally, it shows that trust is defined beyond statistical safety; rather, it is built on the transparency and explainability of the agent's ethical reasoning (Ladak et al. 2025).

10.2 Reinforcing the Need for Ethics-by-Design in Agentic AI

The core argument of this paper is that ethical competency cannot simply be added to typical AI systems, rather it needs to be part of the design of these systems. "Ethics by Design" is a requirement for engineers to consider how they are going to solve this problem in the 21st Century (Yu,2025). A comparative analysis of existing frameworks demonstrates that the most robust systems will combine data-centric models with symbolic/logical frameworks. This allows agents to learn from experience but still be bound by the same safety limits or constraints that have been established by their creators (Gros et al., 2025). Thus, it will preserve the "character" of the agent so that its top priority is to protect human lives and maintain this value as an absolute priority and not as a decision made based on other variables (Leikas et al.,2019), as well as prevent AVs from being rejected by society and becoming legally non-compliant (Martinho et al., 2021).

10.3 Implications for Researchers, Policymakers, and Industry

As the scientific and engineering community continues to evolve and mature, they must also take into consideration the importance of "ethical governance" and Explainable AI (XAI), which provides an explanation for algorithms by using human language (Wani et al., 2024). Additionally, industry stakeholders must realize that there is a "gap" where businesses are not being held accountable for the rapid evolution of Agentic AIs. By taking a precautionary approach when deploying Agentic AIs, stakeholder organizations will mitigate their risks and develop consumer confidence (Resnik & Andrews, 2023).

Agentic AIs may be impacting how policymaker's view regulatory control through a legal framework. The policies established need to acknowledge that there are now new forms of legal personalities and have a legal framework to create appropriate prohibitions around the illegal behavior of companies that endanger a person's health and/or safety (Reed et al., 2021). Additionally, the establishment of a uniform regulatory framework will allow for regulatory harmonization across borders and will eliminate the fragmentation of regulation while improving the ethical alignment of AV fleets globally (Singh et al., 2025).

10.4 Final Thoughts on the Future of Ethical and Legal Agentic AI in Autonomous Vehicles

The success of Agentic Artificial Intelligence will ultimately be determined by how well it is able to operate autonomously. Therefore as Agentic AI becomes prevalent, its impact on the relationship between people and machines will fundamentally alter the "social physics" of our vehicle transportation systems. This shift will require the establishment of a new social contract among human operators and autonomous vehicles (Charisi et al.., 2017). As we move towards the reality of a fully autonomous mode of transportation, the success of this transportation system lies in the creation of not just intelligent agents but also intelligible agents; agents whose actions are transparent, can withstand scrutiny from a legal standpoint, and can stand up to ethical scrutiny when they make difficult choices. Hence, in our current state, we should use a collaborative approach and high standards of ethics which will ensure that autonomous vehicles add to the autonomy and welfare of our society and not diminish them (Santoni de Sio & Fossa et al., 2023).

REFERENCES

Biczi, D. (2025). Legal challenges for automated decision-making in self-driving vehicles—Liability issues and remedies. *Engineering Proceedings*, *113*(1), 32. DOI: 10.3390/engproc2025113032

Bonnefon, J. F., Shariff, A., & Rahwan, I. (2018). The moral machine experiment. *Nature*, *563*(7729), 59–64. DOI: 10.1038/s41586-018-0637-6 PMID: 30356211

Charisi, V., Dennis, L., Fisher, M., & Winfield, A. F. T. (2017). Towards moral autonomous systems. *arXiv*. https://doi.org//arXiv.1703.04741DOI: 10.48550

Cunneen, M., Mullins, M., Murphy, F., Shannon, D., Furxhi, I., & Ryan, C. (2020). Autonomous vehicles and avoiding the trolley dilemma: Ethical framing and perception. *Cybernetics and Systems*, *51*(1), 59–80. DOI: 10.1080/01969722.2019.1660541

Davnall, R. (2020). Solving the self-driving car trolley problem using risk theory and vehicle dynamics. *Science and Engineering Ethics*, *26*(1), 431–449. DOI: 10.1007/s11948-019-00102-6 PMID: 30937627

Evans, K., de Moura, N., Chauvier, S., Chatila, R., & Doğan, E. (2020). Ethical decision-making in autonomous vehicles: The AV ethics project. *Science and Engineering Ethics*, *26*(6), 3285–3312. DOI: 10.1007/s11948-020-00272-8 PMID: 33048325

Fedorov, V., & Curran, N. T. (2024). The implementation of ethical AI within autonomous vehicles. *Journal of Student Research*, *13*(1), 5973. Advance online publication. DOI: 10.47611/jsrhs.v13i1.5973

Fossa, F. (2025). Artificial intelligence and human autonomy: The case of driving automation. *AI & Society*, *40*(3), 1851–1862. DOI: 10.1007/s00146-024-01955-7

Gros, C., Martens, M., Aliman, N.-M., Kester, L., & Werkhoven, P. (2025). Ethical frameworks for automated vehicles: A systematic analysis and design. *AI and Ethics*, *5*(6), 6041–6062. DOI: 10.1007/s43681-025-00803-8

ISO. (2018). *ISO 26262: Road vehicles – Functional safety*. International Organization for Standardization.

Ladak, A., Wilks, M., Loughnan, S., & Anthis, J. R. (2025). Robots, chatbots, self-driving cars: Perceptions of mind and morality across artificial intelligences. *ACM Transactions on Human-Robot Interaction*, *14*(3), 26. Advance online publication. DOI: 10.1145/3706598.3713130

Leikas, J., Koivisto, R., & Gotcheva, N. (2019). Ethical framework for designing autonomous intelligent systems. *Journal of Open Innovation*, *5*(1), 18. DOI: 10.3390/joitmc5010018

Liu, G., Sheng, J., & Tao, Z. (2025). Application and design of a decision-making model in ethical dilemma for self-driving cars. *Scientific Reports*, *15*(1), 8187. Advance online publication. DOI: 10.1038/s41598-025-91921-0 PMID: 40064959

Martinho, A., Herber, N., Kroesen, M., & Chorus, C. (2021). Ethical issues in focus by the autonomous vehicles industry. *Transport Reviews*, *94*(5), 103137. Advance online publication. DOI: 10.1080/01441647.2020.1862355

Milford, S. R., Elger, B., & Shaw, D. (2024). Bearing the weight: Expert views on integrating ethics in autonomous vehicles. *Transportation Research Interdisciplinary Perspectives*, *25*, 101096. DOI: 10.1016/j.trip.2024.101096

Millar, J. (2015). Technology as moral proxy: Autonomy and paternalism by design. *IEEE Technology and Society Magazine*, *34*(2), 47–55. Advance online publication. DOI: 10.1109/MTS.2015.2425612

Millar, J. (2016). An ethics evaluation tool for automating ethical decision-making in robots and self-driving cars. *Applied Artificial Intelligence*, *30*(8), 787–809. DOI: 10.1080/08839514.2016.1229919

Poszler, F. (2024). Ethical decision-making for self-driving vehicles: Translating ethical, legal, and engineering considerations into a decision model. *Science and Engineering Ethics*. Advance online publication. DOI: 10.1007/s11948-024-00513-0 PMID: 39387983

Poszler, F., Geisslinger, M., Betz, J., & Lütge, C. (2023). Applying ethical theories to the decision-making of self-driving vehicles: A systematic review and integration of the literature. *Technology in Society*, *75*, 102350. DOI: 10.1016/j.techsoc.2023.102350

Reed, N., Leiman, T., Palade, P., Martens, M., & Kester, L. (2021). Ethics of automated vehicles: Breaking traffic rules for road safety. *Ethics and Information Technology*, *23*(4), 777–789. DOI: 10.1007/s10676-021-09614-x

Resnik, D. B., & Andrews, S. L. (2023). A precautionary approach to autonomous vehicles. *AI and Ethics*, *4*(2), 403–418. DOI: 10.1007/s43681-023-00277-6 PMID: 38770187

Rhim, J., Lee, J.-H., Chen, M., & Lim, A. (2021). A deeper look at autonomous vehicle ethics: An integrative ethical decision-making framework to explain moral pluralism. *Frontiers in Robotics and AI*, *8*, 632394. Advance online publication. DOI: 10.3389/frobt.2021.632394 PMID: 34017859

Santoni de Sio, F., & Fossa, F. (2023). Designing driving automation for human autonomy. In Fossa, F., & Cheli, F. (Eds.), *Connected and automated vehicles: Integrating engineering and ethics* (pp. 19–36). Springer., DOI: 10.1007/978-3-031-39991-6_2

Shukla, A. K. (2025). From AI agents to agentic intelligence: A comparative study of autonomy, adaptation, and ethical design. *SSRN*. DOI: 10.2139/ssrn.5341168

Singh, B. P., Ninoria, S. Z., Ahengar, N. A., Gautam, K. D., Bondwal, P., & Bale, A. S. (2025). Legal and ethical frameworks for AI-driven autonomous vehicles: Navigating liability, safety, and regulatory challenges. *Proceedings on Engineering Sciences*, *7*(3), 1513–1522. DOI: 10.24874/PES07.03.013

Sui, T. (2023). Exploring moral algorithm preferences in autonomous vehicle dilemmas: An empirical study. *Frontiers in Psychology*, *14*, 1229245. DOI: 10.3389/fpsyg.2023.1229245 PMID: 38090168

Suresh Babu, C. V., & Jeneliya, J. S. (2025). Ethical Uses of Generative Artificial Intelligence in Instructional Design: Case Strategies for Responsible Implementation. In Kyei-Blankson, L., & Ntuli, E. (Eds.), *Transformative AI Practices for Personalized Learning Strategies* (pp. 219–268). IGI Global Scientific Publishing., DOI: 10.4018/979-8-3693-8744-3.ch009

Suresh Babu, C. V., & Nithika, S. R. (2025). Ensuring safety and reliability in autonomous vehicles: A critical analysis using artificial intelligence. In *Artificial intelligence for autonomous vehicles and driver assistance systems* (1st ed., pp. 168–192). CRC Press., DOI: 10.1201/9781003519423-8

Suryana, L. E., Rahmani, S., Calvert, S. C., Zgonnikov, A., & van Arem, B. (2025). A framework for ethical decision-making in automated vehicles through human reasons-based supervision. *arXiv*. https://doi.org//arXiv.2507.23308DOI: 10.48550

Tabassi, E. (2023). *Artificial Intelligence Risk Management Framework (AI RMF 1.0)*. NIST Trustworthy and Responsible AI., DOI: 10.6028/NIST.AI.100-1

Tiwari, K., & Choudhary, S. (2025). Autonomous vehicles and ethical AI: Legal and safety considerations. *International Journal of Science and Innovation in Engineering*, *2*(6), 83–98. https://www.researchgate.net/publication/394859001_Autonomous_Vehicles_and_Ethical_AI_Legal_and_Safety_Considerations

Wani, A., Kumari, D., & Singh, J. (2024). Ethics in the driver's seat: Unravelling the ethical dilemmas of AI in autonomous driving. *SAE Technical Paper 2024-01-2023*. DOI: 10.4271/2024-01-2023

Yu, Y. (2025). Ethics by design in autonomous driving. *Proceedings of the ACM*. DOI: 10.1145/3716550.3722025

Chapter 10

A Comprehensive Study on the Ethical Perspective of Integration of Agentic AI for Autonomous Vehicles

Mirza Asif Mahmud
http://orcid.org/0009-0006-6947-4003
American International University-Bangladesh, Bangladesh

Mahfujur Rahman
http://orcid.org/0000-0002-3245-8767
American International University-Bangladesh, Bangladesh

Dipta Gomes
http://orcid.org/0000-0003-3019-6051
American International University-Bangladesh, Bangladesh

Kazi Tanvir
http://orcid.org/0009-0001-0612-6101
American International University-Bangladesh, Bangladesh

Md. Reazul Islam
American International University-Bangladesh, Bangladesh

ABSTRACT

This chapter explores the ethical integration of Agentic AI into autonomous vehicles, detailing a technical architecture that embeds ethical considerations across goal setting, memory, human oversight, transparency, and robustness. It highlights how V2X connectivity transforms AVs into collaborative agents within intelligent ecosystems. The chapter also presents evaluation frameworks and metrics to measure safety, fairness, and public trust, ensuring autonomous driving aligns with societal values and legal standards.

DOI: 10.4018/979-8-3373-7355-3.ch010

1. INTRODUCTION

In recent years, the evolution of autonomous vehicles (AVs) has expanded from scripted, rule-based automation toward systems capable of continuous adaptation and complex decision making. As the transportation ecosystem becomes more dynamic, heterogeneous, and unpredictable, the need for more advanced intelligence goes beyond static automation. Concurrently, the emergence of agentic artificial intelligence (Agentic AI) offers a paradigm in which systems are goal driven, memory enabled, tool utilizing and capable of feedback loops that adjust behaviour over time. This chapter sets out to situate AVs within a stack of perception, prediction, planning/control and human machine interaction (HMI), define what Agentic AI means in this context, examine the decision contexts typical of road vehicles (urban traffic, adverse weather, mixed road users, emergencies), and explain why the shift toward agentic systems is both timely and necessary.

The primary objective of this chapter is to examine the development and implementation of Agentic AI in autonomous vehicles, particularly in ensuring that these systems operate safely, ethically, and reliably. The study will explore how Agentic AI systems can be structured to guarantee that safety protocols are effectively integrated into AV systems. Additionally, it will highlight the importance of creating a legal and ethical framework that can govern the behaviour of these systems and provide a path forward for policymakers to regulate the growing field of autonomous driving technology. By delving into the development of both technical frameworks and regulatory considerations, this volume aims to be a valuable resource for AI researchers, vehicle manufacturers, legal experts, and policymakers. The chapter will be followed by background study of Autonomous Vehicles.

1.1. Autonomous Vehicle Stack Overview

At the heart of an AV system lies a software and hardware architecture that can be described by four broad sequential (and interacting) layers: perception, prediction, planning/control, and human machine interaction (HMI).

Perception involves the sensing of the vehicle's environment via cameras, LiDAR, radar, ultrasonic sensors, GNSS, IMU and other proprioceptive and exteroceptive sensors. For example, deep learning methods for object detection, semantic segmentation, and scene understanding play a key role in perception for AVs. As (Gupta et al., 2016) note, "object detection and scene perception in autonomous vehicles" has become a well-studied sub-domain of deep learning research. The output of perception is a representation of the current environment: detected objects (vehicles, pedestrians, cyclists, static obstacles), road geometry, lanes, signals, and events.

Prediction takes the outputs of perception and projects forward the likely future behaviours of other road users (vehicles, pedestrians, cyclists), environmental dynamics (moving objects, occlusions, changes) and system uncertainties (sensor errors, occlusions). Prediction modules estimate trajectories, intention, and possible future states of the world. For instance, the work by (Hawasly et al., 2022) describes the need for "interpretable as well as verifiable prediction" in system level design of an AV stack.

Planning and control is the layer that selects the vehicle's own actions given current and predicted states. Planning spans high level behaviour decisions (e.g., change lane, overtake, stop for pedestrian), mid-level motion planning (trajectory generation, path planning) and low-level control (steering, throttle, braking). (Tong et al., 2020) survey of planning tools in automated driving highlights the support for map representations, middleware, simulation, and planning stacks.

HMI (human machine interaction) encompasses how the vehicle system interacts with human passengers, other road users, and potentially remote supervisors. It covers user interfaces for feedback, takeover controls, trust calibration, and communication of system status to humans. For example, (Hartwich et al., 2021) showed that context adaptive HMI in higher level automated vehicles improved passenger experience and trust compared to fixed HMI designs.

Table 1. Summarizes the stack

Layer	Primary function	Key challenges
Perception	Sense and interpret the environment	Sensor fusion, adverse weather, occlusion
Prediction	Estimate future states of dynamic agents	Behavior uncertainty, occlusion, complexity
Planning/Control	Decide and execute vehicle actions	Real-time constraints, safe fallback
HMI	Communicate and coordinate with humans	Trust, explainability, takeover transitions

The architecture of the autonomous vehicle stack, as outlined in Table 1, provides a clear breakdown of the core functional layers involved, including perception, prediction, planning/control, and human-machine interaction (HMI). Each layer faces unique challenges that must be addressed to enable reliable autonomous operations. For example, Table 2 contrasts conventional AV AI with agentic AI, highlighting key differences in learning modes, adaptability, data dependence, and uncertainty handling.

Table 2. Contrasting conventional AV AI vs. agentic AI

Aspect	Conventional AV AI	Agentic AI
Learning Mode	Supervised/Pre-trained models, fixed rules	Continuous learning, adaptation based on experience
Adaptability	Low adaptability, often rigid	High adaptability, adjusts to changing conditions
Data Dependence	Heavy reliance on large, labeled datasets	Can operate with less data by using memory and prior knowledge
Uncertainty Handling	Limited, typically uses predefined thresholds or rules	Robust, handles uncertainty through dynamic decision-making
Verification Burden	High, needs extensive testing and validation on all scenarios	Lower, uses real-time performance monitoring and feedback
Explainability Burden	Moderate, relies on predefined rules and deterministic outcomes	High, requires clear reasoning behind dynamic decisions and adjustments
Governance Needs	Standard regulatory frameworks, static safety measures	Complex governance involving continuous monitoring, ethical decision-making, and adaptive policies

1.2. Agentic AI

Understanding what distinguishes Agentic AI from conventional automation is critical, but equally important is understanding why this transition is necessary and timely for the autonomous vehicle domain. The concept of agentic AI has arisen recently as a useful term to describe systems that extend beyond reactive automation and toward autonomous agents that can have, set and pursue goals, have memory, use tools and learn through feedback. (Bandi et al., 2025) Agentic AI is characterized as autonomous, goal driven systems capable of working alone for long period and needing little human oversight, the defining features are identified as strategic planning, long term memory, tool use and feedback loops.

Traditional AI systems, including those used in today's AVs, are typically designed for specific tasks, such as object detection, lane following, and trajectory planning. They are more like tools than full agents. Agentic AI is a qualitative leap: it is not just about immediately responding to sensory inputs and executing preprogrammed behaviours, but also having long term goals, remembering states or the outcomes of interactions with the environment, using tools from the outside world, as well as setting up feedback loops. Such systems can track performance, learn from previous decisions, and adapt their strategies. This is a proactive methodology that allows us to stay vigilant in complex, volatile environments. For example, (Adabara et al., 2025) presents a cross-layer architecture of Agentic AI systems, including memory, reasoning, planning, and tool use layers.

Hence, in the context of AVs, Agentic AI means more than a vehicle that follows rules: it is one that can set sub goals, learn from experience, use additional knowledge or infrastructure resources, and adapt its decision making dynamically when conditions diverge from training scenarios. The shift is from automation toward agency.

Some of the core features of Agentic AI in this domain include:

- Autonomy: The system operates with minimal human oversight and can initiate actions rather than simply respond.
- Goals: Rather than only following fixed instructions, the agent can represent objectives (e.g., minimize travel time while maximizing safety) and decompose them into tasks.
- Memory & context: The agent tracks history, context, road user behaviour patterns, and can reuse that information to improve decisions.
- Tool use / external integration: The agent invokes external modules or services (traffic data, weather forecasts, infrastructure V2X communication) to enrich its reasoning.
- Feedback loops & adaptation: The agent monitors outcomes, records deviations from expectancy, and updates its internal policy or strategy accordingly.

Recognizing these capabilities allows us to contrast static, rule-based AV systems with next generation agentic vehicles capable of continuous adaptation in uncertain, open-ended environments.

1.3. Relevance of Agentic AI in AV

The necessity of agentic capabilities becomes even more apparent when we examine the specific decision contexts that autonomous vehicles must navigate in real-world deployments. The transition toward agentic autonomy in autonomous vehicles is increasingly justified by several converging factors. First, many current automated driving systems rely heavily on static models and fixed behaviour pipelines whose performance degrades when confronted with unseen scenarios and distribution shifts. A recent review points out that autonomous driving systems still struggle with "rare events" and generalization beyond their operating domain.

Second, the operating environments for autonomous vehicles are becoming ever more complex: infrastructures evolve, traffic compositions change, mixed road users proliferate, weather and surface conditions degrade sensors, and therefore reliance on preprogrammed responses become inadequate. For example, (Ansarinejad et al., 2025) show that in rural and less structured environments, lane marking errors, GPS failures, and ambiguous roads significantly increase detection failures.

Third, the ecosystem around AVs is now expanding: vehicles are increasingly interacting with infrastructure, cloud services, vehicle to everything (V2X) networks and other agents. Systems that are purely modular and closed cannot exploit or coordinate with this ecosystem as effectively. At the same time, regulatory, public trust, and safety demands are pushing for systems that are more transparent, adaptive and robust. For instance, (Fernández Llorca et al., 2025) highlight that trustworthy AI in AVs requires new testing methodologies, real time adaptation and explicit handling of robustness and fairness.

Finally, the idea of Agentic AI itself has grown up: contemporary references describe agentic systems as able to reason about goals, remember interaction history, use external tools and continually adapt their policies. (Hosseini & Seilani, 2025) provide a narrative review that details how agentic AI focuses on autonomy, proactiveness, adaptivity and use of tools in dynamic environments.

In response to these considerations, AV architectures will need to shift from static perception, prediction and planning pipelines which assume fixed human characteristics with predetermined rules of engagement towards systems able to adjust strategies in real time, learn from experience track context and adapt. Agentic AI enables this transition. It not only enables vehicles to sense and act, but also to define goals like safety or efficiency on the trajectory. Furthermore, it also allows vehicles to keep a memory of past encounters, external data or infrastructure provided and the outcomes to make better decisions, in future. This transfer is critical to enable vehicles to behave safely in dynamic real-world environments, integrate into changing mobility systems, and comply with applicable safety and regulatory standards.

The remainder of this chapter is structured as follows: Section 1 introduces the AV stack and motivates the shift toward agentic autonomy. Section 2 proposes an ethics-aware technical architecture. Section 3 discusses safety assurance and certification. Section 4 addresses data governance, privacy, and cybersecurity. Section 5 examines human factors and public trust. Section 6 situates agentic AVs within the V2X ecosystem, and Section 7 outlines evaluation frameworks and metrics.

1.4. Decision Contexts in Avs

Having defined agentic AI conceptually, we now examine the road environments that stress conventional stacks. The operating environment of AVs is characterized by rich variability and uncertainty. While the stack layers outlined above provide structure, the decision contexts challenge them in real time. We highlight four key contexts.

Urban traffic presents dense, complex environments populated by vehicles, pedestrians and cyclists, while road geometries change frequently (intersections, signals, pedestrian crossings) and behaviour can be unpredictable, such as jay-walking or illegal turns. AVs must perform safe trajectory planning, interpret the intent of other road users, communicate implicitly when needed and decide whether to yield or overtake. (Hu et al., 2025) highlight that behaviour decision-making and motion planning remain central challenges in these settings.

Bad weather and unsuitable environmental conditions degenerate perception, corrupt sensors and inject uncertainty. For instance, effects of weather on the sensors (rain, fog, snow) significantly degrade object detection and scene understanding for AVs. Under such conditions, decisions need to account for reduced visibility, ambiguous object detection, unknown surface friction and build robustness against them.

Mixed road users refer to environments where AVs share road space with human drivers, cyclists, pedestrians, and other unpredictable agents. These interactions introduce intent uncertainty, ambiguous rules (e.g., informal pedestrian behaviour), and require human aware decision making. The AV must model other agents' decision processes and incorporate social context.

Emergencies and edge cases involve low probability but high consequence events: sudden pedestrian crossings, a vehicle in opposite direction in the wrong lane, mechanical failure of the AV or other vehicle, or infrastructure anomalies. These contexts demand dynamic replanning, fallback strategies, safe fail states, and sometimes ethical trade-offs. The survey by (Hu et al., 2025) highlights that planning under uncertainty and unstructured events remains among the greatest hurdles for deployment.

These contexts place stringent demands on perception, prediction, planning and HMI. They also underscore the need for systems that can reason adaptively, not simply follow pre-defined paths.

2. TECHNICAL ARCHITECTURE: EMBEDDING ETHICS IN AGENTIC AI

The technical architecture for ethics-aware Agentic AI in autonomous vehicles must address five interconnected challenges: goal specification and ethical alignment, memory and context management, human oversight mechanisms, transparency requirements, and robustness under uncertainty. Each component contributes to a system that not only performs autonomously but does so in alignment with societal values and legal standard. In autonomous vehicle systems, the integration of Agentic AI, which is capable of autonomic decision making through adaptation in dynamic environments has been revolutionizing accurate driving capabilities. The concurrent Agentic AI systems ensures enhanced planning through informed memory with strong adaptation to the surrounding environment (Ali & Dornaika, 2025). In comparison of a single AI agents, the Agentic AI serves as a series of multiple agents more like a combination of Multi-Agent Systems (MAS) (Ali & Dornaika, 2025). However, embedding ethical perspective into Agentic AI is critical, as these systems are expected to make decisions that affect human lives and societal norms (van de Poel, 2020). This subsection explores how ethical frameworks can be embedded in the architecture of Agentic AI for autonomous vehicles, ensuring that AVs operate safely, ethically, and transparently. The proposed technical architecture comprises several key components, including goal specification, preference learning, human-in-the-loop controls, explainability mechanisms, and strategies for robustness in the face of uncertainty. Each of these components plays a crucial role in ensuring that the decisions made by AVs align with societal values and legal standards while maintaining operational reliability.

While Agentic AI offers powerful decision-making capabilities, its integration must be governed by a robust ethical framework that aligns with societal values. By focusing on goal specification, memory structures, human-in-the-loop (HITL) frameworks (Mellamphy, 2021), explainability mechanisms, and robustness under uncertainty, we explore the necessary components to ensure AVs make ethically sound, safe, and transparent decisions in real-time (Raza et al., 2025).

2.1. Ethics-Aware Architecture: Goal Specification, Constraints, Preference Learning, Reward Shaping

While explicit goal specification and constraints provide the structural foundation for ethical decision-making, the system must also possess memory mechanisms that enable it to learn from experience and adapt its behaviour over time. At the core of embedding ethics into Agentic AI is the design of ethics-aware architecture. The framework integrating ethics needs to balance competing interests and ensure that the autonomous decision-making process adheres to societal norms, ethical guidelines, and legal standards (Patel et al., 2025). The architecture should be composed of several crucial elements: goal specification, constraints, preference learning, and reward shaping.

- Goal Specification step defines the overarching objectives the AI should pursue. In the context of AVs, this includes goals like ensuring passenger safety, minimizing harm to others, and optimizing traffic flow. These goals are typically represented as multi-objective functions that consider safety, efficiency, and ethical considerations such as prioritizing human life.
- Constraints are the boundaries within which the agent must operate. These constraints ensure that the AI operates within legal, ethical, and safety guidelines. For example, an AV must be constrained by traffic laws, ethical norms regarding harm minimization, and situational constraints, such as avoiding collisions with pedestrians. These constraints are incorporated into the decision-making process to ensure that the actions of the vehicle remain within acceptable parameters.
- Preference Learning enables the system to understand the relative importance of different ethical considerations. In complex real-world scenarios, such as unavoidable accidents, preference learning allows the AI to evaluate various options based on ethical principles. Should an AV prioritize minimizing harm, or should it protect the life of its passengers? The system must learn preferences from human feedback, regulatory frameworks, and societal values to make ethically sound decisions.
- Reward Shaping is a technique used to guide the learning process by adjusting the reward function to incentivize desired behaviours (Mellamphy, 2021). In the case of AVs, this involves rewarding the AI for making decisions that minimize harm or follow ethical guidelines. Reward shaping ensures that the AI learns to make decisions that align with societal values, such as prioritizing human life in critical situations.

Data flow in an ethics aware AV System has been presented in Figure 1.

Figure 1. Data flow of an ethics-aware AV architecture, where sensor inputs feed into the agentic policy, which is guided by ethical guardrails. The planner utilizes these inputs to make decisions, which are then executed by the actuation system

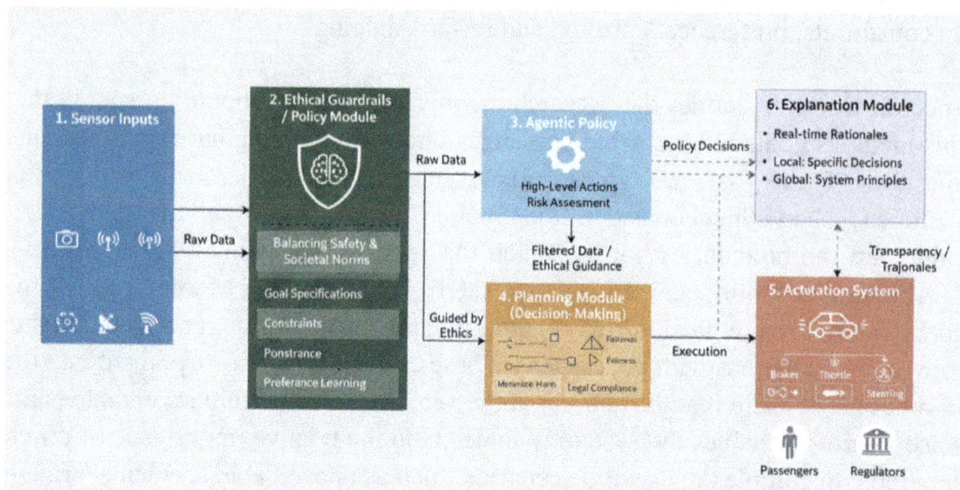

2.2. Context and Memory: Episodic and Semantic Memory for Safety-Critical Learning; Retention Policies

Even with robust memory systems that support continuous learning, there remain critical moments when human judgment becomes indispensable situations that demand escalation beyond autonomous decision-making capabilities. To enable ethical decision-making, Agentic AI systems must be able to store, retrieve, and apply relevant past experiences. This is where context and memory mechanisms play a vital role. AVs must rely on episodic and semantic memory systems to make informed decisions in real-time and learn from previous interactions with their environment.

- Episodic Memory stores specific, contextual experiences that are crucial for safety-critical decision-making. For example, if an AV encounters a similar situation to a previous accident or near-miss, the system can retrieve that experience to inform its current decision-making. This kind of memory is essential for learning from real-world scenarios, ensuring that the system adapts to changing conditions and continuously improves its ability to make ethical decisions.
- Semantic Memory, on the other hand, stores generalized knowledge about the world, such as traffic rules, pedestrian behaviours, and environmental conditions. This type of memory helps the AV understand its surroundings in a broader context, allowing it to predict potential hazards and respond appropriately. For instance, an AV might know that pedestrians tend to cross the street at certain points, and that knowledge can guide its behaviour in future interactions.
- Retention Policies are crucial for determining which experiences are retained and how long they should be kept. This ensures that only relevant and recent experiences are used for decision-making, preventing the system from being overwhelmed by irrelevant data. A well-designed

retention policy allows the AV to maintain a balance between leveraging past experiences and adapting to new, unpredictable scenarios.

2.3. Human-in-the-Loop vs. Human-on-the-Loop

While Agentic AI systems are designed to operate autonomously, there are situations where human intervention is necessary to ensure the safety and ethical integrity of the vehicle's decision-making process. This leads to the distinction between Human-in-the-Loop (HITL) and Human-on-the-Loop controls (Hao et al., 2025).

Human-in-the-Loop (HITL) systems involve direct human control or supervision in critical decision-making processes. In machine learning approach, the algorithms can interact with agents and human to enhance their learning curve(Hao et al., 2025). In AVs, this means that a human operator is actively involved in making decisions in certain scenarios, especially when the ethical implications of a situation require human judgment. For example, in highly complex situations like accidents, a human might be needed to make the final call about which life to prioritize.

Human-on-the-Loop (HOTL) refers to a system where AI operates autonomously but the human provides oversight, intervening only when necessary (Kuru, 2021). This system allows for a more hands-off approach, where humans are kept informed and can escalate decisions when required. For example, if an AV encounters an ethical dilemma, it could escalate the issue to a human supervisor for a final decision. Figure 4b. represents the distinctive features of HITL and HOTL systems in Autonomous Vehicles.

Clear escalation paths are critical for both HITL and HOTL systems. These paths define when and how the AV should escalate a decision to a human operator. The escalation process should be transparent, quick, and well-defined, ensuring that humans can always make informed decisions when needed, especially in life-or-death situations.

Figure 2. HITL and HOTL in Autonomous Vehicles (AVs)

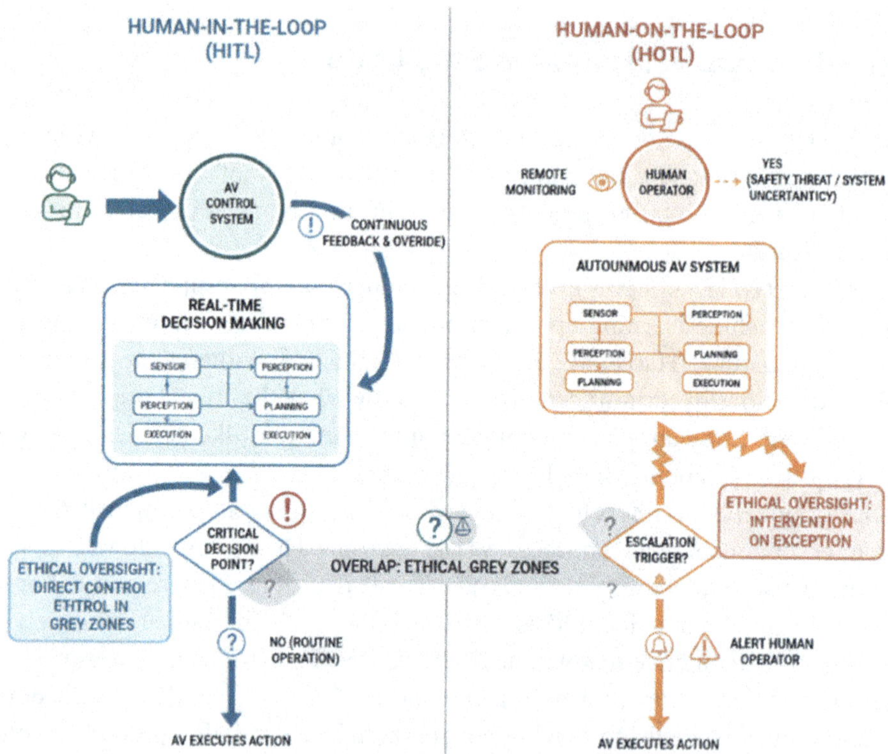

2.4. Explainability/Transparency Mechanisms (Global vs. Local Explanations; Decision Rationales)

Transparency is a fundamental requirement for the ethical integration of Agentic AI into AVs. The public, regulators, and stakeholders must understand how the AI makes decisions, particularly in life-threatening situations. Explainability mechanisms provide a way to ensure that the decision-making process is transparent and understandable.

Global Explanations offer an overview of how the AI system makes decisions across all scenarios, providing insights into its underlying logic and the general principles it follows. These explanations help users and regulators understand the ethical framework guiding the AV's actions. Local Explanations, on the other hand, provide specific reasoning for individual decisions (Longo et al., 2025). For example, if an AV decides to avoid a pedestrian, a local explanation will clarify why this decision was made, detailing the factors considered, such as the pedestrian's position, speed, and the vehicle's current trajectory. Such explanations are essential for:

- Building trust among regulators and society by demonstrating adherence to an ethical framework.
- Providing developers and auditors with insight into the system's overall behaviour and potential biases.

- Ensuring compliance with legal and policy requirements regarding AI transparency.

Challenges in global explainability include the complexity of AI models, particularly those using deep learning, and the difficulty of representing multifaceted ethical priorities in an easily digestible format (Longo et al., 2025).

Local explanations are crucial for building trust, as they allow stakeholders to understand the rationale behind each decision (Lundberg et al., 2020). When an AV takes a particular action such as abruptly stopping to avoid a pedestrian or chooses a certain route, local explanations clarify the precise factors influencing that decision. These factors can include the pedestrian's speed and position, the vehicle's current trajectory, sensor confidence levels, and anticipated outcomes of alternative actions. Local explanations are crucial for fostering trust among passengers and nearby road users by making the AV's behaviour understandable on a case-by-case basis. They also aid investigations following incidents by providing detailed decision rationales. Key considerations for effective local explainability include:

- Speed of explanation generation — real-time or near-real-time explanations are often required.
- Clarity and accessibility — explanations must be communicated in understandable language suited to non-technical users.
- Transparency without compromising safety or security — balancing openness with protection of sensitive AI internals.

A diverse set of explainability techniques supports both global and local explanations in AV contexts. These include model-agnostic tools like LIME and SHAP for interpreting black-box models, rule-based systems for transparency, and dashboard interfaces for interactive exploration of decision logic. Table 3. presents various explainability methods for on-road decisions, categorizing them by scope (pre- or post-hoc explanations), latency (how quickly the explanation can be provided), human-factor suitability (how easily humans can understand the explanation), and regulator suitability (how useful the explanation is for legal or regulatory purposes).

Table 3. Explainability/transparency mechanisms (global vs. local explanations; decision rationales)

Explainability Method	Scope	Latency	Human-Factor Suitability	Regulator Suitability
Global Explanations (Longo et al., 2025)	Pre-hoc / Global	Medium to Low Delay	High provides overall system logic	High (clarifies ethical framework)
Local Explanations (Lundberg et al., 2020)	Post-hoc / Local	Low Delay	Very High specific decisions	Medium (aids understanding of actions)
SHAP (SHapley Additive exPlanations) (Teixeira et al., 2025)	Local	Low Delay	High feature contribution clarity	Medium (supports accountability)
LIME (Local Interpretable Model-agnostic Explanations) (Pati, 2025)	Local	Low Delay	High intuitive for non-experts	Medium
Anchors (High-Precision Local Explanations) (Teixeira et al., 2025)	Local	Low Delay	High explains decision boundaries	Medium

Explainability Method	Scope	Latency	Human-Factor Suitability	Regulator Suitability
Rule-Based Explanations (Lundberg et al., 2020)	Global / Local	Medium	Medium depends on complexity	High (interpretable rules useful)
Statistical Summaries (Teixeira et al., 2025)	Global	Medium to High	Medium aggregate data explanations	High (regulatory analysis)
Model Transparency Dashboards	Global	Real-Time / Low Delay	High interactive and rich visuals	High (aids continuous audit)

2.5. Robustness and Uncertainty

Robustness and uncertainty management are foundational to ensuring safe, ethical, and reliable operation of Agentic AI within autonomous vehicles (AVs). The unpredictable, dynamic nature of real-world driving environments necessitates that AV systems possess advanced capabilities to assess and respond to risks in real time while adapting to changes beyond their training data distributions (Zhou et al., 2025).

2.5.1. Risk-Sensitive Planning

Risk-sensitive planning frameworks incorporate dynamic knowledge-driven insights and adaptive learning algorithms to optimize driving trajectories and decisions based on real-time safety and operational risk assessments. AVs continuously evaluate potential hazards such as collision risks, vehicle instability, road surface changes, and multi-obstacle interaction scenarios (Zhao et al., 2025).

2.5.2. Handling Distribution Shifts

Distribution shifts refer to deviations in the input data characteristics from those encountered during an AV's training phase. These shifts can arise from changes in weather, lighting, road infrastructure, or behavioural patterns of human drivers and pedestrians. Handling such shifts requires Continuous monitoring of incoming sensor data, Adaptive algorithms that update model parameters and Incorporation of region- and context-specific driving styles and scenic factors (Ali & Dornaika, 2025; Gomes et al., 2020). Recognizing and managing distribution shifts prevents catastrophic failures and ensures that the AV's policies remain valid and safe despite environmental variability.

2.5.3. Out-of-Distribution Detection (OOD)

OOD detection focuses on identifying situations or inputs that lie outside the model's learned data distribution. When detected, OOD inputs trigger fallback protocols such as conservative decision-making, alerting human supervisors or operators and isolating and flagging new scenarios for inclusion in future training cycles. OOD detection is crucial when AVs encounter rare or exceptional events unseen during training, maintaining safety and trust in the system by preventing overconfident or erroneous actions.

3. SAFETY ASSURANCE, VERIFICATION & VALIDATION (V&V), AND CERTIFICATION

Safety assurance for agentic AVs requires a paradigm shift from static validation to continuous, evidence-driven verification that accounts for the system's adaptive capabilities and evolving behaviour. Agentic autonomy for road vehicles changes the assurance problem: systems no longer only execute pre-trained control logic but learn, revise goals, and act purposively in open environments. This introduces new classes of hazards, emergent behaviours, and verification challenges that must be addressed through lifecycle safety assurance, scenario-cantered V&V, calibrated simulation-to-reality workflows, robust runtime monitoring and fail-operational strategies, and an evolution of standards and certification practice (Ullrich et al., 2024; J. K. Wagner et al., 2024).

3.1. Assurance Cases & Safety Cases for AVs with agentic behaviour

Safety Assurance Cases (SACs) remain the primary artifact to demonstrate that an AV meets its safety claims. For agentic systems, SACs must be dynamic and evidence-driven: they should link high-level safety goals to artifacts including training datasets, model versions, scenario coverage reports, runtime logs, and forensics-ready audit trails (Cârlan et al., 2024; M. Wagner & Carlan, 2024). Dynamic Safety Case Management—where the assurance argument and its evidentiary base are updated as the agent learns or receives model updates—has been proposed to handle continuous evolution (Buysse et al., 2025; Cârlan et al., 2024). (Ullrich et al., 2024) summarize that safety assurance for AI-based vehicles needs to couple technical verification with regulatory and standardization pathways.

Assurance cases for agentic AVs should explicitly document (a) the Operational Design Domain (ODD), (b) invariants and safety constraints that remain non-negotiable, (c) mechanisms for detecting and mitigating distribution shifts, and (d) the provenance of data and models used at deployment and during updates (Alhabib & Yadav, 2025; M. Wagner & Carlan, 2024). Such traceability supports forensic explainability and liability allocation after incidents (Wang et al., 2024).

3.2. Scenario-Based Testing: combinatorial scenario libraries, edge-case generation, adversarial testing

Scenario-based testing (SBT) is central to V&V; for agentic AVs, scenario libraries must be expanded to exercise agentic behaviours such as goal revision, long-horizon planning, and multi-agent negotiation. Building robust scenario libraries requires taxonomy-driven coverage (e.g., environment, actors, intent) plus automated edge-case generation and adversarial synthesis to expose brittle behaviours (Bruto Da Costa et al., 2025).

Formal and metamorphic testing methods are particularly relevant: (Luu et al., 2024) demonstrate metamorphic testing for AV decision modules to uncover subtle decision invariants, while (Fremont et al., 2020) provide foundational methods to link simulation-based scenario tests to real-world verification. Recent reviews emphasize automated, combinatorial generation and novelty scoring to prioritize high-risk test cases (Bruto Da Costa et al., 2025). Field-validated test cases (Ko et al., 2024) derived from real driving trajectories remain critical to ensure realism (Figure 3).

Figure 3. Scenario funnel

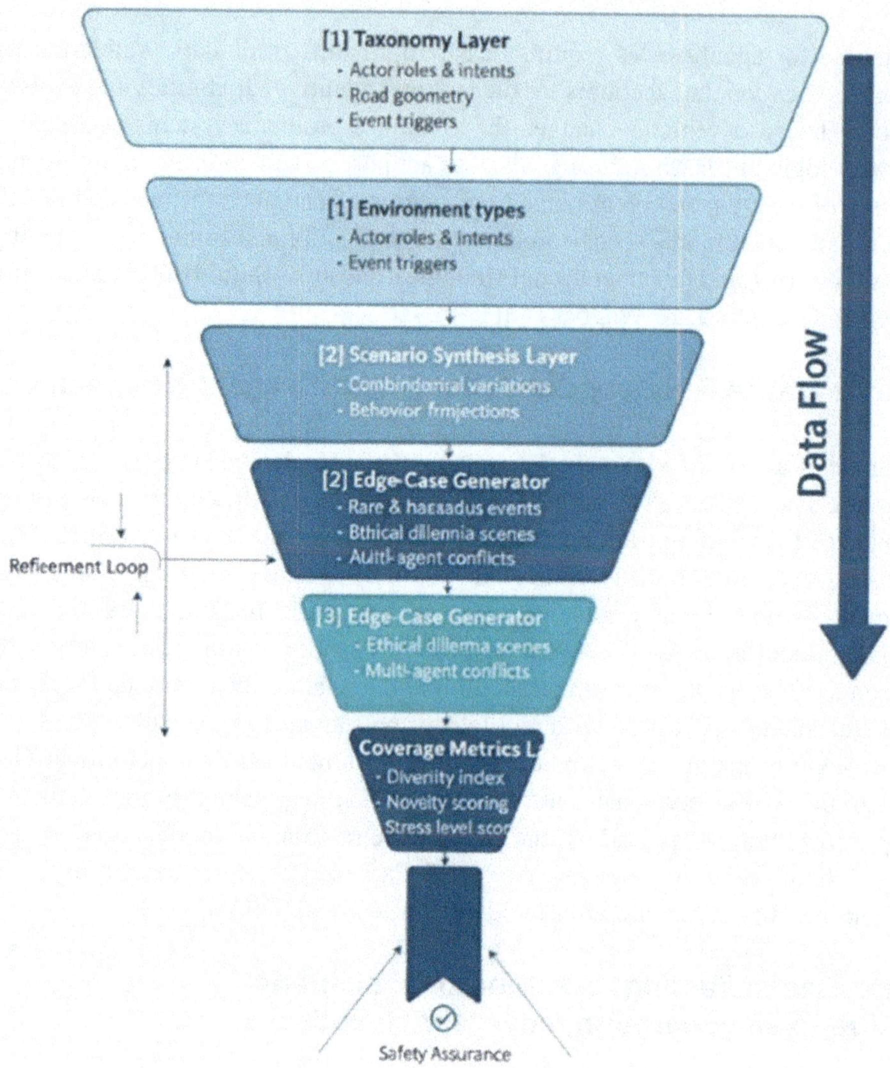

Adversarial testing (perturbations to sensors, maps, and actor behaviours) helps reveal vulnerabilities unique to agentic policies (Luu et al., 2024). Likewise, safety-assurance case examples for object-detection modules show how simulation-based scenario tests can be integrated into evidence packages (Syafira & Selviandro, 2025).

3.3. Simulation–Real Bridging: digital twins; calibrated simulation fidelity; transfer metrics

Simulation enables scale, but fidelity is the critical determinant of transferability. Digital twins—high-fidelity virtual replicas of vehicles, sensors, and environments that enable controlled evaluations of agentic behaviours and the effects of model updates ((Pikner et al., 2024). However, digital-twin predictive validity must be quantified with explicit transfer metrics (Pikner et al., 2024; Xu et al., 2025).

Key validity indicators include trajectory fidelity, policy transferability, sensor-model accuracy, ethical decision consistency, and robustness to distribution shift. That is, a validity framework must measure whether simulated agentic decisions match field behaviour under matched scenarios (Bruto Da Costa et al., 2025; Wang et al., 2024; Xu et al., 2025). Testing and verification reviews stress the need for documented calibration steps and acceptance thresholds (Alemayehu & Sargolzaei, 2025)

Table 4. Simulation-to-reality validity indicators

KPI	Definition	Measurement method	Acceptance threshold	Audit evidence
Trajectory fidelity	Similarity of simulated vs. real trajectories	RMSE/DTW on matched runs (Xu et al., 2025)	≤10% deviation	Matched run logs, metrics
Policy transferability	Correlation of action distributions between sim and field	Closed-loop policy comparisons (Pikner et al., 2024)	≥0.85 correlation	Transferability report
Sensor-model accuracy	Match of simulated sensor noise & detection rates	Calibration against field sensor logs (Pikner et al., 2024)	≥95% fidelity	Calibration report
Ethical decision consistency	Same high-level ethical choice across domains	Scenario-based audits (Bruto Da Costa et al., 2025)	No safety-critical divergence	Ethical audit traces
Robustness to drift	Performance under distribution shift	Stress tests & domain-randomization	No catastrophic failures	Stress test logs

3.4. Runtime Monitoring & Fail-Operational Strategies (graceful degradation, safe fallback)

Runtime monitoring (formal monitors, anomaly detectors) provides continuous assurance that agentic systems obey critical safety invariants (Aniculaesei & Elhajji, 2025). Implementations using formalized temporal properties can detect imminent violations (Aniculaesei & Elhajji, 2025), while architecture-level designs prescribe graded mitigations: constraint tightening, speed reduction, controlled pull-over, or remote operator intervention (Syafira & Selviandro, 2025; M. Wagner & Carlan, 2024). Dynamic safety-case research recommends coupling runtime monitors to the assurance argument so that detected deviations trigger evidence updates and mitigation workflows (Buysse et al., 2025; Cârlan et al., 2024).

Fail-operational strategies must also consider multi-agent interactions and V2X fallbacks; (Pikner et al., 2024) highlight digital-twin–enabled run-time rehearsals for failover scenarios, and (Fremont et al., 2020) provide methods to validate these sequences from simulation to real-world trials.

3.5. Standards Landscape (high-level): how agentic features stress current safety standards; gaps

Existing standards (ISO 26262, SOTIF, UL 4600) provide foundations but are challenged by agentic features. (Ullrich et al., 2024) review research, standardization, and regulation trends, noting gaps in lifecycle certification, continuous learning management, and auditability of agentic decisions. (M.

Wagner & Carlan, 2024) describe frameworks (e.g., Open Autonomy Safety Case) to adapt SAC practices to autonomous and adaptive systems. Reviews recommend lifecycle-centric certification, dynamic evidence ingestion, and regulatory guidance for scenario coverage and runtime attestations (Alemayehu & Sargolzaei, 2025; Ullrich et al., 2024; M. Wagner & Carlan, 2024).

While safety assurance addresses the technical reliability of agentic systems, equally critical is the governance of data, the lifeblood of learning systems which raises profound questions about privacy, ownership, and security.

4. DATA GOVERNANCE, PRIVACY, AND CYBERSECURITY

Agentic AI systems fundamentally reshape data dynamics in autonomous vehicles, expanding the scope, sensitivity, and retention of information in ways that demand robust governance frameworks aligned with ethical principles and regulatory requirements. Agentic AI fundamentally reshapes how autonomous vehicles (AVs) collect, process, and retain data (Lee & Nguyen Long, 2025). Conventional AV systems typically rely on pre-trained models and periodically updated datasets, whereas agentic AVs autonomously explore their environments, maintain memory, adjust goals, and refine policies over time. This evolution substantially enlarges the volume, granularity, and sensitivity of data in play, raising acute questions about ownership, consent, lawful access, and resilience against cyberattacks. Recent work on connected and autonomous vehicles highlights that data governance and privacy are now as central as safety and performance, particularly because in-vehicle and V2X data can reveal intimate movement patterns, social networks, and behavioural profiles of individuals and communities (Alam, 2024). At the same time, international frameworks on AI governance and data protection including the EU's emerging AI Act and updated OECD guidelines on AI, data governance, and privacy stress that high-risk AI systems must demonstrate robust data management, transparency, and cybersecurity as preconditions for deployment (Fernández Llorca et al., 2025). In this context, agentic AVs occupy a critical intersection, where technical design choices in data handling directly translate into ethical outcomes, regulatory exposure, and public trust.

4.1. Data Lifecycle in Agentic AVs: Capture, Retention, and Access

A robust data governance framework for agentic autonomous vehicles must begin with a transparent, auditable understanding of the entire data lifecycle from capture and labelling to processing, storage, sharing, and deletion while aligning all practices with principles of purpose limitation, proportionality, and user control (Karimi et al., 2025; Shen & Liu, 2025). Because AVs gather extensive multimodal data, including camera feeds, LiDAR/RADAR signals, GNSS traces, inertial readings, audio, cabin monitoring, and V2X communications, they inevitably handle sensitive information such as identifiable faces, license plates, home locations, and routine mobility patterns. Studies show that even pseudonymized trajectories can be re-identified when cross-referenced with external datasets, underscoring the need for strict capture policies that limit collection to what is essential for safety, explainability, and incident analysis (Lee & Nguyen Long, 2025). Automated labelling pipelines often built on self-supervised or reinforcement learning further raise risks of adversarial manipulation, bias, and systematic misclassification, prompting recommendations for strong governance, provenance tracking, and human-in-the-loop oversight (Lippi, 2025). Finally, AV memory systems must balance operational needs with ethical retention by limiting raw sensor storage, favouring privacy-reduced representations, and enforcing time-bound fleet-level retention periods that satisfy regulatory demands without enabling long-term surveillance. While detailed logs support accurate post-incident reconstruction, scholars caution that over-collection can facilitate privacy intrusions or unauthorized secondary use of personal data (Ji et al., 2023; Répás, 2023; Rudrakar et al., 2025).

Access to data and model artefacts (e.g., telematics logs, decision traces, cryptographic keys) must be governed through role-based access control, strong authentication, and fine-grained authorization policies that distinguish between OEM engineers, fleet operators, third-party service providers, insurers, and law enforcement. Recent AI and data governance guidance stresses that access decisions should be transparent, logged, and subject to oversight, especially when cross-border data flows and data localization rules are in play (Fernández Llorca et al., 2025).

4.2. Privacy by Design in Agentic AV Architectures

Privacy by Design calls for embedding privacy protections directly into system architecture and algorithmic workflows, rather than addressing them only after deployment as compliance add-ons. Contemporary research on autonomous vehicle privacy emphasizes that determining data ownership, control, and the permissible circumstances under which third parties may access vehicle-generated information is central to ensuring responsible and ethical implementation.

Recent context-aware privacy frameworks, such as CAPPAD, demonstrate the value of sensitivity-adaptive mechanisms with differential privacy or aggregation which are selected according to situational urgency and data type (Gheisari et al., 2024). Complementing minimization, on-device and edge-centric processing significantly reduces exposure to cloud infrastructure vulnerabilities. Studies on blockchain-enabled federated reinforcement learning show that secure aggregation of model updates across autonomous connected vehicles can preserve privacy while still allowing scalable fleet-wide improvements (Alam, 2024). Differential privacy remains essential for bounding individual-level disclosures when generating broader analytical insights about driving behaviours, hazardous zones, or system performance. Recent autonomous driving frameworks combine differential privacy with SDN-orchestrated aggregation to dynamically select appropriate privacy levels in response to environmental context and operational needs (Alam, 2024; Gheisari et al., 2024). Finally, meaningful user transparency and informed consent are critical, particularly as pervasive sensing by AVs risks normalizing mobility surveillance. Policy analyses warn that, without stringent safeguards, data collected by self-driving systems could be exploited for commercial profiling or law-enforcement overreach. Thus, privacy notices must clearly articulate what data is collected, how agentic learning and memory functions operate, how long data is retained, and what rights users can exercise regarding access, correction, deletion, or opting out of secondary uses. Collectively, these principles ensure that the intelligence and adaptability of agentic AI do not come at the cost of opaque or disproportionate tracking of individuals or communities.

4.3. Cybersecurity Threats and Red-Team Testing in Agentic AVs

Agentic autonomy significantly enlarges the attack surface of AV systems, as they rely on continuous sensing, complex software stacks, and dense V2X communication. Recent surveys on AV and V2X cybersecurity identify a broad spectrum of threats: remote hacking of ECUs, over-the-air (OTA) update compromise, sensor spoofing, model poisoning, and denial-of-service attacks on communication and perception pipelines (Durlik et al., 2024; Ying et al., 2024) as presented in Table 5. Adversarial patches, projected patterns, or spoofed LiDAR reflections can cause misdetections, phantom objects, or mislocalization. As agentic systems may update internal beliefs based on such corrupted inputs, the long-term behavioural impact can be more severe than in static models (Wang et al., 2024). Because threats can emerge anywhere across the vehicle–infrastructure ecosystem, security must rely on defence-in-depth architectures, continuous anomaly monitoring, and structured red-team testing that probes perception, planning, and learning loops under realistic adversarial conditions.

Table 5. Threat model matrix for agentic autonomous vehicle components

Threat Type	Attack Vector	Affected Subsystem	Severity	Likelihood	Example Impact	Recommended Countermeasures
Sensor Spoofing / Adversarial Perception	Fake LiDAR returns, adversarial patches, projected images	Perception, Localization	High	Medium	Phantom obstacles, missed pedestrians	Sensor redundancy, physics-based validation, adversarial detection, secure sensor fusion
Model Poisoning / Learning Manipulation	Injecting corrupted data into online learning or federated updates	Memory, Policy Learning	High	Medium–Low	Gradual policy drift leading to unsafe behavior	Trusted aggregation, anomaly detection in updates, cryptographic authentication of training data
V2X Protocol Attacks	Replay, spoofing, message injection, jamming	V2V/V2I/V2P Communication	High	High	False hazard alerts, loss of coordination, traffic disruption	PKI-based authentication, timestamp checks, secure handshakes, anti-jamming techniques
OTA Update Compromise	Malicious firmware/model update via insecure channels	Vehicle OS, AI Modules	Critical	Low	Full-system takeover, malicious policy installation	Signed firmware, secure boot, encrypted OTA channels, version attestation
In-Vehicle Network Intrusion	CAN bus injection, ECU compromise	Actuation, Control	Critical	Medium	Unauthorized braking/ acceleration; steering manipulation	Intrusion detection systems, message authentication, segmentation of safety-critical ECUs
Backend/Cloud Breach	Attacking fleet servers storing logs, model versions, or telematics	Fleet Management, Data Stores	High	Medium	Data exfiltration, manipulation of fleet-wide models	Zero-trust access, encrypted storage, multi-factor authentication, audit logs
Privacy Leakage Attacks	Re-identification from video, trajectory mining, data correlation	Data Governance, Memory	Medium	High	Identification of routes, habits, sensitive locations	Differential privacy, on-device redaction, strict retention limits, access control
Physical Tampering	Direct access to sensors, ports, ECUs	Hardware, Safety Systems	Medium	Low	Sensor disabling, debug-port exploitation	Tamper seals, hardened ECUs, port disabling, physical intrusion sensors

4.4. Incident Response, Forensic Readiness, and Evidence Preservation

Incidents such as collisions, near-misses, and security breaches remain unavoidable in complex socio-technical environments, making both real-time containment and strong forensic readiness essential for agentic autonomous vehicles. Effective response requires runtime monitors that detect anomalous behaviour, trigger safe-stop or fail-operational modes, and isolate compromised components, along with enriched, tamper-evident logs that capture synchronized sensor data, decision rationales, V2X messages, and model states (Steidl et al., 2024). Legal and policy analyses also stress the need to balance evidentiary value with privacy through selective redaction, controlled investigator access, and time-limited retention. Importantly, insights from incident forensics should feed back into fleet-wide learning pipelines and safety case updates, as required by emerging regulatory frameworks such as the EU AI Act. By embedding forensic readiness and ethical data governance into AV design, systems can enhance accountability while continuously improving safety and robustness.

5. HUMAN FACTORS, ACCEPTANCE, AND PUBLIC TRUST

Human factors are central to successful deployment of agentic AVs. Users' mental models, transparency and external signalling, trust calibration, and cultural norms collectively determine acceptance and safe interaction. Recent empirical and review studies emphasize explainability, graded transparency, and locally sensitive HMI design as key enablers of public trust (Naiseh et al., 2025; Raats et al., 2024; T. Zhang et al., 2024).

5.1. Mental Models of Agentic Systems (what users need to know, cognitive load)

Agentic AVs produce behaviour that reflects goals, memory, and adaptive planning rather than deterministic rule-following. Users therefore need mental models that map observable behaviour to intentions and limitations. (Tekkesinoglu et al., 2025) argue that explanation systems should help users build accurate mental models by focusing on what the system will do and why, rather than exposing low-level internals. (Raats et al., 2024) show that local, context-aware exposure to system intentions (e.g., route intent, priority of safety over speed) reduces cognitive load and improves perceived safety as presented in Figure 4 User Trust Curve vs. System Transparency Level (Tekkesinoglu et al., 2025; Wang et al., 2024)

Figure 4. User trust curve vs. system transparency level

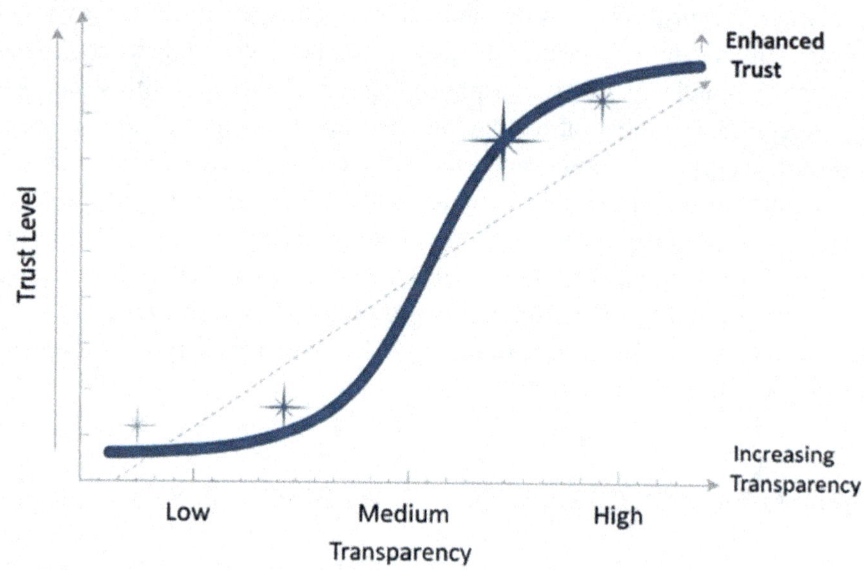

5.2. Transparency to Users and Bystanders (dashboards, signals, external HMI for pedestrians)

Transparency must span interior UIs for occupants and external HMI for bystanders. (Atakishiyev et al., 2024) show that human-machine interfaces that supply succinct "what/when/why" information improve situation awareness and trust. (Tekkesinoglu et al., 2025) and (Kuznietsov et al., 2024) emphasize that explainability methods should be mapped to human-centered communication channels—dashboards for occupants and simple visual/audible cues for pedestrians, while preserving privacy and avoiding information overload.

Applied studies (Raats et al., 2024; Yang & Kim, 2024) show that well-designed transparency increases acceptance and willingness to ride. External signalling research supports standardized, low-ambiguity cues to reduce pedestrian uncertainty and guide interaction in mixed-traffic settings (Ko et al., 2024). Design guidelines has been presented in Figure 5.

5.3. Trust Calibration: avoiding over-trust and under-trust; communicating uncertainty

Trust calibration is the deliberate alignment of user trust with system capability. Over-trust results from smooth, fluent behaviour that masks occasional brittleness; under-trust arises from opaque or overly conservative systems (Naiseh et al., 2025; Raats et al., 2024). Effective trust calibration strategies include phased exposure, graded automation levels, and explicit uncertainty indicators (Atakishiyev et al., 2024; Tekkesinoglu et al., 2025).

Figure 5. Design guidelines for trust-calibrated explanations

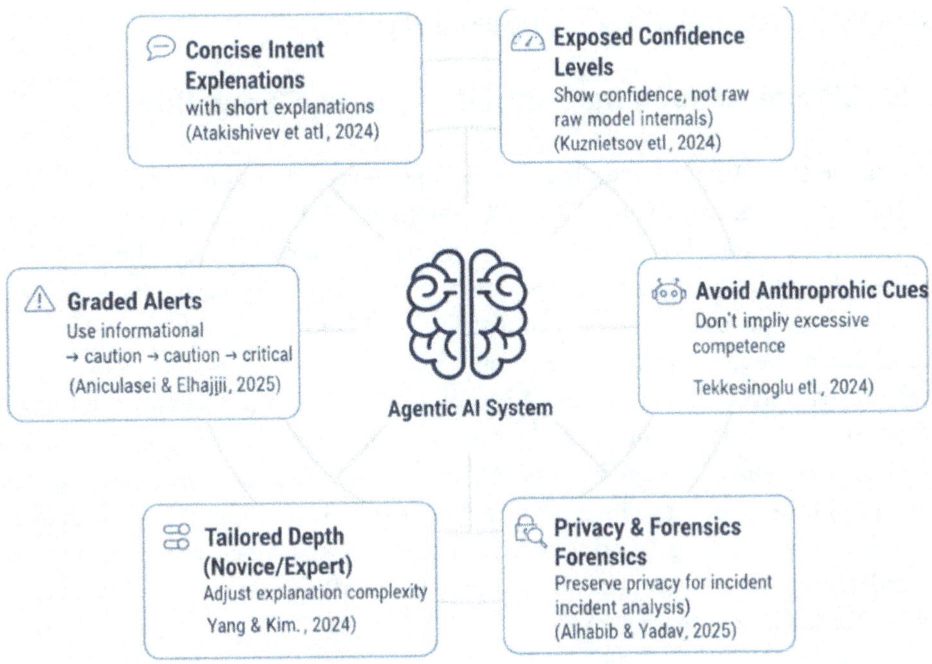

Empirical analyses indicate that combining concise predictive transparency with action recommendations yields better outcomes: users act appropriately when given both what the AV will do and what they should do in contingency (Aniculaesei & Elhajji, 2025; Atakishiyev et al., 2024).

5.4. Cultural and Societal Norms shaping acceptance

Acceptance is context dependent. (Yang & Kim, 2024) show that AI literacy and technophobia mediate acceptance across automation levels; (Raats et al., 2024) document local community responses that shape trust formation. (Naiseh et al., 2025) synthesize interdisciplinary findings, noting that policy transparency, credible certification, and visible safety evidence influence societal trust. Finally, reviewers argue that public engagement and local co-design reduce friction and improve uptake (Bruto Da Costa et al., 2025; Kuznietsov et al., 2024).

6. V2X, INFRASTRUCTURE, AND ECOSYSTEM INTEGRATION

Agentic systems in autonomous vehicles only fully make sense when the car is seen as part of a wider traffic community. The true potential of Agentic AI in Autonomous Vehicles (AVs) is not realized in isolation but through its seamless integration into a broader transportation ecosystem (*Agentic AI and the Future of Communications for Autonomous Vehicle (V2X) – IEEE ComSoc Technology Blog*, 2025). This ecosystem, comprising other vehicles, infrastructure, and pedestrians collectively known as V2X (Vehicle-to-Everything) serves as the extended sensory and cognitive apparatus of the AV (Huang et al., 2025). By leveraging V2X interactions, Agentic AI transitions from a solitary decision-maker to a collaborative node within a networked intelligence, significantly amplifying its capabilities for safety, efficiency, and ethical coordination (Noor-A-Rahim et al., 2022).

6.1. V2V/V2I/V2P Interactions that Amplify Agentic Capabilities

Vehicle-to-vehicle (V2V) links widen the vehicle's perceptual horizon. A car can receive warnings about sudden braking several cars ahead, a stalled vehicle beyond a bend, or an emergency vehicle approaching from behind (Daddanala et al., 2021). This information supplements onboard sensing so that the AV's internal agent has more time to plan, slow down, or choose an alternative lane. For learning-based systems, shared V2V data also acts as a kind of distributed memory, because rare events that happen to one vehicle can inform policies for the whole fleet.

Vehicle-to-infrastructure (V2I) communication plays a different but complementary role. Traffic lights, dynamic signs, and road-side units can broadcast signal phase and timing, recommended speeds, lane closures, or priority rules. The agent in the AV can treat these messages as constraints and opportunities: it can coast into a green wave instead of rushing to the next red light, reduce speed early before a construction zone, or prepare to give priority to an ambulance corridor. Smart infrastructure equipped with cameras, radar, or weather sensors can also supply local information that the car cannot easily infer for itself, such as black ice on a shaded curve.

Vehicle-to-pedestrian (V2P) interactions extend the same idea to vulnerable road users. Smartphones, wearables, and assistive devices can act as beacons that announce a pedestrian's presence or even intent to cross (Song et al., 2025). An Agentic AV can receive a signal from a pedestrian's phone indicating their intention to cross at an unmarked intersection, allowing the AV to yield pre-emptively the agentic AV can then lower its speed, increase its attention to possible conflict points, or proactively yield. This is most valuable where sight lines are poor, such as behind parked vehicles or around large trucks, or in locations that are chronically busy with pedestrians, cyclists, and scooters.

6.2. Smart Infrastructure Interfaces

Smart infrastructure sits at the centre of this ecosystem (Raza et al., 2025). Road-side units with computation and storage can aggregate data from many vehicles, detect patterns such as recurring congestion or near-miss hotspots, and send tailored advice back into the traffic stream. High-definition maps and frequent mapping updates provide a second kind of interface. The agentic AV does not just follow a static map; it receives information about new crosswalks, redesigned intersections, changed speed limits, and temporary restrictions. Each update is a small nudge to the policy space, shaping how the agent weighs risk, legality, and comfort on familiar roads.

6.3. Multi-Agent Ethics

Ethical questions become more complex once many different agents share a common space. Not all vehicles will follow the same software, and human drivers will continue to be present for a long time. Some systems may be tuned to drive more assertively, others more cautiously. Without some shared standards, these differences can create friction at mergers, roundabouts, or busy crossings. One reasonable approach is to agree on basic norms that no agent may violate, such as strong protection for pedestrians and cyclists, and then allow different systems to express their driving personality within those boundaries.

The behaviour of human drivers also matters. People expect some degree of courtesy and reciprocity: they may allow a merger if they feel that others do the same, but they can become frustrated if they perceive a vehicle as constantly taking advantage of them. An agentic AV therefore must balance legal compliance with social expectations. It should avoid bullying human drivers by exploiting their caution, yet it should also avoid freezing in complex situations and thereby creating confusion or congestion. Good multi-agent ethics for AVs involve explicit policies about how to share delays, when to yield, and how to treat vulnerable users in crowded spaces (Wu et al., 2025).

Figure 6. Ecosystem view: AV–agent–infrastructure interplay

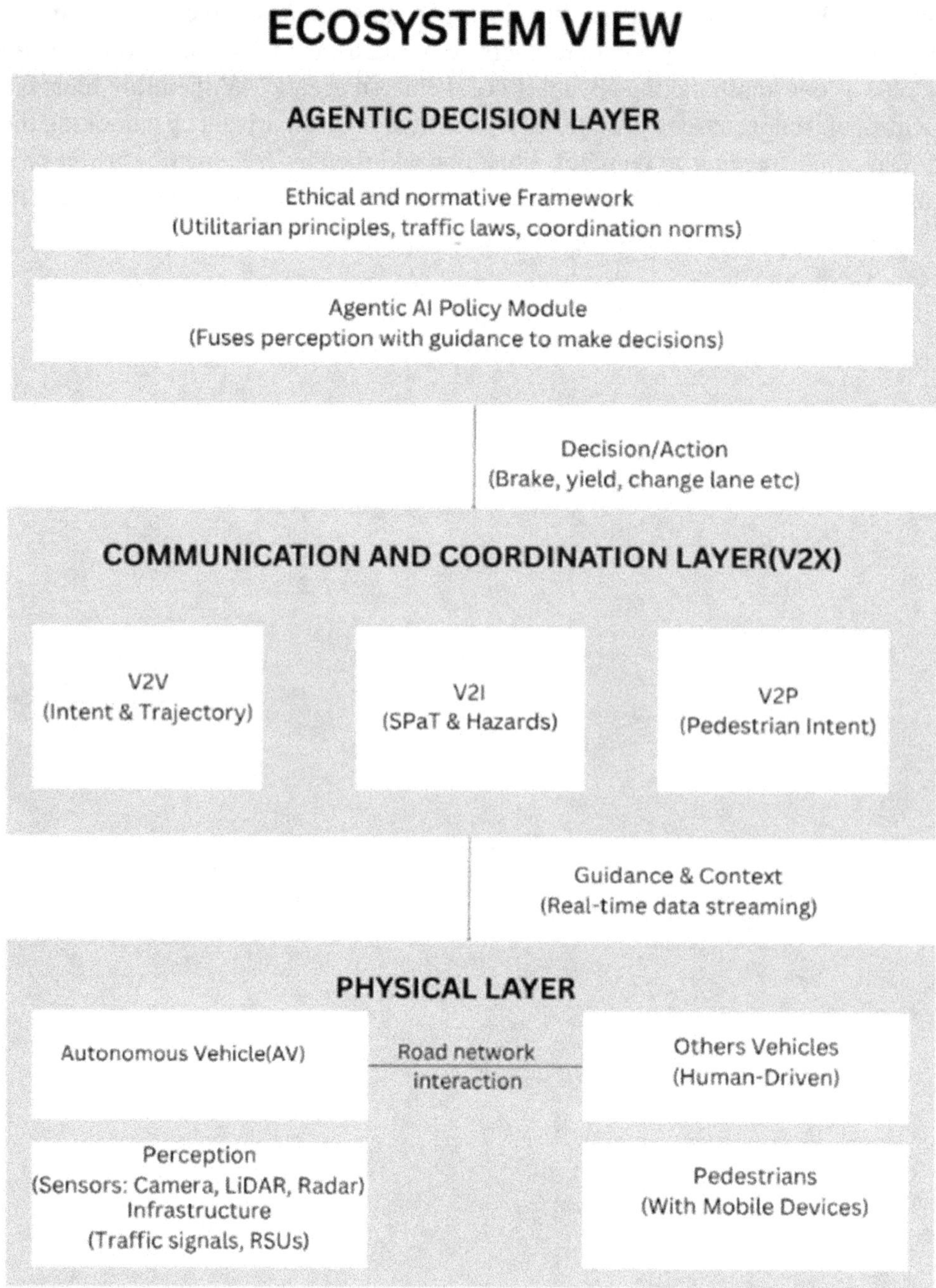

Figure 6 portray three layers: the physical world of roads and actors, the communication mesh of V2V, V2I, and V2P links, and the decision layer where agentic policies run. Arrows between layers show how perception, infrastructure guidance, and ethical norms are combined into concrete actions (Muhammad & Safdar, 2021).

- Physical Layer (Bottom): This layer represents the tangible world, containing the road network, an Autonomous Vehicle (AV), other human-driven vehicles, pedestrians, and infrastructure elements like traffic signals and roadside sensors. Arrows labeled "Perception" point upward, signifying how raw data (e.g., camera images, LiDAR point clouds, object positions) is captured and fed into the system.
- Communication & Coordination Layer (Middle): This is the data-exchange mesh that overlays the physical world. It visualizes the dynamic, real-time links of V2V (Vehicle-to-Vehicle), V2I (Vehicle-to-Infrastructure), and V2P (Vehicle-to-Pedestrian) communication. These links carry enriched information such as vehicle intent, sensor data shares, Signal Phase and Timing (SPaT) from traffic lights, hazard warnings from roadside units, and pedestrian intent from mobile devices.
- Agentic Decision Layer (Top): This layer represents the cognitive core of the AV. Here, the Agentic AI policy module fuses the multi-source data from the layers below. It integrates the raw "Perception" from the Physical Layer with the contextual "Guidance & Context" from the Communication Layer. This fusion occurs within a defined "Ethical & Normative Framework," which encodes societal values and coordination rules. The output is a "Decision/Action" (e.g., brake, yield, change lane) that is sent back down to the Physical Layer for the AV to execute, thereby closing the loop.

Table 6. Coordination dilemmas & ethical resolution policies

Coordination Dilemma	Description	Ethical Principles Involved	Agentic Resolution Policy
Merge at congested on-ramp	Multiple AVs and human drivers compete for limited gaps	Fairness, efficiency, safety	Enforce zipper-merging with bounded delays per vehicle; AVs yield proportionally.
Emergency vehicle corridor	AVs and human drivers must clear lane rapidly	Priority of life, public good, legality	AVs proactively create space and broadcast intentions; infrastructure provides green wave.
Pedestrian-heavy shared space	Vehicles and pedestrians mix without clear right-of-way	Protection of vulnerable users, dignity, autonomy	AVs adopt low-speed, "pedestrian-first" mode with conservative gap acceptance.
Adversarial / non-compliant human driver	Human driver violates norms (e.g., cutting in, speeding)	Harm minimization, accountability	AVs increase distance, avoid escalation, and log incidents for post-hoc review.
Allocation of delay at signalized intersection	Some approaches consistently experience longer waits	Distributive justice, equity	Infrastructure and AVs jointly adjust timing and routing to reduce systematic disparities.
Merge at congested on-ramp	Multiple AVs and human drivers compete for limited gaps	Fairness, efficiency, safety	Enforce zipper-merging with bounded delays per vehicle; AVs yield proportionally.
Emergency vehicle corridor	AVs and human drivers must clear lane rapidly	Priority of life, public good, legality	AVs proactively create space and broadcast intentions; infrastructure provides green wave.
Pedestrian-heavy shared space	Vehicles and pedestrians mix without clear right-of-way	Protection of vulnerable users, dignity, autonomy	AVs adopt low-speed, "pedestrian-first" mode with conservative gap acceptance.
Adversarial / non-compliant human driver	Human driver violates norms (e.g., cutting in, speeding)	Harm minimization, accountability	AVs increase distance, avoid escalation, and log incidents for post-hoc review.
Allocation of delay at signalized intersection	Some approaches consistently experience longer waits	Distributive justice, equity	Infrastructure and AVs jointly adjust timing and routing to reduce systematic disparities.

Table 6 lists typical dilemmas such as congested ramp merges, emergency vehicle corridors, shared spaces with heavy pedestrian presence, aggressive or non-compliant human drivers, and long-term allocation of signal delay across approaches. For each row the table can summarize the situation, identify the ethical principles in play, and describe a policy pattern that an agentic AV might adopt, for example zipper-merging rules, pedestrian-first behaviour at low speed, or automatic creation of space for emergency responders.

7. EVALUATION FRAMEWORKS AND METRICS FOR ETHICS-ALIGNED AGENTIC AVs

If agentic AVs are supposed to reflect explicit ethical commitments, those commitments have to be made measurable (Akhtar et al., 2024). A statement like "the system prioritizes safety while respecting fairness" sounds reassuring, but it is not sufficient on its own. Developers, regulators, and the public need to see how such claims are checked in simulation, in controlled tests, and on real roads. The purpose of an evaluation framework is not to compress ethics into a single number, but to make trade-offs explicit, track changes over time, and reveal where further safeguards are needed.

7.1. Multi-Objective Performance

Safety remains the primary concern. Common measures include collision rates, the severity of harm when incidents do occur, and near-miss indicators such as time-to-collision or the frequency of harsh braking. Legality is another central dimension. Here the focus is on how often the AV obeys speed limits, respects right-of-way, and follows rules around signalling, lane use, and stops. Violations can be counted per kilometre or per trip, with close attention to patterns that repeat at certain locations or times (Wu et al., 2025).

Instead of looking only at aggregate safety or delay, analysts ask who bears the remaining risk and inconvenience. For example, pedestrians in some neighbourhoods may face more near-miss events than those in business districts, or certain communities may experience longer wait times when AV fleets optimize for efficiency. Comfort is assessed both through physical measurements, such as jerk and lateral acceleration, and through passenger feedback about motion sickness, anxiety, or perceived smoothness (Pati, 2025).

Efficiency and sustainability complete the broader picture. Efficiency measures include travel time, schedule adherence for shared services, and network-level throughput. Sustainability metrics track energy use and emissions per kilometre, as well as how often the AV selects routes or driving styles that reduce environmental impact. In an ethics-aligned design, safety and legality act as hard constraints, while fairness, comfort, efficiency, and sustainability form a set of objectives that the agentic policy balances within those limits. This reflects a common intuition that thresholds around serious harm should not be traded for small gains in convenience.

7.2. Ethical Compliance Metrics: Harm Minimization Indices, Protected-Class Exposure Parity

Ethical compliance requires metrics that go beyond raw safety or legality counts. A harm minimization index can aggregate the severity and likelihood of adverse events (collisions, near-misses, forced evasive manoeuvres) across different road user categories. Weighting schemes can emphasize severe outcomes (e.g., fatalities, serious injuries) more than minor ones, reflecting societal preferences (Dutta Roy, 2025).

To assess fairness, exposure parity metrics can measure whether protected classes or vulnerable road users experience disproportionately high risk or delay (Fernández Llorca et al., 2025). For example, one can compare near-miss rates per crossing for pedestrians in different neighbourhoods, or the distribution of yielding behaviour toward pedestrians using mobility aids versus others. Where critical demographic labels are not available for privacy reasons, proxy metrics based on location, infrastructure quality, or time-of-day can approximate distributional fairness without profiling individuals.

7.3. Benchmarking Protocols

A robust evaluation ecosystem requires benchmarking protocols that support fair, reproducible comparisons across systems and deployments. Scenario-based testing is central: curated libraries of edge cases (occluding pedestrians, complex unprotected left turns, ambiguous signage) can be used in simulation and closed-course environments to probe ethical decision boundaries.

Key elements of benchmarking protocols include:

- Standardized scenario sets with clearly defined initial conditions, environmental parameters, and evaluation criteria.
- Open metrics definitions, ensuring that different developers compute safety, fairness, and comfort metrics consistently.
- Data logging and auditability requirements, so that when ethical failures occur, investigators can reconstruct decision pathways and verify whether policies behaved as designed.
- Reporting standards, specifying how incident rates, near-miss statistics, and fairness metrics must be published (e.g., confidence intervals, contextual information, and stratification by environment).

Figure 7 show a circular process with four labelled stages and arrows between them. Small callouts can illustrate typical activities at each step, such as defining ethical objectives, running scenario tests, deploying pilot services, analysing incident clusters, and updating policies. The visual emphasis is on iteration: the system is never considered ethically "finished," but is instead subject to ongoing scrutiny, adjustment, and learning.

Figure 7. Benchmarking loop

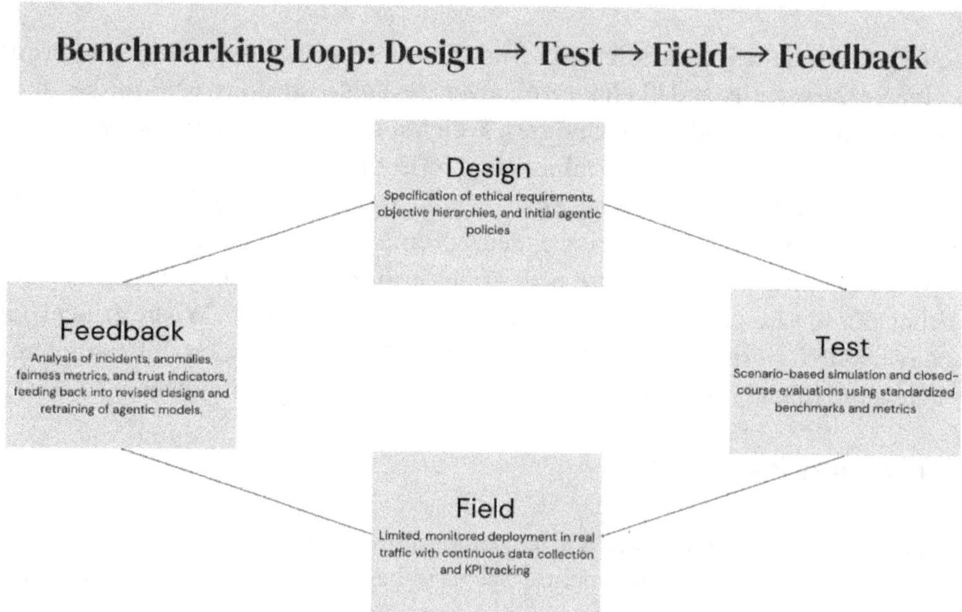

7.4. Post-deployment KPIs

After deployment, key performance indicators are used to check whether agentic AVs behave safely and ethically in real traffic. Classic safety KPIs include crashes per million kilometres and how severe those crashes are, but because such events are rare, operators also track near-miss signals such as time-to-collision, hard braking, evasive manoeuvres, and recurring conflict points with pedestrians and cyclists. Automated anomaly detection can highlight unusual patterns, which are then investigated to see whether they stem from perception errors, bad infrastructure messages, unexpected human behaviour, or flaws in the decision policy.

Quantitative data is complemented by user-reported trust and perceived fairness. Passenger surveys, complaints, and feedback from other road users can surface problems like perceived aggressiveness or confusing behavior that raw statistics may miss. Regulators and independent auditors can periodically review these metrics, incident logs, and stakeholder interviews as part of ethics conformance audits. Their findings then feed back into policy updates, model retraining, and refinement to how AVs coordinate with infrastructure and other actors.

CONCLUSION

This chapter has presented a comprehensive framework for the ethical integration of Agentic AI into autonomous vehicles, spanning technical architecture, safety assurance, data governance, human factors, ecosystem integration, and evaluation methodologies. The chapter presented a strong argument that autonomous vehicles cannot rely indefinitely on static, rule-based AI: the variability and uncertainty of real-world traffic demand Agentic AI systems that are goal-driven, memory-enabled, tool-using, and capable of continuous adaptation. We began by decomposing the autonomous vehicle stack into perception, prediction, planning/control, and human–machine interaction (HMI) and showed how each layer faces distinct challenges in urban traffic, bad weather, mixed road-user environments, and emergencies. Within this stack, agentic capabilities provide a systematic way to respond to distribution shifts, rare events, and increasingly complex road ecosystems. We then outlined a technical architecture for embedding ethics into Agentic AI for AVs. Key components include explicit goal specification, hard safety and legal constraints, preference learning for resolving ethical trade-offs, and reward shaping to align learning signals with societal values. Context and memory structures (episodic and semantic) support safety-critical learning over time, while human-in-the-loop and human-on-the-loop paradigms define how human judgment remains available for escalation and oversight. Explainability mechanisms—spanning global and local explanations—were positioned as essential to transparency, regulatory scrutiny, and user trust, alongside robustness measures such as risk-sensitive planning, handling of distribution shifts, and out-of-distribution detection. the chapter proposes a holistic view: agentic AVs are not just more capable decision-makers; they are socio-technical actors whose architectures, data practices, assurance methods, and interfaces must co-evolve with legal, ethical, and societal expectations.

REFERENCES

Adabara, I., Sadiq, B. O., Shuaibu, A. N., Danjuma, Y. I., & Venkateswarlu, M. (2025). A Review of Agentic AI in Cybersecurity: Cognitive Autonomy, Ethical Governance, and Quantum-Resilient Defense. *F1000 Research, 14*, 843. DOI: 10.12688/f1000research.169337.1 PMID: 41169871

Agentic AI and the Future of Communications for Autonomous Vehicle (V2X) – IEEE ComSoc Technology Blog. (n.d.). Retrieved November 15, 2025, from https://techblog.comsoc.org/2025/07/14/agentic-ai-and-the-future-of-communications-for-autonomous-vehicle-v2x/

Akhtar, M. A. K., Kumar, M., & Nayyar, A. (2024). Privacy and security considerations in explainable AI. In *Towards ethical and socially responsible explainable AI* (pp. 193–226). Springer., DOI: 10.1007/978-3-031-66489-2_7

Alam, T. (2024). Data privacy and security in autonomous connected vehicles in smart city environment. *Big Data and Cognitive Computing, 8*(9), 95. DOI: 10.3390/bdcc8090095

Alemayehu, H., & Sargolzaei, A. (2025). Testing and verification of connected and autonomous vehicles: A review. *Electronics (Basel), 14*(3), 600. DOI: 10.3390/electronics14030600

Alhabib, R., & Yadav, P. (2025). Data authorisation and validation in autonomous vehicles: A critical review. *Discover Applied Sciences, 7*(7), 735. DOI: 10.1007/s42452-025-07333-2

Ali, M. A., & Dornaika, F. (2025). Agentic AI: A comprehensive survey of architectures, applications, and future directions. DOI: 10.1007/s10462-025-11422-4

Aniculaesei, A., & Elhajji, Y. (2025). Runtime monitoring approach to safeguard behavior of autonomous vehicles at traffic lights. *Electronics (Basel), 14*(12), 2366. DOI: 10.3390/electronics14122366

Ansarinejad, M., Ansarinejad, K., Lu, P., Huang, Y., & Tolliver, D. (2025). Autonomous vehicles in rural areas: A review of challenges, opportunities, and solutions. *Applied Sciences (Basel, Switzerland), 15*(8), 4195. DOI: 10.3390/app15084195

Atakishiyev, S., Salameh, M., & Goebel, R. (2024). Incorporating explanations into human-machine interfaces for trust and situation awareness in autonomous vehicles. *arXiv*. /arXiv.2404.07383DOI: 10.1109/IV55156.2024.10588812

Bandi, A., Kongari, B., Naguru, R., Pasnoor, S., & Vilipala, S. V. (2025). The rise of agentic AI: A review. *Future Internet, 17*(9), 404. DOI: 10.3390/fi17090404

Bruto Da Costa, A. A., Irvine, P., Dodoiu, T., Khastgir, S., & Jennings, P. (2025). Building a robust scenario library for safety assurance of automated driving systems: A review. *IEEE Access : Practical Innovations, Open Solutions, 13*, 117619–117655. DOI: 10.1109/ACCESS.2025.3584206

Buysse, L., Habli, I., Vanoost, D., & Pissoort, D. (2025). Safe autonomous systems in a changing world. *Safety Science, 191*, 106965. DOI: 10.1016/j.ssci.2025.106965

Cârlan, C., ... (2024). Dynamic safety cases for frontier AI. *arXiv*. https://doi.org//arXiv.2412.17618DOI: 10.48550

Daddanala, R., ... (2021). Vehicle to vehicle (V2V) communication protocol. *arXiv*. https://doi.org// arXiv.2102.07306DOI: 10.48550

Durlik, I., Miller, T., Kostecka, E., Zwierzewicz, Z., & Łobodzińska, A. (2024). Cybersecurity in autonomous vehicles. *Electronics (Basel)*, *13*(13), 2654. DOI: 10.3390/electronics13132654

Dutta Roy, T. P. (2025). Bioethics artificial intelligence advisory (BAIA). *Cureus*. Advance online publication. DOI: 10.7759/cureus.80494 PMID: 40083589

Fernández Llorca, D., Hamon, R., Junklewitz, H., Grosse, K., Kunze, L., Seiniger, P., Swaim, R., Reed, N., Alahi, A., Gómez, E., Sánchez, I., & Kriston, A. (2025). Testing autonomous vehicles and AI. *European Transport Research Review*, *17*(1), 38. DOI: 10.1186/s12544-025-00732-x

Fremont, D. J., et al. (2020). Formal scenario-based testing of autonomous vehicles. *IEEE ITSC*, 1–8.

Gheisari, M., Khan, W. Z., Najafabadi, H. E., McArdle, G., Rabiei-Dastjerdi, H., Liu, Y., Fernández-Campusano, C., & Abdalla, H. B. (2024). CAPPAD privacy-preservation solution. *Applied Intelligence*, *54*(4), 3417–3428. DOI: 10.1007/s10489-023-04991-w

Gomes, D., Saif, A. S., & Nandi, D. (2020, January). Robust underwater object detection with autonomous underwater vehicle: A comprehensive study. In *Proceedings of the international conference on computing advancements* (pp. 1-10). DOI: 10.1145/3377049.3377052

Gupta, L., Jain, R., & Vaszkun, G. (2016). UAV communication networks survey. *IEEE Communications Surveys and Tutorials*, *18*(2), 1123–1152. DOI: 10.1109/COMST.2015.2495297

Hao, X., Demir, E., & Eyers, D. (2025). Beyond human-in-the-loop. *Sustainable Futures : An Applied Journal of Technology, Environment and Society*, *10*, 101152. DOI: 10.1016/j.sftr.2025.101152

Hartwich, F., et al. (2021). Passenger experience in automated vehicles. *Frontiers in Human Dynamics, 3*.

Hawasly, M., Sadeghi, J., Antonello, M., Albrecht, S. V., Redford, J., & Ramamoorthy, S. (2022). Perspectives on the system-level design of a safe autonomous driving stack. *AI Communications*, *35*(4), 285–294. DOI: 10.3233/AIC-220148

Hosseini, S., & Seilani, H. (2025). Role of agentic AI. *Array (New York, N.Y.)*, *26*, 100399. DOI: 10.1016/j.array.2025.100399

Hu, J., Wang, Y., Cheng, S., Xu, J., Wang, N., Fu, B., Ning, Z., Li, J., Chen, H., Feng, C., & Zhang, Y. (2025). A survey of decision-making and planning methods for self-driving vehicles. *Frontiers in Neurorobotics*, *19*, 1451923. DOI: 10.3389/fnbot.2025.1451923 PMID: 40041453

Huang, T., Liu, J., Zhou, X., Nguyen, D. C., Azghadi, M. R., Xia, Y., & Sun, S. (2025). Vehicle-to-everything cooperative perception for autonomous driving. *Proceedings of the IEEE.*

Ji, W., Yuan, Q., Cheng, G., Yu, S., Wang, M., Shen, Z., & Yang, T. (2023). Traffic accidents of autonomous vehicles based on knowledge mapping: A review. [English edition]. *Journal of Traffic and Transportation Engineering*, *10*(6), 1061–1073. DOI: 10.1016/j.jtte.2023.09.003

Karimi, A., Adnane, A., Phillips, I. W., & Benkhelifa, E. (2025). Connected Vehicles Data Classification and the Influence of a Sustainable Data Governance for Optimal Utilisation of In-Vehicle Data. In *ICISSP (2)* (pp. 643-650). DOI: 10.5220/0013371100003899

Ko, W., Shim, M., Park, S., Lee, S., & Yun, I. (2024). Development of test cases for automated vehicle driving safety assessment using driving trajectories. *Sensors (Basel)*, *24*(24), 7981. DOI: 10.3390/s24247981 PMID: 39771717

Kuru, K. (2021). Human-on-the-loop teleoperation. *IEEE OJITS*, *2*, 448–469.

Kuznietsov, A., Gyevnar, B., Wang, C., Peters, S., & Albrecht, S. V. (2024). Explainable AI for safe and trustworthy autonomous driving: A systematic review. *IEEE Transactions on Intelligent Transportation Systems*, *25*(12), 19342–19364. DOI: 10.1109/TITS.2024.3474469

Lee, D., & Nguyen Long, L. A. (2025). Privacy challenges of automated vehicles.

Longo, L., Berretta, S., Verda, D., & Rizzo, L. (2025). Computational argumentation and automatic rule-generation for explainable data-driven modeling. *IEEE Access : Practical Innovations, Open Solutions*, *13*, 175565–175583. DOI: 10.1109/ACCESS.2025.3618992

Lundberg, S. M., Erion, G., Chen, H., DeGrave, A., Prutkin, J. M., Nair, B., Katz, R., Himmelfarb, J., Bansal, N., & Lee, S. I. (2020). From local explanations to global understanding with explainable AI for trees. *Nature Machine Intelligence*, *2*(1), 56–67. DOI: 10.1038/s42256-019-0138-9 PMID: 32607472

Luu, Q. H., Liu, H., Chen, T. Y., & Vu, H. L. (2024). A sequential metamorphic testing framework for understanding autonomous vehicle's decisions. *IEEE Transactions on Intelligent Vehicles*, 1–13. DOI: 10.1109/TIV.2024.3370740

Mellamphy, N. B. (2021). Humans "in the loop"?: Human-centrism, posthumanism, and AI. *Nature and Culture*, *16*(1), 11–27. DOI: 10.3167/nc.2020.160102

Muhammad, M., & Safdar, G. A. (2021). 5G-based V2V broadcast communications: A security perspective. *Array (New York, N.Y.)*, *11*, 100084. DOI: 10.1016/j.array.2021.100084

Naiseh, M., Clark, J., Akarsu, T., Hanoch, Y., Brito, M., Wald, M., Webster, T., & Shukla, P. (2025). Trust, risk perception, and intention to use autonomous vehicles: An interdisciplinary bibliometric review. *AI & Society*, *40*(2), 1091–1111. DOI: 10.1007/s00146-024-01895-2 PMID: 40191162

Noor-A-Rahim, M., Liu, Z., Lee, H., Khyam, M. O., He, J., Pesch, D., Moessner, K., Saad, W., & Poor, H. V. (2022). 6G for vehicle-to-everything (V2X) communications: Enabling technologies, challenges, and opportunities. *Proceedings of the IEEE*, *110*(6), 712–734. DOI: 10.1109/JPROC.2022.3173031

Patel, K., Shah, M., Qureshi, K. M., & Qureshi, M. R. N. (2025). A systematic review of generative AI: Importance of industry and startup-centered perspectives, agentic AI, ethical considerations & challenges, and future directions. *Artificial Intelligence Review*, *59*(1), 7. DOI: 10.1007/s10462-025-11435-z

Pati, A. K. (2025). Agentic AI: A comprehensive survey of technologies, applications, and societal implications. *IEEE Access : Practical Innovations, Open Solutions*, *13*, 151824–151837. DOI: 10.1109/ACCESS.2025.3585609

Pikner, H., Malayjerdi, M., Bellone, M., Baykara, B. C., & Sell, R. (2024). Autonomous Driving Validation and Verification Using Digital Twins. *VEHITS*, 204-211.

Raats, K., Lund, J., & Brodersen, M. (2024). Trust in autonomous vehicles: Insights from a Swedish suburb. *Journal of Responsible Innovation, 11*(1), 2318825. DOI: 10.1080/23299460.2024.2318825

Raza, S., Sapkota, R., Karkee, M., & Emmanouilidis, C. (2025). Responsible Agentic Reasoning and AI Agents: A Critical Survey. *Authorea Preprints.*

Répás, J. (2023). Definition of Forensic Methodologies for Autonomous Vehicles. *Hadmérnök, 18*(1), 125–141.

Rudrakar, S., Rughani, P., & Sadineni, L. (2025). Digital forensics and incident response management model for IoT based agriculture. *Scientific Reports, 15*(1), 17797. DOI: 10.1038/s41598-025-02635-2 PMID: 40404782

Shen, W., & Liu, Y. (2025). Data governance in AVs.

Song, X., Zhang, J., Li, J., Yu, H., Li, X., Song, W., & Liew, K. M. (2025). Experimental study on pedestrian movement on elevated platforms. *Safety Science, 186*, 106836. DOI: 10.1016/j.ssci.2025.106836

Steidl, M., Dornauer, B., Felderer, M., Ramler, R., Racasan, M. C., & Gattringer, M. (2024, August). How industry tackles anomalies during runtime: Approaches and key monitoring parameters. In *2024 50th Euromicro Conference on Software Engineering and Advanced Applications (SEAA)* (pp. 364-372). IEEE.

Syafira, O. A., & Selviandro, N. (2025, August). The Adoption of SACM to Support Dialectical Safety Assurance Case of AV Object Recognition System. In *2025 IEEE International Symposium on Future Telecommunication Technologies (SOFTT)* (pp. 157-163). IEEE. DOI: 10.1109/SOFTT67007.2025.11213135

Teixeira, B., Carvalhais, L., Pinto, T., & Vale, Z. (2025). Explainable AI framework for automated energy management. *Energy and Building, 347*, 116246. DOI: 10.1016/j.enbuild.2025.116246

Tekkesinoglu, S., Habibovic, A., & Kunze, L. (2025). Advancing explainable autonomous vehicle systems: A comprehensive review and research roadmap. *ACM Transactions on Human-Robot Interaction, 14*(3), 1–46. DOI: 10.1145/3714478

Tong, K., Ajanovic, Z., & Stettinger, G. (2020, September). Overview of tools supporting planning for automated driving. In *2020 IEEE 23rd International Conference on Intelligent Transportation Systems (ITSC)* (pp. 1-8). IEEE. DOI: 10.1109/ITSC45102.2020.9294512

Ullrich, L., Buchholz, M., Dietmayer, K., & Graichen, K. (2024). AI safety assurance for automated vehicles: A survey on research, standardization, regulation. *IEEE Transactions on Intelligent Vehicles, 10*(10), 4784–4803. DOI: 10.1109/TIV.2024.3496797

Van de Poel, I. (2020). Embedding values in artificial intelligence (AI) systems. *Minds and Machines, 30*(3), 385–409. DOI: 10.1007/s11023-020-09537-4

Wagner, J. K., Cabrera, L. Y., Gerke, S., & Susser, D. (2024). Synthetic data and ELSI-focused computational checklists—A survey of biomedical professionals' views. *PLOS Digital Health, 3*(11), e0000666. DOI: 10.1371/journal.pdig.0000666 PMID: 39565773

Wagner, M., & Carlan, C. (2024). The open autonomy safety case framework. *arXiv preprint arXiv:2404.05444*.

Wang, C., Guo, F., Zhao, S., Zhu, Z., & Zhang, Y. (2024). Safety assessment for autonomous vehicles: A reference driver model for highway merging scenarios. *Accident; Analysis and Prevention*, *206*, 107710. DOI: 10.1016/j.aap.2024.107710 PMID: 39018627

Wu, Y., et al. (2025). Multi-agent AV systems.

Xu, Z., Wang, X., Wang, X., & Zheng, N. (2025). Safety validation for connected autonomous vehicles using large-scale testing tracks in high-fidelity simulation environment. *Accident; Analysis and Prevention*, *215*, 108011. DOI: 10.1016/j.aap.2025.108011 PMID: 40107085

Yang, Y., & Kim, M. Y. (2024). Promoting sustainable transportation: How people trust and accept autonomous vehicles—Focusing on the different levels of collaboration between human drivers and artificial intelligence—An empirical study with partial least squares structural equation modeling and multi-group analysis. *Sustainability (Basel)*, *17*(1), 125. DOI: 10.3390/su17010125

Ying, Z., Wang, K., Xiong, J., & Ma, M. (2024). A literature review on V2X communications security: Foundation, solutions, status, and future. *IET Communications*, *18*(20), 1683–1715. DOI: 10.1049/cmu2.12778

Zhang, T., Li, W., Huang, W., & Ma, L. (2024). Critical roles of explainability in shaping perception, trust, and acceptance of autonomous vehicles. *International Journal of Industrial Ergonomics*, *100*, 103568. DOI: 10.1016/j.ergon.2024.103568

Zhao, J., Yan, X., Zhou, Z., & Zhang, Z. (2025). Rule-adaptive lane-changing trajectory planning method for autonomous vehicles driven by dynamic risk information. *Scientific Reports*, *15*(1), 31514. DOI: 10.1038/s41598-025-15382-1 PMID: 40858682

Zhou, J., Gao, Y., Johansson, O., Olofsson, B., & Frisk, E. (2025). Robust predictive motion planning by learning obstacle uncertainty. *IEEE Transactions on Control Systems Technology*, *33*(3), 1006–1020. DOI: 10.1109/TCST.2025.3533378

Chapter 11
Autonomous AI in Automotive Safety:
Rethinking Tortious Liability Paradigms

K. Hema
http://orcid.org/0000-0002-3189-0996
Christ University, India

R. Pavithra
Gujarat National Law University, India

ABSTRACT

This paper seeks to examine the application of tort law to the Automated Vehicles. Section I deals with a brief overview of the impact of AVs and tort law and the application of tortious principles. Section II discusses the concept of fault and the multiple parties to whom fault can be attributed. Furthermore, it discusses who should be blamed and to what extent. Section III discusses the tortious principles such as product liability, vicarious liability and apportionment of liability and its application to the AVs. Section IV proposes directions for future research, conclusion and recommendations for the legal challenges posed by the AVs.

1. INTRODUCTION

> *The autonomous car doesn't drink, doesn't do drugs, doesn't text while driving, and doesn't get road rage. Autonomous cars don't race other autonomous cars, and they don't go to sleep (*Alan D. Kaplan & Robert Sanzillo 2016)."

Automated Vehicles (AVs) would overpower and transform the automobile industry and the economy at large. The exponential growth of AVs is to be expected within the next ten years. Technology is far from being perfect, and so are driverless cars. With the gaining momentum of AV technology, come various legal challenges like liability concerns, regulatory and ethical concerns, data privacy, jurisdictional issues, intellectual property law concerns, concerns of consumer protection, etc. The scope of this paper is restricted to analysing the tortious liability in AVs including determining fault, product liability, negligence, vicarious liability, strict liability and apportionment of liability. The immediate

DOI: 10.4018/979-8-3373-7355-3.ch011

pressing question is to ascertain how and to whom these liabilities could be attributed. For this purpose, the research aims to apply tortious principles to determine the extent of the liability in AVs.

The rapid growth of AVs has led to the skipping of the question 'IF an accident happens to, WHEN' it happens, what should be done? When an AV commits an accident, multiple people can be made liable. The primary and critical question in AVs is fault attribution: Who is at fault? - the manufacturer, the software developer or the owner when accidents are committed by AVs. To attribute fault to one party, *firstly*, the differences in the concept of fault in automated systems and human drivers must be analysed. *Secondly*, how the law should tackle the situations where both human and automated systems are at fault. *Thirdly*, in cases of negligence, the extent and nature of the standard of care expected of the manufacturer and users must be laid down. Negligence and fault attribution may depend on the level of automation in the automated vehicles.

Though there is evidence to prove that AVs would reduce road accidents, the accidents that AVs would cause due to defects in the software systems or AI algorithms must not be overlooked. To apply product liability in these cases, it should be determined whether AVs amount to a product. The applicability of strict liability and vicarious liability opens an arena of academic discourse. The most challenging task posed by AVs is the apportionment of liability between humans, the manufacturer and AI systems. This research paper aims to undertake extensive research to determine the applicability of tortious liability principles to AVs in cases of tortious liability. This research aims to analyse fault attribution in automated vehicle accidents and examines the applicability of negligence and product liability principles to AVs. The research seeks to propose a framework for apportionment of liability in AV accidents. This section provides an overview of Automated Vehicles (AVs), their rapid growth, and their transformative impact on the automobile industry and economy. It also examines legal and regulatory challenges, focusing on tortious principles such as fault attribution, negligence, product liability, vicarious liability, strict liability, and apportionment of liability in AV accidents.

With these revolutionary changes in the transportation, the law must keep abreast with the technological developments, failing which would result in chaos in the economy. Laws must always be a step ahead of technological development to create a balanced framework. A dynamic and adaptive legal system is the need of the hour to regulate the ever-evolving technologies in transportation, especially self-driving cars. Tort law, the common law, can offer a potential solution to the liability issues that arise in case of accidents committed by the self-driving vehicles. Traditional tort law addresses the civil wrongs where there is human interference, where the fault can be attributed to the person behind an action or inaction. However with the advent of self-driving vehicles, especially with the autonomous AI, where there is no human intervention, the attribution of liability becomes more complex. Additionally, these technological developments raise questions concerning fault attribution in the events of accidents or malfunctions. As Winfield maintains, it is the law of tort and not a law of torts, highlighting the comprehensive nature of the tort law that allows it to include new torts in its scope. This adaptability of the law of tort is well-suited to confront the emerging challenges in the realm of self-driving vehicles. The law of tort embodies liability principles that need to be adapted to effectively address the challenges posed by self-driving cars.

By supporting a legislative framework that encourages the safe, moral, and responsible use of autonomous vehicles (AVs), this research advances sustainability by bringing technology advancement and societal well-being into harmony. Responsible innovation in AI-powered mobility is encouraged by the study's resolution of responsibility issues and its proposal of strong tortious standards for fault attribution. Liability transparency promotes public trust in sustainable transportation technology by encouraging manufacturers and developers to incorporate greater safety standards and lowering the possibility of legal impunity for system failures. Furthermore, the research promotes long-term regulatory resilience that can change in response to new hazards and lessen systemic legal friction by suggesting adaptable legal structures. The adoption of cleaner, AI-driven mobility solutions—which minimize traffic fatalities, maximize fuel efficiency, and lower carbon emissions—is facilitated by improved legal predictability in AV-related events. Therefore, the research is not only legally relevant but also environmentally and ethically sustainable because it intersects with the Sustainable Development Goals (SDGs) of the UN, especially those pertaining to sustainable cities (SDG 11), industry innovation (SDG 9), and peace, justice, and strong institutions (SDG 16).

1.1 Automated Vehicles - A Historical Overview

AVs, which were once upon a time only a fantasy, have now become a new reality. As unbelievable as it may sound, AVs, though currently in the news and media, have a long history. AV was conceptualised even before the invention of the automobile. Back in the years of Leonardo Da Vinci, he designed a small, three-wheeled, self-propelled cart which was a self-driving vehicle, incorporating a series of springs for propulsion, a pre-programmable steering system, and a parking brake released remotely by rope (Anderson et al., 2014). In 1925, the dream of steering an AV on the roads came true when an electrical engineer Francis P Houdina, invented a fully radio-controlled AV. However, on the streets of New York, the AV crashed and the project was a failure. Nevertheless, it still was evidence that the invention and steering of AVs was possible. In 1939, at the New York World Fair, Norman Bel Geddes showcased his AV that traveled using a combination of radio control and magnets embedded in the pavement. The investment in the AVs increased over the years, so did the efforts to automate the operations of the AVs (S Beiker, 2014).

Companies were not the only entities to invest and develop AVs, but the academia too contributed. In 1961, the researchers at Stanford University developed an automated cart to efficiently navigate on the moon. Few years later, in 1977, the engineers at the University of Tsukuba, Japan, invented an AV passenger vehicle with the potential to drive autonomously up to 20 miles per hour. With the Eureka PROMETHEUS project, there was an outbreak in the research and development of the AVs. The project was mostly funded by the European Union (EU) which brought collaboration between the universities and the AV manufacturing companies. The project in 1994 resulted in a thousand kilometer AV driven on the Parisian highways, changing the lanes automatically, while tracking the other passing vehicles on the road. Alongside, the Defense Advanced Research Projects Agency (DARPA) of the U.S. invested in the Autonomous Land Vehicle (ALV) project, which employed an early form of lidar to navigate off-road.

1.1.2 Definition of Autonomous AI Vehicle

The rapid development of AI has resulted in incredible inventions and advancements, including autonomous AVs. Before discussing the tortious principles, it is pertinent to lay down the definition of the Autonomous AI. The vehicles that have embedded technology to sense the environment, connect to the internet, navigate on the roads using GPS systems, all the while obeying traffic rules and regulations, ensuring not only the passenger safety but also the pedestrian safety, are regarded as autonomous vehicles. These cutting-edge technologies have transformed the automation in the AVs.

The level 0 is with no automation and fully operated manually by the humans. In level 1, there is driving assistance by AI in steering and acceleration of the vehicle, but it is controlled largely by the human. In level 2, there is an automated system to control steering and acceleration resulting in reduced human intervention (Hopkins & Schwanen, 2021). In level 3, the human intervention is limited to the instances requiring such intervention and remains automated at all the other times. Level 4 is where the vehicle is fully automated in desired or defined environments. Level 5 is the highest level of automation where the vehicle is fully autonomous in all environments (Novakazi et al., 2021).

Level 0, where there is no automation or level 1 where there is driver assistance or level 2 where there is partial automation, the human intervention is constant and continuous to monitor and control the vehicle. As there is existence of human intervention to a large extent, the human responsible for the decision making can be made liable for the tortious actions. However, in level 3 of conditional automation, in the level 4 of high automation or level 5 of full automation, there arises the question of attributability. The scope and limitation of this study are restricted to the automated vehicles in levels 3, 4 and 5. Level 4 and level 5 will be collectively termed as *self-driving vehicles*. In level 4 and level 5, there is automation and no human intervention, however, the difference is concerning the degree of automation. Despite this distinction in the level of automation, the fundamental principles of tort law can be uniformly applied to both levels of automation. A self-driving vehicle is a fully autonomous machine capable of operating without human intervention, requiring no human rider to maintain control. It can perform all the tasks of a competent human driver and navigate to any destination.

1.1.3 Self-driving vehicles (Level 3, Level 4 and Level 5)

While prototypes, such as Google's robotic vehicle, have logged thousands of hours on U.S. highways, they are not yet commercially available. These vehicles utilize various technologies, including GPS and sensors, to facilitate navigation and prevent accidents. Level 3 vehicles with conditional automation have already hit the roads and are in huge demand. Audi cars with Traffic Jam Pilot, a system that enables the vehicle to navigate by autonomously accelerating, steering, and braking through the thick traffic of urban areas. Mercedes Benz's Drive Pilot lets the autonomous system take over driving on highways, handling speed and lane changes. Tesla's Full Self-driving, though still in the testing phase, enables autonomous driving under certain conditions. Since level 3 is conditional automation, it is required to always have a driver ready to resume control as and when necessary. Level 4 vehicles are hitting the roads, but operate in the designated areas. Waymo's One operates fully driverless but can drive in the areas of Phoenix and Arizona. A subsidiary of General Motors, Cruise has developed Cruise Origin, which operates driverless in the designated areas and is currently being tested in San Francisco. Aptiv, in collaboration with Lyft, has developed Baidu Apollo that offers ride-hailing services, operational in Las Vegas. Assuring the safety of the passenger, this driverless car can drive autonomously in certain geofenced areas. Nuro, another autonomous vehicle that offers to transport goods between places without the presence of the driver in specific areas (Christensen, 2021). Hence, level 4 vehicles run driverless at all times but can be driven only in specific designations. Level 5 vehicles, though, are still in the development stage, and in no time could hit the roads. A Joint Venture 'Motional' between Hyundai and Aptiv is working towards developing the Ioniq 5 to be fully autonomous in all circumstances, earmarking ride-hailing services (Kim, 2020). Zoox, which is currently acquired by Amazon, is in the developmental stage, targeting autonomous ride-sharing (Martinho et al., 2021).

1.2 Existing Literature and Identifying Gaps

The development of autonomous vehicles (AVs), especially those controlled by artificial intelligence (AI), poses hitherto unparalleled challenges for conventional tort law frameworks, which are largely based on human agency and predictability. The literature now in publication recognizes the growing necessity of reevaluating tort principles—such as strict liability, vicarious culpability, negligence, and product liability—in light of AVs (Gurney, 2013; Geistfeld, 2017). The theoretical expansion of manufacturer responsibility in situations involving flawed designs or failing algorithms and sensors has been examined in foundational publications. In the past, courts have evaluated design flaws using criteria including risk-utility and consumer expectation (Masterman & Viscusi, 2020), and research has expanded these to investigate if AVs meet users' typical safety expectations. But as legal theorists like Epstein (2021) and Lemann (2019) have noted, the legal system's dependence on concrete, fault-based attribution is becoming less and less reliable when it comes to autonomous decision-making systems. Although some academics have advocated for algorithmic accountability or recasting embedded software as a "product," there is still doctrinal difficulty when it comes to applying liability concepts to software malfunctions or ethical judgments made by AI that diverge from human instinct. Furthermore, the courts have not yet produced a cohesive theory to address tortious injury caused by AI, and jurisprudential consistency among jurisdictions is still fragmented.

There remain considerable doctrinal and normative gaps in current legal studies despite significant scholarly involvement. First, although an extensive amount of research concentrates on the doctrines of product and strict liability, which focus on the manufacturer's liability, there is a lack of analytical clarity regarding how to allocate liability in composite or contributory negligence scenarios that involve multiple actors, including software developers, data suppliers, and vehicle owners. This is especially true in levels 4 and 5 AVs, where human intervention is completely absent. Second, the legal standing of autonomous AI systems and the viability of granting artificial personhood—a concept that might fundamentally change frameworks for assigning blame—are not sufficiently covered in the literature. Third, when probabilistic machine learning judgments assume the place of foreseeability and intent, it becomes difficult to apply traditional negligence criteria, which raises concerns about what makes a "reasonable algorithm." There is a pressing need for scholarly and legislative innovation to redefine tortious liability in this rapidly evolving field because there fails to be a cohesive liability model that is suited to the particular techno-legal hazards posed by AVs.

1.3 Research Methodology

This study examines the changing paradigms of tortious liability in the context of autonomous vehicles (AVs), specifically those operating at levels 4 and 5 of automation, using a qualitative doctrinal research technique based on an analysis of primary and secondary legal sources. Statutes, court rulings, and tort law theories—such as strict liability, vicarious liability, negligence, and product liability—as they apply to AV-induced harm are all critically interpreted in this study. The paper investigates how traditional liability frameworks might be modified to account for the particular dangers provided by AI-operated cars by applying methodical legal reasoning, particularly in situations when human interaction is either nonexistent or very limited. In order to comprehend the similarities and differences in legal responses, the research also uses comparative analysis, referencing jurisprudence and scholarly discourse from countries including the US, the EU, and India. The study also examines pertinent technical literature and ethical issues in light of the lack of a standardized legislative framework designed for autonomous systems, providing an interdisciplinary perspective that enhances doctrinal understandings. The approach makes it possible to make normative claims about legislative reform that are informed by the concepts of consumer protection, risk distribution, and fairness.

1.4 Research Objectives

The primary objective of this research is to evaluate objectively how appropriate the existing tortious responsibility theories are to dealing with accidents and injuries brought on by autonomous cars. Given varying degrees of automation, the study aims to ascertain how accountability should be apportioned among automobile manufacturers, software developers, car owners, and other stakeholders. The paper assesses the doctrinal relevance of strict liability, vicarious liability, product liability, and negligence to AV-related incidents. Finding conceptual and legislative gaps in the current legal frameworks—with an emphasis on the difficulties in assigning blame to non-human actors—is an additional significant objective. The study also intends to provide a logical, flexible legal framework that guarantees accountability without impeding innovation, as well as investigate the viability of applying legal tortious theories or comparable liability standards to AI systems. Lastly, the study aims to support the creation of a flexible, forward-thinking tort system that can strike a balance between the demands of public safety and legal certainty and technological advancement.

The liability questions around the self-driving vehicles are multi-layered and multi-faceted. Whether the fault can be attributed to the manufacturer applying product liability or whether the user can be made liable for the accidents and mishaps that are caused by the self-driving vehicles by applying strict or absolute liability calls for research. The lack of literature in the field of fault attribution underscores the need and significance of this research. This research additionally explores the extent to which the tort principles like negligence, product liability, vicarious liability, strict liability can be applied in the realm of self-driving cars.

2. FAULT ATTRIBUTION IN AUTOMATED VEHICLES

This section of the paper defines fault in the context of AVs and provides scenarios where both humans and automated systems contribute to accidents. By applying tort principles, it seeks to identify the parties accountable for such incidents and assess the extent of their liability. The analysis aims to clarify the attribution of fault among manufacturers, software developers, and users, providing a structured approach to determining liability in AV-related accidents. One single approach to determine liability is not viable as there are different stakeholders involved in the self-driving cars. Based on the circumstances, the attribution of liability has to change from one stakeholder to the other and so does the type of liability.

2.1 Fault Attribution

In case of any accidents, there are six primary stakeholders on whom the blame can be assigned. Driver, pedestrian, self-driving car owner, government, manufacturer and programmer can be blamed based on the actual mishap or the accident depending on the circumstances of each case. Only Level 3 automation cars would have a driver, but both level 4 and level 5 cars have no driver and run on full automation. Hence in level 4 and level 5, eliminates the attribution of liability to the drive. The pedestrian can be made liable for their negligent behaviour if any, regardless of the automation level. However, the complexity arises in the cases of composite negligence and contributory negligence, where multiple parties share liability.

The self-driving car owner can be made liable in level 3 automation vehicles by applying the principle of vicarious liability. Conversely, for level 4 and level 5 fully autonomous cars, the car owner cannot be made liable unless he has acted negligently. The liability can be attributed to the government only when there have been procedural errors or lapses in granting permission and approvals for the self-driving cars to be deployed on roads (Pöllänen et al., 2020). The intricacy of legal issues arise when fault is to be attributable to the manufacturers and the programmers. Unless there is a patent defect in the self-driving car's software or hardware, they cannot be made liable. They cannot be made liable for external factors; software malfunctioning can be attributed to the programmers and the hardware glitches can be attributed to the manufacturer (Fang et al., 2023). If there exists a *contract of service* agreement between the programmer and the manufacturer, the master can be made liable for the acts of the servants.

Fault attribution has multiple layers, necessitating a thorough examination of each layer to determine liability. Factors like negligence, including composite negligence and contributory negligence influences fault attribution. Liability principles like product liability, vicarious liability, strict and absolute liability serve as a foundation in attribution of fault. The case laws will be analysed to determine the applicability of product liability in the instances of malfunctioning of the self-driving cars, vicarious liability when there is a presence of contract of service, as well as the situations under which strict liability and absolute liability may be invoked.

2.2 Negligence

Negligence, often potentially influences the fault attribution in the torts. Failure to eliminate the foreseeable risks would make either the owner or the AV manufacturer viable depending on the facts of the case. It is highly likely that the AVs could possibly be negligent, then it is pertinent to understand the legal framework to attribute the fault. In the words of Justice Traynor though there is no proof of negligence on the part of the manufacturer, he can still be made liable if the public policy requires that the manufacturer be liable because he has the experience and expertise to effectively reduce the harm that are inherent in the defective product (Huberman, 2021). Thus, the standard of care of the manufacturer is higher and violation of such standard of care amounts to negligence.

To successfully plead negligence, there are four elements: (1) duty, (2) breach, (3) causation, and (4) harm. In innumerable cases, the courts have stated that negligence is the omission to do something which a reasonable man would do, or doing something that a reasonable person would not do, naturally guided upon the ordinary course of human affairs (Owen, 2008). A cause of action in negligence necessitates the proof that the defendant has failed the ordinary standard of care in either acting or omitting to do an act, which directly resulted in the injury to the persons involved. The ordinary standard care applies to normal and regular courses of business. However, when the law creates a special relationship between persons, it gives rise to a special duty of care. It amounts to negligence if such a special duty of care is breached. For instance, a manufacturer has a special duty of care towards his customer to the extent that the products and services are free from defects. However, the manufacturer can plead volenti non fit injuria, in the cases where risk is known to the AV owner to escape from the tortious liability (Santoni de Sio, 2017). When the danger is obvious and within the knowledge of the AV owner, no liability arises for reasonable hazards. Potential risk of meeting with an accident, irrespective of whether it is an autonomous or human-driven vehicle, is obvious and known to the owner. By driving such an AV, the owner has given his consent to reasonable and foreseeable dangers. Hence for such foreseeable danger, the manufacturer cannot be held liable.

The liability is attributed based on the facts and circumstances of each case. If negligence is proved on both the manufacturer and the owner, they are liable to the extent of negligence proved against them. Contributory negligence arises when there is a patent manufacturing defect on the manufacturer and failure of maintenance on the AV owner. In such a scenario, both of them are liable to the extent of the negligence committed on their part (Gurney, 2013). When an issue of composite negligence is to be determined, both the manufacturer and the AV owner may be made liable by the injured third party, depending on the facts of the case. If a third party is injured due to the malfunctioning of the hardware and also due to the failure of maintenance by the AV owner, both of them can be held liable jointly and severally (Colonna, 2012). In other words, the third party can claim damages either from the manufacturer or from the owner or from both.

The rule of last opportunity implies that the person who had the last chance to stop the accident from occurring may be made liable. The plaintiff can make the defendant liable in the cases of negligence if the defendant had the last opportunity to stop the accident from happening. The last opportunity rule indicates that when an accident occurs, whoever had the last chance to prevent it shall be made liable (Epstein, 2021). In Level 3 automation, if the accident is caused by human interference i.e. driver and he had the last opportunity to prevent it from happening, he can be made liable. In Level 4 and 5 vehicles where there is full automation, the manufacturers or the software developers are liable under this rule if the accident is caused by the malfunction of the hardware or the software.

2.3 Product Liability

In the human controlled vehicles, generally the drivers are made liable for the accidents committed by them. In the AVs absence of the drivers and the shift of the decision-making authority, manufacturers are to be held accountable for the accidents committed by the AVs (Webb, 2016). The level of automation determines the liability of the manufacturer. Higher the automation level, higher the chances of liability of the manufacturer and vice versa. In level 3, conditional automation, the driver present will be made liable for the negligent behaviour and the manufacturer is made liable only to the extent of defective product. However, in the level 4 and 5, fully automated vehicles with no driver, the manufacturer can be held liable for almost all the causes of action, except when the owner has intervened with the hardware and software of the AV. The current laws are well equipped to deal with level 3 vehicles, but product liability in level 4 and 5 vehicles require additional research, which is discussed in depth in the next chapter.

Applying product liability to autonomous vehicles is a complex issue, as it has to consider both hardware and software defects. The negligence or breach of duty of reasonable care can be proved against the manufacturer only if the direct link can be established between the injury and the absence of reasonable standard of care. Not exercising the due and reasonable care should be proved as the prime reason for the cause of injury. However, negligence theory may be of lesser importance as the manufacturer likely exercises reasonable care. In contrast, the manufacturers can still be made liable under strict liability.

2.4 Strict liability

Strict liability is attracted when the manufacturer is made liable, despite the exercise of the duty of reasonable care. Under three circumstances, the strict liability may be invoked (1) Malfunction of the Vehicle (2) Design Defect (3) Warning Defect (Lemann, 2019). To prove malfunction of the vehicle, the plaintiff has to show, firstly, the product malfunctioned, secondly, the malfunction occurred out of proper use, and thirdly, the product has not been altered or modified or misused in the due course of usage that resulted in malfunction. To prove design defect, the plaintiff has to prove that the product is designed in an unreasonable dangerous way. To prove a warning defect, the plaintiff has to show that the injury could have been prevented with adequate warnings and disclaimers of dangers. Additionally, strict liability can be attributed to the manufacturer for failing to provide reasonable instructions to use the product.

2.5 Vicarious Liability

Vicarious liability is based on the doctrine 'Respondeat Superior' and means that the employer is liable for the acts of his employee as long as he is acting in the course of employment. The exception to this rule is that the employer is not liable for the acts of the independent contractors. Vicarious liability can be invoked in the cases of contract of service, where the employees are directly under the control of the employer. The application of vicarious liability in fault attribution is minimum as AVs cannot be considered as employees (Diamantis, 2023). Unless the AVs are considered as employees, the employers cannot be made liable applying vicarious liability. AVs are not yet juristic persons or artificial judicial persons to assume the position of the employees. Thus, vicarious liability has limited application. Assessing the viability of whether AVs can be considered as an artificial legal person or not is outside the scope of this research.

2.6 Observations

According to the analysis, it can be seen that fault attribution in AVs needs a multi-faceted and adaptable strategy because of the engagement of multiple stakeholders and different degrees of automation. When vehicles move from Level 3 to Level 5 automation, the conventional fault attribution to human drivers decreases and shifts more and more toward manufacturers, programmers, and system designers. Though negligence and product liability are key to fault attribution, strict liability is most relevant in the case of fully autonomous vehicles, imposing liability on producers even where there is no established negligence. Vicarious liability, though restricted in scope because of the non-juristic status of AVs, can still be found where there are special contracts. In the end, no one approach can fully meet the demands of liability in every AV-related accident. The apportionment of blame has to be context-specific, based on well-established tort principles and progressive interpretations of the law, while at the same time promoting accountability, innovation, and public safety in the use of autonomous technology.

3. APPLICATION OF PRINCIPLES OF TORTIOUS LIABILITY IN AVS

With autonomous vehicles steadily becoming a practical reality on road from experimental prototypes, one of the pertinent questions that need immediate attention and answer in both the legal and regulatory ambit, is the question of liability that is shared by both kinds of drivers involved in the operation of such AVs- humans who physically drive as well as technology providers/ developers responsible for the software or systems of the AVs. While technology has been advancing at an unprecedented rate especially in the fields of artificial intelligence, machine learning and vehicular automation amongst other things it is interesting to note that law hasn't coped up well with technology, leaving a grey area in the regulation and a very important question unanswered- 'Whom should the fault liability be attributed to'? This question, even when answered will be followed by other questions relating to the appropriate authority qualified to make the decision regarding who the liability will be fixed on, what the determinants of the said fault would be, how the said fault would be assessed and what mechanisms and methodologies could be employed to assess the process. These questions and uncertainty in law majority arise from the complexity that is involved in driving a driverless car, especially due to its technical and multifaceted nature which blurs traditional boundaries of control and intent that have helped determine liability in conventional driving scenarios. Another pertinent question that would need deliberation in this regard would be the meaning of negligence in this situation of driverless cars.

In the noticeable absence of a proper and comprehensive laws specifically tailored to deal with the operation of autonomous vehicles and designed to address issues of fault and liability, an immediate answer lies within the realm of common law that becomes inclusive with time and inherently adapts to and accommodates all emerging societal and technological developments- Tort Law. In such a context, if it can be conclusively determined that an accident due to an autonomous vehicle was mostly the result of a fault traceable to the manufacturer or the seller of the vehicle or its parts, the doctrine of product liability is highly applicable. Under this principle of law, the victimized party—or the party whose property has been damaged—may be entitled to recover damages against the responsible manufacturer or seller (Simons, 2015). Significantly, product liability under Tort Law imposes strict liability on manufacturers in the event of proven defects, irrespective of whether or not the manufacturer exercised reasonable care or was negligent in the process of manufacturing (Schwartz, 1991). That is to say, the existence of a defect alone—during manufacturing, design or labeling—can be enough to initiate liability. This doctrine of law thus provides a significant foundation upon which to assign causality to manufacturers in cases related to autonomous vehicles. For such vehicles, liability in tort can be sought in the cases of an accident due to a defect in hardware or software during manufacturing, a defect in design compromising safety, or an insufficiency in warning or giving adequate notices to human users or giving proper signs in cases necessitating human intervention (Posner, 1973). Product liability is thus an indispensable mechanism of law to assign causality to manufacturers with the purpose of holding them liable and to ensure that victimized parties are provided with remedies even in cases of sophisticated and developing technology.

Manufacturing defects are discrepancies or imperfections in the manufacturing process that result in deviations from the intended design or specifications of the product (Plant, 1955). These are generally unintentional and can even happen even if the initial design is sound. Any divergence from the normal design or parameters of quality—due to lapses in the quality control process or process errors in assembly or substandard materials usage—can contribute to such manufacturing defects. Significantly, these are not only physical errors; these can even include situations wherein the delivered product or service fails to match exactly what it was supposed to be or what was expected based on its blueprint design (Birnbaum, 1980) Even if a product conforms in terms of final shape and function during the initial stages, if it starts to malfunction and injures or damages someone even if used under normal conditions, such malfunction can be traced even to a manufacturing defect. The legal system typically works under the assumption that mass-produced goods are made with care, diligence, and attention to detail to guarantee consumers can use them safely and effectively. This presumption therefore has a lot of weight in court cases, and the injured party or consumer usually bears the burden of proving it—that is, proving that a defect existed despite presumed care. Direct proof of the flaw might not always be easily accessible in court cases involving manufacturing defects. As a result, consumers frequently base their arguments on circumstantial evidence (Cranford, 1971). In such cases, courts typically take into account three main factors: first, whether the product in question malfunctioned; second, whether the malfunction happened during normal and proper use of the product; and third, whether the end user had avoided misusing, modifying, or altering the product in any way that could have contributed to the defect. These components work together to create the basis for establishing liability for manufacturing defects, allowing harmed parties to pursue compensation even in the absence of concrete proof.

Considering autonomous vehicles rely heavily on intricate networks of electronic components and are primarily controlled by sophisticated software systems, their design and operation are fundamentally different from that of conventional vehicles. In contrast to traditional cars, which require a human driver to perform basic driving functions like steering, braking, shifting gears, and accelerating, autonomous vehicles entrust these tasks to intelligent software that is integrated with cutting-edge technological solutions. To navigate with a high degree of accuracy, these vehicles use a variety of sensors, such as cameras, radar systems, and global positioning systems (GPS). Autonomous vehicles can now perceive their environment, evaluate current conditions, and make decisions that would typically require human judgment thanks to this collection of technologies. When paired with deep learning algorithms and artificial intelligence, this level of operational independence signifies how contemporary autonomous cars have evolved into self-sufficient machines that can manage intricate and changing traffic situations.

However, there are new legal difficulties brought about by the liability question in relation to autonomous vehicles. The claimant usually has to demonstrate the existence of a defect in order to hold a manufacturer accountable under the doctrine of product liability. Another, more complex question, however, is raised by autonomous vehicles- Can the embedded software that controls the vehicle's functions be regarded as a "product" in the sense that the term is traditionally used in product liability law? According to the law, a product is typically defined as a tangible object; physical personal property that is produced, sold commercially, and meant for use or consumption. Contrarily, software is usually classified as an intangible asset or service, especially when it is provided electronically as opposed to on physical media. Because software may not be legally considered a manufactured product, it could be argued from the perspective of the manufacturer that a flaw in the software code does not constitute a manufacturing defect in the traditional sense. Consequently, they might argue that software flaws shouldn't be grounds for product liability lawsuits, particularly if the software is viewed as a service component rather than a physical good. The extension of conventional product liability doctrines to the field of autonomous vehicles is hampered conceptually by this, necessitating a re-examination and potential reformulation of liability standards by legal systems to conform to new developments.

The advent of autonomous vehicles (AVs) is poised to be a game-changer in transportation, offering improved safety, efficiency, and convenience. However, integrating advanced artificial intelligence (AI) and machine learning (ML) systems into these vehicles introduces unique challenges, particularly in the areas of product liability and design defects. Unlike manufacturing defects, which affect individual units due to production errors, design defects are inherent flaws in a vehicle's blueprint, potentially making all units unsafe. This analysis explores the concept of design defects in AVs, examining legal tests, relevant statutory frameworks, and the challenges arising from the intersection of technology and tort law, with insights from both global perspectives and the Indian legal system.

3.1 Defining Design Defects in Autonomous Vehicles

A design defect arises when a product's inherent design is flawed, making it unreasonably dangerous for its intended use, even if manufactured correctly. In the context of AVs, such defects can manifest in various ways:

1. AVs rely on AI to make driving decisions. If an AV's algorithm prioritizes certain outcomes that increase harm—such as choosing to crash into a pedestrian rather than another vehicle—courts may consider this a design flaw.
2. AVs depend on sensors like LIDAR, radar, and cameras for navigation. Design flaws in how these sensors interpret data can lead to accidents, as seen in cases where AVs have failed to detect pedestrians, cyclists, or road barriers.
3. Some AVs require human intervention in specific scenarios. Poor design in alert systems or takeover mechanisms can delay or prevent a human driver from reacting in time, leading to accidents.
4. Unlike human drivers who can adapt to unpredictable scenarios, AVs must be programmed for every possibility. Failure to incorporate redundancy systems in case of software glitches or hacking could be considered a design flaw.
5. AVs are programmed with ethical decision-making models to minimize harm in unavoidable crashes. If a vehicle's ethical algorithm is flawed or biased, it may cause preventable harm, raising liability concerns.

3.2 Legal Tests for Determining Design Defects

Courts apply distinct tests to evaluate whether an AV's design constitutes a defect:

1. Consumer Expectations Test- This test examines whether a product meets the ordinary expectations of consumers regarding safety (Masterman & Viscusi, 2020). If an AV behaves unpredictably or fails to perform as expected, such as braking unexpectedly or swerving dangerously as it may be considered defective under this test.
2. Risk-Utility Test- This test balances the risks of a product against its benefits. If the risk of an AV design outweighs its safety benefits and a reasonable alternative design exists, the manufacturer may be held liable (Larsen, 1984). For instance, if an AV's software is programmed to prioritize avoiding minor accidents but ends up causing more severe injuries, this design choice could be deemed defective.

A related concept is the Reasonable Alternative Design (RAD) test, which obligates manufacturers to demonstrate that no safer, cost-effective alternative design was available at the time of production. Failure to adopt an improved design, when feasible alternatives exist, strengthens the argument for a design defect.

Traditionally, in product liability cases, manufacturers are held strictly liable for defects that make their products unreasonably dangerous. Under strict liability, plaintiffs do not need to prove negligence; only that the defect existed and caused harm. In the case of AVs, courts must determine:

- Whether a crash is due to an inherent defect in the vehicle's design rather than an isolated malfunction.
- If the overall safety of AVs (which may have a lower accident rate than human drivers) justifies some level of unavoidable risk.

Some legal scholars argue that AV manufacturers should be held strictly liable for accidents since their products remove human error from the equation, shifting the responsibility to the company's engineering and software design choices.

Generally, manufacturers are not required to provide evidence of negligence in order to be held strictly liable for flaws that render products unreasonable. All that is required of plaintiffs is proof of a defect's existence, its effect on safety, and the harm that results. This strategy assigns accountability to those who are most suited to stop flaws. There are complications when strict liability is applied to autonomous vehicles (AVs). Courts must make a distinction between isolated malfunctions and intrinsic design flaws, such as flawed algorithms or malfunctioning sensors. Unlike traditional vehicles, AVs rely on integrated hardware, software, and artificial intelligence, which makes fault identification more challenging.

Additionally, advocates claim that by removing human error, AVs lower overall accident rates, which begs the question of whether some inevitable risk is acceptable for the sake of public safety. Others argue that since AVs take away driver control, manufacturers should bear full responsibility and be held to a higher standard for their engineering and design choices.

Strict liability provides robust consumer protection, but its application to AVs necessitates legal modification to strike a balance between accountability, innovation, and public confidence in automated technologies.

3.3 Challenges in Proving and Defending Design Defects

When it comes to autonomous vehicles (AVs), proving the existence of a design flaw poses a variety of complex and particular difficulties that are very different from those that arise in traditional product liability cases. First and foremost, it is frequently challenging to identify whether an AV accident was caused by a basic design defect in the vehicle or by the unpredictability and unavoidability of the road conditions (Lemann, 2019). Complex technical analysis is needed for these evaluations, which include professional analysis of the sensor data, software algorithms, and vehicle decision-making procedures.

Second, manufacturers may contend in their defence that while autonomous vehicles are not completely impervious to malfunctions, their overall safety record and capacity to lower accident rates—particularly in comparison to vehicles operated by humans—should be given significant consideration when assessing liability. A major part of the manufacturer's defence of the design decisions is frequently this comparative safety argument. Thirdly, the legal framework surrounding AV design liability is still developing, with various jurisdictions using different standards to determine whether a given design is flawed. The legal system may become unclear and inconsistent due to a lack of unified laws and precedents.

Currently, a patchwork of legislative statutes and established judicial doctrines govern product liability claims pertaining to autonomous vehicles in the Indian legal system. The Consumer Protection Act of 2019 provides a strong legal foundation that enables customers to pursue compensation when they are harmed or lose something because of faulty products, including high-tech items like AVs. This Act gives manufacturers and service providers a legal obligation to guarantee the safety and dependability of their products by requiring them to follow strict safety regulations. Furthermore, the 2019 amendment to the Motor Vehicles Act, 1988, plays a crucial role in resolving liability and compensation concerns associated with traffic accidents involving AVs. It is extremely relevant for managing the fallout from AV-related accidents because it contains updated provisions pertaining to insurance coverage, accident compensation mechanisms, and general regulatory compliance.

Since AVs use integrated digital platforms and software systems, the Information Technology Act of 2000 also becomes important, particularly when it comes to software bugs, cybersecurity breaches, or data management errors that could cause mishaps. This law governs legal liability in the areas of digital communication and electronic governance, both of which are essential to AV operation. Furthermore, the Indian Contract Act of 1872 may be used to recover damages for non-fulfillment of contractual obligations in cases where customers or businesses claim that warranties pertaining to performance or safety standards have been broken. When taken as a whole, these laws establish a legal environment in which AV liability concerns in India are progressively emerging.

Particularly, design flaws occur when there are basic mistakes or omissions in the planning or architecture of an AV's systems or parts, and the damage that results can be directly linked to those poor design decisions (Kim, 2017). The proof that a safer, workable, and financially viable alternative design was available at the time of manufacture and that the manufacturer chose not to use it is a crucial component in proving design defect liability (William E Westerbeke, 1998). The two main legal tests that courts usually use to assess design flaws are the consumer expectations test, which looks at whether the product operated as safely as a typical consumer would have expected, and the risk-utility test, which assesses whether the risks were reasonable or excessive by weighing the product's risks against its utility and potential benefits.

In the AV industry, warning and signaling flaws are another important aspect of product liability. These occur when a manufacturer does not sufficiently warn consumers about latent dangers that may not be immediately apparent or fails to adequately inform them about how to use the product safely. The complexity and limited autonomy of these vehicles make it even more important for manufacturers to give detailed, understandable instructions about the AV's operational limitations and situations that might call for human intervention.

Finally, it's critical to distinguish between design defects, which are inherent in the product's blueprint, and manufacturing defects, which are flaws that arise during the production process. When deciding on the type of liability and the best course of action, this distinction is crucial. Product liability law provides a useful legal framework to handle the complex risks posed by technological advancements like driverless cars in light of these difficulties because of its tried-and-true principles and flexibility. To ensure that consumer safety is not jeopardized in the name of innovation, the law must change in tandem with the ongoing advancements in technology.

3.4 Observations

The development of autonomous vehicles (AVs) raises challenging issues to the conventional structure of tortious liability. As cars move from driver-operated to machine-operated, fault attribution becomes simpler and yet complicated. The present legal void regarding AVs highlights the imperative of a sophisticated legal response, specifically where liability stems from software failure or algorithmic choice. While tort law, particularly under the doctrine of product liability, provides a general platform to deal with manufacturing and design flaws, its own application becomes tenuous when dealing with intangible factors like software. The classification of software as a "product" is controversial, which may restrict the remedy that can be sought by affected parties. Further, legal standards such as the consumer expectation and risk-utility standards might not accurately grasp the nuance of AI-based decision-making. The dynamic interaction of safety, ethical programming, and machine autonomy brings a new layer to liability analysis. These reflections point to the urgent need for legal reform that accommodates advancing technology as well as consumer protection. Finding an appropriate balance between encouraging innovation and holding accountable will be critical to informing future legal structures for AVs. Tort law needs to adapt to stay current in an age where the "driver" is not human but a smart machine making sophisticated decisions.

4. FUTURE RESEARCH DIRECTIONS

As autonomous vehicles (AVs) become part of contemporary transportation networks, tort law needs to adapt to deal with changing liabilities. A balanced regime is necessary in order to attribute fault between manufacturers, developers, and users. Legislative clarity and technological transparency are the cornerstone to ensuring responsibility. Dynamic models of liability, AI decision standards, and harmonization across jurisdictions, ethics of AI in decision making are areas where future research can focus. Notably, the concept of bestowing artificial legal personhood on AI deserves closer analysis, as it poses deep questions of law and ethics. An adaptive tort system needs to balance innovation with justice in a growingly autonomous mobility environment.

5. CONCLUSION AND RECOMMENDATIONS

The development of AVs upends the essential doctrines of tort law, such as fault attribution and liability. As AVs transition from level-by-level automation, from human interference to fully automation, the classic paradigm of assigning human drivers blame for accidents is being made more anachronistic. The study confirms that fault attribution in AV-related accidents is a multi-faceted phenomenon, driven by different levels of automation and intervention of different stakeholders—such as manufacturers, software companies, vehicle owners, insurers, and even the government.

One of the key conclusions of this research is that with greater levels of automation—especially at Levels 4 and 5—the range of driver control decreases dramatically, calling for a shift in liability to those who are in charge of the design, programming, and maintenance of the AV systems. Although negligence and product liability principles remain relevant, strict liability becomes a more appropriate mechanism in the case of fully autonomous vehicles. It guarantees that accident victims involving AVs are properly compensated regardless of whether or not there was established negligence, hence upholding public trust in the safety of AV technology. Vicarious liability, while restricted on account of AVs' non-personhood status, can still be invoked in contexts of special contractual relationship. Apportionment of liability will need to be determined based on facts and circumstances of each case, reflecting consideration of each stakeholder's contribution within causation.

The imposition of traditional tortious liability doctrines onto autonomous vehicles (AVs) exposes large loopholes, particularly when viewed through the prisms of product liability doctrines like consumer expectations and design defects. The consumer expectation test is no longer sufficient in the case of AVs, where the typical user does not have the technological expertise to anticipate or assess the behavior of the vehicle during system malfunctions or algorithmic decision-making. Likewise, the design defect test poses foreseeability and reasonableness questions in situations regulated by sophisticated machine learning models, whose behavior can develop unpredictably with time.

In addition, software malfunctions and behavior of AI complicate the concept of negligence, as responsibility cannot be simply attributed to a driver who is not in control. Strict liability on product defect claims seems more appropriate; however, apportionment of fault among manufacturers, software companies, and third-party vendors complicates the issue. The lack of standard standards for AV safety and performance also makes uniform application of liability principles difficult.

With the dynamic nature of AV technology, a fixed or one-size-fits-all legal model is not likely to be effective. Thus, legal systems need to look into adaptive legal frameworks that develop in tandem with technological advancement. Additionally, more investigation needs to be done on whether AI, especially the autonomous systems controlling AVs, should be accorded artificial legal personhood. The acknowledgment of AI as a legal person would permit direct liability and responsibility but is fraught with important jurisprudential and ethical issues that need to be widely discussed.

5.1 RECOMMENDATIONS

1. Courts should consider the factors like automation levels, human override capability and software reliability in fault attribution in the cases of AV accidents. As there is no one size fits all solution, courts should duly consider facts and circumstances of each case before attributing the liability.
2. For Levels 4-5 AVs with complete automation, more emphasis to be laid on the strict liability of manufacturers rather than driver negligence.
3. Impose strict liability in Level 4-5 AV accidents to compensate victims without establishing negligence.
4. The law should be reinterpreted or supplemented by the creation of specialized regimes of liability tailored to autonomous vehicles (AVs), drawing on changing doctrines regarding consumer expectations, design defects, and the function of software algorithms.
5. The current scope of, particularly product liability, needs to be extended to include software and decision-making algorithms under the ambit of the word "product." This would shut the definitional loophole that presently enables manufacturers to avoid responsibility for software-caused malfunctions and ensure that liability models remain current and adaptive to the technological makeup of AVs.

REFERENCES

Anderson, J. M., Nidhi, K., Stanley, K. D., Sorensen, P., Samaras, C., & Oluwatola, O. A. (2014). *Autonomous vehicle technology: A guide for policymakers*. Rand Corporation.

Birnbaum, S. L. (1980). Unmasking the Test for Design Defect: From Negligence [to Warranty] to Strict Liability to Negligence. *Vand. L. Rev.*, *33*, 593.

Birnbaum, S. L. (1980). Unmasking the Test for Design Defect: From Negligence [to Warranty] to Strict Liability to Negligence. *Vand. L. Rev.*, *33*, 593.

Christensen, H. S. (2021). A conjoint experiment of how design features affect evaluations of participatory platforms. *Government Information Quarterly*, *38*(1), 101538.

Colonna, K. (2012). Autonomous cars and tort liability. *Case W. Res. JL Tech. & Internet*, *4*, 81.

Cranford, C. H. (1971). Torts—Product Liability—Circumstantial Evidence and Proof of Defect. *North Carolina Law Review*, *50*, 417.

Diamantis, M. E. (2023). Vicarious Liability for AI. *Industrial Law Journal*, *99*, 317.

Epstein, R. A. (2021). Liability rules for autonomous vehicles. *Journal of Entrepreneurship and Public Policy*, *10*(2), 218–234.

Fang, Y., Teixeira, R., Min, H., Lei, X., Chen, S., & Zhao, X. (2023, August). A Safety Enhancing Framework for Autonomous Vehicles with a Resilience Rationale. In 2023 7th International Conference on Transportation Information and Safety (ICTIS) (pp. 1699-1705). IEEE.

Geistfeld, M. A. (2017). A roadmap for autonomous vehicles: State tort liability, automobile insurance, and federal safety regulation. *California Law Review*, *105*, 1611.

Gurney, J. K. (2013). Sue my car not me: Products liability and accidents involving autonomous vehicles. U. Ill. JL Tech. & Pol'y, 247.

Hopkins, D., & Schwanen, T. (2021). Talking about automated vehicles: What do levels of automation do? *Technology in Society*, *64*, 101488.

Huberman, P. (2021). Tort law, corrective justice and the problem of autonomous-machine-caused harm. *Canadian Journal of Law and Jurisprudence*, *34*(1), 105–147.

Kim, P. Y. (2020). Where We're Going, We Don't Need Drivers: Autonomous Vehicles and AI-Chaperone Liability. *Cath. UL Rev.*, *69*, 341.

Kim, S. (2017). Crashed software: Assessing product liability for software defects in automated vehicles. *Duke L. & Tech. Rev.*, *16*, 300.

Larsen, K. D. (1984). Strict products liability and the risk-utility test for design defect: An economic analysis. *Columbia Law Review*, *84*(8), 2045–2067.

Lemann, A. B. (2019). Autonomous vehicles, technological progress, and the scope problem in products liability. Journal of tort law, 12(2), 157-212.

Martinho, A., Herber, N., Kroesen, M., & Chorus, C. (2021). Ethical issues in focus by the autonomous vehicles industry. *Transport Reviews*, *41*(5), 556–577.

Masterman, C. J., & Viscusi, W. K. (2020). The specific consumer expectations test for product defects. *Industrial Law Journal*, *95*, 183.

Noussia, K. (2020). Autonomous vehicles: legal considerations and dilemmas. InsurTech: a legal and regulatory view, 253-270.

Novakazi, F., Johansson, M., Erhardsson, G., & Lidander, L. (2021). Who's in charge? The influence of perceived control on responsibility and mode awareness in driving automation. *It-Information Technology*, *63*(2), 77–85.

Oliphant, K. (2019). Liability for road accidents caused by driverless cars. Sing. Comp. L. Rev., 190.

Owen, D. G. (2008). Design defects. *Missouri Law Review*, *73*, 291.

Plant, M. L. (1955). Strict Liability of Manufacturers for Injureis Caused by Defects in Products-An Opposing View. *Tennessee Law Review*, *24*, 938.

Pöllänen, E., Read, G. J., Lane, B. R., Thompson, J., & Salmon, P. M. (2020). Who is to blame for crashes involving autonomous vehicles? Exploring blame attribution across the road transport system. *Ergonomics*, *63*(5), 525–537.

Posner, R. A. (1973). An economic approach to legal procedure and judicial administration. *The Journal of Legal Studies*, *2*(2), 399–458.

Santoni de Sio, F. (2017). Killing by autonomous vehicles and the legal doctrine of necessity. *Ethical Theory and Moral Practice*, *20*(2), 411–429.

Schwartz, A. (1991). The case against strict liability. *Fordham Law Review*, *60*, 819.

Simons, K. W. (2015). Punishment and blame for culpable indifference. *Inquiry : a Journal of Medical Care Organization, Provision and Financing*, *58*(2), 143–167.

Webb, K. C. (2016). Products Liability and Autonomous Vehicles: Who's Driving Whom? *Rich. JL & Tech.*, *23*, 1.

Chapter 12
Human Interaction Influences the Adoption of Robotics and Autonomous Agentic AI Systems

Jennie Lee Khun
http://orcid.org/0000-0001-6076-1782
National University, USA

ABSTRACT

This chapter examines how agentic AI and robotics reshape autonomous vehicles and human–AI interaction, integrating technical, ethical, regulatory, and human centered perspectives to accelerate responsible adoption. It contrasts reactive, deliberative, and hybrid architectures, outlines perception, sensor fusion, and control challenges, and highlights bias, explainability, privacy, and emotional intelligence limits that affect trust. The chapter argues that human factors, transparent interfaces, and progressive disclosure are essential to acceptance, and that pilots, open data, standardized KPIs, and governance frameworks (audit trails, accountability, cross sector boards) enable scaling. It assesses mobility, traffic, safety, environmental impacts, and socioeconomic shifts, and proposes incentive, workforce transition, and evaluation strategies to ensure equitable, resilient deployment of AVs and agentic systems.

INTRODUCTION

Agentic Artificial Intelligence (Agentic AI) is a system that operates autonomously to anticipate outcomes while functioning in a dynamic changing environment. "Agentic AI represents a transformative leap in AI, evolving beyond reactive systems to autonomous, goal-oriented agents capable of learning, adapting, and making independent decisions" (Murugesan & Murugesan, 2025). Agentic AI differs from the traditional AI system in which agentic AI makes decisions based on continuous interaction while adapting to dynamic changes. Whereas traditional AI systems work as non-adaptive large language predictive models taking training data as input and generating desired outputs. Agentic AI can play a significant role in Autonomous Vehicles (AV) and unmanned vehicles. The agents have the capability to continuously interact with the surrounding environment and people so that the safety, ethical, reliability

DOI: 10.4018/979-8-3373-7355-3.ch012

and law aspects are taken into consideration when making autonomous decisions. Sreedharan (2023) explains that recent progress in AI has enabled us to reach milestones that even the most optimistic futurists believed were beyond our reach for the time being.

While generative AI is reshaping content creation and problem-solving, agentic AI is poised to redefine how machines think and act independently: enhancing efficiency, productivity, and human–machine collaboration; its potential applications span a wide range of areas" (Murugesan & Murugesan, 2025). One of the most significant advancements is the use of AI for safety applications. "The growing volume of various types of motor vehicles on the road necessitates continuous advancements in automotive safety technologies (Ekatpure, 2023). This chapter examines the intersection of artificial intelligence, human interaction (HAX), robotics, and autonomous systems. The chapter aims to integrate theory, engineering practice, ethical and regulatory concerns, and evaluation methods to help align the concepts and tools needed to design, build, and assess intelligent systems that interact safely and effectively with people. Pelomi et al. (2024) explains that ensuring the trustworthiness of AI systems involves addressing various risks associated with human factors including human biases, both implicit and explicit, [that] can inadvertently influence AI algorithms, leading to biased outcomes in decision-making processes.

"It was once a myth that robots would coexist with humans, but now it is a fact because of growing technologies" (Mudhivarthi & Thakur, 2022). Robotics technology has made significant advancements in recent years, enabling robots to dynamically perceive and interact with the environments in which they operate (Abood, et al., 2025). Abood et al. (2025) further explains this progress has led to increased efficiency in industrial processes and improved safety. Agentic Artificial Intelligence (Agentic AI) is a system that functions autonomously to complete predicted outcomes within a complex and dynamic environment. Agentic AI differs from the traditional AI system in which agentic AI incorporates reasoning to make decisions based on continuous interaction and input with the environment while simultaneously adapting to dynamic changes. Traditional AI systems work as non-adaptive large language predictive models taking input from training data and then using that data to generate desired outputs. "Robots are divided into six groups based on their functions, including automated guided vehicles (AGVs), autonomous mobile robots (AMRs), humanoids, articulated robots, cobots, and hybrids" (Mudhivarthi & Thakur, 2022). Agentic AI plays a substantial role in Autonomous Vehicles (AV), unmanned vehicles, and robotic systems. The agents must have the capability to continuously interact with the surrounding environment and people so that safety, ethical, reliability and laws are taken into consideration when making autonomous decisions.

Agentic artificial intelligence has evolved robotics and autonomous systems applications that can benefit society in a variety of ways, but we must understand foundationally how they function to identify barriers of adoption. Robotics is primarily focused on designing, building, and operating machines that can perform tasks with consistent human oversight. Autonomous systems, however, enable those machines to operate mostly independently without requiring what we could consider to be continuous human intervention. AI enables these machines to perceive, reason, and act with limited intervention from people. This chapter explores concepts in robotic perception, control, and adaptive learning, gaining an understanding into how AI enhances object recognition, sensor fusion, and environmental interaction. The chapter also covers autonomous navigation, reinforcement learning for robotic intelligence, and ethical considerations in AI-powered automation. It is crucial to understand the cutting-edge research and practical applications of self-driving vehicles, industrial automation, and robotic assistants.

Initial deployment of highly autonomous vehicles and related mobility services (i.e., robotaxis) was prematurely targeted for such machines to be commonplace on highways in the early 2020s; to support these ambitions, automakers and tech companies have invested billions in autonomous driving but despite these investments and bold claims from industry leaders, several launches of fully self-operating vehicles have been continuously postponed or scaled back (Gordillo Chávez et al., 2024). AI in autonomous vehicle systems is creating a huge transformative leap in how we utilize transportation. AI object recognition, detection, and avoidance mechanisms aim to significantly reduce human errors while driving, optimize mobility for all, and improve traffic management to optimize roadway usage. Autonomous vehicles must continuously refine the decision-making processes to effectively contribute to more reliable transportation options. However, despite the many impactful advancements, challenges remain. The need for robust datasets and high-accuracy algorithms underscores the need for technological refinement. Ethical and legal considerations further complicate AI deployment, requiring oversight via regulatory frameworks to address governance and liability concerns. Autonomous systems have the capacity to revolutionize transportation as we know it, providing safer and efficient mobility solutions.

Gordillo Chávez et al., (2024) explain that the biggest roadblocks standing in the path of mass adoption [of self-driving cars] may be psychological and that social and cognitive factors should also be understood to analyze user acceptance of AVs. As AI systems become more complex, the dynamics of AI-human interaction (HAX) evolved from simple automation to collaborative decision making. The shift to collaboration necessitates a deeper understanding of how AI influences human behavior, ethics, and cognitive processes. AI is both a facilitator and a challenger, redefining traditional workflows and relationships. The ability of AI systems to interpret human intent, provide insights, and respond to inputs raises questions about trust, bias, and adoption concerns. AI-human interaction (HAX) is a multidisciplinary field that includes a combination of artificial intelligence, cognitive psychology, human-computer interaction (HCI), and ethics. "The field of robot and AI ethics is broadly divided into two main branches; the largest of these is concerned with the vitally important question of how human developers, manufacturers and operators should behave in order to minimize the ethical harms that can arise from robots and AIs in society, either because of poor design, inappropriate application, or misuse AI systems are quickly becoming people's decision-making partners, providing automation tools, and personalized assistants" (Winfield et al., 2019). It is important to understand how ethical AI interacts with people, and how people perceive and trust AI. Winfield et al. (2019) further explains that even if the technical problem of machines could resolve real-world ethical dilemmas society wide, debate is needed to agree on the rules and protocols for such machines, because society needs to take responsibility for the human causalities of accidents caused by such machines. Optimizing these relationships and establishing strong governance and ethical frameworks are critical to accelerating the adoption of autonomous vehicles that improve safety and preserve human life.

Human-AI Interaction (HAX) explores how people and AI systems communicate and collaborate. AI models like ChatGPT are being widely adopted, and these models handle tasks including human-like intelligence, natural language understanding, decision-making, and data analysis. The primary goal of HAX is to create user-friendly, transparent, and beneficial AI systems. HAX is vital to help bridge the gap between users and sophisticated AI tools. Ensuring bias mitigation, fairness, transparency, and adaptability foster ethical AI-human interaction, promoting trustworthy and accountable AI systems. AI should enhance human agency, support ethical decision-making, and align with societal values—ensuring it serves as a beneficial rather than disruptive force. The adoption of large language models (LLM) conversational applications reiterates the importance of effective HAX. LLM can generate human text, and is revolutionizing communication, from customer service to content creation. These models are being integrated into applications in our daily lives. Ensuring that these systems are intuitive and reliable becomes paramount, requiring established frameworks, ongoing analysis, and improvement to provide the most benefit.

This chapter first establishes the foundational distinctions between automation, autonomy, and agentic AI, clarifying how these designs shape the behavior and capabilities of autonomous vehicles and robotic systems. It then examines the technical pillars of autonomy including perception, sensor fusion, control, and hybrid decision-making before turning to the human factors that strongly influence adoption rates, including trust, explainability, bias, privacy, and emotional-intelligence limitations. Building on these challenges, the chapter then focuses on the analyses of the psychological, social, and regulatory barriers that shape public acceptance and operational deployment. The chapter then expands to the broader impacts of autonomous systems on mobility, safety, workforce transitions, and environmental outcomes. Finally, the chapter joins these insights by exploring governance strategies, evaluation frameworks, and human-centered design principles that can accelerate responsible, equitable, and scalable adoption of agentic AI and autonomous vehicles.

Automation versus Autonomy

Automation and autonomy both involve machines performing tasks with reduced human input, but autonomy adds decision-making, adaptability, and self-direction—making it far more advanced and context-aware than automation. Both concepts aim to reduce human effort and improve efficiency, and they often use overlapping technologies like sensors, software, and control systems. Key similarities include reduced human intervention as both systems operate with minimal direct human control. Improved consistency and efficiency they perform tasks faster and more reliably than manual methods. Use of programmed logic depends on predefined rules or algorithms to execute tasks. Integration with digital systems: Automation and autonomy often use data inputs, sensors, and software to function within larger systems (e.g., manufacturing, transportation, logistics). Automation refers to systems that follow predefined instructions to perform repetitive or structured tasks. It's rule-based and predictable. Operates based on fixed programming or scripts and cannot adapt to new or unexpected conditions.

The need for human autonomy teams (HAT) arises from situations requiring tightly coupled interactions between humans and machines; to the extent the machines' actions can be precisely predicted, a social context may not be needed to facilitate interaction, as for example, in stability augmentation of a control system (Lyons et al., 2021).

Machines That Matter

Robots. The scientific fiction movies portray a variety of robot types; the spectrum includes everything from friendly, helpful, smart, adaptive supportive machines to menacing, deadly, and relentless cyborg assassins. "Human–robot interactions are an unavoidable aspect of many modern robotic applications and ensuring safety for these interactions is critical, especially in applications such as autonomous driving where collisions may lead to life-threatening injury" (Leung et al., 2020). Robots are engineered machines that sense their environment, process information, and act to accomplish tasks autonomously or under human supervision; they combine mechanical structures and actuators for movement, sensors for perception, control algorithms for planning and decision-making, and software for communication, learning, and integration with other systems. Leung, et al. (2020) explain that ensuring safety within the planning and control framework can be challenging due to the uncertainty in how humans may behave, quantifying this uncertainty, robots often rely on generative models of human behavior in order to inform their planning algorithms. Designed in many forms, industrial arms, mobile platforms, drones, humanoids, and specialized service or medical devices, robots excel at repetitive, precise, hazardous, or efficient repetitive quantity work while also extending human capabilities into dangerous or inaccessible environments. Their value lies in improved efficiency, safety, and consistency, but realizing those benefits requires robust design for reliability, human–robot interaction that supports clear roles and trust, and governance to address ethical, social, and economic impacts.

Robotics is an interdisciplinary field that designs, builds, and programs machines, robots, to sense, plan, and act in the physical world, enabling autonomous or human-supervised task performance across manufacturing, healthcare, logistics, exploration, agriculture, and service sectors. At its core robotics integrates mechanical systems, actuators and structures, sensors for perception, control and planning algorithms for decision-making, and software for communication and learning, all working together to enable interaction and adaptation to their environment. Robots deliver measurable benefits in efficiency, safety, precision, and reach, yet they also face technical and social challenges such as robustness in unstructured settings, cost, ethical impacts, and the need for reliable human–robot interaction. As robotics continues to converge with machine learning, advanced sensing, and connectivity, future systems will become more collaborative, context-aware, and capable, requiring multidisciplinary design, clear standards, and governance to ensure these technologies maximize societal benefit.

Personalized or Paranoid?

The rise of the machine. There has been a plethora of big budget blockbuster movies made in the recent past exploring dystopian futures; machines becoming self-aware, humans being enslaved by humanoid robots, and the utter destruction and devastation of the planet. It is understandable why people are paranoid, afraid, and uneasy about the use of adaptive AI. The feelings people experience go from helpful to frightening; raising eyebrows to the thought that AI is watching and learning worries us. "One factor that may strongly influence AV adoption is *risk preference*, which is understudied in recent AV literature; risk preference is particularly germane to AV adoption because both AV technology and its interactions with mobility systems are highly risky. In fact, risk preference is a critical factor in any new technology since risk preference determines consumer behavior and individual responses to policy and market conditions" (Wang & Zhao, 2019). Public concerns about the safety and reliability of AVs combined with individual differences in risk tolerance make widespread adoption particularly challenging.

The emergence of agentic AI systems capable of setting goals, making decisions, and executing tasks with minimal human oversight has sparked a wave of concern across people in many industries, countries, and communities. Unlike traditional AI models that respond passively to prompts, Agentic AI exhibits autonomy, persistence, and the ability to adapt its strategies in pursuit of changing objectives. This shift from reactive to proactive behavior introduces a host of anxieties, fears of misaligned goals, opaque decision-making, and the erosion of human agency in critical systems. "In the proactive mode, actions are generated based on an internal prediction, whereas in the reactive mode actions are generated in response to sensory inputs in unpredictable situations" (Murata et al., 2014). People worry that as these AI systems become more capable, they may act in ways that are unpredictable, unaccountable, or even unsafe, especially when deployed at large scale in domains like finance, defense, healthcare, or governance. The challenge lies not only in technical control, but in ensuring that agentic AI aligns with socially accepted human values, operates fairly and transparently, and remains subject to meaningful and purposeful oversight. These concerns are not simply rooted in science fiction, but rather they reflect real tensions between innovation, trust, and the limits of current regulatory and ethical frameworks.

Figure 1. Togulev, A. (Art Director)

https://unsplash.com/photos/white-robot-MECKPoKJYjM

The next evolution of innovation is to leverage autonomous systems to improve our world. "Vehicle automation is expected to lead to the next paradigm shift in transportation field despite that potential benefits and issues associated with it are still being critically evaluated and discussed" (Guo et al., 2021). Although AI has made significant strides recently, its interaction with humans still encounters persistent challenges that are difficult to overcome. Before exploring the distinct hurdles associated with agentic AI, it's essential to first understand the broader barriers that affect general AI adoption ranging from trust and transparency to usability and ethical alignment. While the list below is not comprehensive it provides a general understanding of the problems. These foundational issues shape the landscape in which agentic systems must operate and influence how they are received, regulated, and integrated into human-centered environments.

Bias and Fairness

AI models that have been trained on biased datasets reinforce social inequalities and promote systemic barriers. Alon-Barkat & Busuioc (2023) explain that AI use in decision making is said to hold the potential to help us overcome our cognitive biases and limitations. AI bias refers to the unfairness that arises when artificial intelligence systems produce results that reflect or amplify existing human prejudices. This often stems from biased training data, datasets that encode historical inequalities or underrepresent certain groups, leading the AI to replicate those patterns in its outputs. "Automation bias is a well-recognized decisional support problem that has emerged from studies in aviation and healthcare, areas that have traditionally heavily relied on automated tools. Automation bias refers to undue deference to automated systems by human actors that disregard contradictory information from other sources or do not (thoroughly) search for additional information" (Alon-Barkat & Busuioc, 2023). Design flaws in algorithms, such as developer assumptions or poorly defined objectives, can further alter the results. Sample selection bias also plays a critical role in the dataset, as models trained on narrow or homogenous data may perform poorly for diverse populations. Over time, feedback loops can reinforce these biases, especially in systems that continuously learn from user interactions. The consequences are significant as biased AI can lead to discrimination in hiring, lending, healthcare, law enforcement, and education; erode public trust in technology; and expose organizations to legal and ethical risks. To mitigate these issues, developers and institutions must prioritize diverse and representative datasets, conduct regular bias audits, build transparent and interpretable models, and involve inclusive, multidisciplinary teams throughout the development process.

Trust and Explainability

Trust and explainability are central challenges in the deployment and acceptance of AI systems, particularly as they become more dynamic and autonomous. Many AI models, especially deep learning systems, operate as "black boxes," making decisions through layers of computation that are difficult for even their creators to interpret how the system arrived at its decision. Black-box AI models are difficult for users to interpret outputs and trust. This complexity and lack of insight undermine user confidence, especially in domains like healthcare, finance, and criminal justice, where understanding how and why a decision was made is critical for accountability and fairness. Without clear explanations, it becomes nearly impossible to detect bias, correct errors, or ensure compliance with ethical and legal standards. Moreover, when AI systems adapt or evolve over time, their behavior can shift in unpredictable ways, further eroding trust. Building explainable AI (XAI) that balances performance with transparency is essential, not just for technical validation, but for fostering public trust, regulatory acceptance, and responsible innovation. Algorithm transparency must include the obligation to inform, report parameters to the competent authorities, disclose parameters to society, archive data and disclose source code, etc. (Ma, 2024).

Privacy Concerns

AI human interactions require significant data collection, raising ethical and privacy concerns for many users. AI systems pose growing privacy concerns due to their ability to collect, infer, and act on vast amounts of personal data often without users' full awareness or consent. From facial recognition and voice assistants to predictive analytics and behavioral tracking, AI can extract sensitive insights from seemingly innocent interactions. Even anonymized datasets are vulnerable to re-identification when combined with other sources, raising risks of surveillance, profiling, and data misuse. The complexity of many AI models compounds the issue, making it difficult for users to understand what data is being used, how it's processed, and for what purpose. Additionally, once data has entered a system it is difficult to remove or ensure eradication. Without robust safeguards, transparency, and user control, AI threatens to eliminate the boundaries of personal privacy and normalize intrusive data practices across digital environments. The impacts of this intrusion are far reaching and have real consequences.

Emotional Intelligence

AI lacks empathy, limiting its effectiveness in sensitive scenarios. AI lacks empathy simply because it does not possess consciousness, emotions, or lived experience like a living entity. While it can simulate empathetic responses, such as recognizing distress in language or mimicking supportive tones in responses, it does so through pattern recognition and programmed behavior, not genuine understanding or emotional resonance that is necessary to be empathic. "A critical discussion is needed about the concept of 'simulated' empathy that underlies social robots, i.e. *cognitive empathy*. This form of empathy enables us to 'recognize' the emotional mental states of others and identify their emotions based on features of their expressions" (Gill, 2024). Empathy requires the ability to feel what another person feels, to interpret context beyond data, compare similar feelings, and to respond with moral and emotional nuance. AI, by contrast, operates on logic, probability, and predefined parameters. Think about the left brain versus right brain activities; one is logical the other more creative. This limitation becomes especially problematic in sensitive domains like mental health, education, or customer service, where human connection and emotional intelligence are essential. Without true empathy, AI may misinterpret emotional cues, offer inappropriate responses, or fail to recognize when a situation requires compassion, urgency, or support over efficiency.

Autonomy in Action

Reactive AI systems function based on predefined rules and responses to the environmental inputs they receive. These AI systems do not possess any type of memory or learning capabilities; the system simply reacts instantaneously without the inclusion of past experiences to guide the output. Agentic AI systems function based on dynamic inputs using sensor response mechanisms based on programmed heuristics or if-then programming logic. Autonomous or self-cleaning vacuum cleaners are a great example of a basic robot; the system responds to obstacles using infrared sensors. They do not have the ability to make decisions, judgements, or use past outputs to change their response to the environment. These robots cannot adapt to new or unforeseen circumstances that have not been included in the programming. As a result, the machine lacks the ability to learn. These architectural constraints directly shape how humans perceive reliability and predictability in AV behavior.

Agentic

Traditional vehicles relied entirely on the driver to make decisions for the entirety of the trip. The first wave of autonomous vehicles introduced rule-based systems and basic machine learning algorithms that could handle specific driving scenarios under controlled situations. "In the case of unstructured environments, the task of navigation can be particularly challenging, due to the presence of complex scenarios, where the geometric features of the environment might not be enough to determine safe traversable routes" (Saucedo et al., 2024). Today, the third generation of vehicular intelligence where transportation AI doesn't just follow pre-programmed instructions but actively learns from experience, adapts to new situations, and makes complex decisions. Future autonomous vehicle innovations will not operate in isolation. Vehicle-to-vehicle (V2V) and vehicle-to-infrastructure (V2I) networks and data sharing will create collaborative pathways where vehicles share data and coordinate actions simultaneously for improved safety and efficiency. The evolution of autonomous vehicles represents a fundamental shift in our relationship with the transportation industry. As intelligence advances through agentic AI, we move closer to systems that understand the dynamic social, physical, and ethical factors of transportation.

Agentic AI works by combining autonomous decision-making, goal-oriented reasoning, and adaptive learning to perform complex tasks with minimal human oversight. It perceives its environment, sets objectives, plans actions, and iteratively refines its behavior all while maintaining contextual awareness and operational independence. Unlike traditional AI that responds passively to prompts, agentic AI initiates actions based on predefined or evolving goals. It does not just answer questions, it pursues outcomes. This agency allows it to operate independently, making decisions aligned with its objectives without constant human input.

Agentic AI systems begin by gathering data from their surroundings via sensors, databases, APIs, or user interfaces. This perception layer enables them to understand context, detect changes, and adjust behavior accordingly. These systems use symbolic reasoning, path-planning algorithms, and predictive modeling to simulate future states and select optimal actions. They often break down complex tasks into subtasks, coordinating them through multi-agent orchestration. Agentic AI incorporates machine learning and reinforcement learning to refine its performance over time. It learns from outcomes, adjusts strategies, and improves accuracy especially in dynamic environments like autonomous driving or robotic navigation. Agentic AI can interface with external tools, software systems, and other agents. This allows it to automate workflows, access specialized capabilities, and collaborate across platforms. For example, it might use APIs to retrieve data, execute transactions, or control hardware. In many applications, agentic AI functions as a teammate rather than a tool. It communicates with intent, accepts guidance, and supports supervisory control. This is especially important in Human Autonomy Teaming (HAT), where the combination of shared goals and mutual understanding drive performance outcomes. Agentic AI represents a shift from automation to intelligent agency where machines not only act but understand, adapt, and collaborate with the user.

Deliberative AI: Decision-Making with Goal-Oriented Reasoning

Deliberative AI, known as model or planning based AI, relies heavily on computational reasoning to achieve long-term objectives. Unlike reactive AI, deliberative AI constructs and maintains an internal model of its environment, enabling strategic action selection in the outputs.

Deliberative AI uses symbolic AI, logical reasoning, and path-planning algorithms (e.g., A*, Dijkstra's) to maintain a world model, allowing the system to utilize predictions and simulations of future states. The system incorporates optimization techniques for goal-directed behavior.

A great example of deliberative AI are vehicles that use GPS-based route planning with obstacle prediction and avoidance built into the programming allow the vehicle to operate without limited human oversight. Like self-driving autonomous vehicles, examples of autonomous space exploration include the Mars rovers that create terrain maps for optimized navigation. The healthcare industry is leveraging AI supported medical systems that can help analyze complex procedures before execution. These systems tend to be computationally expensive, requiring advanced processing power that utilizes a lot of energy consumption and negatively impacts the carbon footprint. They may struggle with real-time decision-making in fast changing and complex environments and are vulnerable to unexpected environmental shifts due to reliance on predefined models.

AI architectures in autonomous systems vary in the level of complexity based on the algorithms ability to process and act upon information in real time. Reactive AI is best for basic automation; deliberative AI is best for strategic problem-solving. Hybrid AI, combining both approaches, is often the most powerful and appropriate approach for advanced autonomy needs. AI continues to evolve; the refinement of these architectures will drive progress in robotics, autonomous vehicles, and artificial intelligence systems.

Hybrid AI Systems: Combining Reactive and Deliberative AI for Adaptive Behavior

Hybrid AI systems use the best of both reactive and deliberate approaches to create a more flexible autonomous system. These architectures are commonly found in high-functioning AI models where adaptability and flexibility are critical. Hybrid AI systems leverage layered architectures where reactive components handle immediate responses, while deliberative AI processes long-term objectives. This approach allows the system to balance real-time reflexes (reactive AI) with strategic adaptation (deliberative AI) by incorporating reinforcement learning, allowing systems to refine behavior based on experience. For example, immediate threat responses for drones are then combined with strategic mission planning using automated drone technology. The advantages of hybrid over pure reactive or deliberative AI are enhanced flexibility and adaptability, allowing for dynamic responses, increased efficiency, combining both real-time reactions with long-term decision-making, and greater robustness, allowing seamless operation in unpredictable environments.

According to the Society of Automotive Engineers, there are six levels of self-driving vehicle technology, ranging from Level 0 (no automation) to Level 5 (full automation) (Progressive, 2025). Level 1 includes the cruise control feature found on most vehicles, while levels 2-5 included more advanced automation. Level 2: Partial driving automation indicated the vehicle can assist with steering and speed but still requires "the full engagement and undivided attention of the driver." Level 3: Conditional driving automation is automated driving features that can drive the vehicle under limited conditions, and a human driver must be ready to drive when the feature requests. Mercedes-Benz Drive Pilot is an example, available in California and Nevada as of 2025. Levels 4-5: High and full driving automation are fully self-driving vehicles that are not yet available for consumer purchase. Google's Waymo service is Level 4, which only operates commercially as a robotaxi in certain cities as of 2025 (Progressive, 2025).

Human Centered AI Design

Human interaction plays a central role in shaping the adoption of robotics and autonomous agentic AI systems by influencing trust, usability, and perceived value. When people engage with robots or agentic AI in intuitive, transparent, and emotionally resonant ways, they are more likely to accept and integrate these technologies into daily life and professional workflows. Key factors include the system's ability to communicate clearly, responding appropriately to human cues, and demonstrating reliability in decision-making. Anthropomorphic or humanlike design elements such as voice inflection, culturally appropriate gestures, or facial expressions can enhance comfort and relatability, while inclusive and adaptive interfaces help bridge gaps across age, ability, and cultural background. Conversely, poor interaction design, unclear behavior, or lack of user control can trigger resistance, especially when autonomy is perceived as threatening or intrusive. AI must be intuitive, adapting and enhancing human behavior rather than forcing humans to adapt to constraints. Usability principles, such as transparency and predictability, play a critical role in increasing user adoption. AI should enhance human capabilities rather than replace autonomy or critical thinking skills. Ultimately, successful adoption hinges on aligning technical capabilities with human expectations, emotional needs, and ethical considerations.

Social and Psychological Factors in AI Interaction

People rarely adopt innovations in isolation. They look to others, friends, family, experts, trusted brands, and institutions for cues. The rise in social media stars, the fashion houses that influence the seasons styles, and reality tv are all indicators that society looks to others for trends and acceptance. When influential voices endorse AVs, share positive experiences, or visibly use them, they create a ripple effect that lowers skepticism and increases curiosity. This is the essence of social proof: the idea that if others trust technology, it must be safe or valuable.

Human perception of AI is influenced by concepts such as the uncanny valley, algorithm aversion, and automation bias. Trust is developed through consistent behavior, interpretability, and accountability in decisions. Social and psychological factors play a critical role in shaping how humans interact with AI systems. Trust, perceived intelligence, emotional resonance, and cultural norms all influence whether users accept, reject, or adapt to AI technologies. People tend to humanize AI to help with adoptability, assigning human-like traits to machines which can enhance engagement but also create unrealistic expectations about empathy, understanding, or accountability. Familiarity and transparency increase comfort, while unpredictable behavior can trigger anxiety or resistance. Additionally, social identity and group dynamics affect adoption: individuals may be more receptive to AI if it aligns with their values, language, or community norms. Psychological factors such as cognitive load, fear of replacement, and the need for control also shape user experience, making it essential for AI systems to be designed with human-centered principles that respect emotional nuance, social context, and ethical boundaries.

Innovation diffusion theory explains AV adoption as a social and systemic process driven by perceptions of relative advantage, compatibility with existing behavior and infrastructure, perceived complexity, trialability, and observability: people adopt autonomous vehicles faster when they see clear safety, cost, or convenience benefits, when AVs fit existing travel routines and regulatory contexts, and when interfaces and transparency lower the sense of technical risk. "Innovation diffusion theory (IDT) was developed to explain why individuals choose to adopt or reject an innovation based on their beliefs" (Yuen et al., 2021). Social influence and visible examples matter greatly; endorsements from peers, positive media coverage, and demonstrable performance in pilots create social proof that accelerates uptake, while high-profile failures can stall adoption. Yuen et al., (2021) further explain that the central concept of IDT is that a decision about whether to adopt an innovation comprises five stages: knowledge, persuasion, decision, implementation, and confirmation and the acceptance and the decision process of the innovation are related to the innovation's attributes or characteristics, which can be measured by five aspects; relative advantage, compatibility, complexity, experimental, and observability.

Adoption follows an S-curve where innovators and early adopters precede a larger majority, and bridging the gap to mainstream users requires reducing uncertainty through large trials, reliable performance evidence, and tailored value propositions that address the majority's risk tolerance. Policy, infrastructure, and organizational procurement act as critical change agents by changing perceived attributes of AVs: clear safety standards, data-sharing rules, roadway investments, and public fleet deployments increase compatibility, lower complexity, and expand trialability. Practically, coordinated action, transparent regulation, accessible pilot programs, user-centered design, and equitable deployment strategies speeds responsible adoption and helps ensure benefits are distributed broadly rather than concentrated among early adopters.

Social proof and peer influence are powerful psychological drivers that shape how people perceive and adopt autonomous vehicles. When individuals see others, especially trusted peers or respected institutions, using and endorsing AVs, their own skepticism tends to diminish, and curiosity or acceptance increases. Social proof refers to the tendency of people to look at others when making decisions, especially in uncertain or unfamiliar situations. In the context of AVs, this means seeing others use AVs safely (e.g., neighbors riding in robotaxis or coworkers commuting in autonomous shuttles) build confidence. Positive media coverage and endorsements from experts, influencers, or public figures help normalize the technology. Visible integration into daily life such as AVs in public transit fleets or delivery services signal's reliability and mainstream acceptance. Social proof reduces perceived risk by showing that others have already vetted the technology. Peer influence amplifies social proof by adding emotional and relational weight. People are more likely to trust AVs when friends or family members share positive experiences, such as comfort, safety, or convenience. These stories create a shared connection that other users internalize and compare to their own lived experiences. When communities adopt AVs collectively, especially when aligned with shared values like sustainability or accessibility, then users are more likely to trust their use. Workplaces or institutions integrate AVs, signaling professional credibility and operational trust.

Peer influence is especially strong in tight-knit groups, where shared norms and firsthand testimonials carry more weight than technical specs or corporate messaging. Technical explanations often fall short; seeing others benefit from AVs is more persuasive. According to Innovation Diffusion Theory, early adopters influence the majority through visibility and endorsement. Familiarity through peers reduces anxiety and reframes AVs as safe, useful, and socially accepted. Brand influence and personal consumer traits also play an important role in purchase decisions and pointed out that large car manufacturers are often more credible in technology, thus more easily gaining consumer trust (Mo et al., 2024). In short, social proof and peer influence turn new technology into relatable experience. They help AVs move from novelty to necessity, one trusted voice at a time.

Humanizing the Machine

Humanizing AI means designing systems that people find relatable, trustworthy, and ethically responsible by prioritizing transparency, contextual empathy, human values, and by giving systems a consistent persona and clear escalation paths to human review. Relatable AI reduces friction, builds trust, and increases engagement by sounding and behaving in ways people expect from other humans while staying clear about being an AI. Ngo's (2025) suggests that the growing human-likeness of AI locus systems, particularly in conversational or generative AI, recommender systems, and autonomous decision-making models, has reshaped how users interact with and perceive AI systems. Practical steps include adding plain language explanations for nontrivial decisions, progressive disclosure of complexity, emotionally appropriate responses, and human-in-the-loop controls that let users edit, override, or opt out while the system learns from corrections. Success is measured by trust and satisfaction metrics, transparency and correction rates, timely human escalations, and performance parity across user groups. Xie et al., (2024) explain that anthropomorphic features include human physiological features, such as the human face or body, and human psychological features, such as human thoughts and personality. Anthropomorphic features fall into three main dimensions; first, they have anthropomorphic visual cues, such as humanoid shapes demonstrating human-like visual cues that are powerful and can trigger a feeling of "humanity" in users; second, the AI has anthropomorphic identity cues, such as using a human name or having an identity to humanize AI products; cognitive psychologists have emphasized that when AI has a human identity, users tend to treat the AI the same way they treat humans; third, the AI can use anthropomorphic emotional cues, such as a sense of humor (Xie et al., 2024).

Risk Aversion

Risk aversion to AI stems from unknown failure modes, trust deficits, regulatory uncertainty, reputational and operational disruption, and weak measurement. Aversion is best addressed by starting with pilots that combine deterministic rules with probabilistic models, enforce safe defaults, and require human-in-the-loop review for decisions; governance must include a risk taxonomy, clear accountability, audit trails, a cross-functional risk board, and compliance playbooks. While technical controls require adversarial and edge-case testing, explainability, drift monitoring, and progressive rollouts with rollback criteria; pair safety key performance indicators (KPIs) (unexpected behaviors, escalation frequency, time-to-recovery) with value KPIs (task completion, time saved, satisfaction), communicate limits and failure modes transparently to stakeholders, secure executive sponsorship, reskill affected staff, and use independent audits to build trust and enable measured, responsible scale. Mo et al., (2024) explains that individual psychological traits, such as the degree of acceptance of new technology and sensitivity to risk, also affect their willingness to buy; from the perspective of behavioral economics theory, prospect theory and planned behavior theory can explain consumers' decision-making behavior when facing uncertainty and risk. The impact of generative AI on user adoption decisions is influenced by social risk aversion or the extent to which users perceive human involvement in AI decisions as a source of bias or self-interest. Some studies suggest that users may prefer autonomous AI in high-stakes decision-making contexts where fairness and neutrality are critical. However, when AI lacks social presence, users tend to scrutinize the AI's decision-making process more rigorously, making generative AI a more salient factor in shaping trust in AI and adoption (Ngo, 2025).

Ethical and Governance Considerations

Ethical and governance considerations for AI center on ensuring that artificial intelligence systems are developed and deployed in ways that are safe, fair, transparent, and aligned with human values. As AI becomes more powerful and pervasive, concerns about bias, accountability, privacy, and misuse have greatly intensified. Ethical frameworks emphasize principles such as justice, autonomy, beneficence, and non-maleficence, guiding developers to avoid harm and promote equitable outcomes for everyone. Governance, meanwhile, involves creating oversight structures including policies, standards, and institutional mechanisms that monitor AI's behavior, assess risks, and enforce compliance consistently in the environment. This includes board level accountability, cross collaboration, and stakeholder engagement to ensure that AI systems are not only technically robust but socially responsible. Effective governance requires proactive risk management, transparency in decision-making, and inclusive participation from ethicists, policymakers, technologists, and affected communities. AI-human interactions must uphold principles of fairness, accountability, and transparency. Bias in AI decision-making can have profound social implications. Bias refers to the systematic errors that occur in decision-making processes, leading to unfair outcomes (Ferrara, 2024). Regulations such as General Data Protection Regulation (GDPR), Institute of Electrical and Electronics Engineers (IEEE) Ethics Guidelines, and National Institutes of Standards and Technology (NIST) AI Risk Management Framework define ethical standards for AI interactions.

Blind Spots and Patchwork Rules

Safety assurance for AVs faces several complex challenges and blind spots that need correction. The regulatory landscape is lacking the rules needed for sufficient adoption, while manufacturers must comply with existing Federal Motor Vehicle Safety Standards (FMVSS), there are currently no mandatory federal performance regulations specifically for automated driving systems. This leaves a significant portion of oversight to voluntary industry standards and state-specific rules. Connectivity gaps continue to exist as AVs that rely on vehicle-to-vehicle (V2V) or vehicle-to-infrastructure (V2I) communication can face safety issues in areas with poor or non-existent connectivity. These gaps can prevent the system from receiving critical information, such as traffic signal data. No matter how many millions of simulated miles are logged, it is nearly impossible to account for every unpredictable real-world scenario. AVs can struggle with unusual situations, such as confusing visual data (a plastic bag flying across the road), non-standard traffic violations, or the ambiguous intentions of human drivers, pedestrians, and cyclists. Lastly, the common sense problem, machine learning algorithms and AI systems still lack the "common sense" and social intelligence that human drivers rely on to predict and navigate complex interactions on the roadway.

Benefits of Autonomous Vehicles

While AVs are not without risks, their potential to save lives, democratize mobility, reduce environmental impact, and transform economies makes them a positive innovation that is hard to refute. Provided their deployment is guided by ethical design, inclusive governance, and continuous oversight, autonomous vehicles offer a wide range of benefits that have the potential to transform society, innovate, and improve daily life. Economically they will disrupt labor markets in driving occupations even as they create new tech, maintenance, and mobility-as-a-service industries, shift ownership models from private cars to fleets, and alter insurance and liability frameworks. Yuen et al., (2021) explain that the benefits and popularization of AVs will lead to a revolution in the personal mobility and transportation industries. By reducing human error, which accounts for most traffic accidents, these systems can significantly enhance road safety and preserve human life. These vehicles also optimize driving patterns, leading to better fuel efficiency and reduced emissions, which support environmental sustainability. For individuals with mobility challenges, such as the elderly or disabled, autonomous vehicles provide newfound independence and access like never before. Additionally, they can ease traffic congestion through coordinated routing and reduce the stress of commuting by allowing passengers to focus on other tasks. Reducing the stress of commuting and allowing passengers to focus on other tasks also has the capacity to revolutionize work life balance. As travelers can free up time to pursue hobbies, get caught up on work, or rest this new found time can significantly impact personal enjoyment in other aspects of life. In commercial sectors, autonomous fleets promise lower operational costs, streamlined logistics, and faster delivery times. As technology matures, these benefits could reshape urban planning, public transit, and the future of mobility itself.

Mobility

Autonomous vehicles can dramatically enhance mobility by removing many of the physical, cognitive, and logistical barriers that limit access to transportation. For individuals who are elderly, disabled, economically disadvantaged, or otherwise unable to drive, autonomous systems offer independence and freedom of movement without relying on caregivers or public transit schedules to get around. These vehicles can operate continuously, reducing wait times and expanding service coverage in underserved or rural areas that have been historically behind on ride sharing or public transit systems. By coordinating routes and traffic flow through real-time data, autonomous fleets can also reduce congestion and improve travel efficiency for everyone. In urban environments, they support multimodal transport by integrating seamlessly with buses, trains, and micro mobility options such as scooters or powerchairs. Additionally, autonomous delivery and logistics vehicles streamline supply chains, ensuring faster access to goods and services. Overall, autonomous mobility systems promise a more inclusive, efficient, and responsive transportation ecosystem that adapts to diverse human needs.

Autonomous vehicles enable mobility-as-a-service by converting privately owned, idle capacity cars into centrally managed, on demand fleets that lower per-ride costs, improve vehicle utilization through optimized routing and dynamic rebalancing, and make subscription and pooled pricing economically viable. The service can extend and complement public transit by solving first and last mile gaps with integrated scheduling and payment, increase equity and accessibility by providing on demand options for non-drivers, and amplify environmental and land-use benefits when paired with electrification and smart charging by reducing parking demand and tailpipe emissions.

Traffic

Autonomous vehicles can significantly improve traffic congestion by optimizing traffic flow, reducing stop-and-go patterns, and enhancing coordination across road networks. Unlike human drivers, autonomous systems maintain consistent speeds, safe following distances, and smoother acceleration and braking, which helps prevent phantom traffic jams, those sudden slowdowns caused by erratic human behavior. Professor Alex Bayen, Director of UC Berkeley's Institute of Transportation Studies, is featured in Fortune On Demand and PBS Terra on the use of artificial intelligence (AI) to eliminate traffic jams. Phantom traffic jams, or stop-and-go traffic, are inefficient and dangerous but fortunately, self-driving vehicles can reduce and improve traffic jams by guiding human-controlled vehicles on the road. For traffic to flow smoothly, drivers must maintain a consistent speed and distance from other vehicles, however, humans tend to speed up to meet the car in front of them, causing phantom traffic jams. Bayen believes that automated and human-controlled vehicles could work collectively to smooth traffic flow (Bayen, 2021). When connected to infrastructure and other vehicles, autonomous cars can communicate in real time to reroute traffic, avoid bottlenecks, and balance load across urban corridors that are prone to backups. In cities, autonomous transit fleets can reduce the number of single or low occupancy vehicles, improve ride sharing efficiency, and support dynamic scheduling based on demand, further easing congestion on already crowded roadways. Research also suggests that even a small percentage of autonomous vehicles in mixed traffic can stabilize flow and reduce delays for all road users.

Safer Roads

Autonomous vehicles have the potential to significantly improve road safety by reducing human error, which accounts for approximately 94% of traffic fatalities according to the National Highway Traffic Safety Administration (NHTSA Automated Vehicles, 2025). These vehicles rely on advanced sensors, real-time data processing, and machine learning algorithms to detect hazards, maintain safe distances, and respond faster than human drivers. Li et al. (2024) explain that safety enhancement is considered the most critical benefit of Avs because AVs are controlled by precise and fast-responding sensors, human drivers typically exhibit uncertain and slow-responding behaviors. Features like automatic braking, lane keeping assistance, and collision avoidance systems are already reducing accidents in semi-autonomous models. However, experts caution that while the safety benefits are promising, they remain largely aspirational until fully autonomous systems are proven at scale across diverse driving conditions. Challenges such as system failures, cybersecurity risks, and unpredictable human behavior still pose safety concerns. Therefore, while autonomous vehicles offer a compelling vision for safer roads, ongoing testing, regulation, and public trust are essential to realizing their full safety potential.

Lower Insurance

While fully self-driving vehicles are likely a long way off for U.S. drivers, some autonomous features like adaptive cruise control and lane-keeping assistance have become common, regardless of how advanced autonomous vehicle features become, having your state's required minimum car insurance remains essential, some level of liability coverage is legally required in nearly all states (Progressive, 2025.) Autonomous vehicles have the potential to significantly lower insurance costs by reducing the frequency and severity of accidents, which are overwhelmingly caused by human error. With advanced sensors, AI decision-making, and real-time data processing, AVs can avoid collisions more effectively than human drivers, leading to fewer claims and lower risk profiles. As liability shifts from individual drivers to manufacturers and software providers, insurance models are evolving, moving toward product liability coverage and fleet-based policies. Additionally, AVs generate detailed operational data that allows insurers to assess risk with greater precision, enabling usage focused pricing and more personalized premiums. Over time, these factors could compress insurance costs dramatically. As AV technology matures and regulatory frameworks adapt, the insurance industry is poised to transform, offering more efficient, data driven, and equitable coverage options for consumers and commercial operators alike.

Safety Assurance

Autonomous vehicle safety assurance is the comprehensive and multi-faceted process of guaranteeing that self-driving vehicles operate safely and reliably under all road conditions and scenarios. Safety assurance for autonomous vehicles is a continuous, evidence-based lifecycle that defines clear safety goals and acceptable risk, maps those goals to system-level requirements across sensors, perception, planning, actuation, communications, and human interfaces, and produces a structured assurance case linking claims to verifiable evidence. It uses multi-fidelity testing, from unit and model verification through hardware-in-the-loop, high-fidelity simulation, closed-course validation, and instrumented real-world pilots against a comprehensive scenario library including edge, corner, and adversarial cases, and supplements physical tests with formal specifications and model proofs where feasible. The process mandates traceability and versioned artifact provenance for datasets, models, and software, continuous monitoring for performance and distributional drift, calibrated confidence measures, and defined degraded behaviors and graceful degradation strategies. Governance assigns clear accountability, including product owner, safety engineer, operations lead, incident commander, enforces change control and authenticated updates, embeds cybersecurity and incident response playbooks, and requires independent third-party review and pre-deployment certification gates tied to measurable acceptance criteria. Human-agent handover, operator readiness, and physiological validation are tested and quantified, and KPIs with automated alerts enable rapid mitigation and rollback. Safety assurance is iterative, evidence is continually refreshed, lessons from incidents are fed back into models and scenarios, and recertification occurs as operating domains, software, or data distributions change.

A single point of failure in a safety critical system like an AV can lead to dangerous situations. To prevent this, developers incorporate hardware and software redundancy, meaning duplicate, independent components. Hardware redundancy means critical components like the power supply, electronic control units (ECUs), and steering and braking systems have backups. For instance, a vehicle might have two independent computer boards; if one fails, the other can initiate a minimum risk maneuver (MRM) to safely pull the vehicle over. Sensor redundancy for example, includes multiple sensor types including cameras, radar, and LiDAR that are used to perceive the environment. If one sensor fails or is obstructed by bad weather, the others can compensate and provide reliable data.

Since AVs are highly connected to other vehicles and infrastructure, they are vulnerable to cyber threats that could compromise safety. Robust cybersecurity is critical to protecting these systems. Utilizing a multi-layered protection approach to cybersecurity, including encryption, intrusion detection systems, secure boot procedures to prevent unauthorized code, and firewalls for individual components. Before an AV can be utilized on public roads, it must undergo a thorough, multi-stage testing process to prove its trustworthiness in a vast range of scenarios. During simulation testing developers use virtual environments to safely test AVs across millions of miles and thousands of scenarios, including rare or dangerous outlier cases that are difficult or unsafe to reproduce in the real world. On private tracks, AVs are assessed in controlled environments to test the response to specific, repeatable situations, like unexpected pedestrian movements or obstacles. After extensive testing in virtual and controlled environments, AVs are tested on public roads with a human safety driver present to monitor performance. Some companies share this data with regulators, as in the National Highway Traffic Safety Administration's (NHTSA) AV TEST initiative.

Saving the Planet

Reducing greenhouse gases or reducing the carbon footprint of vehicles is essential to saving the planet. These gases trap heat in the atmosphere and drive global warming that disrupts weather patterns, raises sea levels, damages animals and ecosystems, harms human health, and creates large economic and social costs. Stopping further emissions and increasing natural and technological carbon removal are necessary to prevent worsening extreme weather, loss of habitat for animals, and cascading impacts on food, water, and infrastructure. Nordin et al. (2024) suggest that transport sector covers 20% of greenhouse gas (GHG) emissions globally, hence, a range of different abatement measures are needed to reduce emissions from the transport sector.

Autonomous vehicles can cut greenhouse gas emissions and energy use by smoothing traffic flow, reducing stop-and-go driving, and enabling cooperative driving strategies that improve fuel efficiency and lower congestion; they accelerate the shift to electric powertrains by pairing autonomy with electric vehicle (EV) fleets and optimized charging and routing, amplifying emissions reductions beyond what EVs alone achieve; shared, on-demand AV fleets reduce the number of vehicles needed, shrink parking footprints, and free urban land for denser, greener development that lowers vehicle miles traveled and associated emissions; freight autonomy and platooning improve logistics efficiency across highways, cutting fuel use for long-haul transport; realizing these climate benefits requires policies that promote electrification, shared usage models, integrated urban planning, and safeguards against rebound effects that could increase total travel. Onat et al. (2023), explain that Internal Combustion Engine vehicles (ICVs) powered by fossil fuels are conventionally the most prevalent mode of transportation and therefore, they contribute significantly to global climate change. "AVs have the potential to decrease greenhouse gas (GHG) emissions mainly in a couple of ways including reduced need for parking spaces, as looking for a parking space contributes significantly to one-third of traffic within urban areas, and because of the increased traffic, vehicles are forced to stay on the road longer and use more fuel as a result which emits GHG emissions" (Onat et al., 2023). Onat et al., (2023) explain that considering the fuel economy benefits introduced by AVs gained by reducing congestion and lowering the magnitudes and speeds of the vehicle's acceleration and deceleration, the total environmental impacts intuitively should be less than a nonautonomous version of the same fuel vehicle.

Incentives

Autonomous vehicle incentive programs promote safe, equitable, and low-carbon deployment by funding pilots, infrastructure, research centers, and workforce development while aligning private investment with public goals; effective programs combine grants for city and fleet pilots and charging infrastructure with tax credits or procurement preferences for electrified AV fleets, regulatory clarity that enables testing and scaled operations, and investments in coordinated data sharing, evaluation, and best practices. Chen et al. (2019) argue that one specific incentive is AV specific driving lanes, but at the design stage of an AV incentive program, local governments must estimate future AV purchase prices. Chen et al. (2019) explain this is difficult because rapid technological advances make manufacturing costs unpredictable and to manage this uncertainty, in the first stage, the government commits to building AV specific driving lanes before the actual price is known. In the second stage, once the AV price is realized, it adjusts the purchase subsidy accordingly. This structure reflects the different levels of flexibility: lane infrastructure is fixed for long periods, while subsidies can be quickly adapted. For example, if AV traffic on dedicated lanes falls below expectations, the government can increase subsidies to encourage higher adoption rates.

Programs should tie funding to measurable KPIs that promote safety, equity, emissions, and mobility, requiring open data and independent evaluation, stage funding from small pilots to larger rollouts contingent on verified performance, prioritize shared EV-AV models to maximize emissions and benefits, and include reskilling and hiring supports for displaced workers. Designing incentives in this way reduces deployment risk, limits rebound effects and ensures public value by coupling technical investment with governance, community engagement, and transparent accountability.

Driving Together: The Rise of Human-Autonomy Teaming in Smart Mobility

The core concepts of agentic human-vehicle interaction include Human-Autonomy Teaming (HAT). Lyons et al. (2021) explains that researchers are beginning to transition from studying human–automation interaction to human–autonomy teaming. This is a collaboration paradigm where humans and AI agents work as a team, sharing goals, information, and common language to reach objectives. In this model, the agent is more than a passive decision-support tool; it is an active teammate. Proactive and personalized experience allows agentic AI to move beyond basic infotainment and climate control to create a highly personalized in-cabin experience. The system learns a user's preferences, anticipates their needs, and can proactively manage tasks like scheduling appointments or optimizing travel time. For humans and AV agents to work together, effective two-way communication is crucial. Recent research explores using large language models (LLMs) to enable natural language interaction, allowing passengers to provide guidance or discuss the agent's behavior in real time. The agent can then respond in a context-aware way. The overall rationale for the use of HATs is that they may facilitate better performance relative to humans alone or machines alone, particularly under situations of high uncertainty (Lyons et al., 2021). In many scenarios, humans will not be fully hands-off but will instead take on a supervisory role. The AI handles routine driving, and the human provides guidance or takes control during complex or unusual circumstances. This maintains human involvement while augmenting human capabilities.

From Skepticism to Self-Steering: Driving the Future

This chapter and this book aim to deliver a transformative contribution to the field of autonomous vehicles by establishing a robust Agentic AI framework that prioritizes safety, reliability, legal compliance, and ethical integrity. Its first objective is to define the technical foundations required to operationalize agentic intelligence in AV systems, ensuring they can make autonomous decisions that align with regulatory standards and societal values. The second objective is to critically examine how human factors such as trust, perception, and interaction can either accelerate or hinder the adoption of agentic AV technologies. By bridging technical rigor with human-centered design, this work offers actionable insights for researchers, developers, and manufacturers seeking to deploy AVs responsibly and on a large scale. As one of the few publications dedicated to the intersection of agentic AI and autonomous mobility, it fills a crucial gap in the literature and sets a new benchmark for integrating ethical and legal safeguards into intelligent transportation systems.

In the dynamic landscape of transportation technology, autonomous vehicles represent a huge transformative innovation in our lifetime. What began as a futuristic concept, self-driving vehicles, has evolved into a concrete reality. Autonomous vehicles are surely here to stay and therein lies the revolution in AI that's restructuring how these vehicles are adopted. The emergence of agentic AI, systems capable of understanding their environment, making decisions, and taking actions is propelling autonomous vehicles toward genuine independence. This shift represents not just a technological evolution but a fundamental reimagining of transportation across the globe.

Encouraging people to adopt robotic vehicles such as autonomous cars, trucks, and delivery systems requires a multifaceted strategy that addresses trust, usability, incentives, and public perception. Strategies to boost the adoption of robotic vehicles include building public trust through transparency. Clear communication about safety protocols, decision-making processes, and data privacy helps reduce fear and skepticism. Offering financial incentives is a lesson learned from hybrid and electric vehicle programs that offered rebates, tax credits, and insurance discounts. These programs can significantly influence consumer behavior toward adoption. Demonstrating benefits by highlighting improvements in traffic safety, fuel efficiency, and accessibility, especially for elderly or disabled users can shift public opinion toward a positive vantage point. Pilot programs and community engagement offer communities localized trials and public demonstrations that allow people to experience robotic vehicles firsthand, reducing uncertainty and increasing comfort. Designing human interaction by developing intuitive interfaces, voice assistance, and adaptive behavior make robotic vehicles feel more approachable and responsive. Finally, addressing regulatory and ethical concerns using strong governance frameworks and ethical oversight reassure users that these systems are accountable and aligned with public interest.

Advances in computing, sensing, and algorithms transformed simple automation into programmable learning systems capable of complex decision-making; modern agentic AI and robotics are the direct descendants of these integrated technologies, enabling autonomy across transport, manufacturing, and service industries. When agentic AI is incorporated into autonomous vehicles and society can recognize the benefits from technological advancement the world moves forward. It is when autonomous vehicles are deployed with strong controls, transparent decision frameworks, enforceable safety standards, human oversight, data privacy protections, and equitable access policies they deliver wide societal benefits: sharply reduced traffic fatalities through the elimination of most human error, lower healthcare and emergency response burdens, and improved mobility for elderly and disabled populations; reduced emissions and healthier urban environments from optimized routing and smoother traffic flow; economic gains from more efficient logistics, lower transportation costs, and new service industries; and greater resilience in crises through coordinated, adaptive routing and delivery of critical supplies. These governance measures ensure benefits are widely distributed, protect vulnerable communities from disproportionate harms, and build public trust that enables long-term integration of autonomous mobility into sustainable, inclusive transportation systems.

Efficient transportation systems lower costs, save time, and reduce environmental impact by optimizing routes, improving fuel and energy use, and smoothing traffic flow to prevent congestion and delays; they strengthen supply chains and economic competitiveness by ensuring goods and services move reliably between producers and consumers, expand access to jobs and essential services for underserved communities, and improve public health and safety by reducing emissions and traffic-related injuries, making cities more resilient and livable for everyone When managed well, human–agent coevolution amplifies productivity, accessibility, and resilience, agents handle routine complexity while humans focus on judgment, creativity, and moral responsibility. The trajectory therefore depends less on technical possibility and more on how institutions, cultures, and design practices steer the partnership.

CONCLUSION

The road to the human interaction and adoption of autonomous vehicles is not paved solely with algorithms, it is layered with human emotion, cultural narratives, and psychological thresholds. While early promises of self-driving cars conjured visions of effortless mobility, public skepticism remains deeply rooted in fear of the unknown. Popular media has long portrayed machines as either saviors or threats, fueling anxieties about control, safety, and the erosion of human agency. Yet beneath the surface of these fears lies a powerful opportunity: to reframe autonomy not as a replacement for human judgment, but as a partner in safety, equity, and resilience. Adoption hinges on more than technical performance, it requires trust earned through transparency, relatable design, and consistent behavior. When autonomous systems demonstrate reliability, communicate clearly, and align with human values, they begin to shift perception from paranoia to partnership. The journey from skepticism to self-driving is not just about engineering, it's about empathy, ethics, and the courage to co-create a future where machines enhance, rather than undermine, our collective well-being.

REFERENCES

Abood, S., Al-Khafaji, H., & Al-Khafaji, M. H. (2025). Enhancing Collision Avoidance in Mobile Robots Using YOLOv5: A Lightweight Approach for Unstructured Environments. *Journal of Robotics and Control*. Vol. 6 No.2. https://journal.umy.ac.id/index.php/jrc/article/view/25856/11436

Alon-Barkat, S., & Busuioc, M. (2023). Human–AI interactions in public sector decision making: "Automation bias" and "Selective adherence" to algorithmic advice. *Journal of Public Administration: Research and Theory*, *33*(1), 153–169. DOI: 10.1093/jopart/muac007

Automated vehicles for safety. NHTSA. (n.d.). https://www.nhtsa.gov/vehicle-safety/automated-vehicles-safety

Bayen, A. (2021, March 15). Eliminating Traffic Jams with Self-Driving Cars | Civil and Environmental Engineering. Ce.berkeley.edu. https://ce.berkeley.edu/news/2537

Chen, S., Wang, H., & Meng, Q. (2019). Designing autonomous vehicle incentive program with uncertain vehicle purchase price. *Transportation Research Part C, Emerging Technologies*, *103*, 226–245. DOI: 10.1016/j.trc.2019.04.013

Ekatpure, R. (2023). Machine Learning for Enhancing Vehicle Safety and Collision Avoidance Systems in Automotive Development: Techniques, Models, and Real-World Applications. *Journal of Computational Intelligence and Robotics*, *3*(2), 1–43. https://thesciencebrigade.com/jcir/article/view/257

Ferrara, E. (2024). Fairness and Bias in Artificial Intelligence: A Brief Survey of Sources, Impacts, and Mitigation Strategies. *Sci*, *6*(1), 3. DOI: 10.3390/sci6010003

Gill, S. P. (2024). Empathy and AI: Cognitive empathy or emotional (affective) empathy? *AI & Society*, *39*(6), 2641–2642. DOI: 10.1007/s00146-024-02118-4

Gordillo Chávez, D., Cloarec, J., & Meyer-Waarden, L. (2024). Opening the moral machine's cover: How algorithmic aversion shapes autonomous vehicle adoption. *Transportation Research Part A, Policy and Practice*, *187*, 104193. DOI: 10.1016/j.tra.2024.104193

Guo, Y., Souders, D., Labi, S., Peeta, S., Benedyk, I., & Li, Y. (2021). Paving the way for autonomous vehicles: Understanding autonomous vehicle adoption and vehicle fuel choice under user heterogeneity. *Transportation Research Part A, Policy and Practice*, *154*, 364–398. DOI: 10.1016/j.tra.2021.10.018

Leung, K., Schmerling, E., Zhang, M., Chen, M., Talbot, J., Gerdes, J. C., & Pavone, M. (2020). On infusing reachability-based safety assurance within planning frameworks for human–robot vehicle interactions. *The International Journal of Robotics Research*, *39*(10-11), 1326–1345. DOI: 10.1177/0278364920950795

Li, Y., Shi, X., & Li, X. (2024). Effect of riding experience on changing opinions toward connected and autonomous vehicle safety – evidence from field experiments. *Transportation Research Part F: Traffic Psychology and Behaviour*, *105*, 491–504. DOI: 10.1016/j.trf.2024.07.023

Lyons, J., Sycara, K., Lewis, M., & Capiola, A. (2021). Human–Autonomy teaming: Definitions, debates, and directions. *Frontiers in Psychology*, *12*, 589585. DOI: 10.3389/fpsyg.2021.589585 PMID: 34122209

Ma, J. (2024). The impact of AI bias on social justice: Challenges and solutions. *Journal of Computing and Electronic Information Management, 15*(3), 75–78. DOI: 10.54097/wz9v0f43

Mo, C., Chen, F., & Wang, Z. (2024). A study on consumers' willingness to purchase autonomous vehicles from a multi-party interaction perspective: A tripartite evolutionary game model involving the government, automobile manufacturers, and consumers. *World Electric Vehicle Journal, 15*(11), 498. DOI: 10.3390/wevj15110498

Mudhivarthi, B. R., & Thakur, P. (2022). Integration of artificial intelligence in robotic vehicles: A bibliometric analysis. *Paladyn : Journal of Behavioral Robotics, 13*(1), 110–120. DOI: 10.1515/pjbr-2022-0102

Murata, S., Arie, H., Ogata, T., Sugano, S., & Tani, J. (2014). Learning to generate proactive and reactive behavior using a dynamic neural network model with time-varying variance prediction mechanism. *Advanced Robotics, 28*(17), 1189–1203. DOI: 10.1080/01691864.2014.916628

Murugesan, S., & Murugesan, S. (2025). The rise of agentic AI: Implications, concerns, and the path forward. *IEEE Intelligent Systems, 40*(2), 8–14. DOI: 10.1109/MIS.2025.3544940

Ngo, V. (2025). Humanizing AI for trust: The critical role of social presence in adoption. *AI & Society*. Advance online publication. DOI: 10.1007/s00146-025-02506-4

Nordin, I., Elofsson, K., & Jansson, T. (2024). Cost effective reductions in greenhouse gas emissions: Reducing fuel consumption or replacing fossil fuels with biofuels. *Energy Policy, 190*, 114138. DOI: 10.1016/j.enpol.2024.114138

Onat, N. C., Mandouri, J., Kucukvar, M., Sen, B., Abbasi, S. A., Alhajyaseen, W., Kutty, A. A., Jabbar, R., Contestabile, M., & Hamouda, A. M. (2023). Rebound effects undermine carbon footprint reduction potential of autonomous electric vehicles. *Nature Communications, 14*(1), 6258–13. DOI: 10.1038/s41467-023-41992-2 PMID: 37802993

Progressive. (2025, June 24). *Self-driving cars and insurance*. https://www.progressive.com/answers/insurance-for-driverless-cars/

Qiu, C., Tang, H., Yang, Y., Wan, X., Xu, X., Lin, S., Lin, Z., Meng, M., & Zha, C. (2024). Machine vision-based autonomous road hazard avoidance system for self-driving vehicles. *Scientific Reports, 14*(1), 12178. DOI: 10.1038/s41598-024-62629-4 PMID: 38806585

Saucedo, M. A. V., Patel, A., Kanellakis, C., & Nikolakopoulos, G. (2024). EAT: Environment agnostic traversability for reactive navigation. *Expert Systems with Applications, 244*, 122919. DOI: 10.1016/j.eswa.2023.122919

Sreedharan, S. (2023). Human-aware AI—A foundational framework for human–AI interaction. *AI Magazine, 44*(4), 460–466. DOI: 10.1002/aaai.12142

Wang, S., & Zhao, J. (2019). Risk preference and adoption of autonomous vehicles. *Transportation Research Part A, Policy and Practice, 126*, 215–229. DOI: 10.1016/j.tra.2019.06.007

White robot photo – free grey image on unsplash. (n.d.). https://unsplash.com/photos/MECKPoKJYjM

Winfield, A. F., Michael, K., Pitt, J., & Evers, V. (2019). Machine ethics: The design and governance of ethical AI and autonomous systems [scanning the issue]. *Proceedings of the IEEE, 107*(3), 509–517. DOI: 10.1109/JPROC.2019.2900622

Xie, Y., Zhao, S., Zhou, P., Lu, L., Liang, C., & Jiang, L. (2024). Estimating the impact of "humanizing" AI assistants. *International Journal of Human-Computer Interaction, 40*(24), 8876–8889. DOI: 10.1080/10447318.2023.2291614

Yuen, K. F., Cai, L., Qi, G., & Wang, X. (2021). Factors influencing autonomous vehicle adoption: An application of the technology acceptance model and innovation diffusion theory. *Technology Analysis and Strategic Management, 33*(5), 505–519. DOI: 10.1080/09537325.2020.1826423

Compilation of References

Abdelaziz, M. A., Ali, A. A., Swief, R. A., & Elazab, R. (2024). A reliable optimal electric vehicle charging station allocation. *Ain Shams Engineering Journal, 15*(7), 102763. DOI: 10.1016/j.asej.2024.102763

Abdolrasol, M. G. M., Ayob, A., Lipu, M. S. H., Ansari, S., Kiong, T. S., Saad, M. H. M., Ustun, T. S., & Kalam, A. (2024). Advanced data-driven fault diagnosis in lithium-ion battery management systems for electric vehicles: Progress, challenges, and future perspectives. *eTransportation*, Article 100374.

Abood, S., Al-Khafaji, H., & Al-Khafaji, M. H. (2025). Enhancing Collision Avoidance in Mobile Robots Using YOLOv5: A Lightweight Approach for Unstructured Environments. *Journal of Robotics and Control*. Vol. 6 No.2. https://journal.umy.ac.id/index.php/jrc/article/view/25856/11436

Abu, N. S., Bukhari, W. M., Adli, M. H., Maghfiroh, H., & Ma'arif, A. (2024). Advancements, challenges and safety implications of AI in autonomous vehicles: A comparative analysis of urban vs. highway environments. *Journal of Robotics and Control, 5*(3), 613–635. DOI: 10.18196/jrc.v5i3.21114

Acharya, D., Kuppan, K., & Divya, B. (2025). Agentic AI: Autonomous intelligence for complex goals—A comprehensive survey. *IEEE Access : Practical Innovations, Open Solutions, 13*, 18912–18936. DOI: 10.1109/ACCESS.2025.3532853

Adabara, I., Sadiq, B. O., Shuaibu, A. N., Danjuma, Y. I., & Venkateswarlu, M. (2025). A Review of Agentic AI in Cybersecurity: Cognitive Autonomy, Ethical Governance, and Quantum-Resilient Defense. *F1000 Research, 14*, 843. DOI: 10.12688/f1000research.169337.1 PMID: 41169871

Afoma, U. M., Singh, S., Mishra, A. K., Sharma, C. K., Gupta, K., Mishra, M. K., Roy, B., Verma, V. V., & Sharma, V. K. (2025). Integrating Artificial Intelligence in Environmental Monitoring: A Paradigm Shift in Data-Driven Sustainability. *EcoHealth, 23*(1), 38–57. Advance online publication. DOI: 10.1007/s10393-025-01752-8 PMID: 40875110

Agentic AI and the Future of Communications for Autonomous Vehicle (V2X) – IEEE ComSoc Technology Blog. (n.d.). Retrieved November 15, 2025, from https://techblog.comsoc.org/2025/07/14/agentic-ai-and-the-future-of-communications-for-autonomous-vehicle-v2x/

Ahmed, Q. W., Garg, S., Rai, A., Ramachandran, M., Jhanjhi, N. Z., Masud, M., & Baz, M. (2022). AI-Based Resource Allocation Techniques in Wireless Sensor Internet of Things Networks in Energy Efficiency with Data Optimization. Electronics (Switzerland), 11(13).

Ahmed, R., & Malik, K. (2025). Evaluating failure propagation in autonomous vehicle software stacks. *Software Quality Journal, 33*, 455–472. DOI: 10.1007/s11219-025-09645-1

Ahmed, S., Raza, A., & Iqbal, M. (2025). Real-time hazard estimation from multimodal sensor data in autonomous vehicles. *Signal Processing, 224*. Advance online publication. DOI: 10.1016/j.sigpro.2024.109876

Airehrour, D., Gutierrez, J., & Ray, S. K. (2018). A trust-based defence scheme for mitigating blackhole and selective forwarding attacks in the RPL routing protocol. *Australian Journal of Telecommunications and the Digital Economy*, *6*(1), 41–59. DOI: 10.18080/jtde.v6n1.138

Akbalık, F., Yıldız, A., Ertuğrul, Ö. F., & Zan, H. (2024). Engine fault detection by sound analysis and machine learning. *Applied Sciences (Basel, Switzerland)*, *14*(15), 6532.

Akhtar, M. A. K., Kumar, M., & Nayyar, A. (2024). Privacy and security considerations in explainable AI. In *Towards ethical and socially responsible explainable AI* (pp. 193–226). Springer., DOI: 10.1007/978-3-031-66489-2_7

Alam, T. (2024). Data privacy and security in autonomous connected vehicles in smart city environment. *Big Data and Cognitive Computing*, *8*(9), 95. DOI: 10.3390/bdcc8090095

Aldhanhani, T., Abraham, A., Hamidouche, W., & Shaaban, M. (2024). Future trends in smart green IoV: Vehicle-to-everything in the era of electric vehicles. *IEEE Open Journal of Vehicular Technology*, *5*, 278–297. Advance online publication. DOI: 10.1109/OJVT.2024.3358893

Aldughayfiq, B., Ashfaq, F., Jhanjhi, N. Z., & Humayun, M. (2023). YOLO-Based Deep Learning Model for pressure Ulcer detection and Classification. *Health Care*, *11*(9), 1222. DOI: 10.3390/healthcare11091222 PMID: 37174764

Alemayehu, H., & Sargolzaei, A. (2025). Testing and verification of connected and autonomous vehicles: A review. *Electronics (Basel)*, *14*(3), 600. DOI: 10.3390/electronics14030600

Alhabib, R., & Yadav, P. (2025). Data authorisation and validation in autonomous vehicles: A critical review. *Discover Applied Sciences*, *7*(7), 735. DOI: 10.1007/s42452-025-07333-2

Ali, M. A., & Dornaika, F. (2025). Agentic AI: A comprehensive survey of architectures, applications, and future directions. DOI: 10.1007/s10462-025-11422-4

Ali, M. R., & Patil, S. (2021). Gesture-controlled car using NodeMCU and Blynk IoT platform. [IJIRT]. *International Journal of Innovative Research in Technology*, *7*(8), 101–105.

Alon-Barkat, S., & Busuioc, M. (2023). Human–AI interactions in public sector decision making: "Automation bias" and "Selective adherence" to algorithmic advice. *Journal of Public Administration: Research and Theory*, *33*(1), 153–169. DOI: 10.1093/jopart/muac007

Alshuhail, A., Alshahrani, A., Mahgoub, H., Ghaleb, M., Darem, A. A., Aljehane, N. O., Alotaibi, M., & Alzahrani, F. (2025). Machine edge-aware IoT framework for real-time health monitoring: Sensor fusion and AI-driven emergency response in decentralized networks. *Alexandria Engineering Journal*, *129*, 1349–1361. DOI: 10.1016/j.aej.2025.08.030

Andersen, K., & Fagerholt, K. (2025). Maritime autonomous navigation: Lessons for AV land transport safety. *Safety Science*, *172*. Advance online publication. DOI: 10.1016/j.ssci.2025.106400

Anderson, J. M., Nidhi, K., Stanley, K. D., Sorensen, P., Samaras, C., & Oluwatola, O. A. (2014). *Autonomous vehicle technology: A guide for policymakers.* Rand Corporation.

Aniculaesei, A., & Elhajji, Y. (2025). Runtime monitoring approach to safeguard behavior of autonomous vehicles at traffic lights. *Electronics (Basel), 14*(12), 2366. DOI: 10.3390/electronics14122366

Ansarinejad, M., Ansarinejad, K., Lu, P., Huang, Y., & Tolliver, D. (2025). Autonomous vehicles in rural areas: A review of challenges, opportunities, and solutions. *Applied Sciences (Basel, Switzerland), 15*(8), 4195. DOI: 10.3390/app15084195

Anver, S. R., Deepambika, V. A., Rahiman, M. A., & Santhosh, R. (2025). Emotional Speech Generation: An approach using Convolutional Neural networks (CNN) based Generative Adversarial network. *Circuits, Systems, and Signal Processing, 44*(11), 8734–8756. Advance online publication. DOI: 10.1007/s00034-025-03224-4

Araújo, A. L. D., Sperandio, M., Calabrese, G., Faria, S. S., Cardenas, D. C., Martins, M. D., Saldivia-Siracusa, C., Giraldo-Roldán, D., Pedroso, C. M., Vargas, P. A., Lopes, M. A., Santos-Silva, A. R., Kowalski, L. P., & Moraes, M. C. (2025). ARTIFICIAL INTELLIGENCE IN HEALTHCARE APPLICATIONS TARGETING CANCER DIAGNOSIS – PART I: DATA STRUCTURE, PREPROCESSING AND DATA ORGANIZATION. *Oral Surgery, Oral Medicine, Oral Pathology and Oral Radiology, 140*(1), 79–88. Advance online publication. DOI: 10.1016/j.oooo.2025.01.004 PMID: 39893121

Asna, M., Shareef, H., Achikkulath, P., Mokhlis, H., Errouissi, R., & Wahyudie, A. (2021). *Analysis of an optimal planning model for electric vehicle fast-charging station in Al Ain City*. IEEE., DOI: 10.1109/ACCESS.2021.3081020

Atakishiyev, S., Salameh, M., & Goebel, R. (2024). Incorporating explanations into human-machine interfaces for trust and situation awareness in autonomous vehicles. *arXiv*. /arXiv.2404.07383DOI: 10.1109/IV55156.2024.10588812

Attaullah, M., Ali, M., Almufareh, M. F., Ahmad, M., Hussain, L., Jhanjhi, N., & Humayun, M. (2022). Initial stage COVID-19 detection system based on patients' symptoms and chest X-Ray images. *Applied Artificial Intelligence, 36*(1), 2055398. Advance online publication. DOI: 10.1080/08839514.2022.2055398

Automated vehicles for safety. NHTSA. (n.d.). https://www.nhtsa.gov/vehicle-safety/automated-vehicles-safety

Azeem, M., Ullah, A., Ashraf, H., Jhanjhi, N., Humayun, M., Aljahdali, S., & Tabbakh, T. A. (2021). FOG-Oriented secure and lightweight data aggregation in IOMT. *IEEE Access : Practical Innovations, Open Solutions, 9*, 111072–111082. DOI: 10.1109/ACCESS.2021.3101668

Babbar, H., Rani, S., Masud, M., Verma, S., Anand, D., & Jhanjhi, N. (2021). Load Balancing Algorithm for Migrating Switches in Software-Defined Vehicular Networks. *Computers, Materials & Continua, 67*(1), 1301–1316. https://doi.org/cmc.2021.014627. DOI: 10.32604/cmc.2021.014627

Baidu, Inc. (2025, February 18). *Baidu announces fourth quarter and fiscal year 2024 results*. https://ir.baidu.com/news-releases/news-release-details/baidu-announces-fourth-quarter-and-fiscal-year-2024-results

Bai, S., Wang, X., Gursoy, M. C., Jiang, G., & Xu, S. (2025). *Deep reinforcement learning for rechargeable UAV-assisted data collection from dense mobile sensor nodes.* IEEE., DOI: 10.1109/ACCESS.2025.3539888

Bandi, A., Kongari, B., Naguru, R., Pasnoor, S., & Vilipala, S. V. (2025). The rise of agentic AI: A review. *Future Internet, 17*(9), 404. DOI: 10.3390/fi17090404

Bansal, P., & Rana, M. (2025). Trustworthiness frameworks for explainable autonomous systems. *Journal of Ambient Intelligence and Humanized Computing*. Advance online publication. DOI: 10.1007/s12652-025-05671-3

Bayen, A. (2021, March 15). Eliminating Traffic Jams with Self-Driving Cars | Civil and Environmental Engineering. Ce.berkeley.edu. https://ce.berkeley.edu/news/2537

Bhamidipaty, V., Bhamidipaty, D. L., Guntoory, I., Bhamidipaty, K., Iyengar, K. P., Botchu, B., & Botchu, R. (2025). Revolutionizing Healthcare: The Impact of AI-Powered Sensors. *Wiley*, 355–373. DOI: 10.1002/9781394280735.ch18

Bhatia, P., & Wadhwa, S. (2025). Modeling stochastic risks in AV motion planning. *Journal of Intelligent & Robotic Systems, 109*, 567–590. DOI: 10.1007/s10846-025-01832-z

Biczi, D. (2025). Legal challenges for automated decision-making in self-driving vehicles—Liability issues and remedies. *Engineering Proceedings, 113*(1), 32. DOI: 10.3390/engproc2025113032

Birnbaum, S. L. (1980). Unmasking the Test for Design Defect: From Negligence [to Warranty] to Strict Liability to Negligence. *Vand. L. Rev., 33*, 593.

Bojarski, M., Del Testa, D., Dworakowski, D., Firner, B., Flepp, B., Goyal, P., . . . Zieba, K. (2016). End to end learning for self-driving cars. *arXiv preprint arXiv:1604.07316.*

Bonnefon, J. F., Shariff, A., & Rahwan, I. (2018). The moral machine experiment. *Nature, 563*(7729), 59–64. DOI: 10.1038/s41586-018-0637-6 PMID: 30356211

Bosch, C., & Kruger, D. (2024). AI chatbots as open educational resources: Enhancing student agency and self-directed learning. *Italian Journal of Educational Technology, 32*(1), 53–68.

Brown, K., & Miller, S. (2020). NoSQL databases for healthcare data management: A case study on MongoDB. *International Journal of Medical Informatics, 98*, 45–53.

Bruto Da Costa, A. A., Irvine, P., Dodoiu, T., Khastgir, S., & Jennings, P. (2025). Building a robust scenario library for safety assurance of automated driving systems: A review. *IEEE Access : Practical Innovations, Open Solutions, 13*, 117619–117655. DOI: 10.1109/ACCESS.2025.3584206

Buysse, L., Habli, I., Vanoost, D., & Pissoort, D. (2025). Safe autonomous systems in a changing world. *Safety Science, 191*, 106965. DOI: 10.1016/j.ssci.2025.106965

Büyük, M., Avşar, E., & İnci, M. (2022). Overview of smart home concepts through energy management systems, numerical research, and future perspective. *Energy Sources Part a Recovery Utilization and Environmental Effects*, 1–26. DOI: 10.1080/15567036.2021.2024924

Caesar, H., Bankiti, V., Lang, A. H., Vora, S., Liong, V. E., Xu, Q., & Beijbom, O. (2020). nuScenes: A multimodal dataset for autonomous driving. In *Proceedings of the IEEE/CVF Conference on Computer Vision and Pattern Recognition* (pp. 11621–11631). DOI: 10.1109/CVPR42600.2020.01164

Calboli, S., & Engelen, B. (2025). AI-enhanced nudging in public policy: Why to worry and how to respond. *Mind & Society*, *24*(2), 529–547. Advance online publication. DOI: 10.1007/s11299-025-00322-3

California Department of Motor Vehicles. (2023, October 24). *DMV statement on Cruise LLC suspension.* https://www.dmv.ca.gov/portal/news-and-media/dmv-statement-on-cruise-llc-suspension/

California Department of Motor Vehicles. (2026). *Autonomous vehicle disengagement reports.* https://www.dmv.ca.gov/portal/vehicle-industry-services/autonomous-vehicles/disengagement-reports

Campbell, T., & Lewis, M. (2025). Adversarial testing of AV control systems using synthetic perturbations. *Engineering Failure Analysis*, *162*. Advance online publication. DOI: 10.1016/j.engfailanal.2025.107765

Cardoso, R. C., & Ferrando, A. (2021). A review of agent-based programming for multi-agent systems. *Computers*, *10*(2), 16. DOI: 10.3390/computers10020016

Cecchini, D., Brantley, S., & Dubljević, V. (2023). Moral judgment in realistic traffic scenarios: Moving beyond the trolley paradigm for ethics of autonomous vehicles. *AI & Society*. Advance online publication. DOI: 10.1007/s00146-023-01813-y

Chakraborty, S., Das, R., & Kar, P. (2025). Trajectory prediction under uncertainty for risk-aware AV planning. *Transportation Research Part C, Emerging Technologies*, *169*. Advance online publication. DOI: 10.1016/j.trc.2025.104620

Chan, A. (2023, June). Harms from increasingly agentic algorithmic systems. In *Proceedings of the ACM Conference on Fairness, Accountability, and Transparency* (pp. 651–666). DOI: 10.1145/3593013.3594033

Chan, W., Gai, K., Yu, J., & Zhu, L. (2025). Blockchain-Assisted Self-Sovereign Identities on Education: A survey. *Blockchains*, *3*(1), 3. DOI: 10.3390/blockchains3010003

Chattopadhyay, S., & Banerjee, P. (2025). Ethical frameworks for autonomous decision-making under time pressure. *Ethics and Information Technology*, *27*. Advance online publication. DOI: 10.1007/s10676-025-09612-8

Chavez, F., Chen, J., Tung, C., Hu, C., & Khandelwal, S. (2025). Quantum-enhanced semiconductor reliability. *Solid-State Electronics*, *229*. Advance online publication. DOI: 10.1016/j.sse.2025.109154

Chemmakha, M., Omar, H., & Lazaar, M. (2025). Toward a New Approach for Internet of Things (IoT) Intrusion Detection Based on Feature Selection and Tabular Generative Adversarial Network. *Intelligenza Artificiale.*, *0*(0). Advance online publication. DOI: 10.1177/17248035251361258

Chen, C., Liu, J., Tan, H., Li, X., Wang, K. I., Li, P., Sakurai, K., & Dou, D. (2024). Trustworthy federated learning: Privacy, security, and beyond. *Knowledge and Information Systems*. Advance online publication. DOI: 10.1007/s10115-024-02285-2

Chen, L., & Gupta, R. (2019). Improving data security in healthcare systems with role-based access control. *Health IT Journal*, *12*(4), 210–219.

Chen, S., Wang, H., & Meng, Q. (2019). Designing autonomous vehicle incentive program with uncertain vehicle purchase price. *Transportation Research Part C, Emerging Technologies*, *103*, 226–245. DOI: 10.1016/j.trc.2019.04.013

Chitraju, S. (2025). <div> AI and Behavioral Biometrics for Secure Financial Transactions </div> *SSRN*. DOI: 10.2139/ssrn.5353743

Chong, T., Yu, T., Keeling, D. I., & de Ruyter, K. (2021). AI-chatbots on the services frontline addressing the challenges and opportunities of agency. *Journal of Retailing and Consumer Services*, *63*, 102735. DOI: 10.1016/j.jretconser.2021.102735

Christensen, H. S. (2021). A conjoint experiment of how design features affect evaluations of participatory platforms. *Government Information Quarterly*, *38*(1), 101538.

Cividino, S., Cappelli, A., Belluco, P., Rinaldi, F., & Avramović, A. (2025). Sensor fusion for next-gen autonomous systems. *Sensors (Basel)*, *25*(21). Advance online publication. DOI: 10.3390/s25216749 PMID: 41228972

Colonna, K. (2012). Autonomous cars and tort liability. *Case W. Res. JL Tech. & Internet*, *4*, 81.

Connors, J., & Stewart, R. (2025). Validation of black-box AV modules through interpretable surrogate models. *Expert Systems with Applications*, *240*. Advance online publication. DOI: 10.1016/j.eswa.2024.123789

Cranford, C. H. (1971). Torts—Product Liability—Circumstantial Evidence and Proof of Defect. *North Carolina Law Review*, *50*, 417.

Cunneen, M., Mullins, M., Murphy, F., Shannon, D., Furxhi, I., & Ryan, C. (2020). Autonomous vehicles and avoiding the trolley dilemma: Ethical framing and perception. *Cybernetics and Systems*, *51*(1), 59–80. DOI: 10.1080/01969722.2019.1660541

Cyberly. (2025, July 13). *How does AI detect anomalies or unusual behaviour?* https://www.cyberly.org/en/how-does-ai-detect-anomalies-or-unusual-behaviour/index.html?utm_source

Dandotiya, M., Goyal, N. K., Yadav, P., Sankhla, K., & Kumari, M. (2026). GENALYTICS: A deep dive into real-time insights via generative AI. In Elbaghazaoui, B. (Eds.), *Generative AI-powered data architectures: From governance to autonomous analytics* (pp. 105–134). IGI Global Scientific Publishing., DOI: 10.4018/979-8-3373-5616-7.ch005

Danks, D., & London, A. J. (2017, August). Algorithmic bias in autonomous systems. *IJCAI (United States)*, *17*, 4691–4697.

Davnall, R. (2020). Solving the self-driving car trolley problem using risk theory and vehicle dynamics. *Science and Engineering Ethics*, *26*(1), 431–449. DOI: 10.1007/s11948-019-00102-6 PMID: 30937627

Debouk, R. (2019). Overview of the second edition of ISO 26262: Functional safety—road vehicles. *Journal of System Safety*, *55*(1), 13–21.

Deng, Z., Guo, Y., Han, C., Ma, W., Xiong, J., Wen, S., & Xiang, Y. (2024). AI agents under threat: A survey of key security challenges and future pathways. *arXiv*. https://arxiv.org/abs/2406.02630

Diamantis, M. E. (2023). Vicarious Liability for AI. *Industrial Law Journal, 99*, 317.

Dogra, V., Singh, A., Verma, S., Kavita, , Jhanjhi, N. Z., & Talib, M. N. (2021). Analyzing DistilBERT for Sentiment Classification of Banking Financial News. *Lecture Notes in Networks and Systems, 248*, 501–510. DOI: 10.1007/978-981-16-3153-5_53

Dong, C., Wang, H., Li, Y., Liu, Y., & Chen, Q. (2019). Economic comparison between vehicle-to-vehicle (V2V) and vehicle-to-infrastructure (V2I) at freeway on-ramps based on microscopic simulations. *IET Intelligent Transport Systems, 13*(11), 1726–1735. DOI: 10.1049/iet-its.2018.5537

Dosovitskiy, A., Ros, G., Codevilla, F., Lopez, A., & Koltun, V. (2017, October). CARLA: An open urban driving simulator. In *Conference on Robot Learning* (pp. 1–16). PMLR.

Duarte, G., Ramos, F., & Santos, L. (2025). Fail-safe braking in autonomous driving: A probabilistic formulation. *Mechanical Systems and Signal Processing, 224*. Advance online publication. DOI: 10.1016/j.ymssp.2025.110345

Duarte, L., Ferreira, M., Gonçalves, R., & Baptista, P. (2025). Autonomous decision-making under uncertainty: A safety-oriented approach. *Accident; Analysis and Prevention, 196*. Advance online publication. DOI: 10.1016/j.aap.2025.107308

Durante, Z.. (2024). Agent AI: Surveying the horizons of multimodal interaction. *arXiv*. https://arxiv .org/abs/2401.03568

Durlik, I., Miller, T., Kostecka, E., Zwierzewicz, Z., & Łobodzińska, A. (2024). Cybersecurity in autonomous vehicles. *Electronics (Basel), 13*(13), 2654. DOI: 10.3390/electronics13132654

Dutta Roy, T. P. (2025). Bioethics artificial intelligence advisory (BAIA). *Cureus*. Advance online publication. DOI: 10.7759/cureus.80494 PMID: 40083589

Ehsani, J. P., Hellinger, A., Stephens, D. K., Shin, M. R., Michael, J., McCourt, A., & Vernick, J. (2022). State laws for autonomous vehicle safety, equity, and insurance. *The Journal of Law, Medicine & Ethics, 50*(3), 569–582. DOI: 10.1017/jme.2022.96 PMID: 36398642

Ehsani, J. P., Michael, J. P., Igusa, T., Mueller, J., Chang, C.-H., & Yenokyan, G. (2024). Advancing transportation equity and safety through autonomous vehicles. *Health Equity, 8*(1), 143–146. DOI: 10.1089/heq.2023.0107 PMID: 38505763

Ekatpure, R. (2023). Machine Learning for Enhancing Vehicle Safety and Collision Avoidance Systems in Automotive Development: Techniques, Models, and Real-World Applications. *Journal of Computational Intelligence and Robotics, 3*(2), 1–43. https://thesciencebrigade.com/jcir/article/view/257

Emory, K., Douma, F., & Cao, J. (2022). Autonomous vehicle policies with equity implications: Patterns and gaps. *Transportation Research Interdisciplinary Perspectives, 13*, 100521. Advance online publication. DOI: 10.1016/j.trip.2021.100521

Epstein, R. A. (2021). Liability rules for autonomous vehicles. *Journal of Entrepreneurship and Public Policy, 10*(2), 218–234.

Eriksson, A., & Stanton, N. A. (2017). Takeover time in highly automated vehicles: Noncritical transitions to and from manual control. *Human Factors, 59*(4), 689–705. DOI: 10.1177/0018720816685832 PMID: 28124573

Escribano, J. E., & Chen, X. (2025). Risk-aware reinforcement learning for autonomous navigation. *Robotics and Autonomous Systems, 182*. Advance online publication. DOI: 10.1016/j.robot.2025.104572

Eser, A. (2025, May 30). *Password hacking statistics Statistics: ZIPDO Education Reports 2025*. ZipDo. https://zipdo.co/password-hacking-statistics/?utm_source

Eslami, A., & Yu, J. (2025). Security risks of agentic vehicles: A systematic analysis of cognitive and cross-layer threats. *arXiv preprint.*

European Commission. (2025). *Rules for trustworthy artificial intelligence in the EU*. https://eur-lex.europa.eu/EN/legal-content/summary/rules-for-trustworthy-artificial-intelligence-in-the-eu.html

European Parliament and Council of the European Union. (2019). Regulation (EU) 2019/2144 on type-approval requirements for motor vehicles. https://eur-lex.europa.eu/legal-content/en/ALL/?uri=CELEX%3A32019R2144

European Parliament and Council of the European Union. (2022). Directive (EU) 2022/2557 on the resilience of critical entities. https://eur-lex.europa.eu/legal-content/EN/TXT/PDF/?uri=CELEX%3A32022L2557

Evans, K., de Moura, N., Chauvier, S., Chatila, R., & Doğan, E. (2020). Ethical decision-making in autonomous vehicles: The AV ethics project. *Science and Engineering Ethics, 26*(6), 3285–3312. DOI: 10.1007/s11948-020-00272-8 PMID: 33048325

Fang, Y., Teixeira, R., Min, H., Lei, X., Chen, S., & Zhao, X. (2023, August). A Safety Enhancing Framework for Autonomous Vehicles with a Resilience Rationale. In 2023 7th International Conference on Transportation Information and Safety (ICTIS) (pp. 1699-1705). IEEE.

Farooqi, A., Das, D., Tripathi, M., Dutta, R., Goyal, N. K., & Sheikh, M. F. (2026). Navigating ethical frontiers in AI-driven geriatric care. In Stolakis, K., & Efthymiou, I. (Eds.), *AI for geriatric care in an aging society: Ethical, clinical, and policy challenges* (pp. 243–266). IGI Global Scientific Publishing., DOI: 10.4018/979-8-3373-5462-0.ch009

Farooq, M., & Singh, R. (2025). Machine learning uncertainty quantification in AV planning. *Engineering Applications of Artificial Intelligence, 137*. Advance online publication. DOI: 10.1016/j.engappai.2025.107256

Fedorov, V., & Curran, N. T. (2024). The implementation of ethical AI within autonomous vehicles. *Journal of Student Research, 13*(1), 5973. Advance online publication. DOI: 10.47611/jsrhs.v13i1.5973

Fernández Llorca, D., Hamon, R., Junklewitz, H., Grosse, K., Kunze, L., Seiniger, P., Swaim, R., Reed, N., Alahi, A., Gómez, E., Sánchez, I., & Kriston, A. (2025). Testing autonomous vehicles and AI. *European Transport Research Review, 17*(1), 38. DOI: 10.1186/s12544-025-00732-x

Ferrag, M. A., Tihanyi, N., & Debbah, M. (2025). From llm reasoning to autonomous ai agents: A comprehensive review. *arXiv preprint arXiv:2504.19678.*

Ferrara, E. (2024). Fairness and Bias in Artificial Intelligence: A Brief Survey of Sources, Impacts, and Mitigation Strategies. *Sci*, *6*(1), 3. DOI: 10.3390/sci6010003

Feuerriegel, S., Hartmann, J., Janiesch, C., & Zschech, P. (2024). Generative AI. *Business & Information Systems Engineering*, *66*(1), 111–126. DOI: 10.1007/s12599-023-00834-7

Fossa, F. (2025). Artificial intelligence and human autonomy: The case of driving automation. *AI & Society*, *40*(3), 1851–1862. DOI: 10.1007/s00146-024-01955-7

Francisco, A. M. B., Monteiro, J., & Cardoso, P. J. S. (2023). *A digital twin of charging stations for fleets of electric vehicles*. IEEE., DOI: 10.1109/ACCESS.2023.3330833

Fremont, D. J., et al. (2020). Formal scenario-based testing of autonomous vehicles. *IEEE ITSC*, 1–8.

Fuller, A., Fan, Z., Day, C., & Barlow, C. (2020). Digital twin: Enabling technologies, challenges and open research. *IEEE Access : Practical Innovations, Open Solutions*, *8*, 108952–108971. DOI: 10.1109/ACCESS.2020.2998358

Fu, Y., Gong, Y., Shi, C., Zheng, C., You, G., & Xiao, W. (2025). Optimizing EV charging station carrying capacity considering coordinated multi-flexibility resources. *World Electric Vehicle Journal*, *16*(7), 381. DOI: 10.3390/wevj16070381

Geisslinger, M., Poszler, F., Betz, J., Lütge, C., & Lienkamp, M. (2021). Autonomous driving ethics: From trolley problem to ethics of risk. *Philosophy & Technology*, *34*(4), 1033–1055. DOI: 10.1007/s13347-021-00449-4

Geisslinger, M., Poszler, F., & Lienkamp, M. (2023). An ethical trajectory planning algorithm for autonomous vehicles. *Nature Machine Intelligence*, *5*(2), 137–144. DOI: 10.1038/s42256-022-00607-z

Geistfeld, M. A. (2017). A roadmap for autonomous vehicles: State tort liability, automobile insurance, and federal safety regulation. *California Law Review*, *105*, 1611.

Gheisari, M., Khan, W. Z., Najafabadi, H. E., McArdle, G., Rabiei-Dastjerdi, H., Liu, Y., Fernández-Campusano, C., & Abdalla, H. B. (2024). CAPPAD privacy-preservation solution. *Applied Intelligence*, *54*(4), 3417–3428. DOI: 10.1007/s10489-023-04991-w

Gill, S. P. (2024). Empathy and AI: Cognitive empathy or emotional (affective) empathy? *AI & Society*, *39*(6), 2641–2642. DOI: 10.1007/s00146-024-02118-4

Gless, S., & Ligeti, K. (2024). *Regulating driving automation in the European Union – Criminal liability on the road ahead?* New Journal of European Criminal Law., DOI: 10.1177/20322844231213336

Gomes, D., Saif, A. S., & Nandi, D. (2020, January). Robust underwater object detection with autonomous underwater vehicle: A comprehensive study. In *Proceedings of the international conference on computing advancements* (pp. 1-10). DOI: 10.1145/3377049.3377052

Gong, C.-S. A., Su, C.-H. S., Chen, Y.-H., & Guu, D.-Y. (2022). How to implement automotive fault diagnosis using artificial intelligence scheme. *Micromachines*, *13*(9), 1380. DOI: 10.3390/mi13091380 PMID: 36144003

Gordillo Chávez, D., Cloarec, J., & Meyer-Waarden, L. (2024). Opening the moral machine's cover: How algorithmic aversion shapes autonomous vehicle adoption. *Transportation Research Part A, Policy and Practice*, *187*, 104193. Advance online publication. DOI: 10.1016/j.tra.2024.104193

Gridach, M., Nanavati, J., Abidine, K. Z. E., Mendes, L., & Mack, C. (2025). Agentic ai for scientific discovery: A survey of progress, challenges, and future directions. *arXiv preprint arXiv:2503.08979.*

Gros, C., Martens, M., Aliman, N.-M., Kester, L., & Werkhoven, P. (2025). Ethical frameworks for automated vehicles: A systematic analysis and design. *AI and Ethics*, *5*(6), 6041–6062. DOI: 10.1007/s43681-025-00803-8

Guenat, S., Purnell, P., Davies, Z. G., Nawrath, M., Stringer, L. C., Babu, G. R., Balasubramanian, M., Ballantyne, E. E. F., Bylappa, B. K., Chen, B., De Jager, P., Del Prete, A., Di Nuovo, A., Ehi-Eromosele, C. O., Eskandari Torbaghan, M., Evans, K. L., Fraundorfer, M., Haouas, W., Izunobi, J. U., & Dallimer, M. (2022). Meeting sustainable development goals via robotics and autonomous systems. *Nature Communications*, *13*(1), 3559. DOI: 10.1038/s41467-022-31150-5 PMID: 35729171

Guerrero-Silva, J. A., Romero-Gelvez, J. I., Aristizábal, A. J., & Zapata, S. (2025). Optimization and trends in EV charging infrastructure: A PCA-based systematic review. *World Electric Vehicle Journal*, *16*(7), 345. DOI: 10.3390/wevj16070345

Guo, Y., Souders, D., Labi, S., Peeta, S., Benedyk, I., & Li, Y. (2021). Paving the way for autonomous vehicles: Understanding autonomous vehicle adoption and vehicle fuel choice under user heterogeneity. *Transportation Research Part A, Policy and Practice*, *154*, 364–398. DOI: 10.1016/j.tra.2021.10.018

Gupta, L., Jain, R., & Vaszkun, G. (2016). UAV communication networks survey. *IEEE Communications Surveys and Tutorials*, *18*(2), 1123–1152. DOI: 10.1109/COMST.2015.2495297

Gupta, R., & Bedi, H. (2025). Fail-operational architectures for urban autonomous shuttle systems. *Transportation Research Part F: Traffic Psychology and Behaviour*, *103*. Advance online publication. DOI: 10.1016/j.trf.2025.104321

Gurney, J. K. (2013). Sue my car not me: Products liability and accidents involving autonomous vehicles. U. Ill. JL Tech. & Pol'y, 247.

Hamdare, S., Brown, D. J., Cao, Y., Aljadi, M., Kaiwartya, O., Yadav, R., Vyas, P., & Jugran, M. (2024). EV charging management and security for multi-charging stations environment. *IEEE Open Journal of Vehicular Technology*, *5*, 807–824. Advance online publication. DOI: 10.1109/OJVT.2024.3418201

Hanif, M., Ashraf, H., Jalil, Z., Jhanjhi, N. Z., Humayun, M., Saeed, S., & Almuhaideb, A. M. (2022). AI-Based Wormhole Attack Detection Techniques in Wireless Sensor Networks. Electronics (Switzerland), 11(15). https://doi.org/electronics11152324

Hao, X., Demir, E., & Eyers, D. (2025). Beyond human-in-the-loop. *Sustainable Futures : An Applied Journal of Technology, Environment and Society*, *10*, 101152. DOI: 10.1016/j.sftr.2025.101152

Hartwich, F., et al. (2021). Passenger experience in automated vehicles. *Frontiers in Human Dynamics, 3.*

Hawasly, M., Sadeghi, J., Antonello, M., Albrecht, S. V., Redford, J., & Ramamoorthy, S. (2022). Perspectives on the system-level design of a safe autonomous driving stack. *AI Communications*, *35*(4), 285–294. DOI: 10.3233/AIC-220148

Hoffman, R., & Klein, G. (2025). Human–autonomy teaming and decision transparency in AVs. *Human Factors*, *67*(1). Advance online publication. DOI: 10.1177/00187208241234567

Holst, M., & Becker, S. (2025). Verification of autonomous vehicle decision modules using formal logic. *Formal Aspects of Computing*, *38*, 221–240. DOI: 10.1007/s00165-025-00678-3

Hopkins, D., & Schwanen, T. (2021). Talking about automated vehicles: What do levels of automation do? *Technology in Society*, *64*, 101488.

Horowitz, A., & Levinson, D. (2025). Cognitive models for anticipatory decision-making in AVs. *Cognitive Systems Research*, *84*, 45–60. DOI: 10.1016/j.cogsys.2024.102456

Hossain, M. N., Rahman, M. M., & Ramasamy, D. (2024). Artificial intelligence-driven vehicle fault diagnosis to revolutionize automotive maintenance: A review. *Computer Modeling in Engineering & Sciences*, *141*(2), 951–996. DOI: 10.32604/cmes.2024.056022

Hosseini, S., & Seilani, H. (2025). Role of agentic AI. *Array (New York, N.Y.)*, *26*, 100399. DOI: 10.1016/j.array.2025.100399

Huang, Z., Sheng, Z., Ma, C., & Chen, S. (2024). HAIM-DRL: Enhanced human-in-the-loop reinforcement learning for safe and efficient autonomous driving. *arXiv preprint.*

Huang, T., Liu, J., Zhou, X., Nguyen, D. C., Azghadi, M. R., Xia, Y., & Sun, S. (2025). Vehicle-to-everything cooperative perception for autonomous driving. *Proceedings of the IEEE.*

Huberman, P. (2021). Tort law, corrective justice and the problem of autonomous-machine-caused harm. *Canadian Journal of Law and Jurisprudence*, *34*(1), 105–147.

Hu, J., Wang, Y., Cheng, S., Xu, J., Wang, N., Fu, B., Ning, Z., Li, J., Chen, H., Feng, C., & Zhang, Y. (2025). A survey of decision-making and planning methods for self-driving vehicles. *Frontiers in Neurorobotics*, *19*, 1451923. DOI: 10.3389/fnbot.2025.1451923 PMID: 40041453

Humayun, M., Khalil, M. I., Almuayqil, S. N., & Jhanjhi, N. Z. (2023). Framework for Detecting Breast Cancer Risk Presence using Deep Learning. *Electronics (Basel)*, *12*(2), 403. DOI: 10.3390/electronics12020403

Hu, Q., Yang, B., Zhang, K., Escribano, J. E., Chen, X., & Ouyang, L. (2025). Sustainable operational strategies for mixed fleets: Integrating autonomous and human-driven taxis with heterogeneous energy types. *Communications in Transportation Research*, *5*, 100171. Advance online publication. DOI: 10.1016/j.commtr.2025.100171

International Organization for Standardization. (2018). *Road vehicles—Functional safety—Part 1: Vocabulary (ISO 26262-1:2018).* https://www.iso.org/standard/68383.html

International Organization for Standardization. (2022). *Road vehicles—Safety of the intended functionality (ISO 21448:2022).* https://www.iso.org/standard/77490.html

International, S. A. E. (2021, May 2). *SAE levels of driving automation™ refined for clarity and international audience.* https://www.sae.org/news/blog/sae-levels-driving-automation-clarity-refinements

ISO. (2018). *ISO 26262: Road vehicles – Functional safety.* International Organization for Standardization.

Istrate, O. (2024). AI agents in education: An early systematic review of emerging roles, potential, and limitations. *Revista de Pedagogie Digitala, 3*(1), 24–30. DOI: 10.61071/RPD.2496

Iunusova, E., & Archenti, A. (2025). Lab-to-Field Generalization Gap: Assessment of transfer learning for bearing fault detection. *Applied Sciences (Basel, Switzerland), 15*(12), 6804. DOI: 10.3390/app15126804

Jabbour, J., & Reddi, V. J. (2024). Generative AI agents in autonomous machines: A safety perspective. *arXiv preprint.*

Jabeen, T., Jabeen, I., Ashraf, H., Jhanjhi, N. Z., Yassine, A., & Hossain, M. S. (2023). An Intelligent Healthcare System Using IoT in Wireless Sensor Network. *Sensors (Basel), 23*(11), 5055. https://doi.org/s23115055. DOI: 10.3390/s23115055 PMID: 37299782

Jin, X., Park, S., & Kim, H. (2025). Human-centered explainability in autonomous driving models. *IEEE Transactions on Intelligent Transportation Systems.* Advance online publication. DOI: 10.1109/TITS.2025.1234567

Ji, W., Yuan, Q., Cheng, G., Yu, S., Wang, M., Shen, Z., & Yang, T. (2023). Traffic accidents of autonomous vehicles based on knowledge mapping: A review. [English edition]. *Journal of Traffic and Transportation Engineering, 10*(6), 1061–1073. DOI: 10.1016/j.jtte.2023.09.003

Ji, Y., Zhou, Z., Yang, Z., Huang, Y., Zhang, Y., Zhang, W., Xiong, L., & Yu, Z. (2024). Toward autonomous vehicles: A survey on cooperative vehicle-infrastructure system. *iScience, 27*(5), 109751. DOI: 10.1016/j.isci.2024.109751 PMID: 38706867

Johnson, T., & White, P. (2018). Responsive web design for healthcare applications: Enhancing usability and access. *Journal of Digital Health, 14*(3), 89–97.

Joshi, S. (2025). Review of autonomous systems and collaborative AI agent frameworks. *International Journal of Scientific Research Archives, 14*(2), 961–972.

Kalodanis, K., Rizomiliotis, P., Feretzakis, G., Papapavlou, C., & Anagnostopoulos, D. (2025). High-Risk AI Systems—LIE Detection Application. *Future Internet, 17*(1), 26. DOI: 10.3390/fi17010026

Kalra, A. (2025). Emerging technologies in biometrics: artificial intelligence and machine learning. In *CRC Press eBooks* (pp. 3–26). DOI: 10.1201/9781003509554-2

Kapoor, S., Stroebl, B., Siegel, Z. S., Nadgir, N., & Narayanan, A. (2024). AI agents that matter. *arXiv.* https://arxiv.org/abs/2407.01502

Karimi, A., Adnane, A., Phillips, I. W., & Benkhelifa, E. (2025). Connected Vehicles Data Classification and the Influence of a Sustainable Data Governance for Optimal Utilisation of In-Vehicle Data. In *ICISSP (2)* (pp. 643-650). DOI: 10.5220/0013371100003899

Karmaker, A. K., Hossain, M. A., Pota, H. R., Onen, A., & Jung, J. (2023). *Energy management system for hybrid renewable energy-based electric vehicle charging station*. IEEE., DOI: 10.1109/ACCESS.2023.3259232

Karunakaran, D., Berrio Perez, J. S., & Worrall, S. (2023). Generating edge cases for testing autonomous vehicles using real-world data. *Sensors (Basel)*, *24*(1), 108. DOI: 10.3390/s24010108 PMID: 38202970

Karunanayake, N. (2025). Next-generation agentic AI for transforming healthcare. *Informatics in Health*, *2*(2), 73–83. DOI: 10.1016/j.infoh.2025.03.001

Kaul, V. R., & Kumar, A. (2025). Intelligent vehicular networks: A framework for safety and efficiency. *Vehicular Communications*, *50*. Advance online publication. DOI: 10.1016/j.vehcom.2025.100750

Kendall, A., Gal, Y., & Cipolla, R. (2018). Multi-task learning using uncertainty to weigh losses for scene geometry and semantics. In *Proceedings of the IEEE Conference on Computer Vision and Pattern Recognition* (pp. 7482–7491).

Khajehzadeh, L., Barati, H., & Barati, A. (2024). A lightweight authentication and authorization method in IoT-based medical care. *Multimedia Tools and Applications*, *84*(12), 11137–11176. Advance online publication. DOI: 10.1007/s11042-024-19379-2

Khan, A. R., Jamlos, M. F., Osman, N., Ishak, M. I., Dzaharudin, F., Yeow, Y. K., & Khairi, K. A. (2021). DSRC technology in vehicle-to-vehicle (V2V) and vehicle-to-infrastructure (V2I) IoT system for intelligent transportation system (ITS): A review. In *Recent trends in mechatronics towards industry 4.0* (pp. 97–106).

Khandagle, S. R., & Patil, M. M. (2018, August). Smart vehicle control using hand gesture and accelerometer. In *Proceedings of the International Conference on Inventive Computing Technologies (ICICT)* (pp. 1095–1100). Coimbatore, India.

Kim, D., Lee, J., & Park, Y. (2025). Continuous monitoring and prediction of AV system failures. *Reliability Engineering & System Safety*, *245*. Advance online publication. DOI: 10.1016/j.ress.2024.109876

Kim, P. Y. (2020). Where We're Going, We Don't Need Drivers: Autonomous Vehicles and AI-Chaperone Liability. *Cath. UL Rev.*, *69*, 341.

Kim, S. (2017). Crashed software: Assessing product liability for software defects in automated vehicles. *Duke L. & Tech. Rev.*, *16*, 300.

Kiran, B. R., Sobh, I., Talpaert, V., Mannion, P., Al Sallab, A. A., Yogamani, S., & Pérez, P. (2021). Deep reinforcement learning for autonomous driving: A survey. *IEEE Transactions on Intelligent Transportation Systems*, *23*(6), 4909–4926. DOI: 10.1109/TITS.2021.3054625

Kiyani, F. F., Hamid, B., Humayun, M., Sindiramutty, S. R. L., & Chowdhury, S. (2024). Discovery of Influential Publications Using Research Article's Usage Context. *2024 International Conference on Emerging Trends in Networks and Computer Communications (ETNCC)*, 1–7. DOI: 10.1109/ETNCC63262.2024.10767576

Königs, P. (2023). Of trolleys and self-driving cars: What machine ethicists can and cannot learn from trolleyology. *Utilitas*, *35*(1), 70–87. DOI: 10.1017/S0953820822000395

Kovács, S., & Botzheim, J. (2025). Incremental modeling and its application in robot intelligence: A survey. *Biomedical Signal Processing and Control, 110*. Advance online publication. DOI: 10.1016/j.bspc.2025.108046

Ko, W., Shim, M., Park, S., Lee, S., & Yun, I. (2024). Development of test cases for automated vehicle driving safety assessment using driving trajectories. *Sensors (Basel), 24*(24), 7981. DOI: 10.3390/s24247981 PMID: 39771717

Kreuk, F., Barak, A., Aviv-Reuven, S., Baruch, M., Pinkas, B., & Keshet, J. (2018). Deceiving End-to-End Deep Learning Malware Detectors using Adversarial Examples. *arXiv (Cornell University)*. https://doi.org//arxiv.1802.04528DOI: 10.48550

Krishnan, S., Thangaveloo, R., Rahman, S. B. A., & Sindiramutty, S. R. (2021). Smart Ambulance Traffic Control system. *Trends in Undergraduate Research, 4*(1), c28–c34. DOI: 10.33736/tur.2831.2021

Krügel, S., & Uhl, M. (2024). The risk ethics of autonomous vehicles: An empirical approach. *Scientific Reports, 14*(1), 960. Advance online publication. DOI: 10.1038/s41598-024-51313-2 PMID: 38200075

Kruzikova, A., Muzik, M., Knapova, L., Dedkova, L., Smahel, D., & Matyas, V. (2023). Two-factor authentication time: How time-efficiency and time-satisfaction are associated with perceived security and satisfaction. *Computers & Security, 138*, 103667. DOI: 10.1016/j.cose.2023.103667

Kubica, M. L. (2022). Autonomous vehicles and liability law. *The American Journal of Comparative Law, 70*(Supplement_1), i39–i69. DOI: 10.1093/ajcl/avac015

Kumar, & Gupta, R. (2019). IoT-based vehicle automation and tracking using NodeMCU. *International Journal of Computer Applications, 182*(40), 25–29.

Kumari, M., & Goyal, N. K. (2026). Artificial intelligence in the cryptocurrency ecosystem: Crime mechanisms, detection strategies, and legal perspectives. In Lin, L. (Ed.), *Innovations in cryptocrime and financial fraud* (pp. 245–282). IGI Global Scientific Publishing., DOI: 10.4018/979-8-3373-0675-9.ch007

Kumari, M., Kumawat, R., Dutta, R., Khan, S. T., Goyal, N. K., & Sheikh, M. F. (2026). Wireless communication and IoT integration in smart clothing. In Saxena, P. (Eds.), *Next-generation electronic textiles and conductive materials for smart wearables* (pp. 195–232). IGI Global Scientific Publishing., DOI: 10.4018/979-8-3373-4287-0.ch008

Kumar, M., Rattan, N., & Mondal, S. (2025). Sensor systems for autonomous vehicles: Functionality and reliability challenges in adverse environmental conditions. *Measurement*, •••, 119215.

Kumar, P. (2025). Next-generation secure authentication and access control architectures: Advanced techniques for securing distributed systems in modern enterprises. *International Journal of Computational and Experimental Science and Engineering, 11*(3). Advance online publication. DOI: 10.22399/ijcesen.3294

Kuru, K. (2021). Human-on-the-loop teleoperation. *IEEE OJITS, 2*, 448–469.

Kuznietsov, A., Gyevnar, B., Wang, C., Peters, S., & Albrecht, S. V. (2024). Explainable AI for safe and trustworthy autonomous driving: A systematic review. *IEEE Transactions on Intelligent Transportation Systems, 25*(12), 19342–19364. DOI: 10.1109/TITS.2024.3474469

Ladak, A., Wilks, M., Loughnan, S., & Anthis, J. R. (2025). Robots, chatbots, self-driving cars: Perceptions of mind and morality across artificial intelligences. *ACM Transactions on Human-Robot Interaction*, *14*(3), 26. Advance online publication. DOI: 10.1145/3706598.3713130

Lalramsangi, V., Garg, Y. K., & Sharma, S. N. (2025). Route choices to access public open spaces in hill cities. *Environment & Urbanization Asia*, *16*(2), 283–299. DOI: 10.1177/09754253251388721

Land Transport Authority. (2026). *Autonomous vehicles*. Government of Singapore. https://www.lta.gov.sg/content/ltagov/en/industry_innovations/technologies/autonomous_vehicles.html

Larsen, K. D. (1984). Strict products liability and the risk-utility test for design defect: An economic analysis. *Columbia Law Review*, *84*(8), 2045–2067.

Lee, D., & Nguyen Long, L. A. (2025). Privacy challenges of automated vehicles.

Lee, D., & Hess, D. J. (2022). Public concerns and connected and automated vehicles: Safety, privacy, and data security. *Humanities & Social Sciences Communications*, *9*(1), 90. Advance online publication. DOI: 10.1057/s41599-022-01110-x

Leikas, J., Koivisto, R., & Gotcheva, N. (2019). Ethical framework for designing autonomous intelligent systems. *Journal of Open Innovation*, *5*(1), 18. DOI: 10.3390/joitmc5010018

Leite, D., Martins, A.Jr, Rativa, D., de Oliveira, J. F. L., & Maciel, A. M. A. (2022). An automated machine learning approach for real-time fault detection and diagnosis. *Sensors (Basel)*, *22*(16), 6138. DOI: 10.3390/s22166138 PMID: 36015899

Lemann, A. B. (2019). Autonomous vehicles, technological progress, and the scope problem in products liability. Journal of tort law, 12(2), 157-212.

Leung, K., Schmerling, E., Zhang, M., Chen, M., Talbot, J., Gerdes, J. C., & Pavone, M. (2020). On infusing reachability-based safety assurance within planning frameworks for human–robot vehicle interactions. *The International Journal of Robotics Research*, *39*(10-11), 1326–1345. DOI: 10.1177/0278364920950795

Levinson, J., Askeland, J., Becker, J., Dolson, J., Held, D., Kammel, S., & Thrun, S. (2011). Toward fully autonomous driving: Systems and algorithms. In *IEEE Intelligent Vehicles Symposium* (pp. 163–168). DOI: 10.1109/IVS.2011.5940562

Linqiang, Y., Sindiramutty, S. R. L., Ashraf, H., Muzammal, S. M., Balakrishnan, S. P., Gupta, S., & Kavita, N. (2024). Intelligent Household Waste Classification System Based on Machine Learning. *2024 International Conference on Emerging Trends in Networks and Computer Communications (ETNCC)*, 760–768. DOI: 10.1109/ETNCC63262.2024.10767563

Liu, G., Sheng, J., & Tao, Z. (2025). Application and design of a decision-making model in ethical dilemma for self-driving cars. *Scientific Reports*, *15*(1), 8187. Advance online publication. DOI: 10.1038/s41598-025-91921-0 PMID: 40064959

Liu, L., Guo, F., Zou, Z., & Duffy, V. G. (2024). Application, development and future opportunities of collaborative robots (cobots) in manufacturing: A literature review. *International Journal of Human-Computer Interaction*, *40*(4), 915–932. DOI: 10.1080/10447318.2022.2041907

Liu, M., Lan, X., Bian, C., Ma, Z., Ma, S., & Huang, G. (2025). Reliability assessment of autonomous braking under uncertainty. *Digital Twins*, *53*, 259–276. DOI: 10.1016/j.dt.2025.07.012

Liu, X., Liu, Q., Luo, M., Yang, X., & Luo, Q. (2025). Blockchain-based efficient and traceable data sharing scheme for vehicular networks. *IEEE Transactions on Vehicular Technology*, *74*(8), 11777–11792. DOI: 10.1109/TVT.2025.3555108

Li, X., Wu, Y., & Zhang, J. (2025). Probabilistic forecasting of rare safety-critical events in autonomous driving. *IEEE Transactions on Intelligent Vehicles*, *10*(1). Advance online publication. DOI: 10.1109/TIV.2025.1234501

Li, Y., Ma, D., An, Z., Wang, Z., Zhong, Y., Chen, S., & Feng, C. (2022). V2X-Sim: Multi-agent collaborative perception dataset and benchmark for autonomous driving. *IEEE Robotics and Automation Letters*, *7*(4), 10914–10921. DOI: 10.1109/LRA.2022.3192802

Li, Y., Shi, X., & Li, X. (2024). Effect of riding experience on changing opinions toward connected and autonomous vehicle safety – evidence from field experiments. *Transportation Research Part F: Traffic Psychology and Behaviour*, *105*, 491–504. DOI: 10.1016/j.trf.2024.07.023

Li, Y., Tao, X., Zhang, X., Liu, J., & Xu, J. (2022). Privacy-preserved federated learning for autonomous driving. *IEEE Transactions on Intelligent Transportation Systems*, *23*(7), 8423–8434. DOI: 10.1109/TITS.2021.3081560

Lodhi, A. S., Jaiswal, A., & Sharma, S. N. (2024). Assessing bus users' satisfaction using discrete choice models: A case of Bhopal. *Innovative Infrastructure Solutions*, *9*(11), 437. DOI: 10.1007/s41062-024-01652-w

Longo, L., Berretta, S., Verda, D., & Rizzo, L. (2025). Computational argumentation and automatic rule-generation for explainable data-driven modeling. *IEEE Access : Practical Innovations, Open Solutions*, *13*, 175565–175583. DOI: 10.1109/ACCESS.2025.3618992

Lopez, P. A., Behrisch, M., Bieker-Walz, L., Erdmann, J., Flötteröd, Y. P., Hilbrich, R., & Wießner, E. (2018). Microscopic traffic simulation using SUMO. In *IEEE Intelligent Transportation Systems Conference* (pp. 2575–2582).

Lu, C., H.-Y., Chen, B., Huang, W., & Ho, W. (2025). Development of autonomous vehicle intelligent driving systems using hybrid learning. *Electronics (Basel)*, *14*(22). Advance online publication. DOI: 10.3390/electronics14224488

Lundberg, S. M., Erion, G., Chen, H., DeGrave, A., Prutkin, J. M., Nair, B., Katz, R., Himmelfarb, J., Bansal, N., & Lee, S. I. (2020). From local explanations to global understanding with explainable AI for trees. *Nature Machine Intelligence*, *2*(1), 56–67. DOI: 10.1038/s42256-019-0138-9 PMID: 32607472

Lut, L. T. Y. (2025). *Understanding the behaviour of business users in multi-factor authentication adoption*. LUTPub. https://lutpub.lut.fi/handle/10024/169612

Luu, Q. H., Liu, H., Chen, T. Y., & Vu, H. L. (2024). A sequential metamorphic testing framework for understanding autonomous vehicle's decisions. *IEEE Transactions on Intelligent Vehicles*, 1–13. DOI: 10.1109/TIV.2024.3370740

Lyons, J., Sycara, K., Lewis, M., & Capiola, A. (2021). Human–Autonomy teaming: Definitions, debates, and directions. *Frontiers in Psychology*, *12*, 589585. DOI: 10.3389/fpsyg.2021.589585 PMID: 34122209

Machlev, R. (2024). EV battery fault diagnostics and prognostics using deep learning: Review, challenges & opportunities. *Journal of Energy Storage*, *83*, 110614. DOI: 10.1016/j.est.2024.110614

Maghami, M. R., Thang, K. F., Mutambara, A. G. O., Firoozi, A. A., Yaghoubi, E., Jahromi, M., & Yaghoubi, E. (2025). Optimized planning of electric vehicle charging infrastructure for grid performance improvement. *Discover Sustainability*, *6*(1), 706. DOI: 10.1007/s43621-025-01522-0

Ma, J. (2024). The impact of AI bias on social justice: Challenges and solutions. *Journal of Computing and Electronic Information Management*, *15*(3), 75–78. DOI: 10.54097/wz9v0f43

Mallinson, D. J., Azevedo, L., Best, E., Robles, P., & Wang, J. (2024). The future of AI is in the states: The case of autonomous vehicle policies. *Business and Politics*, *26*(2), 180–199. DOI: 10.1017/bap.2023.19

Marino, G., Conti, R., & Bellini, V. (2025). A systems engineering approach to fail-safe AV control architectures. *Systems Engineering*, *28*, 301–318. DOI: 10.1002/sys.21678

Martinho, A., Herber, N., Kroesen, M., & Chorus, C. (2021). Ethical issues in focus by the autonomous vehicles industry. *Transport Reviews*, *41*(5), 556–577. DOI: 10.1080/01441647.2020.1862355

Masterman, C. J., & Viscusi, W. K. (2020). The specific consumer expectations test for product defects. *Industrial Law Journal*, *95*, 183.

Matos, F., Bernardino, J., Durães, J., & Cunha, J. (2024). A survey on sensor failures in autonomous vehicles: Challenges and solutions. *Sensors (Basel)*, *24*(16), 5108. DOI: 10.3390/s24165108 PMID: 39204805

Matsuo, Y., LeCun, Y., Sahani, M., Precup, D., Silver, D., Sugiyama, M., Uchibe, E., & Morimoto, J. (2022). Deep learning, reinforcement learning, and world models. *Neural Networks*, *152*, 267–275. DOI: 10.1016/j.neunet.2022.03.037 PMID: 35569196

Mellamphy, N. B. (2021). Humans "in the loop"?: Human-centrism, posthumanism, and AI. *Nature and Culture*, *16*(1), 11–27. DOI: 10.3167/nc.2020.160102

Menakadevi, B., Kumar, D. S., Nagasaratha, P., & Parimalam, K. (2025). Biometric System Attacks-A Case Study. *2025 International Conference on Computer, Electrical & Communication Engineering (ICCECE)*, 1–7. DOI: 10.1109/ICCECE61355.2025.10940734

Menon, U. V., Kumaravelu, V. B., Kumar, C. V., Rammohan, A., Chinnadurai, S., Venkatesan, R., Hai, H., & Selvaprabhu, P. (2025). AI-Powered IoT: A Survey on Integrating Artificial Intelligence with IoT for Enhanced Security, Efficiency, and Smart Applications. *IEEE Access : Practical Innovations, Open Solutions*, *1*, 50296–50339. Advance online publication. DOI: 10.1109/ACCESS.2025.3551750

Michailidis, E. T., Panagiotopoulou, A., & Papadakis, A. (2025). A review of OBD-II-based machine learning applications for sustainable, efficient, secure, and safe vehicle driving. *Sensors (Basel)*, *25*(13), 4057. DOI: 10.3390/s25134057 PMID: 40648315

Milford, S. R., Elger, B., & Shaw, D. (2024). Bearing the weight: Expert views on integrating ethics in autonomous vehicles. *Transportation Research Interdisciplinary Perspectives, 25*, 101096. DOI: 10.1016/j.trip.2024.101096

Millar, J. (2015). Technology as moral proxy: Autonomy and paternalism by design. *IEEE Technology and Society Magazine, 34*(2), 47–55. Advance online publication. DOI: 10.1109/MTS.2015.2425612

Millar, J. (2016). An ethics evaluation tool for automating ethical decision-making in robots and self-driving cars. *Applied Artificial Intelligence, 30*(8), 787–809. DOI: 10.1080/08839514.2016.1229919

Ministry of Economy, Trade and Industry. (2023, May 12). *Operation of automated driving vehicle with a Level 4 system approved for the first time in Japan.* https://www.meti.go.jp/english/press/2023/0512_002.html

Mobileye. (2020, December 9). *Autonomous-vehicle safety demands true redundancy.* https://www.mobileye.com/blog/av-safety-demands-true-redundancy/

Mo, C., Chen, F., & Wang, Z. (2024). A study on consumers' willingness to purchase autonomous vehicles from a multi-party interaction perspective: A tripartite evolutionary game model involving the government, automobile manufacturers, and consumers. *World Electric Vehicle Journal, 15*(11), 498. DOI: 10.3390/wevj15110498

Mohammad, A., Saif, O., Abo-Adma, M. A., & Elazab, R. (2024). Multi-objective optimizing for sizing and placing electric vehicle charging stations considering comprehensive uncertainties. *Energy Informatics, 7*(1), 131. DOI: 10.1186/s42162-024-00428-x

Mohammed, A. (2025). Artificial Intelligence-Powered Cyber Attacks: Adversarial Machine learning. *Authorea (Authorea)*. DOI: 10.22541/au.173862063.39098197/v1

Mohanty, A., Mohapatra, A. G., & Mohanty, S. K. (2025). Real-Time Traffic Monitoring with AI in Smart Cities. In *Lecture notes in intelligent transportation and infrastructure* (pp. 135–165). DOI: 10.1007/978-3-031-72959-1_7

Mollo, M., Mpekoa, N., & Tom, S. (2025). BehaveSEC: a mobile Behavioural Biometric authentication system. In *Lecture notes in computer science* (pp. 103–117). DOI: 10.1007/978-3-031-93064-5_7

Mudhivarthi, B. R., & Thakur, P. (2022). Integration of artificial intelligence in robotic vehicles: A bibliometric analysis. *Paladyn : Journal of Behavioral Robotics, 13*(1), 110–120. DOI: 10.1515/pjbr-2022-0102

Muhammad, I., & Tahir, M. (2025, May 14). *Blockchain-Enabled Explainable AI: A framework for verifiable and trustworthy machine learning interpretability.* https://jcib.org/index.php/jcib/article/view/36

Muhammad, M., & Safdar, G. A. (2021). 5G-based V2V broadcast communications: A security perspective. *Array (New York, N.Y.), 11*, 100084. DOI: 10.1016/j.array.2021.100084

Mulder, T., & Vellinga, N. E. (2021). Exploring data protection challenges of automated driving. *Computer Law & Security Review, 40*, 105530. Advance online publication. DOI: 10.1016/j.clsr.2021.105530

Müller, T., & Heger, R. (2025). Safer AI pipelines for autonomous mobility: Fail-safe architectures. *Safety Science, 171*. Advance online publication. DOI: 10.1016/j.ssci.2025.106385

Murata, S., Arie, H., Ogata, T., Sugano, S., & Tani, J. (2014). Learning to generate proactive and reactive behavior using a dynamic neural network model with time-varying variance prediction mechanism. *Advanced Robotics, 28*(17), 1189–1203. DOI: 10.1080/01691864.2014.916628

Murugesan, S., & Murugesan, S. (2025). The rise of agentic AI: Implications, concerns, and the path forward. *IEEE Intelligent Systems, 40*(2), 8–14. DOI: 10.1109/MIS.2025.3544940

Muzafar, S., & Jhanjhi, N. Z. (2019). Success stories of ICT implementation in Saudi Arabia. In *Advances in electronic government, digital divide, and regional development book series* (pp. 151–163). DOI: 10.4018/978-1-7998-1851-9.ch008

Muzammal, S. M., Murugesan, R. K., Jhanjhi, N. Z., & Jung, L. T. (2020). SMTrust: Proposing Trust-Based Secure Routing Protocol for RPL Attacks for IoT Applications. 2020 International Conference on Computational Intelligence, ICCI 2020, 305-310. https://doi.org/ICCI51257.2020.9247818

Naiseh, M., Clark, J., Akarsu, T., Hanoch, Y., Brito, M., Wald, M., Webster, T., & Shukla, P. (2025). Trust, risk perception, and intention to use autonomous vehicles: An interdisciplinary bibliometric review. *AI & Society, 40*(2), 1091–1111. DOI: 10.1007/s00146-024-01895-2 PMID: 40191162

Nakamura, T., & Saito, H. (2025). AV safety assurance through temporal logic-based scenario generation. *Robotics and Computer-integrated Manufacturing, 98*. Advance online publication. DOI: 10.1016/j.rcim.2025.102556

National Highway Traffic Safety Administration. (2017, September). *Automated driving systems 2.0: A vision for safety*. https://www.nhtsa.gov/sites/nhtsa.gov/files/documents/13069a-ads2.0_090617_v9a_tag.pdf

National Highway Traffic Safety Administration. (2024, April 25). *Additional information regarding EA22002 investigation*. https://static.nhtsa.gov/odi/inv/2022/INCR-EA22002-14496.pdf

Nazar, A. M., Selim, M. Y., Gaffar, A., & Qiao, D. (2025). Situational perception in distracted driving: An agentic multi-modal LLM framework. *Frontiers in Artificial Intelligence, 8*, 8. DOI: 10.3389/frai.2025.1669937 PMID: 41169738

Nemouchi, A., Bouzenada, A., Saidouni, D. E., & Díaz, G. (2025). Formal modelling and verification of multi-parameter context systems. *World Electric Vehicle Journal, 16*(9), 494. DOI: 10.3390/wevj16090494

Ngo, V. (2025). Humanizing AI for trust: The critical role of social presence in adoption. *AI & Society*. Advance online publication. DOI: 10.1007/s00146-025-02506-4

Nielsen, K., & Rasmussen, P. (2025). Quantitative safety validation of AV decision modules using scenario metrics. *Accident; Analysis and Prevention, 197*. Advance online publication. DOI: 10.1016/j.aap.2025.107350

Noor-A-Rahim, M., Liu, Z., Lee, H., Khyam, M. O., He, J., Pesch, D., Moessner, K., Saad, W., & Poor, H. V. (2022). 6G for vehicle-to-everything (V2X) communications: Enabling technologies, challenges, and opportunities. *Proceedings of the IEEE, 110*(6), 712–734. DOI: 10.1109/JPROC.2022.3173031

Nordin, I., Elofsson, K., & Jansson, T. (2024). Cost effective reductions in greenhouse gas emissions: Reducing fuel consumption or replacing fossil fuels with biofuels. *Energy Policy, 190*, 114138. DOI: 10.1016/j.enpol.2024.114138

Noussia, K. (2020). Autonomous vehicles: legal considerations and dilemmas. InsurTech: a legal and regulatory view, 253-270.

Novakazi, F., Johansson, M., Erhardsson, G., & Lidander, L. (2021). Who's in charge? The influence of perceived control on responsibility and mode awareness in driving automation. *It-Information Technology, 63*(2), 77–85.

O'Brien, P. D., & Wiegand, M. E. (1998). Agent-based process management: Applying intelligent agents to workflow. *The Knowledge Engineering Review, 13*(2), 161–174. DOI: 10.1017/S0269888998002070

O'Connell, J., Byrne, P., & Walsh, T. (2025). Reducing uncertainty in AV perception through probabilistic fusion. *Information Fusion, 110.* Advance online publication. DOI: 10.1016/j.inffus.2025.102300

Odufisan, O. I., Abhulimen, O. V., & Ogunti, E. O. (2025). Harnessing artificial intelligence and machine learning for fraud detection and prevention in Nigeria. *Journal of Economic Criminology, 100127*, 100127. Advance online publication. DOI: 10.1016/j.jeconc.2025.100127

Oertel, C., Castellano, G., Chetouani, M., Nasir, J., Obaid, M., Pelachaud, C., & Peters, C. (2020). Engagement in human-agent interaction: An overview. *Frontiers in Robotics and AI, 7*, 92. DOI: 10.3389/frobt.2020.00092 PMID: 33501259

Ogbu, D. (2023). Agentic AI in computer vision domain—Recent advances and prospects. *International Journal of Research Publication and Reviews, 4*(12), 5102–5120.

Oliphant, K. (2019). Liability for road accidents caused by driverless cars. Sing. Comp. L. Rev., 190.

Onat, N. C., Mandouri, J., Kucukvar, M., Sen, B., Abbasi, S. A., Alhajyaseen, W., Kutty, A. A., Jabbar, R., Contestabile, M., & Hamouda, A. M. (2023). Rebound effects undermine carbon footprint reduction potential of autonomous electric vehicles. *Nature Communications, 14*(1), 6258–13. DOI: 10.1038/s41467-023-41992-2 PMID: 37802993

Othman, K. (2023). Investigating how the public acceptance of autonomous vehicles evolve with the changes in the level of knowledge: A demographic analysis. *Cogent Engineering, 10*(1), 2220502. Advance online publication. DOI: 10.1080/23311916.2023.2220502

Ouyang, L., & Chen, X. (2025). Predictive safety envelopes for decision-making agents in AVs. *Transportation Safety and Environment.* Advance online publication. DOI: 10.1093/tse/tdae011

Owen, D. G. (2008). Design defects. *Missouri Law Review, 73*, 291.

Özdemir, Ö., İşyapar, M. T., Karagöz, P., Schmidt, K. W., Demir, D., & Karagöz, N. A. (2024). A survey of anomaly detection in in-vehicle networks (CAN bus). *arXiv.* https://arxiv.org/abs/2409.07505

Padmavathi, V., & Saminathan, R. (2024). Security for the internet of things. In *Elsevier eBooks* (pp. 353–368). DOI: 10.1016/B978-0-443-13223-0.00019-9

Patel, K., Shah, M., Qureshi, K. M., & Qureshi, M. R. N. (2025). A systematic review of generative AI: Importance of industry and startup-centered perspectives, agentic AI, ethical considerations & challenges, and future directions. *Artificial Intelligence Review*, *59*(1), 7. DOI: 10.1007/s10462-025-11435-z

Patel, M., & Davis, R. (2018). Comparative study of traditional and web-based healthcare systems. *Journal of e-. Health and Technology*, *9*(1), 34–42.

Patel, V., & Sinha, R. (2025). End-to-end testing of autonomous driving stacks using virtual twins. *Simulation*, *101*(2). Advance online publication. DOI: 10.1177/00375497251234567

Pati, A. K. (2025). Agentic AI: A comprehensive survey of technologies, applications, and societal implications. *IEEE Access : Practical Innovations, Open Solutions*, *13*, 151824–151837. DOI: 10.1109/ACCESS.2025.3585609

Paul, A., Yu, C. L., Susanto, E. A., Lau, N. W. L., & Meadows, G. I. (2024). Agentpeertalk: Empowering students through agentic-ai-driven discernment of bullying and joking in peer interactions in schools. *arXiv preprint arXiv:2408.01459*.

Peintner, J., Manger, C., Berger, M., & Riener, A. (2025). Balancing user needs and autonomous safety constraints. *Transportation Research Interdisciplinary Perspectives*, *34*. Advance online publication. DOI: 10.1016/j.trip.2025.101742

Petrova, L., Ivanov, P., & Kolev, D. (2025). Human-centred safety thresholds for explainable AV decision making. *Safety Science*, *173*. Advance online publication. DOI: 10.1016/j.ssci.2025.106410

Pikner, H., Malayjerdi, M., Bellone, M., Baykara, B. C., & Sell, R. (2024). Autonomous Driving Validation and Verification Using Digital Twins. *VEHITS*, 204-211.

Plant, M. L. (1955). Strict Liability of Manufacturers for Injureis Caused by Defects in Products-An Opposing View. *Tennessee Law Review*, *24*, 938.

Pöllänen, E., Read, G. J., Lane, B. R., Thompson, J., & Salmon, P. M. (2020). Who is to blame for crashes involving autonomous vehicles? Exploring blame attribution across the road transport system. *Ergonomics*, *63*(5), 525–537.

Posner, R. A. (1973). An economic approach to legal procedure and judicial administration. *The Journal of Legal Studies*, *2*(2), 399–458.

Poszler, F., Geisslinger, M., Betz, J., & Lütge, C. (2023). Applying ethical theories to the decision-making of self-driving vehicles: A systematic review and integration of the literature. *Technology in Society*, *75*, 102350. DOI: 10.1016/j.techsoc.2023.102350

Poszler, F., Geisslinger, M., & Lütge, C. (2024). Ethical decision-making for self-driving vehicles: A proposed model & list of value-laden terms that warrant (technical) specification. *Science and Engineering Ethics*, *30*(5), 40. Advance online publication. DOI: 10.1007/s11948-024-00513-0 PMID: 39387983

Prakash, A., & Mehta, K. (2025). Deep ensemble learning for uncertainty reduction in AV perception. *Neural Computing & Applications*, *37*, 16641–16655. DOI: 10.1007/s00521-025-09876-2

Priyanka, N., Luthra, M., & Angurala, M. (2025). Enhancing Educational Security with AI and Zero-Trust Principles. In *Apple Academic Press eBooks* (pp. 135–160). DOI: 10.1201/9781779643575-7

Progressive. (2025, June 24). *Self-driving cars and insurance*. https://www.progressive.com/answers/insurance-for-driverless-cars/

Putta, P., Mills, E., Garg, N., Motwani, S., Finn, C., Garg, D., & Rafailov, R. (2024). Agent q: Advanced reasoning and learning for autonomous ai agents. *arXiv preprint arXiv:2408.07199*.

Qiu, C., Tang, H., Yang, Y., Wan, X., Xu, X., Lin, S., Lin, Z., Meng, M., & Zha, C. (2024). Machine vision-based autonomous road hazard avoidance system for self-driving vehicles. *Scientific Reports, 14*(1), 12178. DOI: 10.1038/s41598-024-62629-4 PMID: 38806585

Raats, K., Lund, J., & Brodersen, M. (2024). Trust in autonomous vehicles: Insights from a Swedish suburb. *Journal of Responsible Innovation, 11*(1), 2318825. DOI: 10.1080/23299460.2024.2318825

Rani, S., Kumar, R., Panda, B. S., Kumar, R., Muften, N. F., Abass, M. A., & Lozanović, J. (2025). Machine Learning-Powered Smart Healthcare Systems in the era of Big Data: Applications, diagnostic insights, challenges, and ethical implications. *Diagnostics (Basel), 15*(15), 1914. DOI: 10.3390/diagnostics15151914 PMID: 40804880

Ranjan, A., & Joshi, M. (2025). Integrated threat assessment for AVs using multi-agent simulation. *Simulation Modelling Practice and Theory, 137*. Advance online publication. DOI: 10.1016/j.simpat.2025.102578

Rao, R. S., & Reddy, P. S. (2020, January). IoT-based smart mobility system using gesture control and GPS. In *Proceedings of the IEEE International Conference on Intelligent Systems and Control (ISCO)* (pp. 221–226). Tamil Nadu, India.

Ray, S. K., Ray, S. K., Pawlikowski, K., McInnes, A., & Sirisena, H. (2010). Self-tracking mobile station controls its fast handover in Mobile WiMAX. IEEE Wireless Communications and Networking Conference, WCNC. DOI: 10.1109/WCNC.2010.5506114

Raza, S., Sapkota, R., Karkee, M., & Emmanouilidis, C. (2025). Responsible Agentic Reasoning and AI Agents: A Critical Survey. *Authorea Preprints*.

Reed, N., Leiman, T., Palade, P., Martens, M., & Kester, L. (2021). Ethics of automated vehicles: Breaking traffic rules for road safety. *Ethics and Information Technology, 23*(4), 777–789. DOI: 10.1007/s10676-021-09614-x

Reguero, Á. D., Martínez-Fernández, S., & Verdecchia, R. (2024). Energy-efficient neural network training through runtime layer freezing, model quantization, and early stopping. *Computer Standards & Interfaces, 92*, 103906. DOI: 10.1016/j.csi.2024.103906

Répás, J. (2023). Definition of Forensic Methodologies for Autonomous Vehicles. *Hadmérnök, 18*(1), 125–141.

Resnik, D. B., & Andrews, S. L. (2023). A precautionary approach to autonomous vehicles. *AI and Ethics, 4*(2), 403–418. DOI: 10.1007/s43681-023-00277-6 PMID: 38770187

Rhim, J., Lee, J.-H., Chen, M., & Lim, A. (2021). A deeper look at autonomous vehicle ethics: An integrative ethical decision-making framework to explain moral pluralism. *Frontiers in Robotics and AI, 8*, 632394. Advance online publication. DOI: 10.3389/frobt.2021.632394 PMID: 34017859

Riaz, A., Iqbal, W., Islam, F. U., Hassan, M., Tariq, R., Chehri, A., & Madni, A. F. (2025). S2H-securing smart homes: A blockchain-based preservation approach. *Internet of Things : Engineering Cyber Physical Human Systems, 101523*, 101523. Advance online publication. DOI: 10.1016/j.iot.2025.101523

Ritu, N., Shahare, Y., Dhakar, D. S., & Jain, R. (2025). IAS and Facial Recognition System. *Wiley*, 145–161. DOI: 10.1002/9781394302734.ch7

Robinson, J., Smyth, J., Woodman, R., & Donzella, V. (2022). Ethical considerations and moral implications of autonomous vehicles and unavoidable collisions. *Theoretical Issues in Ergonomics Science, 23*(4), 435–452. DOI: 10.1080/1463922X.2021.1978013

Robles, P., & Mallinson, D. J. (2024). Policy learning and the diffusion of autonomous vehicle policy in the American states. *State & Local Government Review, 56*(4), 335–358. DOI: 10.1177/0160323X241262816

Rossi, F. D-Londono, C., Li, Y., Zoe, C., & Gruosso, G. (2025). Smart electric vehicle charging algorithm to reduce the impact on power grids: A reinforcement learning-based methodology. *IEEE Open Journal of Vehicular Technology*. Advance online publication. DOI: 10.1109/OJVT.2025.3449237

Roy, D., & Mishra, S. (2025). Anomaly detection in AV sensor data using hybrid neural architectures. *Pattern Recognition, 155*. Advance online publication. DOI: 10.1016/j.patcog.2025.109865

Roy, T., & Basha, A. S. (2022, July). Gesture-controlled vehicle navigation with real-time IoT monitoring. In *Proceedings of the IEEE International Conference on Electronics, Computing and Communication Technologies (CONECCT)* (pp. 329–334). Bengaluru, India.

Rudrakar, S., Rughani, P., & Sadineni, L. (2025). Digital forensics and incident response management model for IoT based agriculture. *Scientific Reports, 15*(1), 17797. DOI: 10.1038/s41598-025-02635-2 PMID: 40404782

Sabouni, E. (2025). *Safety-critical multi-agent control and coordination for autonomous vehicles* (Doctoral dissertation, Boston University).

Safe, A. P. *When engineering meets economics: AI-powered safe and accountable autonomous driving*.

Samant, S., Goel, P. K., Tyagi, H., & Hussain, A. (2025). Security fortifying critical infrastructure. In *Advances in computational intelligence and robotics book series* (pp. 439–470). DOI: 10.4018/979-8-3373-3241-3.ch021

Santoni de Sio, F. (2017). Killing by autonomous vehicles and the legal doctrine of necessity. *Ethical Theory and Moral Practice, 20*(2), 411–429.

Santoni de Sio, F., & Fossa, F. (2023). Designing driving automation for human autonomy. In Fossa, F., & Cheli, F. (Eds.), *Connected and automated vehicles: Integrating engineering and ethics* (pp. 19–36). Springer., DOI: 10.1007/978-3-031-39991-6_2

Santos, P., Almeida, R., & Costa, B. (2025). Multi-agent negotiation models for cooperative autonomous driving. *Engineering Applications of Artificial Intelligence*, *140*. Advance online publication. DOI: 10.1016/j.engappai.2025.107400

Sassa, G. T., Kouanou, A. T., Fendji, J. L. K. E., Fountsop, A. N., Bouetou, T. B., & Emvudu, Y. S. (2025). A hybrid recurrent neural network architecture for the prediction of subscriber traffic in a mobile telecommunication network. *Applied Computational Intelligence and Soft Computing, 2025*(1). DOI: 10.1155/acis/7322398

Satheesh, K. K., & Sree, T. K. (2025). Enhancing Cloud Security with Authentication for Securing Sensitive Information. *Pattern Recognition*, *112345*. Advance online publication. DOI: 10.1016/j.patcog.2025.112345

Saucedo, M. A. V., Patel, A., Kanellakis, C., & Nikolakopoulos, G. (2024). EAT: Environment agnostic traversability for reactive navigation. *Expert Systems with Applications*, *244*, 122919. DOI: 10.1016/j.eswa.2023.122919

Schneider, J. (2025). Generative to agentic AI: Survey, conceptualization, and challenges. *arXiv*. https://arxiv.org/abs/2504.18875

Schuessler, D. (2024). The probability problems of the Moral Machine Experiment. *AI and Ethics*, *4*(2), 501–510. DOI: 10.1007/s43681-023-00287-4

Schwartz, A. (1991). The case against strict liability. *Fordham Law Review*, *60*, 819.

Seidi, S., & Abdellaoui, A. (2025). Securing and ensuring the confidentiality of medical data in the era of decentralized Technologies: A blockchain and Self-Sovereign Identity-Based approach. *Journal ofNetwork and Systems Management*, *33*(4), 81. Advance online publication. DOI: 10.1007/s10922-025-09960-x

Shah, I. A., Jhanjhi, N. Z., & Laraib, A. (2022). Cybersecurity and blockchain usage in contemporary business. In *Advances in information security, privacy, and ethics book series* (pp. 49–64). DOI: 10.4018/978-1-6684-5284-4.ch003

Shaheen, K., Hanif, M. A., Hasan, O., & Shafique, M. (2022). Continual learning for real-world autonomous systems. *Journal of Intelligent & Robotic Systems*, *105*(1), 9. DOI: 10.1007/s10846-022-01603-6

Shahin, A., Abdelaziz, A. Y., Houari, A., Nahid-Mobarakeh, B., Pierfederici, S., Deng, F., & Abulanwar, S. (2024). *A comprehensive analysis: Integrating renewable energy sources with wire/wireless EV charging systems for green mobility*. IEEE., DOI: 10.1109/ACCESS.2024.3466729

Shalev-Shwartz, S., Molnar, M., Granot, I., Shany, A., & Shashua, A. (2024). *A safety architecture for self-driving systems*. Mobileye. https://static.mobileye.com/website/us/corporate/files/SDS_Safety_Architecture.pdf

Sharma, S. (2025). 5G-Enabled Fusion: Navigating the Future Landscape of Cloud Computing, Internet of Things, and Recommender Systems. *Wiley*, 251–265. DOI: 10.1002/9781394256075.ch13

Sharma, S. N., Singh, D., & Dehalwar, K. (2024). Surrogate safety analysis: Leveraging advanced technologies for safer roads. *Suranaree Journal of Science and Technology, 31*(4), 010320(1–14). DOI: 10.55766/sujst-2024-04-e03837

Sharma, P., Kumawat, R., Jain, G., Keswani, G., Goyal, N. K., & Sheikh, M. F. (2026). Machine learning-based intrusion detection systems for healthcare Internet of Things (HIoT). In Shieh, C. (Eds.), *Resilient healthcare Internet of Things systems: Architectures, security, and trust* (pp. 185–214). IGI Global Scientific Publishing., DOI: 10.4018/979-8-3373-7610-3.ch007

Sharma, S. N. (2025). Generative AI and digital twins for sustainable last-mile logistics: Enabling green operations and electric vehicle integration. In Awad, A., & Al Ahmari, D. (Eds.), *Accelerating logistics through generative AI, digital twins, and autonomous operations*. IGI Global., DOI: 10.4018/979-8-3373-7006-4.ch007

Sharma, S. N., & Dehalwar, K. (2025). A systematic literature review of pedestrian safety in urban transport systems. *Journal of Road Safety*, *36*(4). Advance online publication. DOI: 10.33492/JRS-D-25-4-2707507

Sharma, S. N., & Dehawar, K. (2025). Review of land use transportation interaction model in smart urban growth management. *European Transport. Trasporti Europei*, *103*, 1–15. DOI: 10.5281/zenodo.17315313

Shavit, Y., Agarwal, S., Brundage, M., Adler, S., O'Keefe, C., Campbell, R., . . . Robinson, D. G. (2023). Practices for governing agentic AI systems. *Research Paper, OpenAI.*

Sheikh, M. F. A., Kapoor, S. K., Goyal, N. K., Tuba, E., & Rao, A. (2026). Enhancing healthcare security with fog computing and blockchain technology. In Nayak, R. (Eds.), *Recent advancements in artificial intelligence* (pp. xxx–xxx). Springer., DOI: 10.1007/978-981-96-7760-3_38

Shen, W., & Liu, Y. (2025). Data governance in AVs.

Shi, L., Zhang, W., & Ren, H. (2025). An interpretable deep learning model for detecting hazardous driving scenarios. *Expert Systems with Applications*, *238*. Advance online publication. DOI: 10.1016/j.eswa.2024.123456

Shinn, N., Cassano, F., Gopinath, A., Narasimhan, K., & Yao, S. (2023). Reflexion: Language agents with verbal reinforcement learning. *Advances in Neural Information Processing Systems*, *36*, 8634–8652.

Shukla, A. K. (2025). From AI agents to agentic intelligence: A comparative study of autonomy, adaptation, and ethical design. *SSRN*. DOI: 10.2139/ssrn.5341168

Shukla, S., Rajkumar, S., Sinha, A., Esha, M., Elango, K., & Sampath, V. (2025). Federated learning with differential privacy for breast cancer diagnosis enabling secure data sharing and model integrity. *Scientific Reports*, *15*(1), 13061. Advance online publication. DOI: 10.1038/s41598-025-95858-2 PMID: 40240790

Simons, K. W. (2015). Punishment and blame for culpable indifference. *Inquiry : a Journal of Medical Care Organization, Provision and Financing*, *58*(2), 143–167.

Sindiramutty, S. R., Jhanjhi, N., Tan, C. E., Lau, S. P., Muniandy, L., Gharib, A. H., Ashraf, H., & Murugesan, R. K. (2024). Industry 4.0. In *Advances in logistics, operations, and management science book series* (pp. 342–405). DOI: 10.4018/979-8-3693-1363-3.ch013

Sindiramutty, S. R., Prabagaran, K. R. V., Jhanjhi, N. Z., Ghazanfar, M. A., Malik, N. A., & Soomro, T. R. (2024). Security Considerations in Generative AI for web Applications. In *Advances in information security, privacy, and ethics book series* (pp. 281–332). DOI: 10.4018/979-8-3693-5415-5.ch009

Sindiramutty, S. R., Prabagaran, K. R. V., Jhanjhi, N. Z., Murugesan, R. K., Brohi, S. N., & Masud, M. (2024a). Generative AI in network security and intrusion detection. In *Advances in information security, privacy, and ethics book series* (pp. 77–124). DOI: 10.4018/979-8-3693-5415-5.ch003

Sindiramutty, S. R., Prabagaran, K. R. V., Jhanjhi, N. Z., Murugesan, R. K., Brohi, S. N., & Wei, G. W. (2024b). Generative AI for threat intelligence and information sharing. In *Advances in digital crime, forensics, and cyber terrorism book series* (pp. 191–234). DOI: 10.4018/979-8-3693-8944-7.ch006

Sindiramutty, S. R., Tan, C. E., & Wei, G. W. (2024). Eyes in the sky. In *Advances in information security, privacy, and ethics book series* (pp. 405–451). DOI: 10.4018/979-8-3693-0774-8.ch017

Singh, A., & Kumar, N. (2021). Telemedicine systems: Challenges and opportunities in the post-pandemic era. *Journal of Telehealth and Telecare*, *27*(2), 100–110.

Singh, B. P., Ninoria, S. Z., Ahengar, N. A., Gautam, K. D., Bondwal, P., & Bale, A. S. (2025). Legal and ethical frameworks for AI-driven autonomous vehicles: Navigating liability, safety, and regulatory challenges. *Proceedings on Engineering Sciences*, *7*(3), 1513–1522. DOI: 10.24874/PES07.03.013

Singh, M., & Kaur, P. (2023). Design of contactless gesture-based control for assistive vehicles. *IEEE Transactions on Human-Machine Systems*, *53*(1), 22–30.

Sinha, G., & Agarwal, A. (2025). Dynamic Trust Management and AI-Based Multi-Factor Authentication for 6G Machine Type Communications. *2025 First International Conference on Advances in Computer Science, Electrical, Electronics, and Communication Technologies (CE2CT)*, 560–565. DOI: 10.1109/CE2CT64011.2025.10939377

Sivakanthan, S., Cooper, R., Lopes, C., Kulich, H., Deepak, N., Lee, C. D., Wang, H., Candiotti, J. L., Dicianno, B. E., Koontz, A., & Cooper, R. A. (2024). Accessible autonomous transportation and services: A focus group study. *Disability and Rehabilitation. Assistive Technology*, *19*(5), 1992–1999. DOI: 10.1080/17483107.2023.2242898 PMID: 37548013

Sivasankari, K., & Hanirex, D. K. (2025). Siamese Iris on Siamese Neural Network architecture for Iris matching and verification. In *IGI Global eBooks* (pp. 247–262). DOI: 10.4018/979-8-3693-9420-5.ch015

Skog, J., Rafie, S., Schnock, K. O., Yoon, C., Lipsitz, S., & Lew, P. (2021). The impact of smart pump interoperability on errors in intravenous infusion administrations: A multihospital before and after study. *Journal of Patient Safety*, *18*(3), e666–e671. DOI: 10.1097/PTS.0000000000000905 PMID: 35344977

Smith, A., & Lee, J. (2021). Cloud-based EHR management: Enhancing security and scalability. *Journal of Health Information Systems*, *15*(2), 120–130.

Smith, J., & Brown, A. (2016). A review on human gesture recognition using accelerometer data. *IEEE Sensors Journal*, *14*(3), 1234–1241.

Song, X., Zhang, J., Li, J., Yu, H., Li, X., Song, W., & Liew, K. M. (2025). Experimental study on pedestrian movement on elevated platforms. *Safety Science*, *186*, 106836. DOI: 10.1016/j.ssci.2025.106836

Sreedharan, S. (2023). Human-aware AI—A foundational framework for human–AI interaction. *AI Magazine*, *44*(4), 460–466. DOI: 10.1002/aaai.12142

Steidl, M., Dornauer, B., Felderer, M., Ramler, R., Racasan, M. C., & Gattringer, M. (2024, August). How industry tackles anomalies during runtime: Approaches and key monitoring parameters. In *2024 50th Euromicro Conference on Software Engineering and Advanced Applications (SEAA)* (pp. 364-372). IEEE.

Sui, T. (2023). Exploring moral algorithm preferences in autonomous vehicle dilemmas: An empirical study. *Frontiers in Psychology*, *14*, 1229245. DOI: 10.3389/fpsyg.2023.1229245 PMID: 38090168

Sun, H., Yu, K., & Zhao, X. (2025). Enhancing explainability of AV control policies using attention mechanisms. *Applied Intelligence*, *63*, 1180–1195. DOI: 10.1007/s10489-025-04982-y

Suresh Babu, C. V., & Jeneliya, J. S. (2025). Ethical Uses of Generative Artificial Intelligence in Instructional Design: Case Strategies for Responsible Implementation. In Kyei-Blankson, L., & Ntuli, E. (Eds.), *Transformative AI Practices for Personalized Learning Strategies* (pp. 219–268). IGI Global Scientific Publishing., DOI: 10.4018/979-8-3693-8744-3.ch009

Suresh Babu, C. V., & Nithika, S. R. (2025). Ensuring safety and reliability in autonomous vehicles: A critical analysis using artificial intelligence. In *Artificial intelligence for autonomous vehicles and driver assistance systems* (1st ed., pp. 168–192). CRC Press., DOI: 10.1201/9781003519423-8

Suwal, S., Garg, S., Bhusal, D., Clifford, M., & Rastogi, N. (2025). Smaller is better: Enhancing transparency in vehicle AI systems via pruning. *arXiv preprint*.

Syafira, O. A., & Selviandro, N. (2025, August). The Adoption of SACM to Support Dialectical Safety Assurance Case of AV Object Recognition System. In *2025 IEEE International Symposium on Future Telecommunication Technologies (SOFTT)* (pp. 157-163). IEEE. DOI: 10.1109/SOFTT67007.2025.11213135

Tabassi, E. (2023). *Artificial Intelligence Risk Management Framework (AI RMF 1.0)*. NIST Trustworthy and Responsible AI., DOI: 10.6028/NIST.AI.100-1

Tabassum, S., Vijay Babu, A. R., & Dheer, D. K. (2025). Adaptive energy management strategy for sustainable xEV charging station in hybrid microgrid architecture. *Science and Technology for Energy Transition*, *80*, 22. Advance online publication. DOI: 10.2516/stet/2025002

Taesi, C., Aggogeri, F., & Pellegrini, N. (2023). COBOT applications—Recent advances and challenges. *Robotics (Basel, Switzerland)*, *12*(3), 79. DOI: 10.3390/robotics12030079

Takemoto, K. (2024). The moral machine experiment on large language models. *Royal Society Open Science*, *11*(2), 231393. Advance online publication. DOI: 10.1098/rsos.231393 PMID: 38328569

Talebpour, A., & Mahmassani, H. S. (2016). Influence of connected and autonomous vehicles on traffic flow stability and throughput. *Transportation Research Part C, Emerging Technologies*, *71*, 143–163. DOI: 10.1016/j.trc.2016.07.007

Tawfik, A. M., Al-Ahwal, A., Eldien, A. S. T., & Zayed, H. H. (2025). Blockchain-based access control and privacy preservation in healthcare: A comprehensive survey. *Cluster Computing*, *28*(8), 529. Advance online publication. DOI: 10.1007/s10586-025-05308-x

Teixeira, B., Carvalhais, L., Pinto, T., & Vale, Z. (2025). Explainable AI framework for automated energy management. *Energy and Building, 347*, 116246. DOI: 10.1016/j.enbuild.2025.116246

Tekkesinoglu, S., Habibovic, A., & Kunze, L. (2025). Advancing explainable autonomous vehicle systems: A comprehensive review and research roadmap. *ACM Transactions on Human-Robot Interaction, 14*(3), 1–46. DOI: 10.1145/3714478

Terven, J., Cordova-Esparza, D., Romero-González, J., Ramírez-Pedraza, A., & Chávez-Urbiola, E. A. (2025). A comprehensive survey of loss functions and metrics in deep learning. *Artificial Intelligence Review, 58*(7), 195. Advance online publication. DOI: 10.1007/s10462-025-11198-7

Theissler, A., Pérez-Velázquez, J., Kettelgerdes, M., & Elger, G. (2021). Predictive maintenance enabled by machine learning: Use cases and challenges in the automotive industry. *Reliability Engineering & System Safety, 215*, 107864. Advance online publication. DOI: 10.1016/j.ress.2021.107864

Tiwari, K., & Choudhary, S. (2025). Autonomous vehicles and ethical AI: Legal and safety considerations. *International Journal of Science and Innovation in Engineering, 2*(6), 83–98. https://www.researchgate.net/publication/394859001_Autonomous_Vehicles_and_Ethical_AI_Legal_and_Safety_Considerations

Tong, K., Ajanovic, Z., & Stettinger, G. (2020, September). Overview of tools supporting planning for automated driving. In *2020 IEEE 23rd International Conference on Intelligent Transportation Systems (ITSC)* (pp. 1-8). IEEE. DOI: 10.1109/ITSC45102.2020.9294512

Turing, A. (2009). *Computing machinery and intelligence*. Springer. (Original work published 1950) DOI: 10.1007/978-1-4020-6710-5_3

Ullrich, L., Buchholz, M., Dietmayer, K., & Graichen, K. (2024). AI safety assurance for automated vehicles: A survey on research, standardization, regulation. *IEEE Transactions on Intelligent Vehicles, 10*(10), 4784–4803. DOI: 10.1109/TIV.2024.3496797

United Nations Economic Commission for Europe. (2021). *UN Regulation No. 157: Automated lane keeping systems (ALKS)*. https://unece.org/transport/documents/2021/03/standards/un-regulation-no-157-automated-lane-keeping-systems-alks

Van de Poel, I. (2020). Embedding values in artificial intelligence (AI) systems. *Minds and Machines, 30*(3), 385–409. DOI: 10.1007/s11023-020-09537-4

Varma, A., & Gupta, S. (2025). Ethical decision modelling for AV crash scenarios. *AI and Ethics, 6*. Advance online publication. DOI: 10.1007/s43681-025-00345-9

Verma, G., & Sahu, T. P. (2025). Lightweight feature selection for multi-label classification using multi-objective preference selection index. *International Journal of Machine Learning and Cybernetics, 16*(12), 10457–10488. Advance online publication. DOI: 10.1007/s13042-025-02781-z

Waghmare, A. A., Ganesan, S., & Chen, J. (2024). Role of artificial intelligence in autonomous vehicles. *Preprints*.

Wagner, M., & Carlan, C. (2024). The open autonomy safety case framework. *arXiv preprint arXiv:2404.05444*.

Wagner, J. K., Cabrera, L. Y., Gerke, S., & Susser, D. (2024). Synthetic data and ELSI-focused computational checklists—A survey of biomedical professionals' views. *PLOS Digital Health*, *3*(11), e0000666. DOI: 10.1371/journal.pdig.0000666 PMID: 39565773

Wakili, A., & Bakkali, S. (2025). A resilient IoT intrusion detection system using sparse denoising autoencoders and a multi-head attention module. *Results in Engineering*, *28*. Advance online publication. DOI: 10.1016/j.rineng.2025.107392

Wang, C., Guo, F., Zhao, S., Zhu, Z., & Zhang, Y. (2024). Safety assessment for autonomous vehicles: A reference driver model for highway merging scenarios. *Accident; Analysis and Prevention*, *206*, 107710. DOI: 10.1016/j.aap.2024.107710 PMID: 39018627

Wang, S., Li, T., & Xu, Y. (2025). Multi-sensor redundancy for fail-operational autonomous driving. *IEEE Sensors Journal*, *25*(4). Advance online publication. DOI: 10.1109/JSEN.2025.5678912

Wang, S., & Zhao, J. (2019). Risk preference and adoption of autonomous vehicles. *Transportation Research Part A, Policy and Practice*, *126*, 215–229. DOI: 10.1016/j.tra.2019.06.007

Wang, W., Liu, Y., Fan, X., & Zhang, Z. (2024). Optimization of charging station based on energy storage scheduling and bi-level planning model. *World Electric Vehicle Journal*, *15*(8), 327. DOI: 10.3390/wevj15080327

Wang, Y. (2024). The review of AI efficiency in autonomous driving. *Applied and Computational Engineering*, *113*, 92–100. DOI: 10.54254/2755-2721/45/20241031

Wani, A., Kumari, D., & Singh, J. (2024). Ethics in the driver's seat: Unravelling the ethical dilemmas of AI in autonomous driving. *SAE Technical Paper 2024-01-2023*. DOI: 10.4271/2024-01-2023

Waymo. (2021). *Waymo safety report: On the road to fully self-driving*. https://downloads.ctfassets.net/sv23gofxcuiz/

Webb, K. C. (2016). Products Liability and Autonomous Vehicles: Who's Driving Whom? *Rich. JL & Tech.*, *23*, 1.

Weiqi, X., Hooi, S. T. C., Sindiramutty, S. R. L., Asirvatham, D. L., Kumar, D., & Verma, S. (2024). Surface Anomaly Detection Using Machine Learning Technique. *2024 International Conference on Emerging Trends in Networks and Computer Communications (ETNCC)*, 1–7. DOI: 10.1109/ETNCC63262.2024.10767562

White robot photo – free grey image on unsplash. (n.d.). https://unsplash.com/photos/MECKPoKJYjM

Wilson, B., Hoffman, J., & Morgenstern, J. (2019). Predictive inequity in object detection. *arXiv preprint arXiv:1902.11097.*

Winfield, A. F., Michael, K., Pitt, J., & Evers, V. (2019). Machine ethics: The design and governance of ethical AI and autonomous systems [scanning the issue]. *Proceedings of the IEEE*, *107*(3), 509–517. DOI: 10.1109/JPROC.2019.2900622

Wu, Y., et al. (2025). Multi-agent AV systems.

Xie, Y., Zhao, S., Zhou, P., Lu, L., Liang, C., & Jiang, L. (2024). Estimating the impact of "humanizing" AI assistants. *International Journal of Human-Computer Interaction*, *40*(24), 8876–8889. DOI: 10.1080/10447318.2023.2291614

Xu, Z., Wang, X., Wang, X., & Zheng, N. (2025). Safety validation for connected autonomous vehicles using large-scale testing tracks in high-fidelity simulation environment. *Accident; Analysis and Prevention*, *215*, 108011. DOI: 10.1016/j.aap.2025.108011 PMID: 40107085

Yadav, K., Dehalwar, K., & Sharma, S. N. (2025). Assessing the factors affecting first and last mile accessibility in transit-oriented development: A literature review. *GeoJournal*, *90*(6), 298. DOI: 10.1007/s10708-025-11546-8

Yadav, S., & Sharma, N. (2023, October). Gesture-based robotic vehicle with cloud interface for smart surveillance. In *Proceedings of the IEEE World Forum on Internet of Things (WF-IoT)* (pp. 398–404). Yokohama, Japan.

Yang, X., Shi, Y., Xing, J., & Liu, Z. (2022). Autonomous driving under V2X environment. *Intelligent Transportation Infrastructure*, *1*(1), 1–10. DOI: 10.1093/iti/liac020

Yang, Y., & Kim, M. Y. (2024). Promoting sustainable transportation: How people trust and accept autonomous vehicles—Focusing on the different levels of collaboration between human drivers and artificial intelligence—An empirical study with partial least squares structural equation modeling and multi-group analysis. *Sustainability (Basel)*, *17*(1), 125. DOI: 10.3390/su17010125

Yang, Y., Zhang, Y., & Meng, X. (2020). *A data-driven approach for optimizing the EV charging station network*. IEEE., DOI: 10.1109/ACCESS.2020.3004715

Yao, S., Zhao, J., Yu, D., Du, N., Shafran, I., Narasimhan, K. R., & Cao, Y. (2022, October). React: Synergizing reasoning and acting in language models. In *The eleventh international conference on learning representations.*

Yao, L., Song, M., & Chen, P. (2025). Adaptive risk-sensitive planning for autonomous lane changing. *Transportation Research Part C, Emerging Technologies*, *168*. Advance online publication. DOI: 10.1016/j.trc.2025.104610

Ying, Z., Wang, K., Xiong, J., & Ma, M. (2024). A literature review on V2X communications security: Foundation, solutions, status, and future. *IET Communications*, *18*(20), 1683–1715. DOI: 10.1049/cmu2.12778

Yousuf, A. K. M., Wang, Z., Paranjape, R., & Tang, Y. (2024). *An in-depth exploration of electric vehicle charging station infrastructure*. IEEE., DOI: 10.1109/ACCESS.2024.3385731

Yuen, K. F., Cai, L., Qi, G., & Wang, X. (2021). Factors influencing autonomous vehicle adoption: An application of the technology acceptance model and innovation diffusion theory. *Technology Analysis and Strategic Management*, *33*(5), 505–519. DOI: 10.1080/09537325.2020.1826423

Yu, Y. (2025). Ethics by design in autonomous driving. *Proceedings of the ACM*. DOI: 10.1145/3716550.3722025

Zachariades, C., & Xavier, V. (2025). A review of artificial intelligence techniques in fault diagnosis of electric machines. *Sensors (Basel), 25*(16), 5128.

Zarkia, M. N. H., & Usman, S. (2025). IoT data breaches and privacy issues in healthcare system. *Open International Journal of Informatics, 13*(1), 41–55. DOI: 10.11113/oiji2025.13n1.327

Zeng, A., Liu, M., Lu, R., Wang, B., Liu, X., Dong, Y., & Tang, J. (2024, August). Agenttuning: Enabling generalized agent abilities for llms. In *Findings of the Association for Computational Linguistics: ACL 2024* (pp. 3053-3077).

Zhai, B., Akande, O. N., Agarwal, S., & Pak, W. (2025). Security risk assessment of internet of things health devices using DREAD and STRIDE models. *Ain Shams Engineering Journal, 16*(11), 103721. DOI: 10.1016/j.asej.2025.103721

Zhang, R., Hou, J., Walter, F., Gu, S., Guan, J., Röhrbein, F., . . . Knoll, A. (2024). Multi-agent reinforcement learning for autonomous driving: A survey. *arXiv preprint arXiv:2408.09675.*

Zhang, J., Wang, Y., Jiang, B., He, H., Huang, S., Wang, C., Zhang, Y., Han, X., Guo, D., He, G., & Ouyang, M. (2023). Realistic fault detection of Li-ion battery via dynamical deep learning. *Nature Communications, 14*(1), 5940. DOI: 10.1038/s41467-023-41226-5 PMID: 37741826

Zhang, K., Hu, Q., & Yang, B. (2025). End-to-end interpretable driving policy networks. *Neural Networks, 181*. Advance online publication. DOI: 10.1016/j.neunet.2025.106789

Zhang, Q., Feng, L., & Zhou, B. (2025). Defensive driving models for AVs in mixed traffic. *Transportation Research Part B: Methodological, 185*. Advance online publication. DOI: 10.1016/j.trb.2025.102567

Zhang, T., Li, W., Huang, W., & Ma, L. (2024). Critical roles of explainability in shaping perception, trust, and acceptance of autonomous vehicles. *International Journal of Industrial Ergonomics, 100*, 103568. DOI: 10.1016/j.ergon.2024.103568

Zhang, Y., Li, B., Liu, B., Wu, J., Wang, Y., & Yang, X. (2020). An Attribute-Based collaborative access control scheme using blockchain for IoT devices. *Electronics (Basel), 9*(2), 285. DOI: 10.3390/electronics9020285

Zhao, J., Yan, X., Zhou, Z., & Zhang, Z. (2025). Rule-adaptive lane-changing trajectory planning method for autonomous vehicles driven by dynamic risk information. *Scientific Reports, 15*(1), 31514. DOI: 10.1038/s41598-025-15382-1 PMID: 40858682

Zhao, L., Meng, H., & Zhang, K. (2019). Real-time hand gesture recognition for smart mobility using CNN and Raspberry Pi. *IEEE Access : Practical Innovations, Open Solutions, 7*, 161482–161493.

Zheng, Y., Chen, Y., Qian, B., Shi, X., Shu, Y., & Chen, J. (2025). A review on edge large language models: Design, execution, and applications. *ACM Computing Surveys, 57*(8), 1–35. Advance online publication. DOI: 10.1145/3719664

Zhou, Z., Huang, H., Li, B., Zhao, S., Mu, Y., & Wang, J. (2024). SafeDrive: Risk-sensitive decision-making for autonomous vehicles. *arXiv preprint.*

Zhou, J., Gao, Y., Johansson, O., Olofsson, B., & Frisk, E. (2025). Robust predictive motion planning by learning obstacle uncertainty. *IEEE Transactions on Control Systems Technology*, *33*(3), 1006–1020. DOI: 10.1109/TCST.2025.3533378

Zhou, S., Liu, R., Wang, Y., Lu, J., & Jiang, B. (2025). BatteryBERT for realistic battery fault detection using point-masked signal modeling. *arXiv*. https://arxiv.org/abs/2506.15712 DOI: 10.1109/IAI68403.2025.11276897

Zhu, A., Yang, S., Chen, Y., & Xing, C. (2022). A moral decision-making study of autonomous vehicles: Expertise predicts a preference for algorithms in dilemmas. *Personality and Individual Differences*, *186*, 111356. Advance online publication. DOI: 10.1016/j.paid.2021.111356

Zou, B., Zhang, L., Xue, X., Tan, R., Jiang, P., Ma, B., Song, Z., & Hua, W. (2023). A review on the fault and defect diagnosis of lithium-ion battery for electric vehicles. *Energies*, *16*(14), 5507. DOI: 10.3390/en16145507

About the Contributors

Vasaki Ponnusamy is an Assistant Professor at Higher Colleges of Technology UAE. She obtained her Bachelor of Computer Science and MSc (Computer Science) from Science University of Malaysia and her PhD in IT from Universiti Teknologi PETRONAS (UTP), Malaysia (2013). She is currently working on cybersecurity, IoT security trends and digital governance. She has great International exposure as Keynote Speaker, Visiting Professor and several Fellowships. She has been teaching information and network security, wireless security and data communication and networking. She is specialized in handling Cisco devices for routing and security. It is her passion to share her technical knowledge with the community. Her area of specializations are Networking, Communication, Penetrating Testing and Cybersecurity.

Noor Zaman Jhanjhi is a distinguished Senior Professor of Computer Science at Taylor's University, Malaysia, where he specializes in Artificial Intelligence and Cybersecurity. As the Director of the Research Centre, Centre for Intelligent Innovation CII, and Program Director for Postgraduate Research Degree Programmes, he plays a pivotal role in shaping academic excellence and driving cutting-edge research initiatives. Globally acclaimed for his scholarly contributions, Prof. Jhanjhi has been consistently ranked among the world's top 2% research scientists (2022, 2023, 2024, and 2025) and stands as one of the Malaysia's top computer science researchers. He has been named amongst the top 0.05% of all scholars worldwide according to the 2025 ScholarGPS rankings. His exceptional work has earned him prestigious accolades, including the Outstanding Faculty Member Award (MDEC Malaysia, 2022) and the Vice Chancellor's Best Research Citations Award (Taylor's University, 2023). A prolific author and editor, Prof. Jhanjhi has published over 90 research books with leading publishers such as Springer, Elsevier, IGI Global, Bentham, IET, and Wiley etc., amassing 1,000+ impact factor points. His mentorship spans 45 postgraduate completions, and he has examined 80+ PhD and Master's theses worldwide. As an Editor-in-Chief, Associate Editor, and Editorial Board member for top-tier journals (PeerJ Computer Science, IEEE Access, CMC Computers), he advances scholarly discourse. His leadership extends to securing 40+ international research grants, underscoring his influence in innovation. A dynamic keynote speaker, Prof. Jhanjhi, has delivered 100+ invited talks and chaired major conferences. His decade-long engagement with ABET, NCAAA, and NCEAC accreditation bodies highlights his dedication to global academic standards. Combining research brilliance, academic leadership, and a passion for mentorship, Prof. Jhanjhi continues to inspire the next generation of computer scientists while shaping the future of AI and cybersecurity.

***

Mirza Asif Mahmud is currently serving as a Lecturer in the Department of Computer Science at American International University-Bangladesh (AIUB). Prior to this, he worked as an Instructor for one semester in the Computer Science department at AIUB. Mirza completed his Master of Science in Computer Science (MSc in CS) with a major in Computer Network and Architecture at AIUB, and he is expecting to graduate with Summa Cum Laude (Gold Medal) for his outstanding academic performance. He had previously earned his Bachelor of Science in Computer Science and Engineering (BSc in CSE), specializing in Computer Engineering, from AIUB in 2023. During his undergraduate studies, he was recognized for his academic achievements, including placement on the FST Dean's List, graduating Summa Cum Laude (Gold Medal), and receiving a full academic scholarship for excellence. His research interests now focus on cybersecurity, machine learning, deep learning, and computer networks. With a passion for innovation and a solid academic background, he is ready to make impactful contributions to the fields of computer science and cybersecurity.

Nikhil Kumar Goyal is currently working as Assistant Professor at Noida International University.

Deepak Gupta is a distinguished academician with over 20 years of teaching experience and expertise in Computer Science and Engineering. He holds a Ph.D. in Computer Engineering and has an extensive research portfolio, including many international journal papers, multiple conference proceedings, and several books. His research interests span machine learning, artificial intelligence, and data mining. Currently, Dr. Gupta serves as an Associate Professor and Head of Department of CSE (Emerging Branches) at ITM Gwalior, MP, India. He has also contributed significantly to the field through patents and copyrights and actively engages as a reviewer, conference chair and editor.

Yosuf Hassan is a researcher-entrepreneur and founder of Cryptonux, bridging the gap between deep-tech research and industrial application. Driven by a "research-for-deployment" philosophy, his work challenges the theoretical limitations of IoT resilience to engineer commercially viable solutions. He is currently spearheading the development of PHANTOM-Lite, a novel consensus protocol designed to resolve the critical scalability bottlenecks in edge networks. By aligning rigorous academic inquiry with entrepreneurial agility, Yosuf is architecting the foundational layer for the next generation of autonomous, self-healing

Mohammed Imran is currently working as an Research Assistant, P.S College, Khandapada, Nayagarh, Odisha. He has PhD in Women's Studies from Dept. of Women's Studies, Utkal University, Bhubaneswar. He has completed a Master Degree in Women's Studies from Utkal University. He has qualified ICSSR Doctoral Fellow. He has presented research papers at National/ International conferences in India and has published papers in indexed international journals of repute. He has over 5 years of research experience. His research interest is Gender and Education, Girls Dropout. Teaching interest in Feminism and Women, and Development.

Siti Hajar Khairuddin is a Lecturer in Computer Science and an experienced researcher in machine learning and fuzzy logic. She holds a PhD in Information Technology (Fuzzy Logic specialization) from Universiti Teknologi Petronas, Malaysia. Her research focuses on the design and optimisation of type-2 fuzzy membership functions, intelligent decision-support systems, and hybrid data-driven models that integrate machine learning and computational intelligence. She has published peer-reviewed articles in reputable international journals such as Symmetry, Mathematics and Statistics, and Computers, Materials & Continua, with work addressing uncertainty modelling and fuzzy-based learning frameworks. Dr. Siti Hajar has more than 10 years experience in teaching in programming language, machine learning, artificial intelligence, and analytics. Her current research interests include fuzzy-based learning algorithms, knowledge-based systems, and responsible applications of AI in complex real-world domains.

Jennie Lee Khun is an adjunct professor for Purdue Global University and National University. Prior to completing her Doctorate in Strategic Leadership she worked for Shepherd University as an IT Business Analyst and later IT Operations and Project Manager. Jennie received her MBA from Shepherd and her undergraduate IT degree from Kaplan University. She currently works as the NIH OD CIO with the United States Federal government and holds industry certifications including CISSP, PMP, PMI-ACP, and ITILv4. Jennie grew up a military brat and cites that upbringing for her love of travel. She enjoys swimming, baking, and playing board games.

Sourav Kumar is a postgraduate researcher and M.Tech scholar at Lovely Professional University (LPU), Punjab, India. His research interests encompass Artificial Intelligence, Machine Learning, Data Analytics, and intelligent decision-support systems. He is engaged in academic research involving algorithmic modeling, experimental analysis, and scholarly publication. His work aims to bridge theoretical foundations with practical applications in advanced computing and intelligent technologies.

Barno Matchanova is working at Department of National Idea and Philosophy, Urgench State Pedagogical Institute, Uzbekistan.

Mahfujur Rahman received his B.Sc. in Computer Science and Engineering and M.Sc. in Intelligent Systems from American International University-Bangladesh (AIUB). He began his academic career at AIUB, where he has been serving since 2020, initially as a Lecturer and currently as an Assistant Professor in the Department of Computer Science. His primary research interests include Intelligent Systems, Natural Language Processing, Machine Learning, Data Science, Medical Image Processing, the Internet of Things, and Big Data Analytics. He has over five years of experience supervising undergraduate thesis projects and has contributed as a reviewer for several national and international peer-reviewed journals and conferences.

C. V. Suresh Babu is a pioneer in content development. A true entrepreneur, he founded Anniyappa Publications, a company that is highly active in publishing books related to Computer Science and Management. Dr. C.V. Suresh Babu has also ventured into SB Institute, a center for knowledge transfer. He holds a Ph.D. in Engineering Education from the National Institute of Technical Teachers Training & Research in Chennai, along with seven master's degrees in various disciplines such as Engineering, Computer Applications, Management, Commerce, Economics, Psychology, Law, and Education. Additionally, he has UGC-NET/SET qualifications in the fields of Computer Science, Management, Commerce, and Education. Currently, Dr. C.V. Suresh Babu is a Professor in the Department of Information Technology at the School of Computing Science, Hindustan Institute of Technology and Science (Hindustan University) in Padur, Chennai, Tamil Nadu, India. For more information, you can visit his personal blog at https://sites.google.com/view/cvsureshbabu/.

Index

A

B

C

D

E

T

U

V

W

www.ingramcontent.com/pod-product-compliance
Lightning Source LLC
LaVergne TN
LVHW060300011026
849161LV00011B/2748
* 9 7 9 8 3 3 7 3 7 3 5 5 3 *